T0135264

Advances in Intelligent Systems and Computing

Volume 1054

The series "Advances in Intelligent Systems and Computing" contains publications on theory, applications, and design methods of Intelligent Systems and Intelligent Computing. Virtually all disciplines such as engineering, natural sciences, computer and information science, ICT, economics, business, e-commerce, environment, healthcare, life science are covered. The list of topics spans all the areas of modern intelligent systems and computing such as: computational intelligence, soft computing including neural networks, fuzzy systems, evolutionary computing and the fusion of these paradigms, social intelligence, ambient intelligence, computational neuroscience, artificial life, virtual worlds and society, cognitive science and systems, Perception and Vision, DNA and immune based systems, self-organizing and adaptive systems, e-Learning and teaching, human-centered and human-centric computing, recommender systems, intelligent control, robotics and mechatronics including human-machine teaming, knowledge-based paradigms, learning paradigms, machine ethics, intelligent data analysis, knowledge management, intelligent agents, intelligent decision making and support, intelligent network security, trust management, interactive entertainment, Web intelligence and multimedia.

The publications within "Advances in Intelligent Systems and Computing" are primarily proceedings of important conferences, symposia and congresses. They cover significant recent developments in the field, both of a foundational and applicable character. An important characteristic feature of the series is the short publication time and world-wide distribution. This permits a rapid and broad dissemination of research results.

** Indexing: The books of this series are submitted to ISI Proceedings, EI-Compendex, DBLP, SCOPUS, Google Scholar and Springerlink **

More information about this series at http://www.springer.com/series/11156

P. Venkata Krishna · Mohammad S. Obaidat
Editors

Emerging Research in Data Engineering Systems and Computer Communications

Proceedings of CCODE 2019

 Springer

Editors
P. Venkata Krishna
Department of Computer Science
Sri Padmavati Mahila University
Tirupathi, Andhra Pradesh, India

Mohammad S. Obaidat
Department of ECE
Nazarbayev University
Astana, Kazakhstan

KAIST
University of Jordan
Amman, Jordan

University of Science
and Technology Beijing
Beijing, China

ISSN 2194-5357   ISSN 2194-5365   (electronic)
Advances in Intelligent Systems and Computing
ISBN 978-981-15-0134-0   ISBN 978-981-15-0135-7   (eBook)
https://doi.org/10.1007/978-981-15-0135-7

This Springer imprint is published by the registered company Springer Nature Singapore Pte Ltd.
The registered company address is: 152 Beach Road, #21-01/04 Gateway East, Singapore 189721, Singapore

Preface

The emergence of data engineering is playing a major role in the context of social networks, data analytics and new-generation communication systems. The use of social networks is growing every day in millions with the introduction of wireless systems and Internet of things. The exponential growth of smart devices helps to create new-generation knowledge-sharing platform. Data analytics has a major role to play in the growth and success of big data and IoT applications. The growth of data has become large, and it is difficult to analyse. This leads to a new paradigm that gives new ecosystem for better living space for humans using the emerging trends in data engineering systems and computer communications. Many researchers depend on the data available on wireless systems and IoT for developing new-generation services and applications. Hence, the emergence of data engineering systems and computer communications needs to be meticulously investigated and analysed to provide solutions to various problems.

This book discusses and addresses the issues and challenges of data engineering systems and computer communications. The editors will seek articles that address different aspects of data analytics consisting of novel strategies based on wireless systems and IoT including machine learning, optimization, control, statistics and social computing. Moreover, this book will investigate how data engineering systems and computer communications is impacted by cutting-edge innovations.

This comprehensive and timely publication aims to be an essential reference source, building on the available literature in the field of data engineering systems and computer communications while developing applications using cloud computing, big data, IoT, etc. The objectives of this book are to identify different issues, suggest feasible solutions to those identified issues and enable researchers and practitioners from both academia and industry to interact with each other regarding emerging technologies related to data engineering systems and computer communications. In this book, we look for novel chapters that recommend new methodologies, system architectures and other solutions to prevail over the limitations of data engineering systems and computer communications.

Computing-, communication- and data engineering-related development continues to emerge, grow and spread as a global phenomenon that spans and connects culture, technology and organization. There is now ever-increasing bandwidth capable of handling media delivery, computing, telephony and so many more applications yet to be developed. Internet audio, video, animation and instant messages flowing over one converged network instaneously create a new platform for communications that will change both how we conduct business and how we handle our personal communications. The product designers and developers have newer challenges of bringing out the end results much faster.

All the said developments are leading us towards a knowledge society which is also getting increasingly mobile. Further, there has been a great change of mindset the way people treat computers, televisions and new media like real people and places, thus necessitating the deployment and usage of 'intelligent environments' and not just the 'computing or communication systems'. The initiatives such as disappearing computers and related research and development will take us finally to the future which is beyond our imagination at present.

In the above context, the theme of CCODE 2019 was decided on 'Computing, Communication and Data Engineering' aiming at bringing experts, researchers and faculty together to discuss the technological advancements in this domain. The theme was well appreciated by academia and industry. It attracted eminent personalities on its International Advisory Board and researchers and practitioners worldwide as resource persons for pre-conference tutorials and for contributed papers from all over the world.

We received over 400 technical papers, out of which domain experts and technical reviewers accepted about 150 papers for publication into the proceedings of the conference as regular, short and poster papers after a thorough review. The international speakers and contributors are from University of Jordan, Jordan; Fordham University, USA; Southeast Missouri State University, USA; Prince Sattam Bin Abdulaziz University, KSA; University of Sfax, Tunisia; Nazarbayev University, Kazakhstan; and Mizan-Tepi University, Ethiopia. Within India, the conference has invited speakers and resource persons from leading institutes/universities such as IITs, NITs, IIITs, VIT, Anna University, Visvesvaraya Technological University and JNTU, Hyderabad, apart from leading ICT companies.

The conference has a total of nine tracks, during 1–2 February 2019, covering dealing with technologies, protocols, design, implementation, applications, methodologies, best practices and case studies. Each technical session shall have a keynote talk, presentation of technical papers and discussions moderated by an expert. The session is expected to bring out a summary of current status of technology and a set of recommendations for future research and development.

The pre-conference tutorials and workshops are very special hands-on technical sessions handled by the domain experts from the industry and subject matter experts from the academia. These sessions have been designed to cater to the needs of budding industry professionals, researchers, faculty members and students.

We take this opportunity to thank our honourable Vice Chancellor (I/C) Prof. V. Uma, Registrar Prof. D. M. Mamatha and Dean of Sciences Prof. A. Jyothi, for motivating and guiding us apart from the generous support in terms of providing infrastructure and resources necessary for hosting the conference. We sincerely thank all our members of the International Advisory Board, invited speakers, resource persons, authors and delegates/participants without whom this international event would have been unconceivable.

We also thank functional/administrative groups of our university who have provided us all the logistic support and their helping hands in numerous activities related to the conference. Last but not least, we thank each one of our colleagues from the Department of Computer Science who have been working untiringly for several months to realize this conference.

Tirupathi, India   P. Venkata Krishna
Astana, Kazakhstan/Amman, Jordan/Beijing, China   Mohammad S. Obaidat

Contents

About the Editors

Dr. P. Venkata Krishna is currently a Professor of Computer Science and Director at Sri Padmavati Mahila University, Tirupati, India. He received his B. Tech in Electronics and Communication Engineering from Sri Venkateswara University, Tirupathi, India, M. Tech in Computer Science & Engineering from REC, Calicut, India and he received his Ph.D. from VIT University, Vellore, India. Dr. Krishna has several years of experience working in the academia, research, teaching, consultancy, academic administration and project management roles. His current research interests include Mobile and wireless systems, cross layer wireless network design, QoS and Cloud Computing. He was the recipient of several academic and research awards such as the Cognizant Best Faculty Award for the year 2009-2010 and VIT Most Active Researcher Award for the year 2009-2010. His biography was also selected for inclusion in the 2009-2010 edition of Marquis Who's Who in Science and Engineering and the Marquis Who's Who in the World, California, USA. He is the editor for ObCom series of International conference proceedings and he is founding member for ObCom International Conference. He has authored over 200 research papers in various national and international journals and conferences. He has produced 17 Ph.D's and 1 MS by research degree under his guidance. He has guided several masters and bachelors projects. He has authored 15 books on Computer Networks and Programming Languages. He is currently serving as editor in chief for International Journal of Smart Grid and Green Communications, Inderscience Publishers, Switzerland and also editor for International Journal of Systemics, Cybernetics and Informatics and Journal of Advanced Computing Technologies. He is an Associate Editor for International Journal of Communication Systems, Wiley. He is senior member several professional societies such as IEEE, ACM, CSI, IE(I) etc.

Prof. Mohammad S. Obaidat received his Ph.D. and M. S. degrees in Computer Engineering with a minor in Computer Science from The Ohio State University, Columbus, Ohio, USA. Dr. Obaidat is currently the Chair and Full Professor of Computer and Information Science at Fordham University, NY, USA. Among his previous positions are Chair of the Department of Computer Science and Director of the Graduate Program at Monmouth University, Dean of the College of Engineering at Prince Sultan University and Advisor to the President of Philadelphia University for Research, Development and Information Technology. He has received extensive research funding and has published Thirty Eight (38) books and over Six Hundreds (600) refereed technical articles in scholarly international journals and proceedings of international conferences. Prof. Obaidat has served as a consultant for several corporations and organizations worldwide. Mohammad is the Editor-in-Chief of the Wiley International Journal of Communication Systems, the FTRA Journal of Convergence and the KSIP Journal of Information Processing. He is also an Editor of IEEE Wireless Communications. Between 1991-2006, he served as a Technical Editor and an Area Editor of Simulation: Transactions of the Society for Modeling and Simulations (SCS) International, TSCS. He also served on the Editorial Advisory Board of several journals and conferences.

Back Pressure Monitoring of Power Plant Condenser Using Multiple Adaptive Regression Spline

Debarchan Basu and Ajaya Kumar Pani

Abstract This research work involves the application of multivariate adaptive regression spline (MARS) for estimating back pressure (p) created in a condenser of a coal-fired thermal power plant. MARS employs the plant load (L) and temperature of cooling water (T) as input variables. The output of the MARS is condenser back pressure $\hat{p}$. The designed MARS-based model gives equations for determination of p. Further, the MARS-generated objective function is optimized by randomized search cross validation. Simulation study shows that the accuracy of the reported MARS model is quite satisfactory for the prediction of back pressure.

Keywords Fault detection · Modelling · Condenser · Multivariate adaptive regression spline · MARS

1 Introduction

In the last decade, data-driven process models have found immense application in process industries for fault detection [1] and quality prediction [2, 3]. Early fault detection is important for safety. For diagnosing equipment failure in a power plant, it might require overhauling and dismantling the equipment, which is cumbersome and might increase further risk of leakage and vibration. Automatic fault detection prevents regular and costly equipment maintenance by continuously detecting problems with performance and brings that to the attention of concerned personnel [4].

The condenser in a thermal power plant is usually a shell and tube heat exchanger. The low-pressure steam coming from low-pressure turbine is passed through the shell side, and cooling water is passed through tubes of the condenser. The steam loses its latent heat and is converted to near saturated water by exchanging heat with cooling water. The entire operation is carried out at

D. Basu · A. K. Pani (✉)
Department of Chemical Engineering, Birla Institute of Technology and Science,
Pilani 333031, India
e-mail: akpani@pilani.bits-pilani.ac.in

© Springer Nature Singapore Pte Ltd. 2020
P. Venkata Krishna and M. S. Obaidat (eds.), *Emerging Research in Data Engineering Systems and Computer Communications*, Advances in Intelligent Systems and Computing 1054, https://doi.org/10.1007/978-981-15-0135-7_1

sub-atmospheric pressure (vacuum) so as to draw steam through the system. This vacuum or back pressure maintained in the condenser is a key parameter affecting the overall efficiency of the power plant. As vacuum (negative pressure) increases inside the condenser, enthalpy of the expanding steam in the turbine drops further. This results in an increase of the amount of available work from the turbine (electrical output) and improvement of plant efficiency. Therefore, continuous monitoring and maintenance of the back pressure value are of paramount importance [5].

Our aim in this work is to design a data-driven model to detect fault in the condenser in power plant. The application of this process model is fault detection of the condenser system. Whenever the difference between predicted pressure by the model and actual pressure indicated by the pressure sensor becomes significant for some period, this indicates a fault. This fault may either be a process fault, i.e. when the pressure of the system is not close to the desired value (indicated by the predicted value) or sensor fault, i.e. the sensor is not giving accurate value of pressure in the system. The process model was developed using the technique of multiple adaptive regression spline, and the simulated values show good accuracy.

2 Methodology

This research employs multivariate adaptive regression splines (MARS) for prediction of back pressure p in the condenser. The data set contains information about the back pressure (in mbar) generated in the condenser, the plant load (in MW) and the inlet temperature of cooling water (in °C) (Table 1). MARS is a flexible, fast and quite accurate simulation technique for both regression and classification applications. The complex nonlinear relationship between process inputs and output can be accurately fitted by a MARS model [6].

2.1 Data Analysis and Selection

The proposed model will predict vacuum (back pressure) created inside the condenser (p) by accepting values of plant load (L) and the cooling water temperature (T). These two variables are input features of the model of steady-state operation. So, the model takes the form

Table 1 Input–output variables for the condenser process

Variables		Description
Inputs	L	Plant load (MW)
	T	Temperature of cooling water (°C)
Outputs	$\hat{p}$	Back pressure (mbar)

$$\hat{p} = f(L, T)$$

The unknown f in the above equation is to be determined so as to ensure that predicted values $\hat{p}$ are as close as possible to the actual values p. The required data are collected from the condensing process of a thermal power plant. The data were collected for one month only and sampled at intervals of 1 h. The data set has 720 instances. The data set has 720 objects containing the values of plant load, temperature (inputs) and condenser back pressure (output). Data were split into training and testing set in 1:1 ratio, using three different random selection techniques. In order to choose the training set, the statistical parameters of the overall original data set and the three training sets were determined. These parameters include maximum, minimum, mean, skewness and standard deviation of both the input and output variables. The set which showed closest resemblance of these parameters to that of the original set was chosen as the training set.

2.2 Modelling

Three types of models were developed for the prediction of back pressure from known values of plant load and temperature. These are linear regression model, polynomial regression model and multivariate adaptive regression spline model. Regression coefficients for linear and polynomial models were determined using least sum of squared error optimization technique. The MARS model was developed from the training data with Python 3.5.

Multivariate Adaptive Regression Splines (MARS): MARS is a multivariate, piecewise regression technique that can be effectively used to describe complex relationship existing between input and output. It is a nonlinear and nonparametric regression method [4]. MARS splits the whole space of input variable into various intervals. The end points of the interval are called "knots". It is done in order to fit a spline function between these knots. The data in each of this interval are represented by a different mathematical equation to relate each sub-region of input variable to the output variable. This relationship is created from a bunch of coefficients and basis functions (BF) which are captured from the regression data [6, 7]. The MARS system is constructed upon piecewise defined linear basis functions as follows:

$$[-(x-t)]^q = \begin{cases} (t-x)^q, & \text{if } x < t \\ 0, & \text{otherwise} \end{cases} \tag{1}$$

$$[+(x-t)]^q = \begin{cases} (x-t)^q, & \text{if } x \geq t \\ 0, & \text{otherwise} \end{cases} \tag{2}$$

where q is the power of the linear basis function, and t represents the knots. The final model is of the form:

$$\hat{y} = \widehat{f(x)} = a_0 + \sum_{m=1}^{M} a_m H_{km}\left(x_{v(k,m)}\right) \tag{3}$$

where y is the output variable, dependent on x which is the input variable, and M is the number of spline functions; a_m and a_0 are the basis function parameters of the function and the model. The term $H_{km}\left(x_{v(k,m)}\right)$ can be defined as follows:

$$H_{km}\left(x_{v(k,m)}\right) = \prod_{k=1}^{K} h_{km} \tag{4}$$

where $x_{v(k,m)}$ is the predictor in the kth of the mth product. When $K = 1$, the model is additive, and when $K = 2$, the model is pairwise interactive.

The MARS algorithm (Fig. 1) consists of two steps:

- *Forward step*: Basis functions are selected to define the final model. Basis functions are added into Eq. (3) to obtain better performance. But due to a large number of basis functions, the model might overfit.
- *Backward step*: To prevent overfitting problem, basis functions which are redundant and contribute least to the model are eliminated from Eq. (3) until the best set is found. For this pruning, a generalized cross validation (GCV) criterion is engaged, which is given by

$$\text{GCV} = \frac{1}{N} \frac{\sum_{i=1}^{N}\left(y_i - \hat{f}(x_i)\right)^2}{\left[1 - \frac{C(B)}{N}\right]^2} \tag{5}$$

Here, N is the number of samples or data points. $\left[1 - \frac{C(B)}{N}\right]^2$ denotes a complexity function; $C(B)$ is the complexity penalty that is defined as $C(B) = (B+1) + dB$. B is the number of basis functions being fit, and d is the cost or penalty associated with each basis function included in the model. It can be also seen as a smoothing parameter. Higher the d, more basis functions will be removed.

2.3 Statistical Error Analysis

The error between actual and predicted values is given by statistical measures of R^2, adjusted R^2 and root mean squared error (RMSE) which are expressed as given below:

$$R^2 = 1 - \frac{\sum_i (y_i - \hat{y}_i)^2}{\sum_i (y_i - \bar{y}_i)^2} \tag{6}$$

$$\text{Adj}.R^2 = 1 - \frac{(1 - R^2)(N - 1)}{(N - p - 1)} \tag{7}$$

$$\text{RMSE} = \sqrt{\frac{1}{N}\sum_{i=1}^{N}(y_i - \hat{y}_i)^2} \tag{8}$$

where y_i = actual value, $\hat{y}_i$ = predicted value, p = total number of regressors in the training model, N = sample size.

3 Results and Discussion

The data set has two inputs, one output and a total of 720 samples. The data set was divided equally having 360 samples each to a training set and a testing or validation set. The division was performed using three techniques of random selection: (1) top half, bottom half (2) alternate elements and (3) generation of 360 random numbers in the range of 1–720 and picking up the samples corresponding to the 360 random numbers. The descriptive statistical parameters of the total data set and the three designed training sets were determined. The training set which had the closest parameter values with the total data was chosen as the data set for model development. The comparison of the designed training set data and total data is presented in Table 2.

3.1 Hyper-parameter Optimization

Parameter optimization or parameter tuning is one of the methods to improve the accuracy of algorithms by getting optimal values of the parameters associated with it. Qualitatively, hyper-parameters, λ, are those parameters in the learning algorithm that influence its accuracy the most. The hyper-parametrization problem is formulated as follows:

Table 2 Statistical parameter of the data set

Statistical parameters		Variables		
		Inputs		Output
		Load	Temperature	Back pressure
Maximum	Total data	408.85	32.7	−874.55
	Training set	408.85	32.7	−874.55
Minimum	Total data	151.52	25.6	−910.94
	Training set	151.52	25.6	−910.94
Range	Total data	257.32	7.1	36.39
	Training set	257.32	7.1	36.39
Mean	Total data	297.66	28.93	−895.86
	Training set	297.69	28.93	−895.84
Standard deviation	Total data	35.26	1.17	6.24
	Training set	35.28	1.17	6.22
Skewness	Total data	−2.34	0.957	0.27
	Training set	−2.34	0.956	0.28

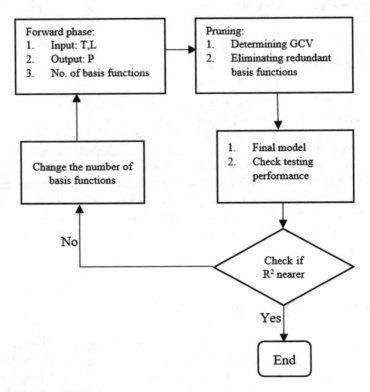

Fig. 1 MARS design algorithm

$$\lambda(\cdot) \approx \mathrm{argmin}_{\lambda \in \Lambda} \mathrm{mean}_{x \in X^{(\mathrm{test})}} \mathcal{L}\left(x; \mathcal{A}_\lambda\left(X^{(\mathrm{train})}\right)\right) \tag{9}$$

where ψ is known as the hyper-parameter response function. The optimization problem reduces to the minimization of $\psi(\lambda)$ over $\lambda \in \Lambda$. Finding the optimal set of λ involves choosing S number of trial points, i.e. $\{\{\lambda^{(i)}, \ldots, \lambda^{(s)}\}$ to estimate the value of response function for each point and return the $\lambda^{(i)}$ that worked best. It is implemented here as random search in scikit-learn library.

3.2 Model Performance

The efficiency of the developed MARS model was assessed by calculating R^2, adjusted R^2 and RMSE. The parameters that were used for model training are given in Table 3.

Forty basis functions were introduced in the forward pass. Out of these, 13 basis functions were pruned in the backward pass. The variation of model accuracy during forward step and backward pruning step is shown in Fig. 2. The final MARS model consists of 27 basis functions. The expression, from Eq. (3), is given as follows:

$$\hat{p} = -894.391 + \sum_{m=1}^{27} a_m B_m(x) \tag{10}$$

The expression for $B_m(x)$ and corresponding a_m is shown in Table 4.

Table 5 presents the prediction accuracies of the linear regression, polynomial regression and the MARS (presented in Table 4) model.

It is observed that the model gives an R^2 value of 0.9174, with a RMSE of 1.79 for the training set. When the model is fit to the testing set, it gives an R^2 of 0.9188 and RMSE of 1.78. The spline model has a better performance than the other two models. Figure 3 presents a graph of actual back pressure values of condenser and the values predicted by the designed MARS model.

Table 3 Parameters of MARS model

Parameters	Values
Maximum functions	10–100
GCV penalty per knot	0, 1–4
Maximum interactions	1–4
Prune	Yes
Parameter controlling the minimum number of data points between points	0–1
Parameter controlling the number of extreme data values of each feature not eligible as knot locations	0–1

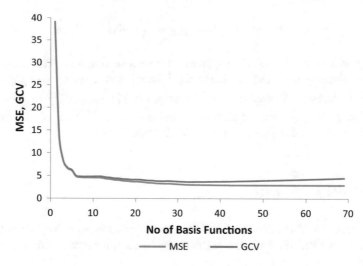

Fig. 2 Variation of model accuracy with respect to the number of basis function of MARS model

Table 4 Basis functions and corresponding coefficients

Basis function	Equation	Coefficient (a_m)
$B_1(x)$	$h(T - 29.0564)$	1.85699
$B_2(x)$	$h(L - 283.182)$	0.0836953
$B_3(x)$	$h(283.182 - L)$	−641.514
$B_4(x)$	$h(T - 28.3168) * h(29.0564 - T)$	44.3432
$B_5(x)$	$h(28.3168 - T) * h(29.0564 - T)$	−20.8766
$B_6(x)$	$h(T - 28.0927) * h(28.3168 - T) * h(29.0564 - T)$	433219
$B_7(x)$	$h(28.0927 - T) * h(28.3168 - T) * h(29.0564 - T)$	−1.0161
$B_8(x)$	$h(27.82 - T) * h(29.0564 - T)$	26.3289
$B_9(x)$	$T * h(T - 28.0927) * h(28.3168 - T) * h(29.0564 - T)$	−15356.6
$B_{10}(x)$	$h(28.5662 - T) * h(T - 28.3168) * h(29.0564 - T)$	−924.984
$B_{11}(x)$	$h(170.633 - L) * h(283.182 - L)$	−5.74788
$B_{12}(x)$	$h(289.07 - L) * h(L - 283.182)$	−1.17447
$B_{13}(x)$	$h(L - 311.669) * h(T - 27.82) * h(29.0564-T)$	0.168671
$B_{14}(x)$	$h(311.669 - L) * h(T - 27.82) * h(29.0564 - T)$	0.0450742
$B_{15}(x)$	$h(T - 27.82) * h(289.07 - L) * h(L - 283.182)$	1.25848
$B_{16}(x)$	$h(27.82 - T) * h(289.07 - L) * h(L - 283.182)$	0.751321
$B_{17}(x)$	$h(T - 28.0927) * h(T - 27.82) * h(29.0564 - T)$	−31.2481
$B_{18}(x)$	$h(28.0927 - T) * h(T - 27.82) * h(29.0564 - T)$	−325.509
$B_{19}(x)$	$h(T - 30.01)$	2.16943

(continued)

Table 4 (continued)

Basis function	Equation	Coefficient (a_m)
$B_{20}(x)$	$h(L - 311.519) * h(311.669 - L) * h(T - 27.82) * h(29.0564 - T)$	−2412.58
$B_{21}(x)$	$h(L - 282.269) * h(283.182 - L)$	−33.081
$B_{22}(x)$	$h(282.269 - L) * h(283.182 - L)$	5.74623
$B_{23}(x)$	$h(281.194 - L) * h(L - 170.633) * h(283.182 - L)$	0.0204301
$B_{24}(x)$	$h(L - 279.98) * h(281.194 - L) * h(L - 170.633) * h(283.182 - L)$	−0.0593202
$B_{25}(x)$	$h(301.145 - L) * h(L - 283.182)$	−0.013041
$B_{26}(x)$	$h(L - 280.27) * h(281.194 - L) * h(L - 170.633) * h(283.182 - L)$	0.105976
$B_{27}(x)$	$L * h(L - 170.633) * h(283.182 - L)$	0.0204305

Table 5 Prediction performance of the three models with test data

Model type	R^2	RMSE
Linear regression	0.815	2.67
Polynomial regression	0.868	2.33
MARS	0.919	1.78

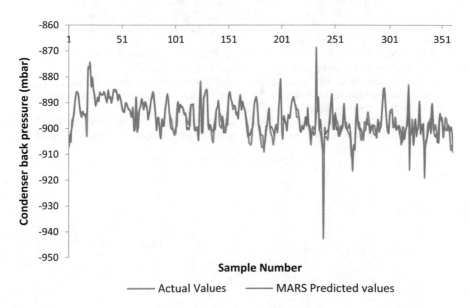

Fig. 3 Actual and MARS-predicted values of condenser back pressure

4 Conclusion

The back pressure existing in the condenser of a thermal power plant is a key feature deciding the power plant efficiency. In this study, multiple linear regression, polynomial regression and multivariate adaptive regression spline-based process models were developed for continuous monitoring of condenser back pressure. Whenever there is a significant deviation between model-predicted pressure and sensor-indicated pressure, a fault in the process or pressure sensor is suspected. Simulation results show that the MARS model has better estimation accuracy than the other two models in predicting back pressure. The accuracy of the MARS model is quite acceptable for use of the model for fault detection and diagnosis of the condensing process in power plant.

Acknowledgements The authors gratefully acknowledge the data provided by NTPC Solapur (Maharashtra, India) for this study.

References

1. Rajakarunakaran, S., Venkumar, P., Devaraj, D., Surya Prakasa Rao, K.: Artificial neural network approach for fault detection in rotary system. Appl. Soft Comput. **8**, 740–748 (2008)
2. Jain, V., Kishore, P., Kumar, R.A., Pani, A.K.: Inferential sensing of output quality in petroleum refinery using principal component regression and support vector regression. In: 2017 IEEE 7th International on Advance Computing Conference (IACC), pp. 461–465 (2017)
3. Morey, A., Pradhan, S., Kumar, R.A., Pani, A.K., Vijayan, S., Jain, V., Gupta, A.: Pollutant monitoring in tail gas of sulfur recovery unit with statistical and soft computing models. Chem. Eng. Commun. **206**, 69–85 (2019)
4. Chen, K.-Y., Chen, L.-S., Chen, M.-C., Lee, C.-L.: Using SVM based method for equipment fault detection in a thermal powerplant. Comput. Ind. **62**, 42–50 (2011)
5. Munoz, A., Sanz-Bobi, M.A.: An incipient fault detection system based on the probabilistic radial basis function network: application to the diagnosis of the condenser of a coal power plant. Neurocomputing **23**, 177–194 (1998)
6. Friedman, J.H.: Multivariate adaptive regression splines. Ann. Stat. **19**, 1–67 (1991)
7. Frank, I.E.: Modern nonlinear regression methods. Chemom. Intell. Lab. Syst. **27**, 1–19 (1995)

An Efficient Image Retrieval System for Remote Sensing Images Using Deep Hashing Network

Sudheer Valaboju and M. Venkatesan

Abstract Due to the huge increase in volumes of remote sensing images, there is a requirement for retrieval systems which maintain the retrieval accuracy and efficiency which requires better learning of features and the binary hash codes which better discriminate the images of different classes of images. The existing retrieval systems for remote sensing images use CNNs for feature learning which fails to preserve the spatial properties of an image which in turn affect the quality of binary hash code and the retrieval performance. This Paper tries to address the above goals by using (1) Extracting Hierarchical features of convolutional neural network and using them to sequential learning to better learn the features preserving spatial and semantic properties. (2) Use lossless triplet loss with two more loss functions to generate the binary hash codes which better discriminate the images of different classes. The proposed architecture consists of three phases: (1) Fine-tuning a pre-trained model. (2) Extracting the hierarchical features of convolutional neural network. (3) Using those features to train the deep learning-based hashing network. Experiments are conducted on a publicly available dataset UCMD and show that when hierarchial convolutional features are considered there is a significant improvement in performance.

Keywords Remote sensing · Image retrieval · Deep learning · Hashing

S. Valaboju (✉) · M. Venkatesan
Department of Computer Science Engineering, National Institute
of Technology Surathkal, Surathkal, Karanataka, India
e-mail: Sudheer.valaboju@gmail.com

M. Venkatesan
e-mail: venkisakthi77@gmail.com

© Springer Nature Singapore Pte Ltd. 2020
P. Venkata Krishna and M. S. Obaidat (eds.), *Emerging Research in Data
Engineering Systems and Computer Communications*, Advances in Intelligent
Systems and Computing 1054, https://doi.org/10.1007/978-981-15-0135-7_2

1 Introduction

The increase in the satellites for remote sensing images and advancements in technology leads to extensive remote sensing images being acquired [1]. To use or manage this data effectively and efficiently, there is a need of retrieval systems which has been a challenge in the field of remote sensing and applications. The CBIR workflow consists of two stages: (1) Image features of query are generated. (2) Similarity measuring is done using the features of query image and the features of images in database.

Due to the enormous databases of remote sensing images, there is a need for faster image retrieval systems. The techniques like the nearest neighbor search for these databases are not practical. The recent work shows that the hashing-based retrieval systems are suitable for large-scale databases.

2 Related Work

The recently proposed hashing-based approximate nearest neighbor search [2] for faster and accurate image search and retrieval in large remote sensing data archives achieves much faster retrieval speed than the conventional retrieval framework which uses similarity measuring between features. But fails to maintain the retrieval efficiency as the hand-crafted features used in this method does not accurately maintain the semantic properties of the image which in turn affect the binary code and ultimately the retrieval performance.

These shortcomings are addressed by the deep hashing neural networks [3–7] which use the deep semantic features generated by the CNNs and map them to binary hash codes. These outperform the hashing-based approximate nearest neighbor search and these methods use cross-entropy as loss which fails to discriminate the images leads to poor generalization.

To address the above problems, deep metric and hash code learning [8] has been proposed which uses the pre-trained inception model to extract the features of the images and use those features for binary hash generation and to better discriminate the images the triplet loss is used. Which outperforms deep hashing neural networks. But in the standard triplet loss, the loss reaches the zero very quickly where the image discrimination is not perfect. And also this framework cannot be directly applied to the images which vary spatially and spectrally (e.g., hyper-spectral images).

To address the above problem, in this paper, we propose a framework which can be extended to the images that vary spatially and spectrally. The approach leverages the hierarchical convolutional features to preserve the spatial properties by sequential learning using LSTM [9] and the lossless triplet loss is used to address the shortcoming of triplet loss.

3 Deep Learning-Based Hashing Network

Deep feature learning with limited labeled samples is a difficult task. To better learn features, we use a pre-trained imagenet-VGG-F model.

The framework consists of three phases:

(1) Fine-tune the model: (i) We freeze five convolutional layers and build the small model say top model containing two dense layers with 256 and 21 neurons (21 being the number of classes of dataset), respectively, and train it using the output of the fifth convolutional layer. (ii) Fine-tune the other convolutional layers one at a time. (2) Extract the hierarchial convolutional features of the network. Use the bilinear interpolation to upsample the image size reduced due to convolution and pooling operations. The existing retrieval systems for remote sensing images use CNNs for feature learning which fails to preserve the spatial properties. Due to Convolution and maxpooling the output of different layers are of different size in padding of zeros is done for making the same size of all hierarchial convolutional features. Then, these hierarchical convolutional features are stacked together to form a feature vector. (3) The generated feature vector is used to train the deep learning-based hashing network. Deep learning-based hashing network is trained using the combination of different losses inspired from [10]. Below given are elaborated details.

Let us say, we have training data $T = i_1, i_2, i_3, \ldots$ in each i having a corresponding label from the set of labels which can be represented as $L = L_1, L_2, \ldots, L_p$. The ultimate aim is to generate feature representations of query image say "f" and map f to k-bit binary hash code. The function can be represented as $f \rightarrow [0, 1]$ k. These hash codes are used as a similarity measure between two images and most similar ones are retrieved in testing phase. Similarity measure between hash codes is done using Hamming distance.

In the first phase, T is used to fine-tune the pre-trained model. And in the second phase, the T is fed to pre-trained model which consists of five convolution layers to obtain five feature vectors, each from each convolution layer can be represented as $cL_1 = [f_i]_0^n, cL_2 = [f_i]_0^n, cL_3 = [f_i]_0^n, cL_4 = [f_i]_0^n, cL_5 = [f_i]_0^n$, and we use bilinear interpolation to upsample and PCA to reduce number of feature maps to make all feature vectors of same dimensions. Now, we stack all these feature vectors to create final feature vector $F = [cL_1, cL_2, cL_3, cL_4, cL_5]$, and F is fed to the deep hashing network for training.

Let LT be lossless triplet loss function used for training the deep hashing network which was pointed out by Marc-Olivier Arsenault [10], which make images of same label come closer and make different images go farther. LT is represented as follows:

$$\sum_{i=1}^{N} \left[-\ln\left(-\frac{(f_i^a - f_i^p)^2}{\beta} + 1 + \epsilon \right) - \ln\left(-\frac{N - (f_i^a - f_i^n)^2}{\beta} + 1 + \epsilon \right) \right]$$

where f_i^a is anchor and f_i^p is positive sample and f_i^n is the negative sample and N is the number of outputs of the network. β is the scaling factor which is set to N.

Let TL be the triplet loss function which can be represented as follows:

$$\sum_{i=1}^{n} \left(0, \|f(g_i^a) - f(g_i^p)\|_2^2 - \|f(g_i^a) - f(g_i^p)\|_2^2 + \alpha \right)$$

where $f\left(g_i^a\right)$ is anchor and $f(g_i^p)$ is positive sample and $f\left(g_i^a\right)$ is the negative sample and α is the scaling factor by default set to 0.2. Inspired from [8], we use two more loss functions. Second loss function aims at maximizing the sum of squared errors between the output of the last layer in the deep hashing-based network and 0.5. Let L_{out} be the second loss function which can be represented as follows:

$$L_{\text{out}} = -\frac{1}{k} \sum_{i=1}^{n} \|f(g_i) - 0.51\|^2$$

where 1 is the k-dimensional vector with all ones and then the third loss function which balances the number of ones and zeros in the binary representations of features.

Let L_{bal} be the third loss function.

$$L_{\text{bal}} = \sum_{i=1}^{n} (\text{mean}(f_g^i)) - 0.5)^2.$$

These losses are combined to form the combined loss.

$$\text{combined loss} = \text{LT} + \lambda_1(L_{\text{out}}) + \lambda_2(L_{\text{bal}}).$$
$$\text{where } \lambda_1 = 0.001 \text{ and } \lambda_2 = 0.1$$

After training is completed, we obtain the binary hash code for every image by binarizing the sigmoid embeddings of the deep learning-based network, i.e., the last layer in the network as follows similar to [10]:

$$b = (\text{sign}(a_n - 0.5) + 1/2)$$

where a_n is the nth activation of the sigmoid layer. With these binary hash codes, the similarity of images is measured using Hamming distance.

4 Experiment Details

The experiments are conducted on the publicly available dataset UCMD which consists of 2100 images from 21 different classes. Each class contains 100 images of size 25 × 256 with spatial resolution 30 cm. The dataset has limited labeled samples which makes a challenge to train the neural network.

No Augmentation of data is done.

A pre-trained VGG-net (Table 1) is used to generate the five intermediate feature representations of the images from five convolutional layer having dimensions [27 × 27 × 256], [13 × 13 × 256], [13 × 13 × 256], [13 × 13 × 256], [13 × 13 × 256], [6 × 6 × 256], respectively. Then, the feature maps of each representation are increased to 256 and size of images are upsampled to have same size in all layers' intermediate features using bilinear interpolation. All these are stacked together to form an intermediate dataset having dimensions [6300 × 27 × 27 × 256].

5 Results

Now, these data are fed into deep learning-based hashing network for training (Table 2).

The network was trained with a small batch size of 90 which are the combination of anchors, positives and negatives as the loss is based on triplet loss. Standard Adam optimization function is used for optimization. The performance of the model is evaluated by the mean average precision (mAP) [8]. 60% of the data is used as the training data and remaining for testing.

Table 1 Architecture of five convolution layers in VGG-F

conv1	64 × 11 × 11, stride 4 × 4, pool 2 × 2, relu
conv2	256 × 5 × 5, stride 1 × 1, pad 2, pool 2, relu
conv3	256 × 3 × 3, stride 1 × 1, pad 1, relu
conv4	256 × 3 × 3, stride 1 × 1, pad 1, relu
conv5	256 × 3 × 3, stride 1 × 1, pad 1, pool 2, relu

Table 2 Deep learning-based hashing network

RNN	512 LSTM cells
RNN	512 LSTM cells
FCL	1024, dropout 0.5, leakyrelu
FCL	512, dropout 0.5, leakyrelu
FCL	256, dropout 0.5, leakyrelu
FCHL	K, sigmoid

A *K*-bit, i.e., 32-bit hash code, is generated which is used for similarity measuring. In the testing phase, different scenarios are considered for each query image and 20 similar images are retrieved based on these mAP obtained. It is observed that lossless triplet loss underperformed and the below results are using triplet loss. (i) Without the hierarchical convolutional features for feature learning and without LSTM, the mAP obtained is 0.41. (ii) With hierarchical convolutional features and LSTM, the mAP obtained is 0.72. The results show that the performance has been improved when considered the spatial properties. And the lossless triplet loss did not perform well on this dataset.

6   Conclusion

In this paper, we show that preserving the spatial properties by using hierarchical convolutional features for sequential learning there is a significant improvement in performance and as a Future development improve the accuracy of the network and extend the network architecture to hyper spectral and multispectral images which vary more spatially and spectrally.

References

1. Chi, M., Plaza, A., Benediktsson, J.A., Sun, Z., Shen, J., Zhu, Y.: Big data for remote sensing: challenges and opportunities. Proc. IEEE **104**(11), 2207–2219 (2016)
2. Demir, B., Bruzzone, L.: Hashing-based scalable remote sensing image search and retrieval in large archives. IEEE Trans. Geosci. Remote Sens. **54**(2), 892–904 (2016)
3. En, S., Crémilleux, B., Jurie, F.: Unsupervised deep hashing with stacked convolutional autoencoders. In: 2017 IEEE International Conference on Image Processing (ICIP), Beijing, pp. 3420–3424 (2017)
4. Li, Y., Zhang, Y., Huang, X., Zhu, H., Ma, J.: Large-scale remote sensing image retrieval by deep hashing neural networks. IEEE Trans. Geosci. Remote Sens. **56**(2), 950–965 (2018)
5. Zhu, H., Long, M., Wang, J., Cao, Y.: Deep hashing network for efficient similarity retrieval. In: AAAI (2016)
6. Cao, Z., Long, M., Wang, J., Yu, P.S.: Hashnet: deep learning to hash by continuation. ArXive-prints arXiv:170200758v4 [cs.LG] (February 2017)
7. Ye, F., Xiao, H., Zhao, H., Dong, M., Luo, W., Min, W.: Remote sensing image retrieval using convolutional neural network features and weighted distance. IEEE Geosci. Remote Sens. Lett.
8. Roy, S., Sangineto, E., Demir, B., Sebe, N: Deep metric and hash-code learning for content-based retrieval of remote sensing images. In: IGARSS 2018—2018 IEEE International Geoscience and Remote Sensing Symposium, pp. 4539–4542 (2018)
9. Lu, X., Chen, Y., Li, X.: Hierarchical recurrent neural hashing for image retrieval with hierarchical convolutional features. IEEE Trans. Image Process. **27**(1), 106–120 (2018)
10. https://towardsdatascience.com/lossless-triplet-loss-7e932f990b24

Tea Leaf Disease Prediction Using Texture-Based Image Processing

Alok Ranjan Srivastava and M. Venkatesan

Abstract Nowadays, Tea is commonly used in India as well as in all over the world. Tea is produced in many states of India, i.e., Assam, West Bengal, Tamil Nadu, Karnataka, and so on. But, production of tea is heavily affected by various diseases and pests. There are various kinds of diseases in tea leaves and various pests that can damage the tea crop and affect the tea production. Tea crop damage is reduced by recognizing the tea leaf diseases in an early stage. After detection of the kind of tea leaf diseases, suitable preventive method can be used to reduce the tea crop damage. For the detection of tea leaves diseases, there are different classification methods. Some classification techniques are random forest classifier, k-nearest neighbor classifier, support vector machine classifier, neural network, etc. After training the dataset with classifier, the image of tea leaf is given as an input, the best possible match is found by the classifier system, and diseases are recognized by the classifier system. This project is going to use various classification techniques to recognize and predict the tea leaves disease which helps us to improve the tea production of India.

Keywords Classification techniques · Dataset · Support vector machine · Neural network · k-nearest neighbor classifier · Random forest classifier

1 Introduction

There are various kinds of diseases in tea leaves. These tea leaves diseases are categorized into two categories: fungal and non-fungal. There are few leaf diseases.

1. **Algal Leaf Spot**: It is a non-fungal disease. Gray, green, and tan-raised spots appear on tea leaves. It causes by algae, high temperature, and humidity. It can be managed by avoiding plant stress and poorly drained sites.

A. R. Srivastava (✉) · M. Venkatesan
Department of Computer Science and Engineering, National Institute
of Technology, Surathkal, Karnataka, India
e-mail: alokranjan71995@gmail.com

© Springer Nature Singapore Pte Ltd. 2020
P. Venkata Krishna and M. S. Obaidat (eds.), *Emerging Research in Data
Engineering Systems and Computer Communications*, Advances in Intelligent
Systems and Computing 1054, https://doi.org/10.1007/978-981-15-0135-7_3

17

2. **Brown Blight and Gray Blight**: It is a fungal disease. First, some small pale yellow spots start to appear on young leaves. These spots grow and turn into brown or gray. To avoid this, we have to reduce plant stress.
3. **Blister Blight**: There are some small pinhole-size spots on young leaves. These spots become larger, transparent, and light brown with time. It is a fungal disease. It can be managed by removal of affected leaves and spraying the Bordeaux mixture.

There are various other tea leaf diseases like horse hair blight and pests like tea aphids and spider mites that affect the tea crop production. Above is the brief introduction of tea leaves diseases and pests [1] (Figs. 1 and 2).

This paper proposes an idea to detect these tea leaf diseases using texture feature extraction and classification techniques.

(a)   (b)

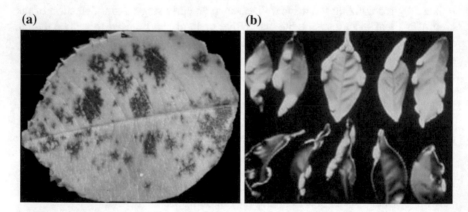

Fig. 1 **a** Algal leaf spot and **b** Blister blight

(a)   (b)

Fig. 2 **a** Brown blight and **b** Gray blight

2   Literature Survey

Karmokar et al. [2] have proposed a tea leaf disease recognizer that combines feature extraction and neural network ensemble for training. First step is the acquisition of tea leaf images by digital camera. Next step is the pre-processing of the dataset. In this step, images are converted into threshold images. Next step is the feature extraction to get matrix representation of the extracted image. Training of these dataset is done by negative correlation algorithm. The negative correlation algorithm is one of the methods to combine the predictions of the various independent trained neural networks. Next step is the testing that is recognizing the tea leaf disease.

Kaur et al. [3] have proposed a classification system to classify soybean leaf diseases into three different categories, i.e., downy mildew, septoria leaf blight, and frog eye. First step is the acquisition of dataset. The pre-processing of acquired images is done in next step. Then, background removal is done to get clear leaf image. Next step is the segmentation. Next step is the feature extraction. A number of texture features and color features can be used for feature extraction. Next step is the classification. SVM is used for classification in this paper [3].

The segmentation process is required to get clear leaf image so that feature extraction can be done. Badnakhe et al. [4] have proposed the Otsu's threshold and the k-means clustering for identification of leaf diseases. The extracted feature values are less for k-means clustering and give better results.

Choudary et al. [5] have proposed a methodology to predict accurate disease groups for plant leaf images. They have performed k-means clustering for segmentation. They have used texture features to train neural network [5, 6].

Hossain et al. [7] have proposed a system which is based on support vector machine classification model. The tea leaf images are converted into gray-scale images, and GLCM is used for texture feature extraction. SVM-based classification gives an accuracy of 93% in predicting tea leaf diseases [7–10].

Gulhane et al. [11] have proposed the techniques of identifying cotton disease based on the pattern of disease and extracted feature of the image. ANN method gives 85–91% of exact disease prediction depending upon the quality of acquired images and pre-processing process. It is found that morphological features give better performance [11–14].

Revathi et al. [15] have proposed computer vision system for the recognition of the cotton crop diseases from RGB images. They have proposed enhanced particle swarm optimization (PSO) feature selection method. The proposed PSO method uses skew divergence method and features like texture, edge, and color to extract the features. They have compared the pre-existing classifiers (i.e., support vector machine, back propagation neural network, fuzzy with edge CYMK color feature) [15].

Jha et al. [16] have proposed tea leaf disease detection technique. First step is the feature extraction. In this step, the RGB images are converted into HSV color space for feature extraction. The features calculated by the mean and variance of the

Gabor filtered images. The image descriptor SIFT is applied for image-based matching and recognition for feature extraction. The features extracted from HSV-Gabor and SIFT are applied to probabilistic neural network (PNN), which is used to map any input pattern to number of classification [16].

Khirade et al. [17] have proposed an image processing technique to identify and classify diseases of plant leaf. First step is the acquisition of image dataset. Next step is the pre-processing of the acquired images. Next step is the segmentation of image. Then, in the next step, perform feature extraction and then k-mean clustering. The next step is Otsu threshold algorithm. For feature extraction, Sachin D. Khirade et al. proposed color co-occurrence matrix using texture-based analysis. For classification, they proposed artificial neural network and back propagation network [17].

3   Research Methodology

This paper focuses on tea leaf disease classification. This research proposes a classification system in which actual prediction is done on the basis of majority vote of the different classifier system.

3.1   Dataset Acquisition

A large sample of tea leaf images is collected from High Field Tea Estate, Coonoor, Tamil Nadu, using a digital camera. The sample contains different types of tea leaf, some are diseased (algal, blister blight, and gray blight), and some are healthy.

3.2   Image Pre-processing

This phase is divided into three steps: Image Normalization, Image Segmentation and Image Augmentation. In first step, tea leaf image is normalized, i.e., its size is made constant. In second step, a python script is used for background removal and extraction of area of interest in tea leaf image. In third step, the dataset is expanded using different methods. Additional images are created by rotating the images 90°, 180°, and 270°, by mirroring each rotated image horizontally and vertically (Figs. 3 and 4).

(a)   (b)

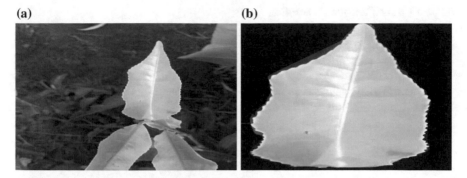

Fig. 3 **a** Raw leaf image and **b** Segmented image

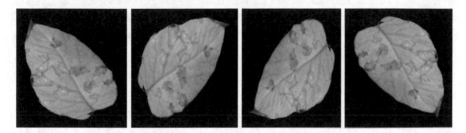

Fig. 4 Augmented tea leaf image

3.3 Feature Extraction

3.3.1 Texture Feature Extraction Using Gray-Level Co-occurrence Matrix

Texture is an important aspect that can help in image classification problem. Texture is an important property that measures the variation of intensities in nearby pixels of a surface. All tea leaf samples are converted into gray-scale images. Gray-level co-occurrence matrix (GLCM) is created for all tea leaf images. GLCM is a statistical way of examining texture, i.e., the relationship in the intensities of nearby pixels. The GLCM is calculated by how often a pixel with certain intensity value occurs in a specific spatial relationship to another one with some other fix values [18]. By using GLCM, texture features (haralick features) are extracted [19]. A total of 13 texture features are extracted using GLCM. These features are contrast, energy, correlation, variance, inverse difference moment, entropy, sum variance, sum entropy, sum average, difference variance, difference entropy, maximal correlation coefficient, and information correlation that are evaluated using gray-level co-occurrence matrix.

3.3.2 Color Feature Extraction

Color is an important feature. All tea leaf images are converted into RGB images. The mean and standard deviation of all three components (R, G, B) are computed. Therefore, a total of six more features are extracted. Contour area and contour perimeter are also evaluated as features. Hence, a total of 21 features (13 texture, 6 color, contour area and contour perimeter) are extracted for each tea leaf image.

3.4 Feature Selection and Feature Scaling

A total of 21 features of a tea leaf image are extracted. But, some features are irrelevant. A set of relevant features are selected by doing the statistical analysis of these 21 features. A total of 15 features are selected. The relevant features are scaled between zero and one so that each feature is considered equally for classification.

3.5 Classification System

Image classification is a technique in which a particular image is classified into pre-defined category. In the case of tea leaf images, tea leaf is classified into one of the four pre-defined class (healthy, algal, blister and gray). This paper proposes a classifier system in which four different classifiers are used, and majority voting of these classifiers is used for classification. The classification system consists of support vector machine classifier, decision tree classifier, random forest classifier, and Ada-boost classifier. These are the four classifiers that are used to train the tea leaf image dataset. Individual prediction on test dataset of each classifier is considered, and majority voting is taken as actual prediction.

When a new tea leaf image is given for classification, its features are extracted and given to the each of the trained classifier for prediction, and each classifier predicts its class based on the trained dataset. The actual prediction is given by the majority vote.

4 Results and Discussion

A set of 1680 tea leaf images (i.e., healthy and diseased tea leaf images) are used as training dataset. Another set of 336 tea leaf images are used as testing dataset.

Then, 15 features of tea leaf (9 texture feature and 6 color features) have been considered to train each of the four classifiers—support vector machine, decision tree classifier, random forest classifier, and Ada-boost classifier. The four texture features (correlation, information correlation, maximal correlation coefficient, entropy), contour area and contour perimeter are found not to show much variation among healthy, algal spot, blister blight, and gray blight using statistical analysis. Therefore, these features are not taken account in the classification process. Removal of these features speeds up the classification system and does not degrade the accuracy of classification system (Table 1).

The set of 1680 tea leaf images are used as training dataset. Each of the four classifier models is trained using this dataset. Another set of 336 tea leaf images are used as testing dataset. Table 2 shows the true and false classification of the testing dataset based on the majority voting of support vector machine, decision tree, random forest, and Ada-boost classifier.

According to the above classification, a set of metrics is evaluated for each class. Some metrics like recall, precision, F1-score, and support are used for proposed classification system to describe its performance (Table 3).

Table 4 shows the corresponding accuracy for each class obtained by the proposed classification model. The overall average accuracy obtained by the classification model is 96.43%.

Table 1 Different feature (texture and color) of healthy, algal spot, blister blight, and gray spot tea leaf images

Leaf/features	Healthy	Algal spot	Blister blight	Gray spot
Energy	0.5385	0.05340	0.7284	0.1737
Contrast	26.4150	20.3550	17.6103	29.8940
Variance	3392.90	859.40	1467.88	8034.254
Inverse difference moment	0.85090	0.5737	0.9072	0.6386
Sum average	62.0603	73.0456	22.0687	170.786
Sum variance	13545.1	3417.22	5853.91	32107.12
Sum entropy	2.9018	6.2770	1.69567	5.59594
Difference variance	0.0025	0.0013	0.0030	0.001280
Difference entropy	1.32460	2.5702	0.9651	2.47543
Mean blue	30.3730	22.9856	4.4274	57.4124
Mean green	43.0080	38.0888	12.4957	89.1394
Mean red	7.46444	38.2120	10.5223	87.9125
Standard deviation blue	58.4200	19.9370	20.5223	62.8278
Standard deviation green	80.0784	29.8681	42.6445	93.9897
Standard deviation red	25.2183	32.6855	37.6790	91.8901

Table 2 Classification table of 336 testing leaf images

Predicted/actual	Healthy	Algal spot	Blister blight	Gray spot
Healthy	296	0	0	0
Algal spot	12	9	0	0
Blister blight	0	0	5	0
Gray spot	0	0	0	14

Table 3 Set of metrics for each class

Metric/class	Precision	Recall	$F1$-score	Support
Healthy	0.96	1.00	0.98	296
Algal spot	1.00	0.43	0.60	21
Blister blight	1.00	1.00	1.00	5
Gray spot	1.00	1.00	1.00	14
Average/ total	0.97	0.96	0.96	336

Table 4 The accuracy of the classification system

Class	Success rate (%)
Healthy	100
Algal spot	43
Blister blight	100
Gray spot	100

5 Conclusion and Future Scope

A classification system is developed for the detection of three kind of tea leaf diseases based on the majority voting of the support vector machine, decision tree, random forest, and Ada-boost classifier in this research. The overall accuracy of this classifier system is 96.43%. This tea leaf disease detection helps the industry to enhance its tea crop quality and quantity.

The segmentation process can be improved by using texture-based segmentation. The use of different kind of neural networks in the proposed ensemble model is also the future scope of this research to achieve more accurate result.

References

1. https://plantvillage.psu.edu/topics/tea/infos
2. Karmokar, B.C., Ullah, M.S., Siddiquee, M.K., Alam, K.M.R.: Tea leaf disease recognition using neural network ensemble. Int. J. Comput. Appl. **114**(17), 27–30 (2015)
3. Kaur, S., Pandey, S., Goel, S.: Semi-automatic leaf disease detection and classification system for soybean culture. Inst. Eng. Technol. Image Process. J. **12**(6), 1038–1048 (2018)
4. Badnakhe, M.R., Deshmukh, P.R.: Infected leaf analysis and comparison by Otsu threshold and k-means clustering. Int. J. Adv. Res. Comput. Sci. Softw. Eng. **2**(3) (2012)
5. Choudhary, G.M., Gulati, V.: Advance in image processing for detection of plant diseases. Int. J. Adv. Res. Comput. Sci. Softw. Eng. **5**(7) (2015)
6. Mainkar, P.M., Ghorpade, S., Adawadkar, M.: Plant leaf disease detection and classification using image processing techniques. Int. J. Innov. Emerg. Res. Eng. **2**(4) (2015)
7. Hossain, M.S., Mou, R.M., Hasan, M.M., Chakraborty, S., Abdur Razzak, M.: Recognition and detection of tea leaf's diseases using support vector machine. In: 2018 IEEE 14th International Colloquium on Signal Processing & Its Applications (CSPA 2018), 9–10 March 2018, Penang, Malaysia. IEEE, New York (2018)

8. Arivazhagan, S., Newlin Shebiah, R., Ananthi, S., Vishnu Varthini, S.: Detection of unhealthy region of plant leaves and classification of plant leaf diseases using texture features. Int. Agric. J.—CIGR J. (2013)
9. Gavhale, K.R., Gawande, U., Hajari, K.O.: Unhealthy region of citrus leaf detection using image processing techniques. In: International Conference for Convergence of Technology—2014. IEEE, New York (2015)
10. Meena Prakash, R., Saraswathy, G.P., Ramalakshmi, G.: Detection of leaf diseases and classification using digital image processing. In: 2017 International Conference on Innovations in Informations, Embedded and Communication Systems. IEEE, New York (2018)
11. Gulhane, V.A., Gurjar, A.A.: Detection of diseases on cotton leaves and its possible diagnosis. Int. J. Image Process. (2011)
12. Kaur, R., Singla, S.: Classification of plant leaf diseases using gradient and texture feature. ACM (2016)
13. Jhuria, M., Kumar, A., Borse, R.: Image processing for smart farming: detection of disease and fruit grading. In: 2013 IEEE Second International Conference on Image Information Processing (ICIIP-2013). IEEE, New York (2014)
14. Anand, R., Veni, S., Aravinth, J.: An application of image processing techniques for detection of diseases on Brinjal leaves using k-means clustering method. In: 2016 Fifth International Conference on Recent Trends in Information technology. IEEE, New York (2016)
15. Revathi, P., Hemlatha, M.: Cotton leaf spot diseases detection utilizing feature selection with skew divergence method. Int. J. Sci. Eng. Technol. (2014)
16. Jha, S., Jain, U., Kende, A., Venkatesan, M.: Disease detection in tea leaves using image processing. Int. J. Pharma Bio Sci. (2016)
17. Khirade, S.D., Patil, A.B.: Plant leaf disease detection using image processing. In: 2015 International Conference on Computing Communication Control and Automation. IEEE, New York (2015)
18. http://matlab.izmiran.ru/help/toolbox/images/enhanc15.html
19. Haralick, R., Shanmugam, K., Dinstein, I.H.: Textural features for image classification. IEEE Trans. Syst. Man Cybern. **3**, 610–621 (1973)

Analyzing Significant Reduction in Traffic by Using Restricted Smart Parking

Amtul Waheed, Jana Shafi and P. Venkata Krishna

Abstract With the growth of population specifically in market areas and industrial areas, the traffic-related consequences such as congested roads, finding vacant parking space, traffic congestion at major traffic points in the town, etc. are also increasing particularly at office hours, business hours, and peak hours. Drivers of the vehicle just keep on circling in parking lot to find the vacant parking space. Due to this, people are facing a lot of problems like delay in traveling time, getting frustrated as well as health hazard caused due to vehicular fuel pollution. To overcome this problem, many parking management systems have been introduced which helps in providing information about exact real-time car parking space availability to drivers. In this paper, we are considering real-time smart parking system for collecting data of smart parking lot occupancy and parking restrictions on each lot from field sensor devices and then forwarded to servers for processing as well as usage for application which are on Google Play store and iOS users from App store. The main objective of this analysis is to help end users such as citizens and parking manager to ease traffic congestion and reduces time by using real-time bay sensor data and application to show driver exact location in parking lot. The focus of this paper is to analyze statistics of maximum stay of each car in parking lot, influencing traffic in parking restriction lot depending on the weekdays.

Keywords IoT · Smart parking · Smart traffic management

A. Waheed (✉) · J. Shafi · P. Venkata Krishna
Sri Padmavati Mahila Visvavidyalayam, Tirupati, Andhra Pradesh, India
e-mail: w_amtul@yahoo.com

J. Shafi
e-mail: janashafi09@gmail.com

P. Venkata Krishna
e-mail: pvk@spmvv.ac.in

© Springer Nature Singapore Pte Ltd. 2020
P. Venkata Krishna and M. S. Obaidat (eds.), *Emerging Research in Data Engineering Systems and Computer Communications*, Advances in Intelligent Systems and Computing 1054, https://doi.org/10.1007/978-981-15-0135-7_4

1 Introduction

With the increase of population in metropolitan cities, the rise of transport vehicular is also increasing. This leads to a shortage of availability for parking space due to the scarcity of available land [1].

A location used for parking of vehicles which can be structured or unstructured, paid or unpaid is defined as parking space. Parking is finding and occupying vacant space for keeping vehicle at a bay. A parking lot is defined as a cluster of parking location. Most popular traditional or manual parking system uses tokens and coins. The disadvantage of this parking system is it does not provide accurate details about any specific parking space and vehicle parked. As such systems are manual, it relies on person in charge for counting vehicles parked in parking space, number of free parking spaces, maximum number of parking spaces, and number of available spaces. There can be many variations as all the calculations are done manually. Due to this, it is not counted as smart parking system.

According to study conducted worldwide, 30–50% of the drivers search for free parking space [2–5] and spends 3.5–14 min to find a parking space [6, 7]. This leads to frustration, accidents, traffic congestion, lost professional opportunities and also increases in air pollution.

The study emphasis on the locations where there is scarcity of parking spaces and increase in traffic congestions [6]. Congestions occur mostly at peak traffic hours and traffic density is high [8, 9]. To reduce traffic and parking-related issues, field of smart parking service systems is introduced which is a part of intelligent transportation systems (ITS).

Smart technologies and smart parking sensors such as ultrasonic sensors, magnetometers, and multiagent systems are used to detect and gather parking occupancy information which helps drivers in navigating toward free parking space [10]. Smart parking is defined as parking vehicle with the use of applications and technologies for assistance. Many online smart parking applications are available which demonstrate efficient navigational directions to reserve empty parking spaces [11].

Many researches are conducted to improve parking efficiency at closed parking lot, restricted parking lots which are paid parking lots and offer reservation of parking spaces. By using smart parking services, parking space can be reserved by online payment through mobile application or web application. Once payment is done, parking space is reserved and the user will be provided with a reference code which can be used for authorization. And the vehicle can directly enter the parking lot and occupy a vacant or allotted parking space.

There is a very uncommon type of parking lot that is restricted parking lots in which parking spaces or areas that are designated for specific purposes and which does not support reservation, and but are available for a limited amount of time for a short period. Therefore, still, there are research breaches to improve parking efficiency at restricted parking lots.

Here, this paper focuses on smart parking tools include technologies, sensors, and application for improving parking efficiency and identifying parking occupancy information analysis for smart parking systems. Data are collected from large commercial shopping precinct smart parking deployment. We start our work along with some preliminary analysis discussing smart parking sensors, technologies, and applications used in real-time parking system. Then we analyze streaming of data process mechanism between smart parking system and users. Last through real datasets, our final considerations are analyzed and concluded by focusing on how maximum stay of each car in parking lot influences traffic in parking restriction lot depending on the weekdays.

2 Smart Parking Tools

Smart parking tools include technologies, sensors, and application for improving parking efficiency and identifying parking occupancy information.

2.1 Smart Parking Sensors

Sensors are widely used tools in smart parking area. There are many different types of sensors which help in detecting parking occupancy information. Below are some the mostly used sensors in smart parking system.

1. **Active infrared sensor.** These types of sensors are deployed at indoor closed parking lots as these sensors are very sensitive to environmental changes such as rain or snow and are not suitable for open parking lots. These sensors detect vehicle or object by the amount of energy reflected by the emitted infrared energy of sensors [12, 13]. High investment is needed for maintenance of these sensors.
2. **Passive infrared sensor**. These types of sensors work on basis of variation in energy such as when vehicle occupies a parking space, sensors sense a change in energy and detect occupancy [12, 13]. These sensors are very sensitive to environmental changes such as rain or snow; there is a possibility to show inaccurate result under certain environmental conditions. These sensors are positioned underground or on the ceiling which requires high investment for both procurement and maintenance of sensors. Mostly, these types of sensors are suitable for closed parking lots or inside building.
3. **Ultrasonic sensor**. These sensors are very sensitive to environmental changes such as rain or snow; therefore, these sensors are deployed on the ceiling and are suitable for indoor parking lots when compared to open parking lots [14]. These sensors detect objects based on reflected energy emitted by sound waves between 25 and 50 kHz. These sensors are placed on top of every parking space

to get parking occupancy status. Deployment and maintenance of this type of sensor are expensive but the sensors are available for low cost.

4. **Inductive loop detectors**. These detectors work on the principle of electro-magnetism using underground wiring system to detect the presence of a vehicle [12]. These types of sensors are used to keep the count of number of vehicles enter and exit from the parking lots to check the availability of parking space. These are mostly deployed at entrance and exit of indoor parking lots. Accuracy of these detectors is very high in counting the vehicles parked in multiple commercial parking lots.

5. **Parking guidance systems**. Parking guidance system uses inductive loop detectors or visual camera at entrance or exit of parking lot to calculate the vehicle in a parking lot and later displayed on the display screen near the parking lot from which driver can see and decide the parking space to occupy [15, 16]. Anyhow, driver has to find the parking space to occupy by viewing the display screens; they do not direct the driver to a particular parking space [14]. Investment for this installation is very less and also suitable for open parking lots.

6. **Radio-frequency identification (RFID)**. This tool can be used for both indoor and closed parking lots. Radio-frequency identification (RFID) is used to identify the vehicle by providing radio-frequency tags for identification. Antenna and transceiver are installed at the entrance of parking lots to identify the tag and then allowing vehicle to occupy a parking lot [17]. Open parking lot is freely available and so for this reason, RFID is not appropriate to deploy. Anyhow, it neither directs driver for finding vacant parking space nor does it provide individual parking occupancy status.

7. **Magnetometer**. These types of sensors are installed in both open and closed parking lots as they are not sensitive to environment. These sensors use elec-tromagnetic field to detect the vehicle by detecting the change in the field. These sensors are placed under the surface in order to be in close proximity to the vehicle [12]. To calculate occupancy of parking spaces, sensors have to be placed under every parking space. However, the investment for installation and maintenances of these sensors are very expensive on bulky scale.

8. **Microwave radar**. These tools are suitable for both open and closed parking lots as they are not sensitive to environment. Uses dual microwave radars for both moving and stationary vehicles, as it does not support to detect stationary objects; based on the reflected signal, it estimates the velocity of the moving target. These are deployed under the surface for vehicle detection. However, they are expensive to install and maintain these microwave radars on a large scale as should be placed in every parking space to detect parking occupancy status [18].

2.2 Smart Parking Technologies

Foremost technologies for occupying vacant parking space and providing intelligent parking system are summarized in this section.

1. **Global positioning system**. GPS is in smart parking system based on navigation directed by smart parking management choosing shortest or optimal route from the current location of driver for occupying a vacant parking space. Anyhow, GPS alone cannot collect information regarding parking spaces. Occupancy of parking spaces is estimated using navigation provided using GPS to the estimated parking space and historical occupancy information [19]. GPS services are inclined to errors when the signals are blocked because of walls inside the building or tall towers and underground building. GPS is suitable for outdoor open parking lots as there are very less chances of signal blocking when compared to indoor parking lots. And also, its accuracy is completely dependent on the signal availability through satellite.

2. **Machine vision**. In machine vision, visual camera is used for LRP-license plate recognition to identify parking lot occupancy. Cameras are placed near the entry of parking lot. Based on LRP, number of vehicles entered and exited are monitored which helps in counting vacant parking spaces. This system does not provide the occupancy status of parking spaces. In order to attain continuous monitoring of parking lot, video of higher bandwidths should be broken into images at regular intervals and frame rate [20]. These visual cameras for parking lots can cover large area in parking lot and investment expenditure is also very minimal [21].

3. **Vehicular ad hoc networks**. This type of technology is based on wireless communication devices offering services such as antitheft and smart parking. Vehicles are equipped with onboard units (OBU), parking lots with roadside units (RSUs) and both are registered with trusted authority [22]. As soon as a vehicle approaches the parking lot installed with RSUs, navigational information to the vacant parking space will be provided to OBU. These devices are suitable for both open parking lot and closed parking lot as it is not environmental sensitive. Anyhow, installation and maintenance are very expensive.

4. **Multiagent systems**. These are user-friendly parking systems which allow user according to their preferences to select parking space using mobile applications or web applications. Here, multiple mediums are used like visual camera, sensors, mobile applications, algorithms, etc. The user will also receive navigational information to reach the parking space [23]. These systems are suitable for both open and closed parking lots, and therefore, expenditure would be dependent on the usage of technology to identify occupancy status of parking spaces.

5. **Neural networks**. A neural network is based on brain nervous system for data processing. It has been evolved in various types over the years like fuzzy, neural network, feedforward, convolutional neural networks, and fluid neural network.

It is used in identifying license plates in real-time videos [24, 25]. Deep learning is a branch of machine learning which uses neural networks in object detection and classification. There is another evolving technology such as convolutional neural networks which would take images as input and is more efficient in analyzing images. In a recent study, convolutional neural networks were used along with machine vision to capture parking occupancy information efficiently [26]. This technology would function as an efficient tool in data processing while it is not involved in real-time data capturing. Therefore, it is suitable for open and closed parking lots with minimal expenditure.

6. **Fuzzy logic**. Fuzzy logic is based on multivalued logic in data processing and can also be used to develop forecasting models. Like neural networks, fuzzy logic can also be used in multiagent systems. Accuracy of this model is very high when combined machine vision and sensor technologies. These technologies are suitable for both open and closed parking lots and expenditure is also minimal. Fuzzy logic supports autonomy in providing information on the availability of parking spaces.

2.3 Smart Parking Applications

There are many free smart parking applications available on web and mobile stores of android or iOS which can be accessed by internet and smartphones. These applications help driver in taking decision for occupying a vacant parking space. These applications consider user preferences and choices while allotting parking slots. Smart parking applications serve as decision support systems in occupying available parking spaces as shown in Fig. 1.

Smart parking applications can improve efficiency in the following ways,

- Helps driver in allotting parking slot.
- Navigate driver to parking lot.
- Booking and acknowledging the driver to a parking lot.
- Guide and booking particular parking slot and navigating the driver to a specific parking space using navigational information.

2.4 Smart Parking System Model

Development and implementation of a smart parking system allow drivers to find and reserve the vacant parking locations. System determines parking status from the sensor networks deployed in parking locations. Results are to obtain reduction in search time of parking slot in parking lot. A block diagram of the smart parking system approach is shown in Fig. 2 and working model of smart parking system architecture is presented in Fig. 3.

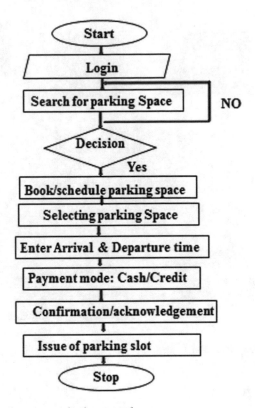

Fig. 1 Smart application communication control

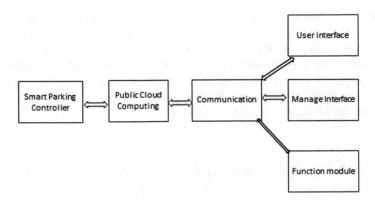

Fig. 2 Block diagram of smart parking system approach

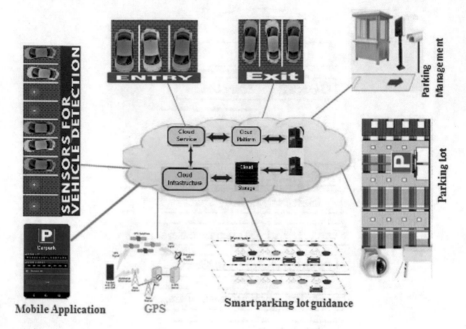

Fig. 3 Smart parking system model

3 Methodology

3.1 Python

With advancements in programming languages, python has become most popular and dynamic programming language for building web frameworks and by its large and active scientific computing community. Python programing language is widely exploring for data visualization, data computing, and data analysis and interactivity for domain-specific open sources. Python's extended libraries such as pandas, NumPy, matplotlib, IPython, and SciPy are extended alternative for data manipulation task for building data-centric application.

3.2 Analyzing Open Datasets Using Pandas in a Python

Open datasets are powerful analysis tool that connects multiple datasets to derive new perceptions and can be used by anyone without any restrictions or permissions, which are freely available to access, modify, and store. Most of the common providers of open datasets are academic institutions, government, and public agencies sharing data such as environmental, economic, census, and health datasets according to region or country-specific. In python, a panda is one of the effective and efficient data analysis packages which are very flexible and easy to load, index, classify, and group data.

1. Accessing dataset in python.
2. Open the notebook called analyze open datasets with pandas dataset.
3. Loading data from open datasets into a python notebook using pandas LINK-TO-DATA string.
4. Join the two different datasets using pandas merge function.
5. Check the data, run a first check of the data with describe().
6. Data analysis, create a scatter plot, and relate the values.
7. Customize the look by defining the font and figure size and colors of the points with matplotlib.

4 Data Collection

4.1 Current Used Datasets

Currently, we are accessing datasets from google dataset library for smart parking lots and smart parking restrictions. Details and references for each dataset can be found in the Google dataset web page, wiki, and source code. Licensed under Creative Commons Attribution 4.0 International, Variables measured is Parking Lot. Smart parking datasets are collected from Manuka shopping precinct and by using smart mobile application available in Google Play store and iOS user from App store for helping in simplifying traffic congestion and reducing travel time. The smart parking restrictions dataset shows the parking restrictions on each lot and the smart parking lot dataset shows the locations and describes each lot.

5 Observation

5.1 Dataset–1 (Smart Parking Lot)

Observation of smart parking lot dataset shows the correlation between lot code and maximum stay of vehicle on each parking lot as shown in Fig. 4.

- Parking lot code from 500 too 700 is fully occupied due to residential area.
- Parking lot code 200 and 300 is slightly fewer occupied due to street lane.
- Parking lot code 100 and 400 is least occupied due to street lane.

Fig. 4 Correlation between lot code and maximum stay of vehicle on each parking lot

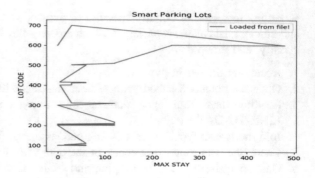

Fig. 5 Sunday is holiday, so traffic and parking are least whereas restricted parking is static

Sunday

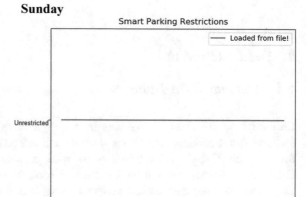

Fig. 6 Monday is highly crowded

Monday

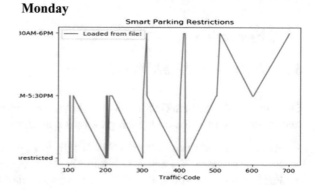

Fig. 7 Tuesday is highly crowded

Tuesday

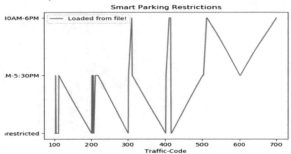

Fig. 8 Wednesday is highly crowded

Wednesday

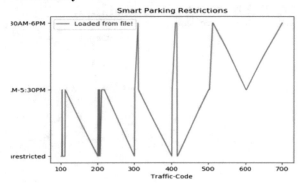

Fig. 9 Thursday is highly crowded

Thursday

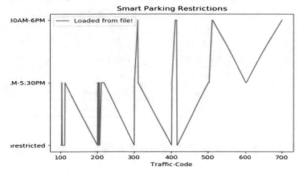

Fig. 10 Friday is highly crowded

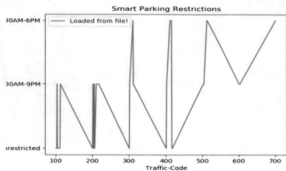

Fig. 11 Saturday least crowded due to weekend

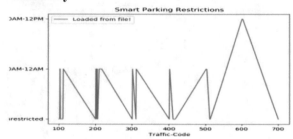

Table 1 Demonstrates the parking and traffic variation on weekdays and weekends

S. No	Days	Status	Observation	Figures
1	Sunday	Static/minimal	Holiday	Figure 5
2	Monday	Fully crowded	Working day	Figure 6
3	Tuesday	Fully crowded	Working day	Figure 7
4	Wednesday	Fully crowded	Working day	Figure 8
5	Thursday	Fully crowded	Working day	Figure 9
6	Friday	Fully crowded	Working day	Figure 10
7	Saturday	Least crowded	Weekend	Figure 11

5.2 Dataset–2 (Smart Parking Restrictions)

Observation of smart parking restriction dataset shows the correlation between peak hours of weekdays and traffic code of smart parking restriction lot.

The observation and analysis of dataset show that traffic code which lies between 300 and 500 is mostly fully occupied between 7:30 AM and 9:00 PM weekly on working days; this shows that unrestricted parking lots minimal on weekdays when compare to weekends and holidays (Table 1).

6 Conclusion

With the increase in the use of smart parking system, latest advancement in technologies is also gradually increasing to improve facilities provided to smart cities user. By the use of advanced technology in smart parking management system, user can get rid of traffic congestions, wasting time for finding vacant parking space especially at peak hours, and also health hazard caused due to vehicular fuel pollution making life easier. This paper analyzes smart parking tools, technologies, and application used in smart parking management system for open and closed parking lots along with their installation expenditure and environment conditions. Also analyzes real-time datasets for smart parking lot and smart parking restrictions on each lots which can help user and parking manager to ease traffic congestion and reduces time by using real-time bay sensor data and application to show driver exact location in parking lots by analyzing statistics of maximum stay of each car in parking lot, influencing traffic in parking restriction lot depending on the weekdays.

Advance research can be done by using deep learning and multiagent systems which help to provide real-time parking occupancy information along with navigational directions to the available parking space in an open and closed parking lot. Automated vehicles are the proposed future research in the field of smart parking that can be experimented to automated robotic parking valets, self-parking, and automated vehicle parking.

References

1. Chinrungrueng, J., Sunantachaikul, U., Triamlumlerd, S.: Smart parking: an application of optical wireless sensor network. In: International Symposium on the Applications and Internet Workshop, Japan, Hiroshima. IEEE, New York (2007)
2. Polak, J., Vythoulkas, P.: An assessment of the state-of-the-art in the modelling of parking behaviour. Report to Transport Research laboratory, Issue 753 (1993)
3. Boltze, M., Puzicha, J.: Effectiveness of the parking guidance system in Frankfurt am main. Parking Trend Int. 27–30 (1995)
4. White, P.: No Vacancy: Park Slopes Parking Problem And How to Fix It, in Internet. http://www.transalt.org/newsroom/releases/126. 7 Jan 2016 (2007)
5. Gallivan, S.: IBM global parking survey: drivers share worldwide parking woes technical report. Technical report, IBM (2011)
6. Shoup, D.C.: Cruising for parking. Transp. Policy 13(6), 479–486 (2006)
7. Polycarpou, E., Lambrinos, L., Protopapadakis, E.: Smart parking solution for urban areas. In: 2013 IEEE 14th International Symposium and Workshops on a World of Wireless, Mobile and Multimedia Networks (WoWMoM), Madrid, Spain (2013)
8. Geroliminis, N., Daganzo, C.F.: Existence of urban-scale macroscopic fundamental diagrams: some experimental findings. Transp. Res. B, Methodol. 42(9), 759–770 (2008)
9. Geroliminis, N., Sun, J.: Properties of a well-defined macroscopic fundamental diagram for urban traffic. Transp. Res. B, Methodol. 45(3), 605–617 (2011)
10. Yang, Z., Liu, H., Wang, X.: The research on the key technologies for improving efficiency of parking guidance system. In: Intelligent Transportation Systems, Proceedings of 2003. IEEE, Shanghai, China (2003)

11. Kotb, A.O., Shen, Y.C., Zhu, X., et al.: Iparker—A new smart car-parking system based on dynamic resource allocation and pricing. IEEE Trans. Intell. Transp. Syst. **17**(9), 2637–2647 (2016)
12. Shaheen, S.: Smart parking management field test: a bay area rapid transit (bart) district parking demonstration. Institute of Transportation Studies, Davis, CA, USA (2005)
13. Mouskos, K., Boile, M., Parker, N.A.: Technical solutions to overcrowded park and ride facilities. New Jersey Department of Transportation, New Jersey, NY, USA (2007)
14. Kianpisheh, A., Mustaffa, N., Limtrairut, P., et al.: Smart parking system (SPS) architecture using ultrasonic detector. Int. J. Softw. Eng. Appl. **6**(3), 51–58 (2012)
15. Idris, M., Tamil, E.M., Noor, N.M., et al.: Parking guidance system utilizing wireless sensor network and ultrasonic sensor. Inf. Technol. J. **8**(2), 138–146 (2009)
16. Waterson, B., Hounsell, N., Chatterjee, K.: Quantifying the potential savings in travel time resulting from parking guidance systems-a simulation case study. J. Oper. Res. Soc. **52**(10), 1067–1077 (2001)
17. Rahman, M.S., Park, Y., Kim, K.-D.: Relative location estimation of vehicles in parking management system. In: 11th International Conference on IEEE Advanced Communication Technology, ICACT 2009, Phoenix Park, South Korea (2009)
18. Bao, X., Zhan, Y., Xu, C., et al.: A novel dual microwave Doppler radar based vehicle detection sensor for parking lot occupancy detection. IEICE Electron. Express **14**(1), 1–12 (2017)
19. Pullola, S., Atrey, P.K., El Saddik, A.: Towards an intelligent GPS-based vehicle navigation system for finding street parking lots. In: IEEE International Conference on Signal Processing and Communications, ICSPC 2007, Dubai, United Arab Emirates (2007)
20. Enríquez, F., Soria, L.M., Álvarez-García, J.A., et al.: Existing approaches to smart parking: an overview. In: International Conference on Smart Cities, Malaga, Spain (2017)
21. Ichihashi, H., Notsu, A., Honda, K., et al.: Vacant parking space detector for outdoor parking lot by using surveillance camera and FCM classifier. In: IEEE International Conference on Fuzzy Systems, FUZZ-IEEE 2009, Jeju Island, South Korea (2009)
22. Lu, R., Lin, X., Zhu, H., et al.: An intelligent secure and privacy-preserving parking scheme through vehicular communications. IEEE Trans. Veh. Technol. **59**(6), 2772–2785 (2010)
23. Bilal, M., Persson, C., Ramparany, F., et al.: Multi-agent based governance model for machine-to-machine networks in a smart parking management system. In: IEEE International. Conference on Communications, Ottawa (2012)
24. Rahman, C.A., Badawy, W., Radmanesh, A.: A real time vehicle's license plate recognition system. In: Proceedings of IEEE Conference on Advanced Video and Signal Based Surveillance, Miami, FL, USA (2003)
25. Villegas, O.O.V., et al.: License plate recognition using a novel fuzzy multilayer neural network. Int. J. Comput. **3**(1), 31–40 (2009)
26. Amato, G., Carrara, F., Falchi, F., et al.: Deep learning for decentralized parking lot occupancy detection. Expert Syst. Appl. **72**, 327–334 (2017)

Implementation Study and Performances Evaluation of an 802.11ad Model Under NS-3

Nouri Omheni, Imen Bouabidi, Faouzi Zarai and Mohammad S. Obaidat

Abstract IEEE 802.11ad is an amendment to the current WLAN 802.11 standard. The goal of this amendment 802.11ad is to deliver lower latency and higher data throughput results away from more crowded spectrum bands by operating in the 60 GHz millimeter wave band. Nevertheless, as of today, limited simulation models for this new amendment are conducted. In this chapter, we present a detailed implementation model of this standard under NS-3 network simulator environment. In addition, a series of simulation experiments have been detailed to study the performance of this new technology in terms of throughput, jitter and latency performance metrics.

Keywords IEEE 802.11ad · NS-3 · 60 GHz millimeter wave band · Modeling and simulation · Performance evaluation

N. Omheni (✉) · I. Bouabidi · F. Zarai
New Technologies and Telecommunications Systems Research Unit,
Department of Telecommunications, National School of Electronics
and Telecommunications of Sfax, Sfax, Tunisia
e-mail: nouri.omheni@gmail.com

F. Zarai
e-mail: faouzifbz@gmail.com

M. S. Obaidat
ECE Department, Nazarbayev University, Astana, Kazakhstan
e-mail: m.s.obaidat@ieee.org; msobaidat@gmail.com

King Abdullah II School of Information Technology, University of Jordan,
Amman, Jordan

University of Science and Technology, Beijing, China

The Amity University—A Global University, Noida, India

© Springer Nature Singapore Pte Ltd. 2020
P. Venkata Krishna and M. S. Obaidat (eds.), *Emerging Research in Data Engineering Systems and Computer Communications*, Advances in Intelligent Systems and Computing 1054, https://doi.org/10.1007/978-981-15-0135-7_5

1 Introduction

The IEEE 802.11ad standard allows multi-gigabit wireless communications in the 60 GHz band [1]. The Wi-Gig specification was delivered to the IEEE 802.11ad and was confirmed in May 2010 as the basis for the 802.11ad project. The first current standards for the wireless network (802.11a and b) were designed mainly to meet the requirements of a laptop in the home and office, and later to permit connectivity in airports, hotels, and shopping centers [2]. Their key function was to offer a link to a broadband connection for Web activities. Nevertheless, at the same time, new usage models with the need for higher throughput have been recognized; data sharing between home and small office connected devices and wireless printing as examples. The key goal of 802.11ad is to allow faster connection in dense and ultra-fast deployment environments in homes, which is partially achieved because 60 GHz has a shorter range, and therefore, other networks are much less likely to intercept and interfere with the connection [3].

In general, a network simulator is made up of a wide range of technologies and network protocols and helps users build complex networks from basic building blocks such as clusters of nodes and links. NS-2 has been widely used in the research community [4, 5]. More than one hundred people and organizations have contributed to the simulation code, and the use of the simulator is still referenced in many network research works. However, a major gap in NS-2 is its scalability in terms of memory usage and simulation runtime.

The NS-3 simulator aims to be a replacement to NS-2, written in C++, python and object-oriented version of Tel (OTel), to try to overcome its limitations (the use of multiple interfaces on a node) [5, 6]. It can be used on Linux/Unix, OS X (Mac) and Windows platforms. Its development first began in July 2006, and it is funded by institutes such as the University of Washington, Georgia Institute of Technology and the ICSI Center for Internet Research. The first public and stable version was published in June 2008. The NS-3 developers decided that the simulation architecture needed to be completely reworked. From this perspective, the NS-2 experience must be associated with advances in programming languages and software engineering. The idea of backward compatibility with NS-2 was dropped from the start. This act frees NS-3 from inherited NS-2 constraints and allows the construction of a simulator that is well designed from the beginning. The major problem of the NS-3 simulator and all other network simulators is that it does not allow simulation of new standards and technologies.

To our knowledge, there is a very limited number of works in the literature that have studied the implementation of the new 802.11ad standard under NS-3. That is why in this chapter, we conduct a detailed study of the implementation of this standard under NS-3.

In addition, a series of simulations have been made to study the performance of this new technology in terms of throughput, jitter and latency. The rest of the chapter is organized as follows. In Sect. 2, we conduct a theoretical study of the IEEE802.11ad amendment, and we detail and criticize some related works.

Section 3 describes the implementation steps of the amendment. In Sect. 4, we present our simulation results. Finally, Sect. 5 concludes the chapter.

2 Theoretical Study of IEEE 802.11ad and Related Works

2.1 IEEE 802.11ad

The Wi-Gig specification explains the physical (PHY) and medium access control (MAC) layers based on IEEE 802.11 [1] as shown in Fig. 1.

2.1.1 Physical Layer (PHY)

The IEEE 802.11ad physical layer allows having three different modulation techniques: Spread Spectrum Modulation, Single Carrier (SC), and Orthogonal Frequency Division Multiplex (OFDM).

Each PHY technique has a distinct purpose and a packet structure. Figure 2 shows these structures.

All types of the physical layer have different contents in their package structures. All packet structures contain the preamble, the header, and the payload. In the physical layer, each version supports a different transmitter to another.

(a) *Transmitter of the physical layer of SC (signal carrier)*: This 802.11ad physical layer mode supports data rates ranging from 385 to 4620 Mbps, which is dependent on MCS (Modulation and Coding Scheme). The basic packet structure is the same in low power SC and SC versions, but 16QAM is not used in low power SC mode to save power. The transmitter of the SC PHY layer consists of a scrambler, an LDPC encoder (Low-Density Parity Check),

Fig. 1 Architecture of the IEEE 802.11ad amendment

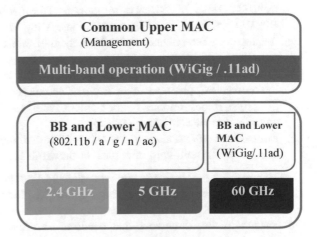

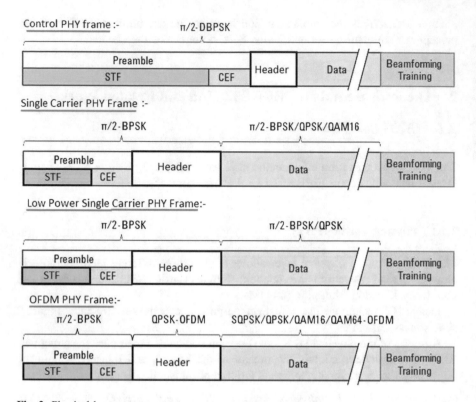

Fig. 2 Physical layer structures

a repetition block (2X), a CP insertion, and a modulation before the packet is formed according to the structure mentioned above. No spectrum configuration is defined. After the baseband processing, the I/Q packet is modulated using the RF bearer and transmitted into the air. About 12 modulation coding schemes from MCS1 to MCS12 are supported. These modulation schemes include BPSK, QPSK, or 16QAM modulations. They use RF support at the center frequency of the channel and with a symbol rate of about 1.76 Gsymbol per second.

(b) **Transmitter of the physical layer of SC with low power (signal carrier)**: This low power SC physical layer transmitter consists of scrambler, RS encoder (224,208), block encoder, block interleavers, CP insertion, and modulation types ($\pi/2 - $ BPSK, $\pi/2 - $ QPSK).

(c) **Transmitter and receiver of the OFDM physical layer**: The OFDM transmitter consists of the following modules as shown in Fig. 3.

The repetition for header is applied to the 802.11ad header part. The same header data information is available three times in the 802.11ad frame. This helps to correct errors at the receiver. This is necessary because the header is the most critical information in the framework that needs to be protected. The mapping module

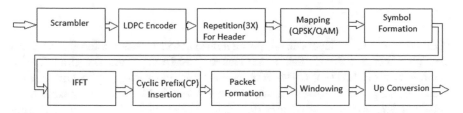

Fig. 3 OFDM TX

converts bits into complex symbols. Various modulation schemes such as SQPSK, QPSK, 16-QAM, and 64-QAM are used according to MCS. The symbol formation module takes care of the formation of OFDM symbols. The OFDM 802.11ad physical layer version uses 16 pilot subcarriers, 1 DC support, 336 data subcarriers, and 159 guard subcarriers. This is formed according to the IFFT structure. The IFFT (Inverse Fast Fourier Transform) module converts frequency domain data symbols to time-domain data symbols. The cyclic prefix insertion is the insertion of the guard interval. About 25% of the overhead is used in the OFDM 11ad physical layer. In other words, the 802.11ad standard uses a cyclic prefix equal to 1/4.

The different steps applied to the signal at the OFDM receiver are as summarized below:

- The conversion and unlocking of First Down are applied to the received packet at the OFDM PHY 11ad receiver. The next forward synchronization is performed as described in the next steps before the demodulation and decoding are performed.
- Windowing is detected by threshold detection. The basic algorithm should be implemented in a way that differentiates the noise and the real 11ad WLAN packet.
- The following time difference is estimated and corrected at the sample resolution level.
- The following frequency offset estimation and correction are performed.
- After performing FFT, complex channel response coefficients are found using the frequency domain channel estimation method using preamble sequences (either STF or CEF, or both).
- Now, the channel equalization is performed using estimated channel response coefficients for each of the OFDM packet symbols.
- After equalizing the channel using the channel response, the phase rotation is performed using estimated phase rotation utilizing driver subcarriers embedded in the OFDM symbol. This is essentially called phase disconnection
- A symbol deformation is performed, which removes the drivers, DCs, and guard subcarriers from each of the OFDM symbols.
- The carrier demapping is performed on the data subcarriers according to the needs according to MCS. It can be QPSK, 16QAM, or 64QAM.

2.1.2 MAC Layer

IEEE 802.11ad MAC layer can support a hybrid access technique, the Carrier Sense Multiple Access with Collision Avoidance (CSMA/CA) and the time-division multiple access (TDMA). CSMA/CA is used for a burst-like application, such as Web browsing due to lower average latency, while TDMA is more desirable for video transmission because of its better quality of service and its efficiency. The rate of the MAC layer is determined by the amount of information bits exchanged between the MAC transceivers and the time required to provide the information successfully.

The 802.11ad specification supports two types of modulation and coding, which offer different advantages: (1) Orthogonal frequency-division orthogonal multiplexing (OFDM), which supports communication over longer distances with longer delay spreads, providing more flexibility in managing obstacles and reflected signals. In addition, OFDM allows the highest transmission speeds up to 7 Gbps. (2) Signal carrier (SC) usually results in lower power consumption, so it is often suitable for small and low-power portable devices.

2.2 Related Works

A few number of works were interested in the implementation and simulation of the new Wi-Fi standards, especially mmwave technologies. Authors in [7] described a model for IEEE 802.11ad implemented using NS-3. It is an interesting work, nevertheless, enormous tasks remain to be done, especially the addition of the different techniques and mechanisms defined in the IEEE 802.11ad amendment. A similar study was presented by [8]. Authors of this work provided a design for simulating IEEE 802.11ad in NS-3. Particularly, they described in their implementation different techniques of beamforming present in the amendment. However, the implementation details are not publicly available. Authors in [9] exploit an IEEE 802.11ad PHY layer to create multi-gigabit links in data centers using NS-3. In their implementation, they use data rates provided for both SC and OFDM PHY layers in the amendment. Additional information pertaining to modeling and simulation of wireless networks can be found in [10, 11].

3 IEEE 802.11ad Implementation

To install NS-3, we need to have a Linux interface, and this is either done by installing Linux directly on the machine or using a virtual machine (Virtual Box or VMware), so we can use Ubuntu as Linux interface.

For the implementation of the IEEE 802.11ad model under NS-3, we added the different classes required in the different models under the directory "ns-allinone-3.26 /ns-3.26 /src /wifi." To add these classes, we used the C++ programming language. Moreover, we used the Eclipse tool. The latter must be configured because the NS-3 platform does not support it. Eclipse is an integrated development environment (IDE) used in computer programming and Java [12]. It contains a basic workspace and an extensible plug-in system to customize the environment. Eclipse is mainly written in Java, and its main use is to develop Java applications. The development environments include Eclipse Java Development Tools (JDT) for Java and Scala, Eclipse CDT for C/C++ and Eclipse PDT for PHP, among others.

The current Wi-Fi model in NS-3 supports different IEEE 802.11 specifications such as a /b /e /g /n /ac with precise implementation of the MAC layer. The Wi-Fi model can be divided into four layers as follows:

- Wi-Fi Channel: An analytical approximation of the physical medium on which data is transmitted (i.e., air in the case of Wi-Fi), consisting of propagation loss and delay models.
- Wi-Fi PHY: The PHY part of the protocol supports sending and receiving packets and determining the loss due to interference.
- Mac Low: It implements RTS/CTS/DATA/ACK transactions, Distributed Coordination (DCF) and Enhanced Distributed Channel Access (EDCA), packet queues, fragmentation, retransmission, and frequency control.
- Mac High: It provides certain features of the Mac Layer Management Entity (MLME) based on the underlying network that it supports, such as the Base Infrastructure System (BSS) or BSS-independent.

Figure 4 describes the architecture of the Wi-Fi model. Modules in gray are modified to support the IEEE 802.11ad standard. In our implementation, we have added new features at the physical layer and Mac layer.

The physical layer's functionalities already exist in the "WiFi-phy" and "yans-WiFi-phy" models. In this layer, we configured the different MCS (Modulation and Coding Scheme, MCS0 to MCS 24) that are supported by the standard. Figures 5 and 6 illustrate configuration and additions excerpts of the MCS under NS-3.

The features of the MAC layer exist in "Wi Fi-Mac" and "Wi Fi-mode" models. In this layer, we have configured the different interframe timers such as SIFS and DIFS.

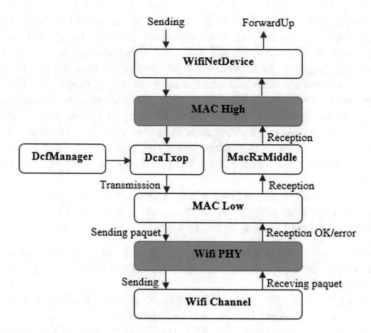

Fig. 4 Wi-Fi model architecture

```
void
YansWifiPhy::Configure80211ad_OFDM (void)
{
  NS_LOG_FUNCTION (this);
  m_channelStartingFrequency = 60e3;  //60.000 GHz
  SetChannelWidth (2160); //2160 MHz

  m_deviceRateSet.push_back (WifiPhy::GetOfdmRate7GbpsVHT ());

  m_deviceMcsSet.push_back (WifiPhy::GetVhtMcs13_OFDM ());
  m_deviceMcsSet.push_back (WifiPhy::GetVhtMcs14_OFDM ());
  m_deviceMcsSet.push_back (WifiPhy::GetVhtMcs15_OFDM ());
  m_deviceMcsSet.push_back (WifiPhy::GetVhtMcs16_OFDM ());
  m_deviceMcsSet.push_back (WifiPhy::GetVhtMcs17_OFDM ());
  m_deviceMcsSet.push_back (WifiPhy::GetVhtMcs18_OFDM ());
  m_deviceMcsSet.push_back (WifiPhy::GetVhtMcs19_OFDM ());
  m_deviceMcsSet.push_back (WifiPhy::GetVhtMcs20_OFDM ());
  m_deviceMcsSet.push_back (WifiPhy::GetVhtMcs21_OFDM ());
  m_deviceMcsSet.push_back (WifiPhy::GetVhtMcs22_OFDM ());
  m_deviceMcsSet.push_back (WifiPhy::GetVhtMcs23_OFDM ());
  m_deviceMcsSet.push_back (WifiPhy::GetVhtMcs24_OFDM ());

  m_bssMembershipSelectorSet.push_back (VHT_PHY);
}
```

Fig. 5 MCSs adding

```
WifiMode
WifiPhy::GetVhtMcs1_SC ()
{
  static WifiMode mcs =
    WifiModeFactory::CreateWifiMcs ("VhtMcs1_SC", 1, WIFI_MOD_CLASS_VHT_SC);
  return mcs;
}
WifiMode
WifiPhy::GetVhtMcs2_SC ()
{
  static WifiMode mcs =
    WifiModeFactory::CreateWifiMcs ("VhtMcs2_SC", 2, WIFI_MOD_CLASS_VHT_SC);
  return mcs;
}
WifiMode
WifiPhy::GetVhtMcs3_SC ()
{
  static WifiMode mcs =
    WifiModeFactory::CreateWifiMcs ("VhtMcs3_SC", 3, WIFI_MOD_CLASS_VHT_SC);
  return mcs;
}
```

Fig. 6 MCSs configuration

4 Simulation and Performances Study

4.1 Simulation Parameters

In the simulated scenario, an access point (AP) and a Wi-Fi station (STA) are connected to a WLAN infrastructure. We have simulated the following parameters: the UDP rate, the jitter, and the latency for each rate value, which depends on the MCS value (1–12 for SC and 13–24 for OFDM), the bandwidth (2.16 GHz), and the guard interval (long or short). PHY throughput is constant during the full simulation run. Simulations' parameters are illustrated in Table 1.

Table 1 Simulations parameters

Mobility model	Constant position mobility model
Packets size	1472 bytes
Manager	Constant rate Wi-Fi manager
Propagation speed	Constant speed propagation delay model
Propagation model	Friis propagation loss model
Guard interval	SGI, LGI
MCS	$1 \rightarrow 12$ (SC), $13 \rightarrow 24$ (OFDM)
Simulation time	20 s

4.2   Simulation Results

4.2.1   Throughput

Figures 7 and 8 show the evolution of the throughput for the different MSCs for both SC and OFDM modes. We observe that the practical rate of the IEEE 802.11ad standard is almost equal to the half of the theoretical rate. For example, with the SC modulation in the MCS-12, the theoretical throughput can reach 4.6 Gbps. However, in practice, the flow rate does not exceed 2.5. To increase throughput, the 802.11ad amendment allows using a technique to reduce inter-symbol interference called the guard interval. The short guard interval called SGI (Short Guard Interval), and the long guard interval called LGI (Long Guard Interval).

The throughput variation for the two guard intervals; short and long is depicted in Fig. 9. We can observe that the throughput with LGI is greater than that with SGI. This can be justified by the fact that with SGI, there is a chance of the existence of a higher inter-symbol interference than the LGI. For example, in the MCS-12 with SGI, the throughput reaches 2597 Gbps. However, with LGI, it reaches 2.85 Gbps. For the identical MCS, it reaches 2582 Gbps without SGI and LGI.

Figure 10 describes the throughput evolution in MCS (OFDM), SGI, and LGI. We can note the same conclusions as the two previous curves. The dissimilarity is that for this scenario we used OFDM modulation for MCS 13 to MCS 24. This method allowed offering higher throughput than SC modulation. For example, in the MCS-24, the throughput reaches 3.35 Gbps with the use of a short guard interval. However, it reaches 3.678 Gbps with a long guard interval and for the same MCS. While without SGI and LGI, the throughput is around 3.3 Gbps only.

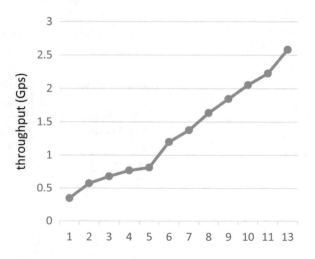

Fig. 7 Throughput versus MCS (SC)

Fig. 8 Throughput versus MCS (OFDM)

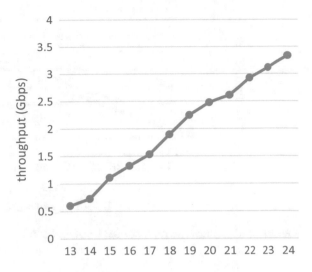

Fig. 9 Throughput versus MCS (SC), SGI, and LGI

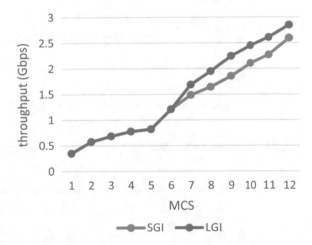

4.2.2 Jitter

In this scenario, we have characterized the jitter evolution as a function of MCS for both SC and OFDM modulation. From graphs shown in Figs. 11 and 12, we observe that the jitter with the SC modulation is better than with the OFDM modulation. For example, the smallest jitter value with SC in the MCS-12 is equal to 9.024 ms, while with OFDM in the MCS-24, it is in the order of 3.47 ms.

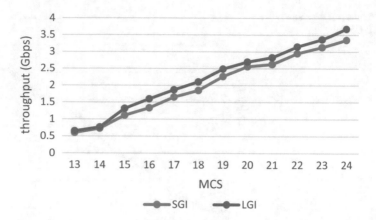

Fig. 10 Throughput versus MCS (OFDM), SGI, and LGI

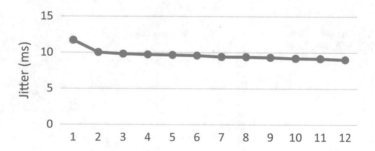

Fig. 11 Jitter versus MCS (SC)

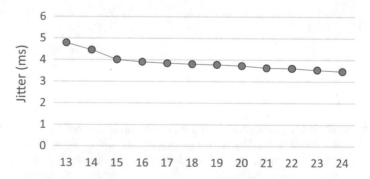

Fig. 12 Jitter versus MCS (OFDM)

4.3 Latency

In this third scenario, we have illustrated the latency evolution as a function of MCS for the two types of modulation SC and OFDM. From Figs. 13 and 14, we notice that the average value of latency does not exceed 10 ms. This value justifies the high flows offered by this standard. We also note that for OFDM modulation, we

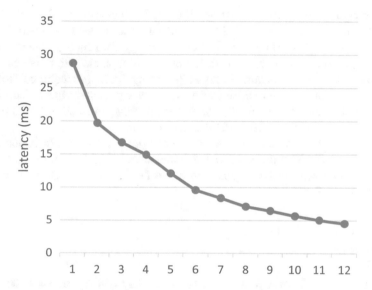

Fig. 13 Latency versus MCS (SC)

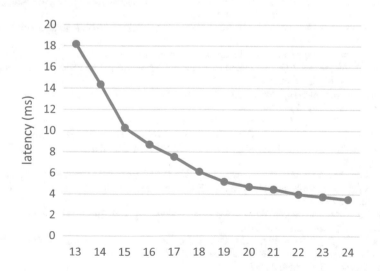

Fig. 14 Latency versus MCS (OFDM)

have always obtained a lower latency than SC modulation. For example, in the first MCS with SC modulation, the latency value is equal to 28.7 ms, and with the OFDM modulation, it is equal to 18.15 ms.

5 Conclusion

In this chapter, we described a comprehensive implementation study of an IEEE 802.11ad model. Then, we conducted a deep evaluation through a series of scenarios under the NS-3 simulator. The considered performance evaluation metrics are latency, jitter, and throughput. In the proposed implementation, we added different classes, which are required under NS-3 using the Eclipse computer tool. The Wi-Fi model that has been added under NS-3 has can support the major of IEEE 802.11 amendments like b/g/n/ac... with a precise implementation of the physical and MAC layers of the IEEE 802.11ad amendment.

As upcoming work, we will ameliorate the proposed model to support other mechanisms defined by the 802.11ad standard such as the frame aggregation. We also think to implement further amendments in the family of 60 GHz technologies such as the IEEE 802.11ay.

References

1. IEEE 802.11ad, Amendment 3: Enhancements for Very High Throughput in the 60 GHz Band. IEEE 802.11 Working Group (2012)
2. IEEE 802.11-2012, Part 11: Wireless LAN Medium Access Control (MAC) and Physical Layer (PHY) Specifications, IEEE std. (March 2012)
3. Maslennikov, R., Lomayev, A.: Implementation of 60 GHz WLAN channel model. Technical report, IEEE (2010)
4. http://www.isi.edu/nsnam/ns/ns-documentation.html-nsmanual
5. Obaidat, M.S., Boudriga, N.: Fundamentals of Performance Evaluation of Computer and Telecommunication Systems. Wiley (2010)
6. https://www.nsnam.org/docs/tutorial/ns-3-tutorial.pdf,ns-3. Tutorial Release ns-3-dev (November 26 2017)
7. Assasa, H., Widmer, J.: Implementation and evaluation of a WLAN IEEE 802.11ad Model in ns-3. In: Proceedings of the Workshop on ns-3, WNS3 '16, pp. 57–64 '(2016)
8. May, M.: Modelling and Evaluation of 60 GHz IEEE 802.11 Wireless Local Area Networks in ns-3 (2014)
9. Halperin, D., Kandula, S., Padhye, J., Bahl, P., Wetherall, D.: Augmenting data center networks with multi-gigabit wireless links. In: Proceeding of the ACM SIGCOMM Conference (2011)
10. Obaidat, M.S., Nicopolitidis, P., Zarai, F.: Modeling and Simulation of Computer and Network Systems. Elsevier (2015)
11. Obaidat, M.S., Papadimitiou, G.I. (eds.): Applied System Simulation: Methodologies and Applications. Kluwer (Springer) (2003). ISBN: 1-4020-7603-7
12. http://www.eclipse.org/eclipse/

Distributed Trust-Based Monitoring Approach for Fog/Cloud Networks

Wided Ben Daoud, Amel Meddeb-Makhlouf, Faouzi Zarai and Mohammad S. Obaidat

Abstract The fog computing paradigm presents an extension of the cloud computing in order to more ensure the requirements of Internet of things (IoT). Hence, the growth of fog-IoT system and their openness to their socioeconomic environment has led to new needs in terms of security. There are many damages resulting from security breaches and violations. In order to address these challenges, this article proposes a novel framework to further enhance the data security and privacy and the system performance, for the IoT users trying to connect to fog network. In fact, a trust-based monitoring strategy is suggested to avoid system's damages and attacks, and also to maintain low latency, because it is easier to avoid damage resulting from security breaches than to wait for a violation and to try to restore and repair the damage caused. The findings of our evaluation indicate that the proposed monitoring process is an efficient solution for system supervision, and detection and prevention attacks, with a fast processing time.

Keywords Access control · IoT · Fog computing · Security · Monitoring · Trust · Risk

W. B. Daoud (✉) · A. Meddeb-Makhlouf · F. Zarai
(NTS'Com) Research Unit, ENET'COM, University of Sfax, Sfax, Tunisia
e-mail: wided.bendaoud@gmail.com

A. Meddeb-Makhlouf
e-mail: amel.makhlouf@enetcom.usf.tn

F. Zarai
e-mail: faouzifbz@gmail.com

M. S. Obaidat
Department of ECE, Nazarbayev University, Astana, Kazakhstan
e-mail: msobaidat@gmail.com

University of Jordan, Amman, Jordan

© Springer Nature Singapore Pte Ltd. 2020
P. Venkata Krishna and M. S. Obaidat (eds.), *Emerging Research in Data Engineering Systems and Computer Communications*, Advances in Intelligent Systems and Computing 1054, https://doi.org/10.1007/978-981-15-0135-7_6

1 Introduction

Cloud computing that can provide on-demand resilient computing resources is seen as an appropriate way of efficiently processing the massive information generated by IoT devices [1]. Yet, cloud computing for IoT also carries two concerns: large amount of resource demand and high transmission latency. Hence, the exchange of the huge amount of data from IoT devices to the cloud needs an enormous amount of time, energy, and bandwidth. Moreover, long-distance transmission may produce a longer delay. For dealing with those problems, the fog computing is proposed. As a result, fog computing can significantly reduce the time processing and enhance and optimize the resource utilization. Besides, since the fog node is closer to the IoT devices, the end-to-end delay can be reduced [2].

Potentially, usually, the first challenge for these environments is the security aspect. Hence, in fog computing, data exchange is important in terms of volume and authenticity; however, the traditional application of network-level security is unlikely to be adequate due to the highly need of a distributed security architecture.

Additionally, by using wireless communications to interconnect nodes, this can make the system vulnerable to attacks, such as sniffing, spoofing, and denial-of-service (DoS) attacks [2]. Besides, the fog network is a recent trend within, security and privacy protection persist an understudied [3].

To address the above problems of existing systems, we based our work on a monitoring process. In fact, to supervise the overall traffic between the cloud and the fog computing, which are combined to serve the IoT devices, collaborative controllers are introduced. Controllers implement the proposed monitoring process that evaluates and updates the trust level of connected users to protect the network from attackers during user access. Moreover, if the trust level is maintained during user access, then a certificate is delivered to this user as a trust authorization for the next access.

This paper is organized as follows. Section 2 presents the problem statement for our work. Section 3 describes the proposed monitoring-based trust assessment scheme. Then, in Sect. 4, we introduce the results of the model's evaluation. We conclude our work by a conclusion in Sect. 5.

2 The Problem Statement

Information systems are exposed to attacks achieved by malicious users. This exposes the data to the risk of damaging and destruction. Users need legitimate and appropriate access to effectively perform their work. However, at any time, they may abuse their privileges accidentally or not [4].

These attacks can have a negative impact on the organization's reputation and productivity, leading to significant losses in revenue and customer.

Systems in fog can stop attacks if access control methods succeeded in reacting when a user performs inappropriate actions for their normal work functions. Moreover, these attacks could be avoided if users are monitored to identify suspicious activities and adapt them to negative changes in user's behavior [5].

Therefore, we need to design and implement techniques and methodologies that are adaptive to the dynamic behavior of fog and IoT environments. Thus, we propose a security model, where agents operating at the fog nodes offer intelligent supervision services, and monitoring strategies.

3 Proposed Distributed Trust-Based Monitoring (DTM) Approach

The intention of the present research begins from a previous work [1, 6]. Indeed, anomalies and suspicious task have been distinguished. But we believed that there exists a correlation between two or more threats and that can lead to other threats. These threats have to be resolved before damaging and destructing the entire network.

The monitoring function is a basic solution to the security management system. There are four steps in control loop that may be deployed in some management systems (Fig. 1). These steps are: monitoring, analyze, planning, and execution, where the control loop cannot be carried out deprived of the monitoring function [7].

The monitoring security system has to be highly available, as needed to protect data exchange and supervise the user's behavior. Yet, such a temporal failure during the monitoring procedure affects the global functioning of the connected systems (cloud, fog, and IoT).

Therefore, we need to design and implement techniques and methodologies that are adaptive to the dynamic behavior of fog and cloud computing. Therefore, we propose a security model, where agents operating at the fog nodes and cloud layers offer intelligent data access services, supervision services, and monitoring strategies.

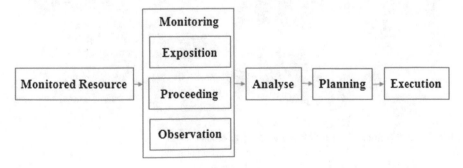

Fig. 1 DTM overall process

Furthermore, defining such a highly secure and available monitoring system becomes more challenging because of the large heterogeneity, the massive distribution, and the highly dynamic feature of the cloud/Fog/IoT networks.

3.1 DTM Architecture

In this paper, we based our work on the monitoring process. To supervise the overall traffic between cloud and fog computing, which are combined to serve the IoT devices, collaborative controllers are introduced. Then, the supervision reports are shared between different systems, leading to a more effective use of the network infrastructure. As illustrated in Fig. 2, the system based on multiple agents is able to manipulate dynamic configurations, heterogeneity, and volatility offer a promising approach to address these requirements.

Because the monitoring process is running to supervise connected users, we assume that the access procedure of the user is done successfully. Our main goal is to maintain the security of data stored or exchanged between the cloud or fog system on one side and the users on the other side. That is, the monitoring agent is placed at the fog nodes and data centers. The purpose of this deployment is to control the activities of the user and capture the anomalies to stop the malicious actions. In fact, our proposed system is proactive; it reacts before the attack is done.

By monitoring the behaviors of each user, if he had completed his missions without a deviation of the behavior, based on the output of the monitoring, the

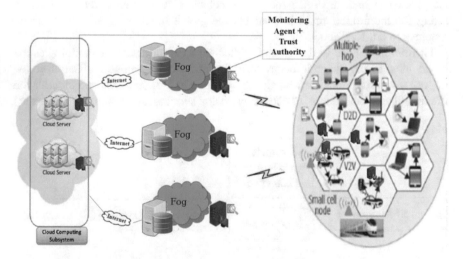

Fig. 2 Architecture of the DTM in wireless fog–cloud

Trust Level	ID_User	ID_Fog	ID_Cloud	Period of access	TA sign

Fig. 3 Trust authorization (auth_tr) format

system gives him a trust authorization. In his next access, this user does not repeat all the procedure of the access, since he has this trust authorization, quite simply, he must present it. The concept of trust authorization is presented as a security certificate indicating the trust level of each user. This certificate contains the following fields (as in Fig. 3):

Where:

- **Trust level**: indicating the trust level of each user. It is composed of three fields: lev_trust_fog/lev_trust_cloud. At each network layer of our architecture, a trust level is allocated to the user in order to be authenticated (or not) in front of the next layer.
- **ID_User**: indicating the identity of this user.
- **ID_Fog, ID_Cloud**: indicating the identifier of the element of the layer that has been accessed by this user and which has already done the monitoring and calculated the trust.
- **Period of access**: indicating the period during which the user has the right to access the fog/cloud without going through the procedure of access, while taking into account the field level of trust. In fact, if the access period is still valid, but the level of trust is decreased, the certificate will be revoked automatically, the access permission will be rejected, and this user must return to the initial phase of the access request.
- **TA sign**: indicating the signature of the Trust Authority, which is a component that is integrated with the monitoring agent.

3.2 The Monitoring Process

For the purpose of collaboration, the monitoring process reports the user behaviors and sends them to the fog nodes. In addition, the fog node is responsible for filling the field of the certificate, the level of the trust of this user. As soon as the user reaches another fog node, he presents his certificate to have the required amount of resources. Within the fog, the monitoring is functional until the end of the activity of the user. This procedure is repeated in the cloud as well as for each user. Figure 4 represents the whole steps of the proposed monitoring system, where Auth_Tr is the trust authorization and Tr is the trust level of such a user.

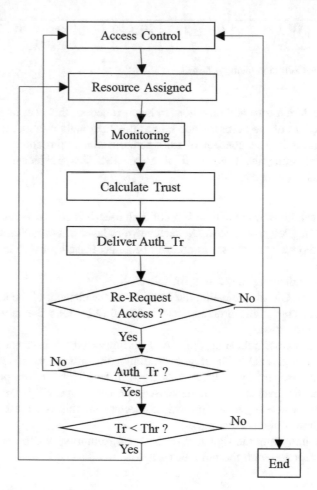

Fig. 4 Flowchart of the proposed system

3.3 Monitoring System Specification

To overcome the limit of existing strategies of security that are not adapted to the fog/cloud wireless environment, mostly due to the openness and the size of the network, the monitoring agents are placed on the fog nodes and in the cloud infrastructure level. The purpose of these agents is to monitor the behaviors of users who are already authenticated by the fog/cloud network. Several users enter the network legally, but they change their behaviors afterward and do attack and malicious activities.

Proposed monitoring agents are implemented within fog nodes, using the basic properties of agents such as autonomy, proactivity, negotiation, and knowledge. View the elasticity of fog and cloud; they can expand and retreat at the request of users or applications. This property is very important for reliable execution of the monitoring process. For that, we model the monitoring process as a detection of a deviation of the user behavior. This needs two phases:

- The initialization phase, where a learning process is executed to determine the "normal" behavior of the connected users. For that, we propose to use different metrics, like the duration of the connection to the system, the number of requested resources, and the type of the used applications. The initialization phase is required to fix the value of thresholds.
- The work phase, where any deviation about this behavior is considered as an attack and an alert is generated by the system to decrease the trust value. In fact, if the metric(s) exceeds the related threshold, the trust Tr will be decreased depending on the type of the metric and the gap between the metric and its threshold [see (1)].

$$Tr = f(\text{trust\_metric}, \text{threshold}) \tag{1}$$

4   Expected Results

4.1   Achieved Security Issues

The access control-based trust monitoring resource allocation system on IoT–fog network is proposed to increase the security and the availability of the network. Moreover, it is able to achieve the following issues:

Scalability: Having hundreds of billions of connected devices and nodes forming the IoT use a variety of fog services as new wireless access technologies to enable and facilitate communication between these nodes and overcome compute and storage limits and improve performance. In our proposal, monitoring agents are able to control and supervise a large number of resources.

Resilience to network changes/failures: The IoT–fog wireless environment is an open network due to its distributed infrastructure. That is, the user's information are vulnerable to network failures mostly at the edge level (fog, IoT). The proposed system has a global view of the network to observe and control the activities of every element in the network and to update trust according to the supervised metrics. This ensures the deactivation action in case of detecting changes in the user's behaviors.

4.2 Performance Evaluation: Usage Case: e-Health Applications

In this section, we choose a use case that necessitates time-sensitive tasks and present it to validate the proposed process for supporting health care. The health information is very sensitive, so any patient who wants to access to their health data and control their health histories should be trusted [8]. Currently, real-world user has not many experiences with fog/cloud-based health monitoring, due to the sensitivity level of the data stored, also due to the network latency and length of response time. This new computing paradigm, if our proposed access control-based monitoring scheme, is applied successfully; for e-health monitoring, it will have countless potential to accelerate the finding of early predictors to enable rapid and efficient care decision.

Therefore, for the purpose of discussion, we evaluate the delay produced when accessing to the fog node. We evaluate the performance of the suggested access control-based monitoring agents, based on time analysis. Therefore, we have built our evaluation methods using mathematical analysis.

Figure 5 provides an overview of the scheme's time factors. The information presented in the figure below corresponds on two time categories. The first represents the processing time (PT) required for the network fog layer to supervise the activities of the user, compute the risk, and deliver the auth_Tr for them. The second category represents the propagation delay (PD) produced by the system, which depends on the speed of propagation and other conditions.

In following, we present the different length of time indicating as in Fig. 5:

- $(t_0 - t_1)$: Delay (PD$_{UF}$) is the propagation delay between an IoT user and fog node.
- $(t_1 - t_2)$: Time (PTF) required for the fog node to achieve an access control process.

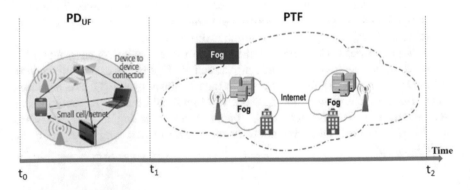

Fig. 5 Time factors of the trust monitoring phase

To improve the quality of service and consumers, it is required to have a fast response time. In our scheme, the risk assessment and the delivery of auth_Tr should be done with the minimum of time. We determine the response time using Eq. (2):

$$\text{Response time} = PT + PD \tag{2}$$

The latency between end users and the fog nodes that are near to them (noted $L_{\text{User-Fog}}$) is comprised between 10 and 100 ms. As illustrated in Fig. 6, cloud infrastructure is the farthest network elements in terms of distance and network latencies.

In our work, we add the security aspect beyond a wireless fog computing based on IoT architecture. We know that there are always a compromise between the security and the quality of the services allowed by each network element. Yet, our monitoring scheme offers high method for access control, authentication, attack detection and prevention, high scalability, and data allocation with low latency.

Moreover, we measure the latency produced when implementing our suggested scheme.

We argue that the monitoring process takes place in the background; its delay is included with the time system. Equation (3) should be modified as follows:

$$\text{Response Time} = PD_{\text{UF}} + PTF + L\_\text{cert} \tag{3}$$

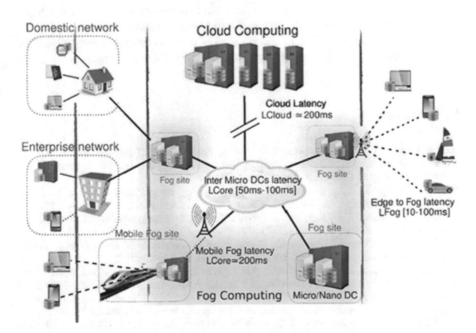

Fig. 6 Overview of cloud and fog latencies

where L_cert is the latency added by our system after the creation and the delivering of the auth_tr to users. In our case, $L\_cert = 0.509$ µs.

All in all, the implementation of the proposed security scheme just produces a small latency in some cases along the user's access. The findings of this evaluation indicate that our monitoring system can be an efficient solution for system supervision, and detection and prevention attacks, with a fast processing time. This is because that the trust computation, the delivering, or the verification of the trust certificate do not consume long time.

5 Conclusions

In IoT and fog computing, there are a huge amount of data that might be executed and manipulated. These environments support multiple connectivity and faster transmission of data. The cooperation between these paradigms requires a robust security system to cope with expected attacks. To this end, we suggested a distributed and cooperative scheme for risk-based monitoring. Moreover, we introduced the concept of certificates in order to ensure high security level and lowest latency level. The proposal achieved data security, data confidentiality, user trust, user revocation, and attack prevention and detection. Our architecture offers high scalability and best resource utilization with low latency based on resource available. The findings of our analysis indicate that our scheme provides proactive security scheme under ultra-trustworthiness and low-latency overhead of 0.509 µs, which is promising regarding to the achieved security issues.

For future work, we plan to more develop the cooperation process between monitoring agents and to simulate our proposal for a deep evaluation.

References

1. Ben Daoud, W., Meddeb-Makhlouf, A., Zarai, F.: A model of role-risk based intrusion prevention for cloud environment. In: 2018 14th International Wireless Communications and Mobile Computing Conference, pp. 530–535 (2018)
2. Ni, J., Zhang, K., Lin, X., Shen, X.S.: Securing Fog computing for internet of things applications: challenges and solutions. IEEE Commun. Surv. Tutorials 20(1), 601–628 (2018)
3. Abbas, N., Zhang, Y., Taherkordi, A., Skeie, T.: Mobile edge computing: a survey. IEEE Internet Things J. 5(1), 450–465 (2018)
4. Alrawais, A., Alhothaily, A., Hu, C., Cheng, X.: Fog computing for the internet of things: security and privacy issues. IEEE Internet Comput. 21(2), 34–42 (2017)
5. Gutierrez-Aguado, J., Alcaraz Calero, J.M., Diaz Villanueva, W.: IaaSMon: monitoring architecture for public cloud computing data centers. J. Grid Comput. 14(2), 283–297 (2016)
6. Ben Daoud, W., Meddeb-Makhlouf, A., Zarai, F., Obaidat, M.S., Hsiao, K.-F.: A distributed access control scheme based on risk and trust for fog-cloud environments. In: Proceedings of 15th International Conference on E-business Telecommunications, vol. 1, ICETE, pp. 296–302 (2018)

7. Abderrahim, M. et al.: A Holistic Monitoring Service for Fog/Edge Infrastructures: A Foresight Study (2017)
8. Giri, D., Maitra, T., Obaidat, M.S.: An efficient fog based secure data transmission of healthcare sensors for e-medical system. In: Proceedings of IEEE GlobeCom 2017, Singapore (Dec 2017)

Artificial Intelligence-Based Technique for Detection of Selective Forwarding Attack in RPL-Based Internet of Things Networks

Vikram Neerugatti and A. Rama Mohan Reddy

Abstract Internet of Things (IoT) is an upcoming technology in computing networks, which uses the novel and constrained technologies like IPv6 over Low-powered wireless personal area networks (6LoWPAN), Routing protocol for low power and Lossy networks (RPL), etc. The 6LoWPAN is a compressed version of the IPv6 protocol with the integration of the wireless personal area network, which plays a major role to connect the things (any objects around us such as chair, refrigerator, light, etc.) in the IoT-based network. The RPL is a distance-based network routing protocol that used to establish a route to send packets in the 6LoWPAN-based IoT networks. As the 6LoWPAN and RPL protocols are compressed/constrained versions, will leads to many attacks in IoT networks. The few attacks in the RPL protocol are wormhole, black hole, sink hole, sybil, rank, selective forwarding, various denial of service attacks, etc. This paper focused on the detection technique, for the selective forwarding attack, which occurs in the 6LoWPAN based RPL protocol. This attack will disrupt the routing path in IoT network by selectively forwarding the few packets and dropping remaining all the packets between the nodes in the network. To mitigate this attack, a novel artificial intelligence-based detection technique was proposed in this paper. This proposed technique was implemented in the ContikiCooja simulator with the Sky motes and the results shown the efficiency of the proposed system in terms of the delay, packets delivery ratio, and in detection of selective forwarding attack.

Keywords Internet of Things · RPL attacks · 6LoWPAN · IDS · Detection technique · Artificial intelligence

V. Neerugatti (✉) · A. Rama Mohan Reddy
Department of CSE, Sri Venkateswara University, Tirupati, Andhra Pradesh, India
e-mail: vikramneerugatti@gmail.com

A. Rama Mohan Reddy
e-mail: ramamohansvu@yahoo.com

© Springer Nature Singapore Pte Ltd. 2020
P. Venkata Krishna and M. S. Obaidat (eds.), *Emerging Research in Data Engineering Systems and Computer Communications*, Advances in Intelligent Systems and Computing 1054, https://doi.org/10.1007/978-981-15-0135-7_7

1   Introduction

In 1990, the Mark weiser put a concept called ubiquitous computing [1]. This computing says that the computing will be everywhere. The Internet of Thing (IoT) also a ubiquitous, and everything can be connected to the Internet with the unique address. The IoT was coined by kevin Ashton in 1999 at Massachusetts Institute of Technology for the work of Auto-ID center [2]. IoT was possible with the advent and open wireless technologies like RFID, Bluetooth, Wi-Fi, embedded systems, sensors, actuators, telephonic data services, etc. This leads to novel integrated framework called Sensor-Internet-Actuator framework. This created the novel environment called smart environment [3]. Figure 1 shows a Gartner hype cycle (August 2018) for the emerging trends in technologies. This cycle says that the IoT platforms will be more than 10 years.

The IoT has more scope and it has vast applications in almost every fields like transportation, logistics, healthcare, monitoring the environmental parameters, etc., and with this interesting applications has open issues like standardization activity, addressing and network issues, security and privacy, etc. [4]. In [5], author given the directions to pursue research in the area of IoT with the different sub-areas like massive scale, architecture and dependencies, creating big data, robustness, openness, and security. The major challenge was a security in IoT. IoT has various layers, which shown in Fig. 2.

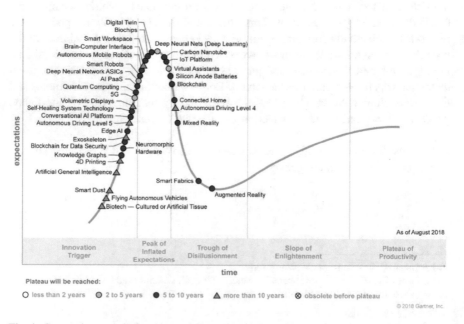

Fig. 1 Gartner hype cycle for the emerging technologies

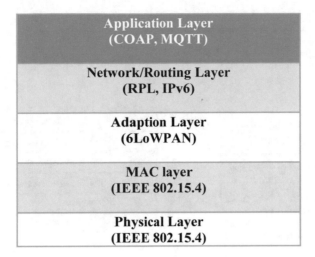

Fig. 2 IoT layers with protocols [6]

Every layer should provide security in terms of the integrity, confidentiality, non-repudiation, and authentication. The IoT is a IPv6-based low-power wireless personal area networks (6LoWPAN), and the routing will be done based on the routing protocol for Low Power and Lossy Networks (RPL) protocol. Security in this RPL was very crucial and plays a major role in the IoT Networks [6]. The RPL arranges the nodes in the network as a destination-oriented directed acyclic graph (DODAG) shown in Fig. 3.

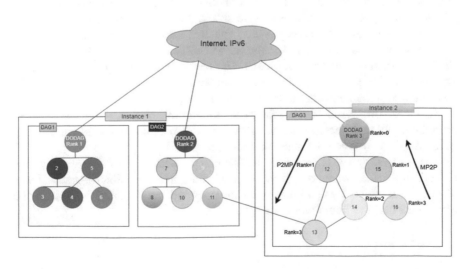

Fig. 3 RPL with DODAG's [7]

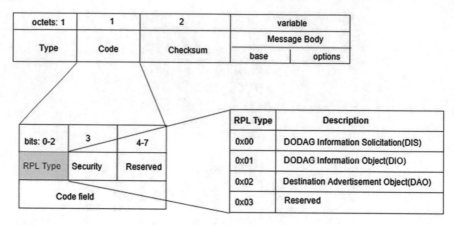

Fig. 4 RPL control messages [7]

In RPL, the nodes may be a border node, which will be a root node in DODAG; Router node, to forward the packets; and the Host node is an end node. RPL has the control messages to create and transfer the packets [7]. The control messages are shown in Fig. 4.

RPL can support the communications in three ways. Many to one communication, one to many communication, and one to one communication. To maintain the route, the nodes may be in two modes called the storing mode and non-storing mode. The border router always will be the storing mode and remaining nodes may be either storing or non-storing mode [7].

The RPL has various attacks like selective forwarding attack, sinkhole attack, sybil attack, hello flooding attack, wormhole attack, clone ID attack, blackhole attack, spoofing attack, denial of service attack, etc. [8, 9]. The current paper was focused on selective forwarding attack, and this attack will disrupt the network routing paths in the RPL, here the attacker will selectively forward the packets and remaining packets may dropped [8, 10]. The selective forwarding attack may be in various forms like, may be, one node in network, two or more nodes in network, and nodes may be consecutive or nonconsecutive [11].

To detect this types of the selectively forwarding attacks, proposed a novel artificial intelligence-based schema called artificial intelligence-based packet drop ratio (AIPDR) by using the neighborhood information.

The next section was organized as follows. In Sect. 2, related work was discussed, in Sect. 3 proposed system was discussed, in Sect. 4 implementation was discussed, in Sect. 5 the results were incorporated, and finally, in Sect. 6 the conclusion and the future work was discussed.

2 Related Work

In [10], authors implemented the selective forwarding attack in the cooja simulator with the Contiki RPL. After implementing the attack they noticed that packs was dropping, so their applied the self-healing mechanism for 24 h of time and noticed that still the packets were dropping and concluded that it requires a detection and prevention scheme for this attack.

In [11], authors implemented an intrusion detection system (IDS) to detect the attack like sinkhole, selective forwarding, and spoofing attack. This IDS was primarily focused on the RPL routing attacks and can be extended to the various other routing attacks.

In [12], authors discussed about the various selective forwarding attacks based on the count of the nodes like single, two consecutive, surrounding and nonconsecutive. Hers this also discussed the possible prevention and detection approaches.

In [13], authors discussed and proposed a detection technique for the selective forwarding attack. The proposed detection method was based on the support vector machines. Proposed the attack model and detection approach and shown the results in terms of the true and false alarms of the attack detection.

3 Proposed System

To detect the selective forwarding attack, here proposed a novel artificial intelligence-based detection technique called artificial intelligence-based packet drop ratio (AIPDR) by using the neighborhood information. This technique will work based on the principle of the packet drop rate (PDR). The PDR will be calculated for every node in the destination-oriented directed acyclic graph (DODAG) of RPL by calculating the below formula

$$PDR = P_{in} - P_{out}$$

where 'P_{in}' = incoming Packets and 'P_{out}' = Outgoing Packets.

Here, the technique is decentralized. The nodes and the border router node of DODAG should work mutually. The nodes in the IoT network will be artificial intelligence based. It means the nodes will work based on the environmental conditions in the network. Here, the environmental condition was the PDR value. Here, the PDR value may be negative, zero, or positive. Whenever the PDR value of the node was other than zero, that particular node was considered as an attack node. It indicates that the packets were dropped or selectively forwarding from that node. This kind of nodes may be one or more. Once detected that kind of malicious nodes, then automatically it will be removed from the DODAG. This is how we can detect the attack nodes of type selective forwarding attack in RPL-based IoT networks. The proposed technique was shown in the Algorithm 1.

Algorithm 1: Proposed Detection process of the Selective forwarding Attack

/* where 'X' is Border router node, 'y' is other nodes, 'Z' is Selective forwarding attack node 'PDR' is a Packet drop rate of one node,   PDR = P_{in}– P_{out}Outgoing Packets, 'P_{in}' isincoming Packets and 'P_{out}' is Outgoing Packets */

0: Start
1: root node 'X' broad cast DIO (DODAG ID, Objective Function, rank=0)
2: other node 'Y' receives the message DIO
3:          calculates rank 'R' based on DIO
4 nodes 'Y' sends (unicast) the DAO to node 'X' with the PDA value
3: node 'X' performs
4:        based on PDA value
5:                  if PDA = = 0
6:                            node 'X' sends DAO-ACK to 'Y'
7:              else
8:                            node 'X' removes the node (Selective forwarding attack node 'y'=='Z')
9: then node 'Y' adds the node 'X' as it parent
10: Node 'Y' multicasts the DIO message to other nodes.
11: end

4   Implementation

The proposed algorithm is implemented in both the border router nodes, attack nodes, and normal nodes. All the nodes in DODAG will check the PDR value. The border router node will check the PDR value of every node that sent by other nodes. Based on that PDR value, the malicious nodes will be detected. Once the malicious node detected it will be removed from DODAG.

In the cooja, 30 sky motes are used. In all the 30 motes, the artificial intelligence code was injected to calculate the PDR value. Among 30 sky motes, one sky mote is considered as RPL border router node. Where proposed algorithm was injected in one of the node and it considered as RPL border router node. Other few nodes among 29 nodes were considered as the malicious nodes with selective forwarding attack (selective forwarding attack code was injected) and remaining nodes have normal nodes (normal code without attacks). After simulating with this setup the results are shown in next section.

5   Results

From the implementation setup that discussed in the Sect. 4, results were drawn and shown in Figs. 5, 6, 7, 8, 9, and 10 (Table 1).

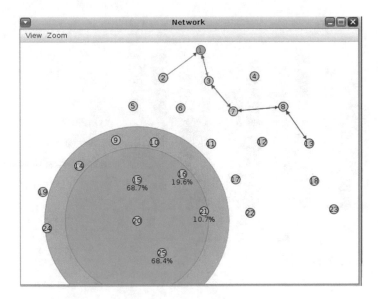

Fig. 5 Sky motes in COOJA simulator

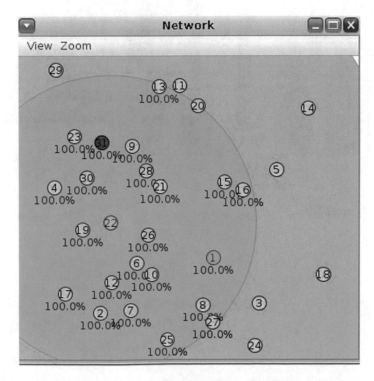

Fig. 6 Sky motes in COOJA simulator with the normal attack and proposed codes

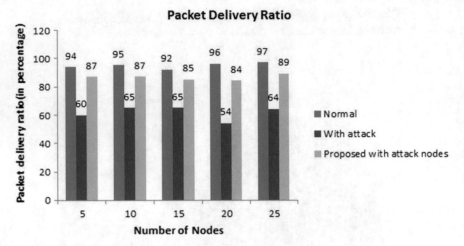

Fig. 7 Packet delivery ratio

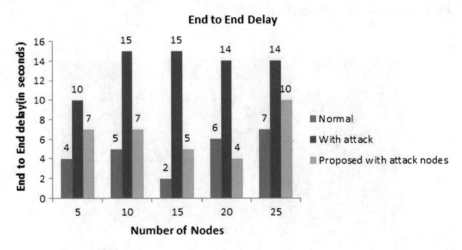

Fig. 8 END to END delay

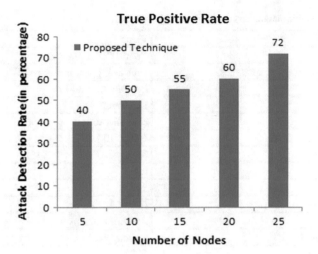

Fig. 9 True positive rate

Fig. 10 False positive rate

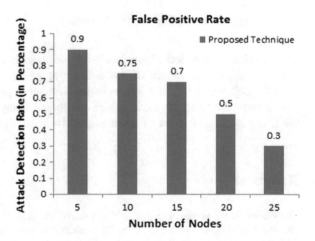

Table 1 Simulation parameters in ContikiCooja simulator

Parameter	Value
Simulator	ContikiCooja Simulator
Radio medium model distance loss	Unit disk graph medium (USGM)
Range of nodes	$R\times$ and $T\times$: 100 m
Mote type	Sky motes
Duty cycle	contikiMAC
Size of deployment area	100×100 m
Number of nodes	30
Number of sinks	0.1
Number of malicious nodes	5–30%
Physical layer	IEE 802.15.4
MAC layer	ContikiMAC, lpv6
Network layer	contikiRPL
Transport layer	UDP
Objective function	Hop count and ETX, MRHOF

6 Conclusion and Future Work

The novel detection technique based on the neighborhood information with the artificial intelligence nature of the nodes was proposed and implemented in the ContikiCooja simulator. The implemented simulation parameters are tabulated. The results were drawn in the terms of the end-to-end delay, packet delivery ratio, and the detection rate in terms of the true positive and false positive rate.

References

1. Weiser, M.: The computer for the 21st century. Sci. Am. **265**(3), 94–105 (1991)
2. Mattern, F., Floerkemeier, C.: From the Internet of Computers to the Internet of Things. From active data management to event-based systems and more, pp. 242–259. Springer, Berlin, Heidelberg (2010)
3. Gubbi, J., et al.: Internet of Things (IoT): a vision, architectural elements, and future directions. Future Gener. Comput. Syst. **29**(7), 1645–1660 (2013)
4. Atzori, L., Iera, A., Morabito, G.: The internet of things: a survey. Comput. Netw. **54**(15), 2787–2805 (2010)
5. Stankovic, J.A.: Research directions for the internet of things. IEEE Int. Things J. **1**(1), 3–9 (2014)
6. Granjal, J., Monteiro, E., Sá Silva, J.: Security for the internet of things: a survey of existing protocols and open research issues. IEEE Commun. Surv. Tutorials **17**(3), 1294–1312 (2015)
7. Gaddour, O., Koubâa, A.: RPL in a nutshell: a survey. Comput. Netw. **56**(14), 3163–3178 (2012)
8. Pongle, P., Chavan, G.: A survey: attacks on RPL and 6LoWPAN in IoT. In: 2015 International Conference on Pervasive Computing (ICPC). IEEE, New York (2015)

9. Mayzaud, A., Badonnel, R., Chrisment, I.: A taxonomy of attacks in RPL-based internet of things. Int. J. Netw. Secur. **18**(3), 459–473 (2016)
10. Wallgren, L., Raza, S., Voigt, T.: Routing attacks and countermeasures in the RPL-based internet of things. Int. J. Distrib. Sens. Netw. **9**(8), 794326 (2013)
11. Raza, S., Wallgren, L., Voigt, T.: SVELTE: real-time intrusion detection in the internet of things. Ad Hoc Netw. **11**(8), 2661–2674 (2013)
12. Khandare, P., Sharma, Y., Sakhare, Y: Countermeasures for selective forwarding and wormhole attack in WSN. In: 2017 International Conference on Inventive Systems and Control (ICISC). IEEE, New York (2017)
13. Kaplantzis, S., et al.: Detecting selective forwarding attacks in wireless sensor networks using support vector machines. In: 2007 3rd International Conference on Intelligent Sensors, Sensor Networks and Information. IEEE, New York (2007)

Decision Tree-Based Fraud Detection Mechanism by Analyzing Uncertain Data in Banking System

Neelu Khare⊙ and P. Viswanathan⊙

Abstract As of now, enormous electronic information stores are being kept up by banks and other money-related organizations. Information mining advancement gives the region to get to the right information at the right time from massive volumes of information. Data classification is an established issue in AI and information mining. In regular choice (decision) tree investigation, a normal for a tuple is either supreme or partial. The choice tree calculations are utilized for dissecting solid and numerical information of uses. In the surviving techniques, they play out the extended model of choice tree examination to help information tuple having factual characteristics with uncertainty characterized by discretionary pdf. Along these lines, we proposed an improved novel choice tree for the two information, speaking to the development and a framework utilized for AI and information mining to strengthen the requirement of the financial undertaking. This paper expects to evaluate the utilization of strategies for choice trees to aid the trepidation of bank extortion. The choice trees aid this work of choosing the characteristic that will build up a superior exhibition in determining the odds of bank fraud.

Keywords Fraud detection · Decision making · Machine learning · Decision trees · Data mining

N. Khare (✉) · P. Viswanathan
School of Information Technology and Engineering, VIT University,
Vellore, Tamil Nadu 632014, India
e-mail: neelukh.29@gmail.com

P. Viswanathan
e-mail: viswatry2003@ieee.org

© Springer Nature Singapore Pte Ltd. 2020
P. Venkata Krishna and M. S. Obaidat (eds.), *Emerging Research in Data Engineering Systems and Computer Communications*, Advances in Intelligent Systems and Computing 1054, https://doi.org/10.1007/978-981-15-0135-7_8

1 Introduction

Fraud characterized as illegitimate or felonious trickiness pointed to happen in budgetary or individual addition [1]. The two primary components to maintain a strategic distance from fakes and misfortunes because of fake exercises are extortion aversion and fraud location frameworks [2]. Fraud counteractive action is the proactive instrument with the objective of handicapping the event of extortion. Fraud location frameworks become an integral factor when the fraudsters outperform the extortion anticipation frameworks and begin a false exchange. Classification is a built-up issue in AI and information mining. Addressed a great deal of planning data tuples, all possessing a class mark and being addressed by a component vector, the job is to algorithmically collect a model that foretells the class name of a subtle test tuple subject to the tuples component vector. A champion among the most standard course of action models is the decision tree shown. Choice trees are unmistakable in light of the fact that they are convenient and clear. Models can be in like manner, and be expelled from choice trees adequately. Various figurines, for instance, C4.5 and ID3 have derived for the advancement of decision tree [3–5]. It figurines the most grasped part and used in a wide extent , for instance, medicinal determination, extortion location, picture acknowledgment, logical tests, and a FICO score of credit candidates.

In customary decision tree gathering, a segment (a quality) of a tuple is either supreme or numerical. For the last referenced, precise and positive point regard is regularly acknowledged. In various applications, regardless, data defenseless-ness is ordinary. The estimation of an incorporate/quality is thusly best gotten not by lone point regard, but instead by an extent of characteristics offering to rise to plausibility scattering [6–8]. An essential strategy to manage data defenseless-ness is to extricate likelihood movements by diagram estimations, for instance suggests and vacillations. We describe this procedure as equalizing. A different philosophy to examine the all-out knowledge passed on by the likelihood of appointments is to develop a choice tree. We describe this technique as a decentralized base. Therefore, we consider the issue of mounting choice tree derivations on data with mysterious numerical characteristics. The objectives are

- To streamline a count for developing decision trees from mysterious information using the decentralized-based procedure.
- To examine whether the decentralized-based methodology could incite a greater order precision differentiated and the averaging approach and
- To develop the speculative misrepresentation location strategy framework at pruning methods remains resolved that can basically increase the computation inclination of AI-based figuring.

2 Related Work

2.1 Application of Banking Sector Data Mining

There are different zones in which information mining can be utilized in money-related segments like client division and productivity, credit examination, anticipating instalment default, showcasing, deceitful exchanges, positioning ventures, advancing stock portfolios, money the board and gauging activities, high hazard credit candidates, most productive credit card customers and cross selling. Certain models, where keeping money industry, have been using the information mining innovation successfully as pursues.

2.2 Extortion Detection

Extortion discovery is the acknowledgment of indications of extortion where no earlier doubt or inclination to extortion exists [9, 10]. As per The American Heritage word reference, second school release, extortion is characterized as a trickiness purposely rehearsed in request to anchor out of line of unlawful gain. Extortion recognition alludes to identification of criminal exercises happening in business associations, for example, banks, Mastercard issuing associations, protection organizations, portable organizations, and securities exchange. The vindictive clients may be the real clients of the association or may act like a client (too known as wholesale fraud). Monetary organizations particularly managing an account divisions pursues essentially two methodologies toward deciding the misrepresentation designs, online exchange check, and Offline exchange Check [11]. For this reason, the organizations buy and keep up information stockrooms of approvals what's more, politically exposed persons information documents from consistence and anti-money laundering arrangement and information suppliers like The Office of Foreign Assets Control (*OFAC*) of the US.

2.3 Risk Analysis in Decision Tree

As learning is ending up increasingly, more synonymous to riches creation and as a methodology plan for contending in the commercial center can be no superior to anything the data on which it is based, the significance of learning and data in the present business can never be viewed as an exogenous steward to the business [12, 13]. Associations and people approaching the correct data at the correct minute have more noteworthy odds of being effective in the age of globalization and ferocious challenge.

Information mining is the way toward separating learning, avoided vast volumes of crude information. The information must be new, not self-evident, and one must most likely use it. Information mining has been characterized as the nontrivial extraction of understood, already obscure, and conceivably important data from raw data [14–16]. It is the exploration of extricating valuable data from huge databases. Information mining is one of the errands during the time spent learning revelation from the database [17, 18]. The means, the steps associated with Knowledge disclosure are

- Data selection: The information significant to the investigation is chosen and recovered from different information areas.
- Data pre-preparing: In this stage, the procedure of information cleaning and information joining is finished.
- Data cleaning: It is otherwise called information purging; in this stage, clamor information and unimportant information are expelled from the gathered information.
- Data integration: In this stage, various information sources, regularly heterogeneous, are joined in a typical source.
- Data transformation: In this stage, the chosen information is changed into structures, suitable for the mining system.
- Data mining: It is the essential advance in which smart strategies are connected to separate possibly valuable examples. The choice is made about the information mining strategy to be utilized.
- Interpretation and Evaluation: In this progression, intriguing examples speaking to information are distinguished dependent on given measures. The found learning is outwardly displayed to the client. This basic advance uses perception strategies to enable clients to get it.

The computerization of money-related tasks, availability through World Wide Web and the help of robotized programming has totally changed the essential idea of business and the way the business activities are being done, and the financial area is not an exemption to it [19–21].

3   Proposed Model

Data mining can add to doing what needs to be done issues in banking and record by finding precedents, causalities, and connections in business information and market costs that are not quickly obvious to administrators in light of the way that the volume data is too much immense or is delivered too quickly to screen by authorities.

A. Classification

Gathering is the most ordinarily associated data mining method, which uses a ton of pre-portrayed advisers for developing a model that can arrange the quantity of occupants in records and free to move around at will. Blackmail area and credit chance applications are particularly fitting to this kind of examination. This approach frequently uses decision tree course of action computations. The data gathering process incorporates learning and portrayal.

B. Fraud Detection

While overseeing banks, the customers and the banks get the chances of falling a straightforward prey to the cheats. Thusly, both the social occasions wish to be secure while overseeing each other. The data mining techniques can help them with distinguishing and therefore keep away from fakes. The data mining techniques will help the relationship with concentrating on the accessible assets of separating the customer data in order to recognize the precedents that can incite swindles.

C. Decision trees

A decision tree is both a data addressing structure and a system used for data mining and AI. The method "partition to conquer" is used in decision trees, which includes breaking the issue into less intricate issues and less requesting to understand. In addition, frameworks associated in a particular portion of a tree can be associated recursively.

A decision tree is made out of the going with parts:

1. Node—Contains a preliminary of a trademark
2. Branch—Contains a response to every trademark
3. Leaf—Each leaf is connected with a class
4. Rule—Each course from the root to a leaf looks at to a portrayal rule.

Directly, the request is to make sense of which attribute will be the primarily picked. We should pick first the ones that have the best information. Decision trees use the possibility of entropy to test how much information has a trademark.

From the information theory field, we can describe entropy as an extent of inconsistency of a variable. By techniques for this thought, it is possible to evaluate whether a quality is unequivocally or not an average one. In case, there are n possible messages with proportionate probability, by then the probability p for each one is $1/n$, accordingly:

$$\log(p) = \log(n).$$

By and by, for an allotment of probabilities $P = (p_1, p_2, \ldots, p_n)$:

$$I_P = -(\{P_1\_1\} * \log(\{P_1\_1\}) + \{P_2\_2\} * \log(\{P_3\_2\}) + \cdots + P_n * \log P_n$$

If P is (0.5, 0.5) by then I_P is 1. And If P is (0.67, 0.33), by then I_P is 0.92.

In these points of reference, we can see that when the transport of the probability is higher, by then we have better information as for this variable. This is the basic property at the period of selection of properties in a decision tree.

The entropy is used to evaluate the haphazardness of the variable to predict the class. The expansion of information evaluates the decline of entropy realized by dividing models as shown by the estimations of the picked property. We portray the expansion of information as seeks after:

$$\text{Expansion(quality)} = I(\text{trademark}) - I(\text{express characteristics}).$$

4 Credit Card Data Structural Format

The first information of the timespan, a year, used to shape the preparation set has around 22 million Visa exchanges. The conveyance of this information regarding being ordinary or deceitful is exceptionally skewed. A year, time frame is utilized to assemble our example incorporated 978 fake documents and around 22 million typical items among a proportion of concerning 1:22,500. Thus, to enable the models to learn the two kinds of silhouettes, we utilized stratified testing to under example the genuine records to a significant number. We attempted examples with various genuine/misrepresentation proportions. In any case, all the information having a place with the timeframe of next a half year, which incorporates around 11,344,000 exchanges but 484 are fakes, and straightforwardly incorporated into the examination set. Every one of the exchanges in the examination set is obtained by the classifier strategies. The information appropriations of the preparation and experimental set are given in Table 1.

4.1 Decision Tree-Based Fraud Analysis Mechanism

In addition, this suspicion has any kind of effect among the expense also, the need for each deceitful exchange. At the end of the day, distinguishing a deceitful exchange submitted with a card having a bigger accessible usable point of confinement spares in excess of a fake one submitted with a card having a littler accessible usable breaking point. In this way, the need for the identification of the main deceitful exchange is higher than that of the second one. In this way, each bogus negative has an alternate fraud cost and the execution of the model ought to

Table 1 Data set distribution w.r.t. classes

Sets	Type	Record count in population	Record count in sets
Training set	Normal	22,000,000	8802
	Fraud	978	978
Test set	Normal	13,644,000	13,644,000
	Fraud	484	484

be assessed over a cost-delicate metric, for example, the recently characterized metric DSLR, which is level of the aggregate measure of spared accessible usable points of confinement rather than the measurements in view of the quantity of cheats recognized as given in Eq. 2.

$$\text{DSLR} = N - \left(\frac{\sum_{j=1}^{k} (C_{FN})_j}{\sum_{i=1}^{f} (C_{FN})_i} \right) \times 100 \tag{2}$$

The all-out fraud cost in the instance of relegating the exchanges as ordinary, C_N, is determined to be the entirety of the accessible usable farthest point of each false records $C_{(FN)j}$ in that hub. Allocating a real exchange as extortion just outcomes in a perception cost which is the same for each real exchange (CFP). Thus, in this strategy, just the fraud costs are utilized for tree acceptance and arrangement.

Accept that there are f fake records and n typical (real) records those falling into a hub where N and $(N = f + n)$ gives the complete number of records in this hub, C_P and C_N can be determined as given beneath in bello equations:

$$C_N = \sum_{i=1}^{f} (C_{FN})_i \tag{3}$$

$$C_P = n \times C_{FN}$$

The traditional decision tree techniques use contamination measures to pick the part trait and the split esteems. ID3 utilizes entropy and data gain, while the successor C5.0 utilizes gain proportion and C and RT utilizes Gini record for contamination estimation. For a two-class case, anticipated data (entropy) and Gini Index can be determined as given in belle equation:

$$\text{(i)} \quad \text{Entropy} = \sum_{i=1}^{2} (-p_i \times \log_2 (p_i))$$

$$\text{(ii)} \quad \text{Gini} = 1 - \sum_{i-1}^{2} (-p_i)^2 \tag{4}$$

4.2 Architectural Design

In order to perform our proposed algorithm, we required a data set on fraud, on which we will apply classification using WEKA tool and get the job of attribute selection done. Data set attributes and data set descriptions are updated in Table. Location: where the credit cardholder will be using the card.

Table 2 Data set description

No.	The attribute	Description	Data type
1	Credit history	Previous history of customer credit	Nominal
2	Purpose	The loan purpose	Nominal
3	Gender	Male or female	Nominal
4	Credit amount	The amount of credit	Numeric
5	Age	Customer age	Numeric
6	Housing	Rent, own or for free	Nominal
7	Job	Is the customer has a job	Nominal
8	Class	The class of loan good/bad	Nominal

- Product: the kind of products that the user will purchase.
- AT amount: the average possible transaction amount.
- n transaction: Number of possible transactions that the cardholder will make.
- t amount: the actual transaction amount.
- Transaction: the current number of transactions used for validation.
- Fraud: classifies whether the transaction can lead to fraud (yes) or not (no).

For each of these attributes, we need to find the information gain and take this as the basis to make the decision tree (Table 2).

5   Decision Tree Algorithm

We utilized negative tree pruning calculation, which maintains a strategic distance from the need of a different test informational index. We initially set the quantity of accurately or miss-effectively arranged tuples at each leaf hub dependent on preparing tuple. At that point, the accompanying calculation is utilized to prune the tree, which is illustrated in Fig. 1.

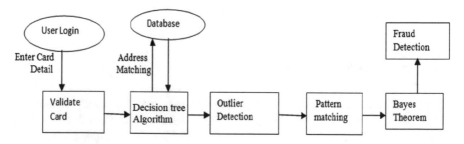

Fig. 1 Flow of fraud identification mechanism

Algorithm 1: Decision tree Algorithm

---

1 **Input:** Node is the root node of the tree.
2 **Output:** Return the pruned version of the tree.;
3 initialization;
4 void Prune(node)
5 Step 1. If node is leaf then return;
6 Step 2. Find the majority class in sub-tree (in this case node) and set it to MajorityClass
7 Step 3. Calculate the number of mis-correctly classified tuples at each leaf with MajorityClass
8 and set it to a variable noMisLeaf
9 noMisLeaf = noMisLeaf + 0.5 (1/2 for single node i.e leaf);
10 Step 4. Calculate the number of mis-correctly classified tuples at node (sub-tree) and set it to
11 noMisSubTree.
12 noMisSubTree = noMisSubTree + NT/2.0; (where NT is total number of leaf in sub-tree)
13 Step 5. Set Total = noMisLeaf + noMisSubTree;
14 Step 6. Calculate standard error using
15 StandardError = Math.sqrt((noMisSubTree *(Total - noMisSubTree))/Total);
16 Step 7. Check for pruning condition
17 IF noMisSubTree+ StandardError ¿= noMisLeaf THEN
18 //Pruned the tree i.e replaced sub-tree with leaf.
19 Set node level to MajorityClass, node child to null (means convert it to leaf), add noMisLeaf,
20 to node and return;

---

6 Experimental Results

Figure 2 demonstrates the WEKA execution and parameters initialization process. Figure 2a shows the initialization of WEKA tool parameters. It enables seven different parameters to process the execution of decision tree. Those values are used to find the received data foundation along with comparison of initialized predefined values.

A huge bit of the time-arrangement informational collection is constants and invalid fields. Separating these fields out in the beginning, time can fundamentally diminish the preparing time and improve the model exactness. The experimental decision tree is illustrated in Fig. 3. The 11% of records were solitary esteemed traits and 9% of records were an invalid esteemed and was regarded as measurably inconsequential. These fields are expelled from the objective information table to lessen the registering time for guaranteeing the displaying procedure quicker.

Subsequently to applying arrangement, information mining system calculations, which acquired the outcomes from the three analyses and in the wake of looking at the accurately grouped case percent, we find that the best calculation for credit order. It is best since it has high exactness and low mean supreme blunder as appeared in the outcome. Likewise, it is able to order the examples effectively than alternate systems. The analyses have been completed a few times and in each times, the preparation and test sets measure have been changed (80% preparing 20% test set, 60% preparing 40% test, and 70% preparing 30% test) and we acquire a similar outcome that the choice tree calculation is best in grouping credits to great and awful advance. This model helps bank chief to distinguish extortion utilizing machine-learning applications by foreseeing that if the exchange will lead the bank to chance or not and bolster leader to settle on composing choices.

```
=== Run information ===

Scheme:weka.classifiers.trees.RandomTree -K 0 -M 1.0 -S 1
Relation:      frodtest-weka.filters.unsupervised.attribute.Remove-R1-8
Instances:     29
Attributes:    7
               location
               product
               atamount
               ntransaction
               tamount
               transaction
               fraud
Test mode:evaluate on training data

=== Classifier model (full training set) ===


RandomTree
==========

tamount < 13750 : No (16/0)
tamount >= 13750
|    tamount < 23250 : Yes (8/0)
|    tamount >= 23250
|    |    atamount < 17500 : Yes (2/0)
|    |    atamount >= 17500 : No (3/0)

Size of the tree : 7

Time taken to build model: 0.01 seconds
```

(a) WEKA phase1

```
=== Evaluation on training set ===
=== Summary ===

Correctly Classified Instances      29               100      %
Incorrectly Classified Instances     0                 0      %
Kappa statistic                      1
Mean absolute error                  0
Root mean squared error              0
Relative absolute error              0        %
Root relative squared error          0        %
Total Number of Instances           29

=== Detailed Accuracy By Class ===
```

	TP Rate	FP Rate	Precision	Recall	F-Measure	ROC Area	Class
	1	0	1	1	1	1	Yes
	1	0	1	1	1	1	No
Weighted Avg.	1	0	1	1	1	1	

```
=== Confusion Matrix ===

 a  b   <-- classified as
10  0 | a = Yes
 0 19 | b = No
```

(b) WEKA phase2

Fig. 2 Decision tree execution for fraud detection algorithm

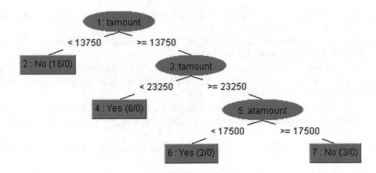

Fig. 3 Experimental decision tree analysis

7 Conclusion

The determination of this work is a choice tree that offers the ability to make arrangements with the assistance of scientific ideas, explicitly of entropy, examined in data hypothesis, permitting a calculation to scientifically figure the haphazardness of variable in regards to the conceivable decisions, in this way, diminishing the trouble of definitely achieving the objective choice. In any case, we ought to most likely make little choice trees so the objective is come to in a couple of inquiries and as fast as could be expected under the circumstances. In particular, when the bases are numeric, they can produce enormous trees that have a troublesome investigation. Situations that require fast reactions, similar to bank extortion logs, can not be utilized with applications that have a high postponement. In view of the execution of our proposed choice tree, and the test results on an example genuine database, we infer that the choice trees with criteria for information mining help in the choice making, particularly in the treatment of vast information. With this, we can conclude that data mining and decision trees can be used in real-world applications. WEKA is a great tool for classification. Weka is used in the concept of decision trees in developing a website that checks whether a transaction can be considered fraud by banks or not. Better work can be done in the future to develop better use of this concept.

References

1. Save P., Tiwarekar, P., Jain, K.N., Mahyavanshi, N.: A novel idea for credit card fraud detection using decision tree. Int. J. Comput. Appl. **161**(13), 6–9 (2017)
2. Sahin, Y., Bulkan, S., Duman, E.: A cost-sensitive decision tree approach for fraud detection. Expert Syst. Appl. **40**, 5916–5923 (2013)
3. Leite, R.A., Gschwandtner, T., Miksch, S., Gstrein, E., Kuntner, J.: Visual analytics for event detection: focusing on fraud. J. Vis. Inf. (2019)

4. Jurgovsky, J., Granitzer, M., Ziegler, K., Calabretto, S., Portier, P., He-guelton, L., Caelen, O.: PT US CR. Exp. Syst. Appl. (2018)
5. Agarawal, T.I., Swami, A.N.: Database mining: a performance perspective. IEEE Trans. Knowl. Data Eng. **5**(6), 914–925 (1993)
6. Chau, M., Cheng, R., Kao, B., Ng, J.: Uncertain data mining: an example in clustering location data. In: Proceedings of Pacific-Asia Conference Knowledge Discovery and Data Mining (PAKDD), pp. 199–204 (April 2006)
7. Chen, J., Cheng, R.: Efficient evaluation of imprecise location-dependent queries. In: Proceedings of International Conference on Data Engineering (ICDE), pp. 586–595 (April 2007)
8. Dalvi, N.N., Suciu, D.: Effecient query evaluation on probabilistic databases. VLDB J. **16**(4), 523–544 (2007)
9. Nierman, A., Jagadish, H.V.: ProTDB: probabilistic data in XML. In: Proceedings of International Conference on Very Large Data Bases (VLDB), pp. 646–657 (August 2002)
10. Janikow, C.Z.: Fuzzy decision trees: issues and methods. IEEE Trans. Syst. Man Cybern. Part B **28**(1), 1–14 (1998)
11. Mitchell, T.M.: Machine Learning. McGraw-Hill (1997); Senthil Vadivu, P., et al.: Int. J. Wireless Commun. Netw. Technol. **1**(1), 27–30 (2012)
12. Umanol, M., Okamoto, H., Hatono, I., Tamura, H., Kawachi, F., Umedzu, S., Kinoshita, J.: Fuzzy decision trees by fuzzy ID3 algorithm and its application to diagnosis systems. In: Proceedings of IEEE Conference on Fuzzy Systems, IEEE World Congress Computational Intelligence, vol. 3, pp. 2113–2118 (June 1994)
13. Fayyad, U.M., Irani, K.B.: On the handling of continuous-valued attributes in decision tree generation. Mach. Learn. **8**, 87–102 (1992)
14. Kohli, N.: Banking Software Marketing, Sales and Service Contacts. Strategic Information Technology Ltd, 2016. Web (12 March 2017)
15. The Hindu: The Hindu Business Line: From mine to shine. The Hindu Business Line, 2003. Web (13 March 2017)
16. Bhasin, M.L.: Data mining: a competitive tool in the banking and retail industries, pp. 588–594. The Chartered Accountant (2006)
17. Chitra, K., Subashini, B.: Fraud detection in the banking sector. In: Proceedings of National Level Seminar on Globalization and its Emerging Trends (2012)
18. Bhambri, V.: Application of data mining in banking sector. IJCST **2**(2), 199–202 (2011)
19. Paige, S.: Building Classification Models: ID3 and C4.5. Temple University, 2007. Web (15 March 2017)
20. Michie, D., et al.: Quinlan, JR C4. 5: Programs for Machine Learning (1993)
21. Morris, C.: Building Classification Models: ID3 and C4.5. Stanford University, 2003. Web (21 March 2017)

Numerical Weather Analysis Using Statistical Modelling as Visual Analytics Technique

Govindan Sudha, Muthuraman Thangaraj and Suguna Sangaiah

Abstract This research work focuses on analysing weather data, described by plenty of attributes, creating a high-dimensional data. After fetching the real-time observations recorded at the city on interest, the data is pre-processed for the purpose of feature selection; this also ensures data correctness, consistency and prepares data for analysis. Statistical model arrives at the aggregate of the insights observed in the empirical data. The statistical model delivers the parameters which are used to estimate the future weather attributes. Substitution of results in underlying statistical model fitness equation, the predictions can be obtained.

Keywords Numerical weather forecasting · Regression · Correlation

1 Introduction

The life in this earth is beautiful and peaceful; undoubtedly, our daily routine has the influence of environment. The word 'Environment' may include many factors like weather, mental and physical health of human and productivity of non-living things involved in the routine. Among the external factors, weather plays the key role. Weather and climate study are ever ending and ever-growing research thrust, and many researchers are doing their research in it. This is because, predicting the weather is not a trivial one. It is also not possible to say—'The weather follows the same pattern as it was in a year back or last week'. Since it includes many weather-influencing factors, predicting weather of 'How will be the next day/week/

G. Sudha · S. Sangaiah (✉)
Sri Meenakshi Government Arts College for Women (A), Madurai, India
e-mail: kt.suguna@gmail.com

G. Sudha
e-mail: g.sudha79@rediff.com

M. Thangaraj
Department of Computer Science, Madurai Kamaraj University, Madurai, India
e-mail: thangarajmku@yahoo.com

© Springer Nature Singapore Pte Ltd. 2020
P. Venkata Krishna and M. S. Obaidat (eds.), *Emerging Research in Data Engineering Systems and Computer Communications*, Advances in Intelligent Systems and Computing 1054, https://doi.org/10.1007/978-981-15-0135-7_9

month?' always yields a marginal value on the weather attribute. Statistical model is successful when the data set has voluminous observations. Linear regression is the basic statistical method which builds a line with known values on X, Y plane. This yields the future values based on intercept of the line on X- and Y-axis. Multiple linear regression is suitable if a number of attributes or variables are greater than two. The intercept values produced by regression model always come up with residual error, i.e. degree to which the estimate's variation is over the actual one.

2   Survey

Jaroensutasinee et al. [1] have developed integrated online weather data analysis by collecting the weather data from 18 automatic weather data servers and eight loggers of Thailand that handled high-dimensional data and provided summary index. Valipur [2, 3] has discussed the different weather, climatic conditions due to evaporation and radiation. Biswas [4] has developed weather prediction with recurrent neural networks. Pathan [5] has contributed the statistical data model applicable to climate science and tested using ANOVA. Anik et al. [6] developed ICT for the farmers by delivering short-term forecasting to save their crops. Diehl et al. [7–9] have proposed visual comparison, spatiotemporal data and probabilistic weather forecasting, respectively. The objective of Shabariram et al. [10, 11] work is to predict maximum rainfall, upcoming rainfall and maximum rain storm location based on hourly and overall storm parameters and discuss challenges in the weather forecasting domain. Liu et al. [12] have discussed about interactive visualization derived through machine learning and artificial intelligence. Arumugam et al. [13–15] have surveyed visual analytic techniques in weather forecasting and medical disease acute level analysis domain.

3   Proposed Work

This research workflow has three phases (i) pre-processing (ii) model building (iii) insight obtained through model and visualizations. The architecture is given in Fig. 2.

3.1   Pre-processing

The weather observations recorded at every 10 min interval at 'Madurai' City are the data set. This district has 19 attributes such as serial number, station, date, time, rain hourly cumulative (mm), rain day cumulative since 0300 UTC (mm), temp (in Celsius), temp min hourly (in Celsius), temp max hourly (in Celsius), temp day min–max (in Celsius), relative humidity (in %), wind direction (in degrees), wind speed (in knot), sea level pressure (in hpa), mean sea level pressure (in hpa/gpm), sunshine (hh.mm), battery (volts) and GPS. This data set has to be cleaned and processed so that we can build the regression model. The following steps are applied to clean the data set.

I. Observe for helpful attributes which actually depicts the weather-oriented observations. Remove the non-helpful attributes. In our domain, we identified attributes such as serial no., gps, district, station since we are considering the single city of interest.

II. Check for empty values at every observation of each attribute. If it is derivable with the collaborative attributes, then obtain the missing value through the application of formula, which relates them. For an example, the Missing value of an observation can be derived if mean attribute exists.

III. If the device is not in working condition, the recordings for the failure period are indicated by some special, abnormal values like 99999.99, −1, NA. Identify and omit such attribute or records.

IV. Normalize the attribute values which have very high ranges. Normalization actually helps in improving the processing speed in later stages. Care must be taken while formulating the normalization formula; it must work accurately in reverse fashion and yield the original attribute value.

After the pre-processing, the underlying data set has date, time, minimum temperature, maximum temperature, precipitation (rain) amount recorded hourly, cumulative precipitation amount per day, min_max_temperature, wind speed, wind direction, humidity, sea level pressure and mean sea level pressure. Few sample observations in the original data set are shown in Fig. 1.

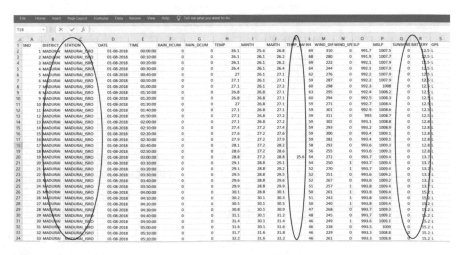

Fig. 1 Few observations as per original data set to be considered for pre-processing stage

3.2 Statistical Model Construction

Now, the data set is ready for statistical model building. The algorithm following the architecture (Fig. 2) builds statistical model.

I. Testing of relationship between the attributes.

  a. Refine the data set with required fields/records which is sufficient.
  b. Perform correlation between every pair of attributes.
  c. Create the frame of pair of attributes along with the degree of correlation either positive or negative correlation.
  d. Select the set that contains the pair of attributes having least correlation. They are basically independent attributes.

II. Relationship testing by constructing linear model.

  a. Build the linear model by choosing a pair of attributes from the list of attributes. The equation is

$$Y = \beta_0 + \beta_1 X_1 + \varepsilon \tag{1}$$

  The value of β_0 (intercept) and β_1 (coordinate of X_1 in the line fitted by the model) shows influence between response (Y) and selected independent attribute (X_1). ε is the error in the model.
  b. Repeat the steps II-a and II-b for every pair of attributes and record the degree of relationship between attributes observed through linear regression.

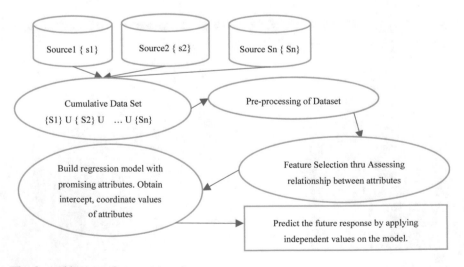

Fig. 2 Architecture of proposed work

III. Select the set of attributes on observing the result of step I and II. Construct the multiple regression model.

a. Construct the multiple liner regression model based on the formula.

$$Y = \beta_0 + \beta_1 X_1 + \beta_2 X_2 + \cdots + \beta_n X_n + \varepsilon \qquad (2)$$

b. Observe the intercept and coefficient values of respondent (Y) variable over independent variables $X_1, \ldots, X_n$.

IV. Substitute the values of intercept, coefficients of attributes in regression formula to obtain future prediction of Y on known (given) $X_{1, \ldots, X_n}$ after a stable model has been constructed; model can be tested using well-known statistical tests like ANOVA, P-Test and F-Test in order to assess the stability with underlying attribute set.

4 Experimental Results and Discussion

Courtesy to Indian Meteorology Department (IMD) [16], this work uses the data set fetched from IMD archive. The statistical model and visualizations are implemented using open-source statistical tool R. Here, our response (prediction) variable is the amount of rain (technically, precipitation) expected for the underlying known weather attributes.

4.1 Data Set

This experiment has been experimented using two data sets, having observations recorded in different time periods at the city 'Madurai'.

I. Data set-1: The data set having 19 attributes and 4202 number of observations recorded at every 10 min interval from 01-08-2018 to 31-08-2018.
II. Data set-2: The data set having 19 attributes and 12,923 number of observations recorded at every 10 min interval from 01-06-2018 to 31-10-2018.

4.2 Correlation

After performing the pre-processing steps, the data set refined to have ten attributes which are essential for analysis. A number of observations are same as the original data set. The correlation between the attributes is shown in Fig. 3.

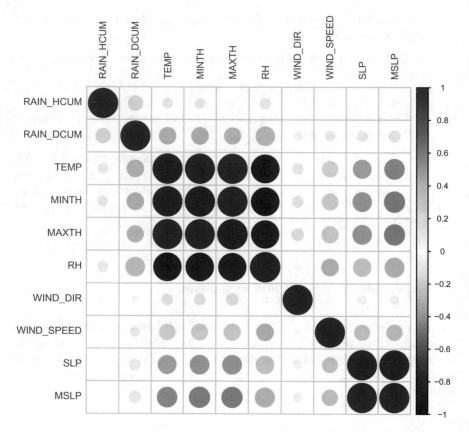

Fig. 3 Correlation plot of weather data set attributes (data set-1: 1 August 1 to 31 August 2018)

This correlation visualization shows that every attribute has 100% correlation with itself. Our response attribute has minimum level of positive correlation with relative humidity, negligible quantity of correlation with wind speed and mean sea level pressure. It also proves that precipitation attribute has very less negative correlation with wind direction.

4.3 Linear Regression Modelling

Linear regression model is built on response attribute with one independent attribute, and the results are plotted for further observations to demonstrate the influence of the independent attribute over precipitation.

The results bring the observation that minimum and maximum temperature have same impact on respondent attribute, and we conclude that minimum temperature can be used as temperature-related component in prediction since it has higher

Table 1 Linear regression between response (precipitation) and independent variables (1 August to 31 August 2018)

Model factors	Independent attributes						
	Relative humidity	Mean sea level pressure	Sea level pressure	Min. Temp.	Max. Temp.	Wind direction	Wind speed
Residual error (4200 of freedom)	5.87	6.35	6.37	6.07	6.11	6.3	6.3
F-stat	777.2	47.49	27.39	452.9	393.6	2.15	24.25
P-value	2.2e−16	6.3e−12	1.74e−07	2.2e−16	2.2e−16	0.14	8.8e−07

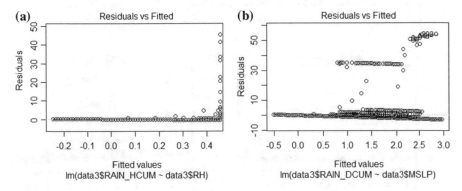

Fig. 4 Linear model of data set-I, **a** (hourly precipitation $\sim$ humidity) versus residual, **b** (cumulative day precipitation $\sim$ mean sea pressure) versus residual

F-stat value. It is observed that humidity also has higher influence in precipitation. The results are tabulated as in Table 1, and the visualizations are given in Fig. 4.

Similarly, the linear regression model is built on three months of observations recorded in 10 min interval from 1 June 2018 to 31 August 2018 (data set-2). The proposed work on data set-2 shows improvement in the following criteria.

- The statistical model has improved in its stability; observed by F-stat value.
- Standard residual error is reduced compared to model constructed by data set-1.
- P-value is less compared to the previous model constructed with data set-1.

The results of data set over the model are tabulated in Table 2, and few visualizations are given in Fig. 5.

Table 2 Linear regression between response (precipitation) and independent variables recorded from 1 June 2018 to 31 August 2018 (data set-2)

Model factors	Independent attributes						
	Relative humidity	Mean sea level pressure	Sea level pressure	Min. Temp.	Max. Temp.	Wind direction	Wind speed
Residual Error (12,923 of freedom)	1.005	5.95	5.92	5.6	5.7	6.0	5.9
F-stat	202.6	266.6	182.2	1487	1314	28.27	122.9
P-value	<2.2E −16	<2.2E −16	<2.2E −16	<2.2E −16	<2.2E −16	1.07E −07	<2.2E −16

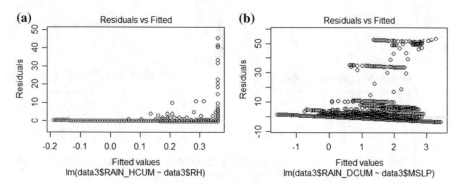

Fig. 5 Linear model of data set-2, **a** (precipitation $\sim$ humidity) versus residual, **b** (precipitation $\sim$ mean sea level pressure) versus residual

4.4 Multiple Linear Regression Modelling

Multiple linear regression model is built on more than one independent attribute which are promising after proceeding correlation and linear regression with response attribute—precipitation; as discussed earlier (Sects. 4.2 and 4.3). The results of the multi-regression models on the set of independent attributes over precipitation are tabulated in Table 3, and few visualizations are given in Fig. 6.

Table 3 Multiple linear regression between response (precipitation) and set of independent variables using observation recorded from 1 August 2018 to 31 August 2018 (data set-1)

Model factors	Independent set of attributes				
	RH + MinT	RH + MinT + WindSpeed	RH + MinT + wind dir.	RH + MinT + Wind dir + RainH
Residual Error (4200 of freedom)	5.8	5.8	5.8	5.6
F-stat	410.6	274	279.2	290.7
P-value	2.2E−16	<2.2E−16	<2.2E−16	2.2E−16

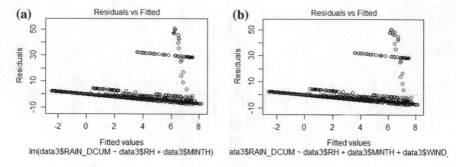

Fig. 6 Multi-regression model of data set-1, **a** (precipitation ∼ humidity + Min. Temp.) versus residual, **b** (precipitation ∼ humidity + Min. Temp. + wind speed)

The obtained results comprise the impact of the multiple linear regression with 2, 3 and 4 attributes, respectively. The model produces the intercept and corresponding coefficients that can be substituted on the respective multi-regression formula for given set of independent attributes in order to forecast/predict the future precipitation.

– Multiple regression model with data set-2 (three months weather observation):

The accumulated, three months weather observations recorded in 10 min interval are used to build multiple regression model on various set of attributes and their performance are listed in Table 4. The proposed work shows improvement in criteria discussed as in linear regression, such as standard residual error, F-stat and P-value model descriptors. The results are tabulated in Table 4, and few visualizations are given in Figs. 7, 8 and 9.

These results show that model build with three months observations (data set-2) has significant results in terms of reliability, stability, handling at most 25% variations on input when compare to models build with single month observations (data set-1).

5   Conclusion

This work concludes that the feature selection described by our work and multi-regression models provide the better results in terms of stability, error rate, accepting variations than simple linear models. Thus, multi-regression models are capable of providing better predictions on underlying empirical data sets; the success rate of the model is directly proportional to the number of observations in empirical data. The result shall be improved if the observation includes attributes like cloud describing information, sunshine describing parameters, pressure at various atmospheric levels and visibility.

Table 4 Multiple linear regression between response (precipitation) and set of independent variables observed from 1 June 2018 to 31 August 2018 recorded at every 10 min interval

Model factors	Independent set of attributes				
	RH + MinT	RH + MSLP	RH + MinT + WindSpeed	RH + MSLP + SLP	RH + MinT + Wind dir + RainH
Residual Error (12,923 of freedom)	5.4	5.5	5.4	5.4	5.3
F-stat	1268	1235	851.9	846.2	810.7
P-value	<2.2E−16	<2.2E−16	<2.2E−16	<2.2E−16	<2.2E−16

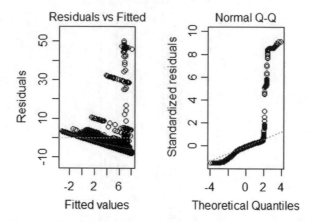

Fig. 7 Data set-2: multi-regression model (precipitation ∼ RH + Min. Temp.) versus residual (Note: June month has no rain, and August has high rain fall observation)

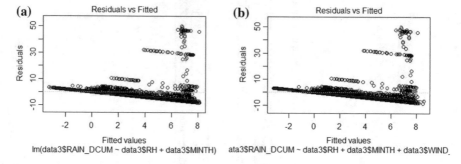

Fig. 8 Multiple linear regression model fitness with data set-2, **a** (precipitation ∼ humidity + Min. Temp.) versus residual, **b** (precipitation ∼ humidity + Min. Temp + wind speed) versus residual

Fig. 9 Data set-2's multi-regression model (RH + Min. Temp. + Wind Dir.) versus residual (Note: June month has no rain, and August has high rain fall observation)

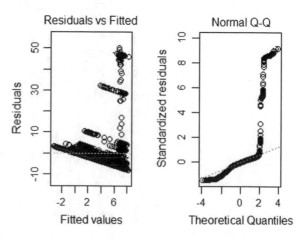

References

1. Jaroensutasinee, K., Pheera, W., Jaroensutasinee, M.: Online Weather Data Analysis and Visualization Tools for Applications in Eco informatics. Springer, Heidelberg (2013)
2. Valipur, M.: Study of different climatic conditions to assess solar radiation in reference crop evaporation equations. Arch. Agronomy Soil Sci. **61**, 679–694 (2014)
3. Valipur, M.: Analysis of potential evaporation using limited weather data. Appl. Water Sci. (2014)
4. Biswas, S.K., Sinha, N., Purkayastha, B., Marbaniang, L.: Weather prediction by recurrent neural network. Int. J. Intell. Eng. Inform. **2**(2/3) (2014) (InderScience Enterprises Ltd.)
5. Pathan, M.B.: Modeling data from climate science using introductory statistics. J. Climatol. Weather. Forecast. (2015). http://dx.doi.org/10.4172/2332-2594.1000129
6. Anik, M.T.A., Dhali, K.P., Ferdous Hossain, S.M., Johara, F.T.: Spatial data visualization methodologies in ICT4D research. In: 18th International Conference on Computer and Information Technology (ICCIT). IEEE, New York (2015)
7. Diehl, A., Pelorosso, L., Delrieuax, C., Saulo, C., Ruiz, J., Groller, M.E., Bruckner, S.: Visual analysis of spatio-temporal data: applications in weather forecasting. Euro Graph. Vis. (Euro VIS) **34**(3) (2015)
8. Samuel Quinan, P., Meyer, M.: Visually comparing weather features in forecasts. IEEE Trans. Vis. Comput. Graph. **22**(1) (Jan 2016)
9. Diehl, A., Pelorosso, L., Delrieux, C., Matkovic, K., Ruiz, J., Groller, M.E., Bruckner, S.: Albero: A visual analytics approach for probabilistic weather forecasting. Comput. Graph. Forum **36**(7) (2017). https://doi.org/10.1111/cgf.13279
10. Shabariram, C.P., Kannammal, K.E., Manojprabakar, T.: Rainfall analysis and rain storm prediction using map reduce framework. In: 2016 International Conference on Computer Communication and informatics (ICCI-2016). IEEE, New York (2016)
11. Liu, S., Wang, X., Liu, M., Zhu, J.: Towards Better Analysis of Machine Learning Models: A Visual Analytics Perspective. Elsevier B.V. Visual Informatics (2017). http://dx.doi.org/10.1016/j.visinf.2017.01.006
12. Anatharajan, T.R.V., Abishek Hariharan, G., Vignjith, K.K., Jijendiran kushmita, R.: Weather monitoring using artificial intelligence. In: 2016 International Conference on Computational Intelligence and Networks. https://doi.org/10.1109/cine.2016.26
13. Sudha, G., Sangaiah, S.: A survey on contribution of visual analytics in health care domain (April 19, 2018). In: 2018 IADS International Conference on Computing, Communications & Data Engineering (CCODE). Available at SSRN: https://ssrn.com/abstract=3165309 © Elsevier
14. Arumugam, G., Sangaiah, S., Sudha, G.: Weather analysis and prediction: a survey with visual analytic perspective. Int. J. Res. Anal. Rev. UGC Approved Research Journal ISSN 2349-5138, E- ISSN 2348-1269 (July 2018)
15. Jain, H., Jain, R.: Big data in weather forecasting: applications and challenges. In: 2017 International Conference on Big Data Analytics and Computational Intelligence (ICBDACI). IEEE, New York, 978-1-5090-6399-4/17/
16. Indian Meteorology Department

Recommendation System Based on Optimal Feature Selection Algorithm for Predictive Analysis

Malaichamy Vithya and Suguna Sangaiah

Abstract The Internet business model with predictive analysis provides efficient way of accessibility to customers with less resources and minimum expenses. User access log files are considered in this work for predictive analysis. But the log files contain lots of unwanted information and not in a form for predictive analysis. So, the web logs need to be pre-processed and further handled by the recommendation system for predictive analysis. In the processing phase, unwanted details are removed; generated user sequences, session sequences and complete user access path patterns are also generated. In the proposed recommender system, significant feature set is considered using our proposed feature selection algorithm to reduce the computational cost and to improve the quality in the predictive analysis process. In this paper, vizhamurasu news web server user access logs are considered for predictive analysis. Our proposed feature selection algorithm performance is analyzed using the performance metrics. The performance analysis shows that our proposed algorithms significantly improve the prediction accuracy than the existing.

Keywords Web logs · Recommendation system · Feature selection · Predictive analysis · User access pattern

1 Introduction

In recent years, E-commerce data set are much larger that huge data set may contain thousands of instances and each of which may be characterized by thousands of features. The massive amount of high-dimensional social and E-commerce data

M. Vithya · S. Sangaiah (✉)
Sri Meenakshi Government Arts College for Women (A), Madurai, Tamil Nadu, India
e-mail: kt.suguna@gmail.com

M. Vithya
e-mail: vithyagopal20@gmail.com

© Springer Nature Singapore Pte Ltd. 2020 105
P. Venkata Krishna and M. S. Obaidat (eds.), *Emerging Research in Data
Engineering Systems and Computer Communications*, Advances in Intelligent
Systems and Computing 1054, https://doi.org/10.1007/978-981-15-0135-7_10

presents great challenges to deal with traditional data mining and machine learning task [1, 2]. Feature selection is a process commonly used in machine learning. It is used to reduce the dimensionality before applying any data mining techniques such as classification, association rules, clustering and regression. Feature selection process selects the subset of original features, without loss of useful information. This helps to remove unrelated and redundant features for reducing data dimensionality. The main aim of feature selection is to find out minimal feature or best subset from our problem domain while retaining high accuracy in signifying the least number of original features [3]. This improves the learning performance and reduces the computational time, enhances result clarity, offer less amount of storage requirements, avoids over fitting, increases the speed of mining algorithm and decreases training times [4]. In general, feature selection algorithms are categorized as wrapper, filter and embedded algorithms. The filter methods used geometric measures for feature selection. In wrapper model, some encoded learning algorithm used to discover the relevant features. Embedded methods combine the qualities of filter and wrapper methods. In this paper, filter model is used to improve the accuracy of recommender system based on user access frequency. The paper is structured as follows. An overview of related reviews is given in Sect. 2. In Sect. 3, the framework of proposed architecture and working principles of proposed feature selection algorithm are discussed. The performance evaluation discussed in Sect. 4. Finally, concluded the work discussed in Sect. 5.

2 Literature Review

In [3, 4] discussed different feature selection methods and algorithms. They presented a case study on different performance metrics and its applications. In [5] conducted an experiment on Mumbai engineering college to enroll in 2014 using 17 attributes from academic and personal records. Their main aim was to find out which feature (attribute) to force to admit student in engineering college using feature selection algorithm. They applied four different classification algorithms in selected significant features and concluded results. In [6] used relief-f feature selection algorithm to find which features highly cause the cotton damage using Weka tool. In [7] conducted experiment to predict best feature from high dimensionality data set using monte carlo feature selection algorithm. They test weather selected features are significantly related to class or not using Z-test for performance measures. Finally, they proved Z-test was an efficient method to identify the significant features. In [8] compared six filter feature selection algorithms and reveals the best method on educational data set. They observed which algorithm expected minimum computational time, constructional time and high accuracy. In [9, 10] conducted experiment on two real-world signed social networks data and demonstrated the effectiveness of their proposed FSNet algorithm. They proved signedFS significantly improves the clustering performance. In [11–13] presents a novel proposed framework for unsupervised personalized feature selection (UPFS) to find out a subset of shared features and their instances specific discriminative features in each instance. They experimented their framework in real-world data sets and showed the effectiveness.

3　Proposed System

In our earlier work [14], proposed algorithm with parallel computing and Hbase for cleaning the irrelevant and unwanted details from vizhamurasu news Web site user access logs. In [15] proposed the algorithm for user identification, session identification and path completion and generated the complete user access path sequences. In this paper, proposed the feature selection algorithm for recommender system to provide efficient predictive analysis. Figure 1 shows the proposed framework to identify significant features from user access patterns using feature selection algorithm.

3.1　Feature Extraction

The pre-processed user access pattern [14, 15] of vizhamurasu news Web site contains huge instances and attributes. The access context is used to personalize the user from their usage in this Web site. Usage data has huge volume and veracity, so

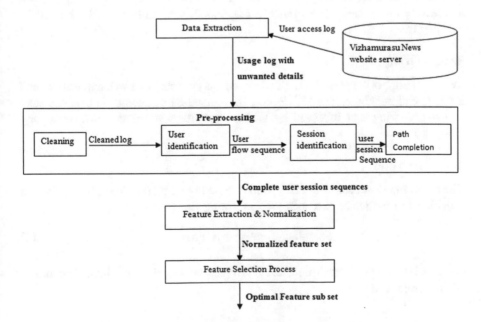

Fig. 1 Proposed architecture for feature selection process

it needs more computation as well as memory space to generate access context. The main aim of the framework is to identifying significant features for identification of a user access behavior with minimum effort and less computation time. Some features are extracted from a user access logs such as IP address, URL, date and time, query string, spending time, average time, page weight, inter-visiting time, page size, burst of visit, user link hierarchy, episode and total of number of visitors.

IP Address

An Internet protocol address is a numerical label assigned to each device connected to a computer network that uses the Internet protocol for communication. It is a unique address that identifies a machine on the Internet or local network, used to recognize who's requested the Web site; the sample IP address is 106.66.169.221.

URL

A uniform resource locator (URL) is the address of a source on the Internet. A URL indicates the location of a resources as well as the protocol used to access it, and it contains some information such as resource, location of the server (domain name), port number on the server, the location of the resources in the directory structure of the server and a part identifier (optional). The sample URL is http://vithya. vizhamurasu.com/sports.php.

Date and Time

The time and date request was received. That is the time that the server finished processing the request. The format is [day/month/year] Aug 9 17. It denoted by number from 1 to 31.

Query String

A query string is the part of a URL where data is accepted to a web application and/ or back-end data base. The HTTP protocol is stateless by design, so data can store via query string that followed by URL. http://vithya.vizhamurasu.com/category. php,id=5.

Spending Time

User how many seconds/minutes spend in a particular page? The spending time is calculated to particular user using the following Eq. (1).

$$\text{Spending Time} = \max_{T}(\text{Page}_{\text{visited}}) \qquad (1)$$

Max is denoted by maximum, and T is represented by time in minutes/seconds of visited page to user.

Average Time

The average time is calculated by finding the time duration between two consecutive page visits. It is calculated by using the following Eq. (2).

$$\text{Average Time} = T(\text{Page}_{\text{Visited}}(n)) - T(\text{Page}_{\text{Visited}}(m)) \tag{2}$$

Here, m and n are consecutive page visits by a particular user.

Inter-visiting Time

It is calculated with the time difference between the sessions. Inter-visiting is calculated for a particular user using the formula in Eq. (3).

$$\text{Inter-visiting Time } (T) = \sum_{i=1}^{n} (T(s2)) - (T(s1)) \tag{3}$$

Here, $s1$ and $s2$ are sessions in minutes.

Page Weight

Particular page was how many times visited by a user. It is calculated by sum of occurrence of a page in same or different user sessions. The following equation is used to find the page weight Eq. (4).

$$\text{Page Weight} = \sum_{i=1}^{n} \text{Page Visited} \in \text{User} \tag{4}$$

Total Number of Visitors

The weight of how many users visited a particular page. To find the total number of visitors the following Eq. (5).

$$\text{Total No. of Visitor} = \sum_{i=1}^{n} \text{User} \in \text{Page} \tag{5}$$

Burst of Visit

Burst of visit is the request coming into the site due to some popular events occurring on the site. Burst of visit time is calculated by sum of how many times the page seen by a particular user using Eq. (6).

$$\text{Burst of Visit} = \sum_{i=1}^{n} \text{Page}_{\text{Popular\_Events}} \tag{6}$$

Here, n denotes the maximum of the visited time.

User Flow Hierarchy (Hyperlink)

The user flow hierarchy (F) refers to the user's sequence flow of the visited pages for different sessions. The sample of single session flow sequence for a user is. http://vithya.vizhamurasu.com/sports.php,   http://ithya.vizhamurasu.com/cricket. php, http://vithya.vizhamurasu.com/economics.php.

Episode

Episode calculation is based on user flow hierarchy result. Equation (5) is used to find the total number of occurrences of the same user flow (F) with different sessions of visited pages.

$$\text{Episode} = \begin{cases} \sum_{i=1}^{n} W\big(F(i)_{\text{Session}}\big), n \; \forall \; \text{No. of Same Occurence} \\ \text{Otherwise } 0; \end{cases} \tag{7}$$

Page Size

Page size is referred to the size of the page, which is the block of stored memory. A page size includes all of the files that are used to create the web page. These web page or files are in any format like HTML, PHP, etc. The experimental web pages are in PHP, and the page size ranges between 10 K to 200 KB.

3.2 Illustration of Proposed Feature Extraction from a User Access Log

Table 1 shows the extracted feature set.

The proposed system extracted 13 features from different user's access log information. Table 2 shows few sample features with their values of different user access information.

3.3 Normalization of Feature Vector

In Table 2, attributes of features are in different formats, so these feature values are normalized into fixed-length vectors of numbers before applying feature selection algorithm. Normalization is the process of generalizing a set to same type of consistent. It is used to achieve the efficient computation of the feature selection with minimized redundancy. The proposed normalization is one type of "One hot encoding". It is a process by which categorical variables are converted into a form that could be provided for feature selection process to do a better job in prediction.

Table 1 Feature fusion set

S. No.	Feature set number (FSN)	Description
1	FSN-1	IP address
2	FSN-2	URL
3	FSN-3	Query string
4	FSN-4	Date & time
5	FSN-5	Inter-visiting time
6	FSN-6	Page weight
7	FSN-7	Total number of users
8	FSN-8	Burst of visit
9	FSN-9	Page size
10	FSN-10	Spending time
11	FSN-11	User hierarchy
12	FSN-12	Episode
13	FSN-13	Average time

Table 2 Sample feature set before normalization

IP	URL	Query string	Date and time	Inter-visiting time in minutes	Page weight	Total number of visitors	Burst of visit	Page size (KB)	Spending time in minutes	User hierarchy	Episode	Average time in minutes
106.66.69.221	http://vithya. vizhamurasu.com/ sports.php	Id = 10	Aug 9 17:07:01 2017	3	9	15	1	90	43	Sid-1: sports → cricket → politics	7	8
64.233.173.23	http://vithya. vizhamurasu.com/ politics.php	Id = 20	Sep 15 3:45:59 2017	5	19	25	3	60	35	Sid-3: Politics → reall-estate → economic	4	16
106.77.172.107	http://vithya. vizhamurasu.com/ economics.php	Id = 25	Oct 25 8:49:55 2017	10	29	27	2	69	28	Sid-6: economic → real-estate-	6	24

The words need to be encoded as integer or floating point values for the input to machine learning algorithm. The proposed normalization feature vector algorithm is shown below:

```
Input: FA=Feature Set
Output: OP = Normalized Set
A=read Feature Array (FA)
N= Total Row (A)
for (i=0;i<n;i++)
{
CIP[i] =ConvertIP (A[i]);
CURL[i] =ConvertURL (A[i]);
CQS[i] =ConvertQueryString (A[i]);
OP= {CIP[i], CURL[i], CQS[i]}
return OP;
}
Input: ip=IP_Address
Output: No = Hash Key
Sub ConvertIP (String ip)
{
if (exists(ip))
{
No=exist_no; // No denotes the Hash Key of Ip address
}
else
{
  No=Max (No) +1;
}
return No;
}
Input: URL
Output: No = Hash Key
Sub ConvertURL (String url)
{
    if (exists (url))
{
  No=exist_no; // No denotes the Hash Key of Ip address
}
else
{
No=Max (No)+1;
}
return No;
}
Input:QS=Query_String
Output: No = Hash Key
Sub ConvertQueryString(String QS)
{
  i f(exists(QS))
{
No=exist_no; // No denotes the Hash Key of Ip address
}
else
{
No=Max(No)+1;
}
return No;
}
```

Table 3 Hash table of IP address

Key	Value
1	192.168.16
2	223.223.29.30
3	168.98.23.98

Table 4 Hash table of URL and query string

Key	Value (URL)	Value (query string)
1	http://vithya.vizhamurasu.com	Id = 3
2	http://vithya.vizhamurasu.com/index.php	Id = 5
3	http://vithya.vizhamurasu.com/category.php	Id = 7

Table 3 and 4 show the hash table with normalized values of features IP address, URL and query sting.

The extracted features set are generalized to enhance the accuracy of recognition process. After normalization, the features are represented as vector by combining the various extracted feature set for different user access log patterns and shown in Table 5.

3.4   Feature Selection

During this study, the dimension reduction techniques for high-dimensional sample spaces are investigated. The feature selection algorithms assist in finding only those features that are really important for identifying user access patterns, and dimensionality reduction typically improves the accuracy of recognition of user pattern with saving memory and time consumptions.

3.4.1   Proposed Feature Selection Algorithm

User access features are extracted from pre-processed user usage logs. Then the feature fusion set is generalized using normalization. The normalized feature set is employed to select the significant features for enhance the predictive model of the user access behavior. Thus, to enhance the predictive model, a proposed feature selection algorithm known to be Feature Reduction and Selection Algorithm (FRSA) is depicted as follows.

Table 5 Sample feature vector set after normalization

IP	URL	Query string	Date and time	Inter-visiting time	Page weight	Total number of visit	Burst of visit	Pagesize (KB)	Spending time	User hierarchy	Episode	Average time
1	8	4	24	5	9	15	1	90	43	$5 \rightarrow 6 \rightarrow 7 \rightarrow 8$	7	8
5	12	9	15	5	19	25	3	60	35	$12 \rightarrow 15 \rightarrow 4$	4	16
9	15	25	12	10	29	27	2	69	28	$13 \rightarrow 16 \rightarrow 18$	6	24

Algorithm: Feature selection algorithm

Input: D_{Log}- Log Dataset

N-Number of Log Attributes

Output: A_{Subset} – Attribute Subset

1:    Obtain the User Web Log Lfrom D_{Log}
2:    Read the set of log attributes from the log L
3:    Select n attributes from number of log attributes N
4:    T=Compute Threshold (N,D_{Log})
5:    For each i = 1 to n of log attributes
6:    For all subsets n_{sub} from N
7:    If InConCal(n_{sub},D_{Log}) <– T // inconsistency is up tothreshold
8:    {
9:    Choose n_{sub} as significant attribute
10:   Return n_{sub}
11:   }
12:   Else
13:   {
14:   If InConCal(n_{sub},D_{Log}) >T// inconsistency is greater than threshold
15:   N_{sub}=n-n_{sub} // remove one feature at a time
16:   }
17:   Return N_{sub}

   }

Sub InConCal (S,D)

  { Obtain the attribute subset S and Dataset D
1:    Compute number of Occurrence i^{th}attribute of S in D
2:    For j = 1 to k instances of distinct feature combination in S
3:    Compute S_{Incon} = $\dfrac{|D_j|-|M_i|}{N}$ // consistency of majority instance of S and Dataset D

4:    Return S_{Incon}
   }

The algorithm explains that obtained web log information of the users is denoted as D_{Log}. From the collection of web log information, the set of attributes are fetched from the each log. The set of attributes are processed to generate a random subset, n_{sub}, from N attributes in every log. Then the computation of threshold value T is performed with the total number of log attributes with respect to the log dataset D_{Log}. The randomly generated subsets are employed to select the significant attributes among the subset. If the number of attributes n of S is less than the threshold T, the data D_{Log} with the attributes prescribed in S is checked against the inconsistency criterion, and the attribute is selected as significant. If the inconsistency criteria are greater than T, then the attribute is removed from the subset of the log attribute set. Then construct a selected feature vector (SFV) to make accurate personalization for the user access prediction model.

Table 6 illustrates the selection of significant feature subset based on the proposed feature selection (FRFS) procedure for the user access pattern recognition.

Algorithm	Selected feature set number
FRFS	1, 2, 4, 5, 6, 7, 10, 11, 12, 13

Table 6 The selected feature vector

4 Implementation and Experimental Results

The proposed architecture of correlation-based subsidy feature selection algorithm is implemented using Hadoop. Hbase server, eclipse and executed on an ubuntu 4.6 operating system. The test data are the vizhamurasu web site user access logs.

4.1 Comparative Study

The proposed algorithm compared with two existing feature selection algorithm based on two performances metrics such as accuracy and computation time. The effectiveness of the feature selection process is usually measured by the Precision, Recall and F-measure; these metrics are calculated by the following formulae

$$\text{Precision} = \frac{(\text{Relevant features}) \cap (\text{Retrieved features})}{\text{Total No. of Feature Set}}$$

$$F = 2 * \frac{(\text{Precision} * \text{Recall})}{\text{Precision}} + \text{Recall}$$

Figures 2 and 3 show the result for comparison of the proposed feature selection algorithm (FRFS) with two existing feature selection algorithm (Correlation feature selection [5] and Relief $-F$ [6]) by evaluating performance metrics such as accuracy and execution with respect to the values of Precision, Recall and F-Measure. Proposed FRFS outperforms with good accuracy of 92% than existing methods. The accuracy comparison results are depicted in (Fig. 2).

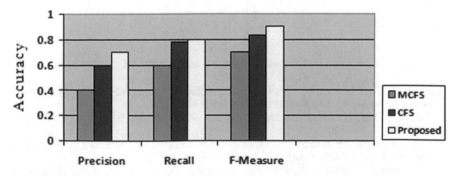

Fig. 2 Comparison of precision, recall and F-measure accuracy between existing and proposed

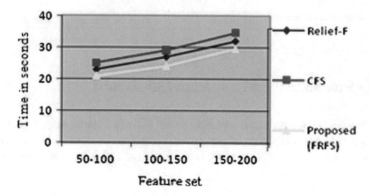

Fig. 3 Comparison of execution time between existing and proposed algorithm

Figure 3 shows that the proposed method (FRFS) consumes less time than other existing with various set of features.

5 Conclusion

The system concludes that the enhancement of the predictive analysis of the recommendation based on the user access behavior is carried out with the dimensionality reduction of obtaining web log files. The main goal of this work is to identify the significant feature. The feature selection algorithm reduces the feature set without loss of information for further clustering/classification process. The proposed algorithm (FRFS) focused on generating significant feature subset for the user access behavior identification. The obtained significant attributes are employed in predicting and recommending the access results to the users. The proposed algorithm out performed with less computation and high accuracy than other existing algorithms.

References

1. Cheng, K., Li, J., Liu, H.: Unsupervised Feature Selection in Signed Social Networks. August 13–17, ACM (2017)
2. Bornelov, S., Komorowski, J.: Selecting of Significant Features using Mote Carlo Feature Selection. Springer International Publishing Switzerland (2016)
3. Ladha, L., Deepa, T.: Feature selection methods and algorithms. Int. J. Comput. Sci. Eng. (IJCSE) 3(5) (May 2011)
4. Sutha, K., Jebamalar Tamilselvi, J.: A review of feature selection algorithms for data mining techniques. Int. J. Comput. Sci. Eng. (IJCSE) 7(6) (June 2015)
5. Mital Doshi, S., Chaturvedi, K.: Correlation based feature selection (CFS) technique to predict student performance. Int. J. Comput. Netw. Commun. (IJCNC) 6(3) (May 2014)

6. Francisca Rosario, S., Thangadurai, K.: Relief: feature selection approach. Int. J. Innov. Res. Develop. **4**(11) (October 2015)
7. Zundong Li, J., Lin, J.: Challenges of feature selection for big data analytics. IEEE Intell. Syst. **32**, 9–15 (2017)
8. Ramaswami, M., Bhaskaran, R.: A study on feature selection techniques in educational data mining. J. Comput. **1**(1) (December 2009)
9. Cheng, K., Li, J., Liu, H.: Unsupervised Feature Selection in Signed Social Networks. In: KDD '17'. ACM, New York (August 13–17, 2017)
10. Liu, H., Motoda, H.: Computational Methods of Feature Selection. CRC Press (2007)
11. Li, J., Cheng, K., Wang, S., Morstatter, F., Robert, T., Tang, J., Liu, H.: Feature Selection: A Data Perspective. Arxiv (2016)
12. Durgabai, R.P.: Feature selection using ReliefF Algorithm. Int. J. Adv. Res. Comput. Eng. **3** (10) (October 2014)
13. Li, J., Wu, L., Dani, H., Liu, H.: Unsupervised Personalized Feature Selection. Association for the Advancement of Artificial Intelligence (2018)
14. Vithya, M., Suguna, S.: Optimal algorithms for preprocessing of server side web logs using parallel computing and HBASE. Comput. Commun. Data Eng. Ser. **1**(1) (2018)
15. Vithya, M., Suguna, S.: Parallel computing techniques for predictive perfecting of server side pages. In: International of Conference of Big Data Analytics and Intelligent of technology. (February 2018)

Improving the Accuracy of Prediction of Plant Diseases Using Dimensionality Reduction-Based Ensemble Models

A. R. Mohamed Yousuff and M. Rajasekhara Babu

Abstract In many real-world applications, different features can be obtained and how to duly utilize them in reduced dimension is a challenge. Simply concatenating them into a long vector is not appropriate because each view has its specific statistical property and physical interpretation. Many dimensionality reduction methods have been developed to identify this lower-dimensional space and map data to it, reducing the number of predictors in supervised learning problems and allowing for better visualization of data relations and clusters. However, the plethora of dimensionality reduction techniques provides a variety of nonlinear, linear, global, and local methods, and it is likely that each method captures different data features. Ensemble methods have achieved much success in supervised learning, from Random Forest to AdaBoost. Ensembles exploit diversity and balance bias, variance, and covariance to achieve these results is likely that disparate dimensionality reduction methods will enhance diversity within a dimensionality reduction-based ensemble. AdaBoost and Random Forest are popular ensemble methods which are widely used for classification of target variables. Major problem with ensembles like AdaBoost and Random Forest is that they perform worse when dimensionality of data is high. Random Forest is the predictor ensemble with a set of decision trees that grow in randomly selected subspaces of data. The proposed research work aims to improve the performance of classification tasks on diseased plants by exploring t-distributed Stochastic Neighbor Embedding (t-SNE) based Ensemble Models. The infected and healthy plant images are subjected to deep learning model to produce their corresponding image embedding. The high dimensional data with thousands of features is then reduced to a smaller number of features dataset by the state-of-the-art t-SNE algorithm. The significant feature dataset is then given as input to the ensemble models to perform the prediction task.

A. R. Mohamed Yousuff · M. Rajasekhara Babu (✉)
School of Computer Science and Engineering, VIT University, Vellore, India
e-mail: mrajasekharababu@vit.ac.in

A. R. Mohamed Yousuff
e-mail: mohamedyousuff.ar2018@vitstudent.ac.in

© Springer Nature Singapore Pte Ltd. 2020
P. Venkata Krishna and M. S. Obaidat (eds.), *Emerging Research in Data Engineering Systems and Computer Communications*, Advances in Intelligent Systems and Computing 1054, https://doi.org/10.1007/978-981-15-0135-7_11

Keywords Dimensionality reduction · *t*-SNE · Deep learning model · Ensemble models

1 Introduction

Humans are capable of visualizing two-dimensional and three-dimensional space or data in form of tables and graphs. Real-time data such as plant disease images, x-ray images, MRI scan, and many other medical images is classified under high dimensional data. The dimension of the data should be reduced to extend that the important features within data should also be preserved. The curse of dimensionality affects classification, clustering, and visualization of data [1]. The *t*-distributed Stochastic Neighbor Embedding (*t*-SNE) algorithm can effectively visualize high dimensional data by insuring feature. Each data point present in the dataset is considered, optimized, and clustered on the two-dimensional or three-dimensional map, which helps to understand the correlation between each data point. The *t*-SNE technique is adapted to reduce the high dimensional plant disease data to two-dimensional data for better classification task and identification of disease [2]. The hyperspectral images and spatial-spectral images have huge dimensions which result in pose collinearity and ill-determination problems. Multivariate type of dimensionality reduction via regression (DRR) makes it suitable to maintain the properties of principal polynomial analysis (PPA) and principal component analysis (PCA) while handling high dimensional remote sensing data [3].

Discriminant information present in the high dimensional images produced by infrared imaging spectrometer is preserved and utilized in dimensionality reduction via regression (DDR). Discriminating information of intermanifold structure and sparse reconstructive relations present in the image dataset is maintained and intrinsic information is also preserved after dimensionality reduction using sparse discriminant embedding (SPE) [4]. Hyperspectral image (HSI) and medical images classification task is made difficult because of its high dimensional nature. HIS is converted into data cubes and further divided into small tensors. Based on the relations among the small tensor, they are clustered and finally projected on low-rank subspace in order to get a proper feature space [5].

The intrinsic geometric structure present in image data can be exploited using manifold learning but this method fails to perform better in case if the data is corrupted. To overcome this problem, manifold learning is combined with low-rank sparse representation to obtain optimal features [6]. Dimensionality reduction methods based on the graph give more promising results but they are more sensitive to noise and hence graph structure changes because of additive noise. Sparse graph and low-rank graph are used to discover hidden local and global structures present in the images but fail in determining the intrinsic information present in the images. Low-rank graph along with patch tensor-based sparse representation helps to exploit the intrinsic information in the images and also reduce the dimensionality with low processing cost [7].

Graph structures are constructed on the given image dataset to exploit the underlying distribution of data and dimensionality reduction is considered as a separate process. Unification of both these processes avoids a lot of limitations and this can be achieved using multiple locality-constrained graph optimizations for dimensionality reduction (MLGODR) algorithm. The distribution of image data and its projection is done on low-dimensional subspace using a matrix can be learned at the same time by implementing MLGODR algorithm [8].

2 Related Works

The infected plant leaves images are preprocessed and subjected to disease segment identification and feature extraction then K-means algorithm and genetic algorithm are used to cluster and classify the plant disease. Nature-inspired algorithm like bacterial foraging optimization (BFO) is used to train the radial basis function neural network (RBFNN) on plant disease dataset to get the infected regions of leaves [9]. A mobile application is used to capture the image of an infected leaf from the plant. The image is then subject to preprocessing to remove the background of the image using the EM algorithm for Gaussian mixture (EMGM). The gray level co-occurrence matrix is computed and made to identify the texture and features using Haralick texture feature method. The extracted features are then given as input to feedforward neural networks in order to perform the classification task [10].

The input images are converted into the RGB matrix which is fed as input to the convolution neural networks (CNN) with the rectified linear unit as activation function (ReLU). The output of CNN is further given as input to the learning vector quantization (LVQ) algorithm. LVQ is heuristic-based classification algorithm which learns the data in Kohonen layer and classifies the given infected input images into its corresponding category [11]. The infected mango leaves dataset was collected and set of rules for classification of the disease has been laid down by getting field expert knowledge. The input data is then subjected to processing using the J48 algorithm implemented in weka 3.8, to identify normal leaf, anthracnose, and algal spot disease found in the mango plant leaves [12].

Manual data collection process was done on 40 different species of peanuts affected by tomato spot wilt disease. Later, images of diseased plants are captured using quadcopter. Orthomosaic 5 channel image creation was initiated after giving the quadcopter images as input, and then image cropping, rotation, and channel separation were done. The plots in the images are isolated using false color image creation. Finally, masked index images are created which are further subjected to evaluation using the correlation of manual data by computing percentage and pixel count varying threshold [13].

3 Proposed Methodology

The input images from four categories are converted into image embedding, i.e., numerical representation of images, using widely popular deep learning model called Visual Geometry Group (VGG19). The image embedding is a high dimensional dataset of the size 4095 features, respectively. The high dimensional image embedding is then subjected to a classification task using Random Forest and AdaBoost-based ensemble models and accuracies are measured. The same high dimensional image embedding is then subjected to the t-SNE algorithm to reduce their dimensions, and then again classification task is performed using Random Forest and AdaBoost-based ensemble models and accuracies are measured. It has been found that the accuracy of the classification of diseased leaves has improved after dimensionality reduction.

3.1 Dataset Description

The apple leaf disease dataset used in this paper was acquired from publicly available Kaggle dataset repository. Apple leaf diseases such as black rot, scab, and cedar rust were addressed in this paper. All the images used have a size of 256×256 pixels. The detailed description of the categories present in the datasets is given in Table 1.

3.2 Visual Geometry Group (VGG19)

The best performing convolution neural networks (CNN) were used to build this model. Preprocessing of the dataset is achieved by computing mean RGB value and negating it from each pixel. In order to capture the notion of directions such as left, right, up, and down, the images are meant to pass through filters with the small receptive field of 3×3 present in layers of CNN. The stride is fixed at 2 and 2×2 pixel window is meant for Max-pooling. A pile of CNN layers is followed by three fully connected (FC) layers with the same configuration throughout all networks. The

Table 1 Dataset category details

Apple leaf diseases	Black rot	Scab	Cedar rust	Healthy
Number of images	497	504	220	1316

first two FC contains 4096 channels, whereas the third one contains 1000 channels. The softmax layer is configured as the last layer. Rectification nonlinearity (ReLU) is configured for all hidden layers [14]. VGG19 model is trained with 1.3 million images, validated with 50,000 images, and finally tested with 100,000 images. The top-one and top-five error were used to evaluate the performance of the model.

3.3 Student-t Distributed Stochastic Neighbor Embedding (t-SNE)

The data points with many features probably yield high dimension Euclidean distance among them. The process of converting the Euclidean distance into conditional probability is called stochastic neighbor embedding. t-SNE algorithm takes high dimensional dataset $\mathcal{D} = \{d_1, d_2, d_3, \ldots, d_n\}$ and converts into two-dimensional or three-dimensional data points $\varepsilon = \{e_1, e_2, e_3, \ldots, e_n\}$ that can be visualized using a scatter plot. The similarity between data point d_x and d_y can be given as conditional probability, $p_{y|x}$. The conditional probability value is high for nearby data points, whereas it is infinite for distributed data with large distance of separation between them. The conditional probability is given by Eq. 3

$$p_{y|x} = \frac{\exp(-\|d_x - d_y\|^2 / 2\sigma_x^2)}{\sum_{m \neq x} \exp\left(-\|d_x - d_m\|^2 / 2\sigma_x^2\right)} \tag{1}$$

The variance of Gaussian which is centered on data point d_x is denoted as σ_x. In t-SNE, using student-t distribution joint probabilities q_{xy} can be computed using Eq. 1 with 1 degree of freedom.

$$q_{xy} = \frac{\left(1 + \|e_x - e_y\|^2\right)^{-1}}{\sum_{m \neq l} \left(1 + \|e_m - e_l\|^2\right)^{-1}} \tag{2}$$

The gradient of the Kullback–Leibler divergence between P and the student-t-based joint probability distribution Q (computed using Eq. 1) is given by Eq. 2

$$\frac{\delta C}{\delta e_x} = 4 \sum_y (p_{xy} - q_{xy})(e_x - e_y)\left(1 + \|e_x - e_y\|^2\right)^{-1} \tag{3}$$

Algorithm 1: Simple version of t-distributed Stochastic Neighbor Embedding.

Input: data set $\mathcal{D} = \{ d_1, d_2, d_3, \ldots, d_n\}$
Input: parameters of cost function: perplexity perpx,
Input: optimization parameters: number of iterations I, rate of learning L,
      momentum $M(i)$
Output: low-dimensional data representation $\varepsilon^{(I)} = \{e_1, e_2, e_3, \ldots, e_n\}$.
Begin:
    calculate pairwise affinities $p_{y|x}$ with perplexity perpx (using equation 3)
    set $p_{xy} = \dfrac{p_{y|x} + p_{x|y}}{2n}$
    Initialization: $\varepsilon^{(0)} = \{e_1, e_2, e_3, \ldots, e_n\}$ from $N(0, 10^{-4}T)$
    for $i = 1$ to I **do**
        calculate low-dimensional affinities q_{xy} (using equation 1)
        calculate gradient $\dfrac{\delta C}{\delta \varepsilon}$ (using equation 5)
        set $\varepsilon^{(i)} = \varepsilon^{(i-1)} + L \dfrac{\delta C}{\delta \varepsilon} + M(i) (\varepsilon^{(i-1)} - \varepsilon^{(i-2)})$
    end
end

3.4   Ensemble Models

Decision trees may result in high variance which causes their output to be delicate
to the specific training data. This variance can be reduced by building multiple
models from samples of the training data and this is called bagging. Random Forest
improves the concept of bagging by finding the correct feature used to construct
trees, apart from building trees on the samples of the training dataset. Thus, performance can be improved. The best split point for the given dataset is found by
decision trees using a greedy selection method but the trees are more vulnerable to
high variance so they need pruning process. The trees in bagging based on decision
tree create similar trees because of similar split points resulting in similar predictions. These split points can be forced to change for each tree using the Random
Forest algorithm [15]. Since the trees generated by Random Forest are different
from each other, the prediction also gives better performance that is the strong
reason behind the usage of Random Forest as an ensemble.

    A number of weak classifiers are ensembled to create a strong classifier; this
method is called as boosting. The AdaBoost algorithm is one among the boosting
algorithms available. In this method, a model is built on the training dataset and the
second model is created to rectify the errors in the first model [16]. Similarly,
models are summed up until maximum model limit is reached or the prediction task
is perfectly achieved.

4 Results and Discussion

Apple leaf dataset with 2537 images, respectively, under four categories of diseases is converted into image embedding using deep learning model VGG19. The image embedding is further subjected to a classification task using ensemble models such as Random Forest and AdaBoost. Later, the high dimensional image embedding is subjected to the dimensionality reduction process using the *t*-SNE algorithm. The dimensionality-reduced data is forwarded to the classification task and the results are displayed in Table 2.

Table 2 Apple leaf dataset classification results

Ensemble method	Area under curve (AUC)	Classification accuracy (CA)	*F*1 score	Precision	Recall
Random Forest (WODR)	0.982	0.903	0.901	0.903	0.903
Random Forest (ADR)	0.985	0.950	0.950	0.951	0.950
AdaBoost (WODR)	0.855	0.812	0.813	0.813	0.812
AdaBoost (ADR)	0.955	0.942	0.942	0.942	0.942

WODR—Without dimensionality reduction
ADR—After dimensionality reduction

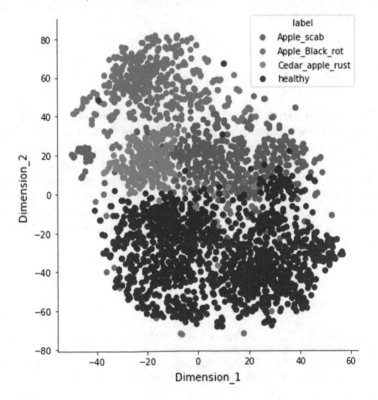

Fig. 1 Apple leaf dataset *t*-SNE visualization

Table 3 Ensemble models confusion matrix for apple leaf dataset

Predicted

Actual		Black rot	Scab	Cedar rust	Healthy	Σ
	Black rot	**462-RF**	16-RF	05-RF	14-RF	497-RF
		468-AB	12-AB	02-AB	15-AB	497-AB
	Scab	10-RF	**458-RF**	11-RF	25-RF	504-RF
		14-AB	**454-AB**	13-AB	23-AB	504-AB
	Cedar rust	00-RF	07-RF	**206-RF**	07-RF	220-RF
		03-AB	10-AB	**200-AB**	07-AB	220-AB
	Healthy	04-RF	27-RF	00-RF	**1285-RF**	1316-RF
		11-AB	35-AB	03-AB	**1267-AB**	1316-AB
	Σ	476-RF	508-RF	222-RF	1331-RF	**2537-RF**
		496-AB	511-AB	218-AB	1312-AB	**2537-AB**

RF Random Forest and *AB* AdaBoost
Highlighted elements in the table 3 represent the diagonal elements of the confusion matrix and
higher numbers in diagonal matrix of the table indicates better classification

The low dimensionality visualization of t-SNE algorithm for VGG19 image embeddings belonging to apple leaf dataset is given in Fig. 1.

The confusion matrices for the ensemble models after dimensionality reduction are displayed in Table 3.

5   Conclusion

The diseased plant leaves image samples from apple plant are classified using ensemble models without reducing the dimensions of the features. The parameters like AUC, CA, $F1$ Score, Precision, and Recall are measured. The diseased leaves dataset is made to reduce their dimensions and classification task is done with the same ensemble models. It is found that the parameters especially classification accuracy (CA) had a noticeable improvement. The image dataset contains only a few thousand images; in case of many plant diseased sample images as a training dataset, the best performing CNN can also be used to perform the classification task in the future.

References

1. van Der Maaten, L., Postma, E., Van den Herik, J.: Dimensionality reduction: a comparative. J. Mach. Learn. Res. **10**, 66–71 (2009)
2. van der Maaten, L., Hinton, G.: Visualizing data using t-SNE. J. Mach. Learn. Res. **9**, 2579–2605 (2008). Available: http://www.jmlr.org/papers/v9/vandermaaten08a.html

3. Laparra, V., Malo, J., Camps-Valls, G.: Dimensionality reduction via regression in hyperspectral imagery. IEEE J. Sel. Top. Signal Process. **9**(6), 1026–1036 (2015). Available: http://ieeexplore.ieee.org/document/7089196/

4. Huang, H., Yang, M.: Dimensionality reduction of hyperspectral images with sparse discriminant embedding. IEEE Trans. Geosci. Remote Sens. **53**(9), 5160–5169 (2015). Available: http://ieeexplore.ieee.org/document/7090989/

5. An, J., Zhang, X., Jiao, L.C.: Dimensionality reduction based on group-based tensor model for hyperspectral image classification. IEEE Geosci. Remote Sens. Lett. **13**(10), 1497–1501 (2016). Available: http://ieeexplore.ieee.org/document/7536213/

6. Xie, L., Yin, M., Yin, X., Liu, Y., Yin, G.: Low-rank sparse preserving projections for dimensionality reduction. IEEE Trans. Image Process. **27**(11), 5261–5274 (2018). Available: https://ieeexplore.ieee.org/document/8410623/

7. An, J., Zhang, X., Zhou, H., Feng, J., Jiao, L.: Patch tensor-based sparse and low-rank graph for hyperspectral images dimensionality reduction. IEEE J. Sel. Top. Appl. Earth Obs. Remote. Sens. **11**(7), 2513–2527 (2018). Available: https://ieeexplore.ieee.org/document/8386807/

8. Zheng, C., Zhao, R., Liu, F., Kong, J., Wang, J., Bi, C., Yi, Y.: Dimensionality reduction via multiple locality-constrained graph optimization. IEEE Access **6**, 54,479–54,494 (2018). Available: https://ieeexplore.ieee.org/document/8472215/

9. Chouhan, S.S., Kaul, A., Singh, U.P., Jain, S.: Bacterial foraging optimization based radial basis function neural network (BRBFNN) for identification and classification of plant leaf diseases: an automatic approach towards plant pathology. IEEE Access **6**, 8852–8863 (2018). Available: http://ieeexplore.ieee.org/document/8289411/

10. Kamble, J.K.: Plant disease detector. (UNAV) (2018). Available: https://ieeexplore.ieee.org/document/8529612/

11. Sardogan, M., Tuncer, A., Ozen, Y.: Plant leaf disease detection and classification based on CNN with LVQ algorithm. (UNAV) (2018). Available: https://ieeexplore.ieee.org/document/8566635/

12. Trongtorkid, C., Pramokchon, P.: Expert system for diagnosis mango diseases using leaf symptoms analysis. (UNAV) (2018). Available: https://ieeexplore.ieee.org/document/8376496/

13. Patrick, A., Pelham, S., Culbreath, A., Holbrook, C.C., De Godoy, I.J., Li, C.: High throughput phenotyping of tomato spot wilt disease in peanuts using unmanned aerial systems and multispectral imaging. IEEE Instrum. Meas. Mag. **20**(3), 4–12 (2017). Available: http://ieeexplore.ieee.org/document/7951684/

14. Simonyan, K., Zisserman, A.: Very Deep Convolutional Networks for Large-Scale Image Recognition (2014). Available: http://arxiv.org/abs/1409.1556 [cs]

15. Zhao, X., Wu, Y., Lee, D.L., Cui, W.: iForest: interpreting random forests via visual analytics. IEEE Trans. Vis. Comput. Graph. **25**(1), 407–416 (2019). Available: https://ieeexplore.ieee.org/document/8454906/

16. Zhang, C., Cai, Q., Song, Y.: Boosting with pairwise constraints. Neurocomputing **73**(4–6), 908–919 (2010). Available: https://linkinghub.elsevier.com/retrieve/pii/S0925231209003749

Generic Architecture for Ubiquitous IoT Applications

Shivani, Satyam Srivastava and M. Rajasekhara Babu

Abstract In this research work, we study the ubiquitous IoT-based applications, systems and technology amongst areas like healthcare, personal assistance, social interactions, agriculture, environmental conditions, traffic management and home control. We further explore to seek out the common trends, framework and design methodology followed in these systems in order to propose an omnipresent generic architecture which can be used unanimously across all these domains along with their respective plug-ins to achieve their specific functionalities. We go further and verify our proposed architecture against the functionalities of various IoT applications. Hence, the final output of this research is a Generic Architecture for Ubiquitous IoT Applications, which can be used as a base architecture to build upon by specifying domain-oriented functions.

Keywords IoT · Ubiquitous · Application · Framework · Architecture · Generic · Healthcare · Agriculture · Transportation · Climate · Traffic management · Environment monitoring

1   Introduction

IoT or the Internet of things is one of the latest trends which are gaining serious attention by developers and users alike. It involves the extension of internet connectivity from standard devices like laptops, desktops, smartphones and phablets to more non-conventional non-traditional devices and things like fans, microwaves, trees, soil, cars and other everyday objects in our surrounding environment.

Shivani · S. Srivastava · M. Rajasekhara Babu (✉)
Vellore Institute of Technology, Vellore, Tamil Nadu, India
e-mail: mrajasekharababu@vit.ac.in

Shivani
e-mail: shivani.twr94@gmail.com

S. Srivastava
e-mail: srivastavasatyam94@gmail.com

© Springer Nature Singapore Pte Ltd. 2020
P. Venkata Krishna and M. S. Obaidat (eds.), *Emerging Research in Data Engineering Systems and Computer Communications*, Advances in Intelligent Systems and Computing 1054, https://doi.org/10.1007/978-981-15-0135-7_12

The definition of IoT has changed over the years primarily because of multiple technologies converging into the field such as machine learning and real-time analytics. The devices and sensors attached to these things utilize embedded technology and Internet to interact and communicate with the environment and each other.

The things in IoT can range from being a vehicle to a person with blood pressure monitor, i.e. objects being assigned IP address and have the capacity and ability to transfer and collect the heterogeneous data over the network without any manual interference.

This enables the devices to be controlled remotely from anywhere. This connectivity in turn helps us capture more data from more places promising greater ways to increase efficiency, security and safety.

2 Application Areas of IoT

IoT is standing at the dawn of its usage and has just started invading the various fields of applications and shown promising results for more endeavours to follow. IoT has been invading areas like agriculture, healthcare, logistics, insurance, transportation, environment monitoring, traffic management, home control, social interaction, smart manufacturing and business applications, and the list is large.

Ubiquitous presence of IoT applications is the future, and some of its implementations apart from the common application areas like smart homes and space are mentioned below.

2.1 Healthcare

This is the sector which is deeply impacted by the extensive use of IoT and promises a better future for IoT and people. Various existing applications and researches show that IoT is primarily being used for personal and public health monitoring and surveillance with use of various sensors and physiological data and also used by hospitals and other such enterprises to monitor physical health data of large-scale patients and people and keep a watch on them.

Researches [2–5] show that the primary data collected from healthcare-oriented IoT devices are:

- Blood pressure
- Heart rate
- Pulse rate
- Body fat percentage
- Blood sugar levels

- Calories burnt
- Other physical activities.

These services aim to monitor and alert the person if anything is above or below the alarming level and monitor the person's overall health status over a period of time providing them with feedback, suggestions and data which gives the user a better control over their health. These applications also help in better monitoring of patients.

2.2 Transportation

This field is quite a major area for IoT to work in, and it is highly entangled with other areas such as climate, weather and environment monitoring. Researches [6–8] show that the primary data collected from transportation and traffic-oriented IoT devices is:

- Light sensing
- Visibility sensing
- Congestion sensing
- Temperature
- Humidity
- Snow/rain level
- Parking space monitoring.

These services aim to provide a better control over traffic and make the transportation safer and easier by processing this data.

2.3 Environment

Monitoring of environment is one of the key applications of IoT. It helps us in modelling city-wide or much larger area wise inspection mechanisms which provide us with data that keeps a check on the local environment.

Researches [9–11] show that the primary data collected from devices deployed in environment is:

- Rain level
- Snow level
- Water level
- Humidity level
- Ground vibrations
- Radiation level
- Atmospheric particle level
- Forest fire detection.

This data helps us in monitoring the quality of our environment and alert/alarm us in case of any mishappening or the various levels crossing their permissible marks.

2.4 Agriculture

This is the field which can greatly benefit from the use of IoT and data from various other fields like environment monitoring [12, 13].

Various applications show that the primary data collected for and used by agricultural industry is

- Soil moisture levels
- Soil nutrients level
- Fertilizer levels
- Monitoring irrigation water
- Weather forecast information like rain and snow
- Prediction of sudden changes in temperature
- Natural disaster warning like floods.

Agricultural industry is the most important. Still, it is the one lacking resources in terms of technological advancements and implementations. IoT applications can and are helping them bloom at a faster pace.

3 Architecture

With the help of modern wireless sensors and technologies, we propose a generic architecture for ubiquitous IoT applications for the development of IoT applications. The model as shown in Fig. 1 aims at providing a standard and an idea and approach to developers of such applications. By thorough inspection of all the IoT application areas and their working architecture, we propose this generic architecture which aims at:

- Facilitating the application development by providing a generic model.
- Unify the development process and make it easy for upcoming developers to approach ubiquitous IoT.
- Provide a base framework so that the developers can add plug-ins as per their requirement for the specific application field.

As shown in Fig. 2, we label the main components of this architecture as:

- Connection and communication module
- Heterogeneous database
- Processing

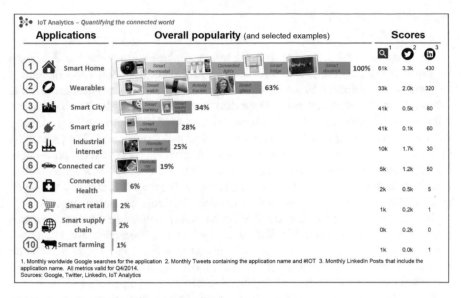

Fig. 1 Analysis of IoT application areas [1]

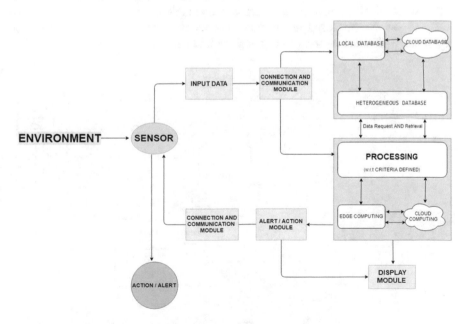

Fig. 2 Proposed architecture for IoT applications

- Alert/action module
- Display module.

This generic architecture involves the sensor capturing or sensing the data from its surrounding environment and forwarding this data. The connection module is responsible for connection and communication between the application and the sensor. Then, this data acts as the input data and enters the system. The system can either store it in heterogeneous database and then send it for processing or directly send it for processing based on the nature of data and requirement. The processing unit can request and retrieve the data from the database and perform the processing. Based on this, the data is either displayed or triggers an action. This action is then identified in the alert/action module, and based on the output, it is either displayed or sent as a command to the sensor which then performs the action (Fig. 3).

For example, a fitness tracker that tracks many physiological data measures blood pressure, then sends this data to the system via connection and communication module, which is then sent to the heterogeneous database to store it for future monitoring and simultaneously sent for the processing where it is matched against the criteria of normal range and displayed to the user via display module. If it is above or below a certain threshold, then it is sent to the action/alert module where it is defined to send a beep alert to the user, and this module sends this request of action to the device which in turn performs this action.

Based on the proposed architecture and its flow chart, we also propose an algorithm which can be used to develop the architecture from scratch.

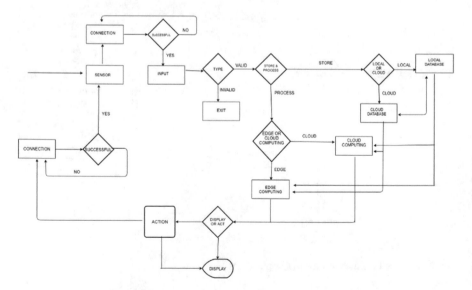

Fig. 3 Activity flow chart of proposed architecture

The algorithm is written as below:

1. Start
2. Sensor Connects to Application
3. If (connection = TRUE)

   a. Input received
   b. IF (input = valid)
   c. IF (input to be stored)
   d. THEN STORE to LOCAL or CLOUD Database
   e. ELSE IF (input to be processed)
   f. THEN EDGE or CLOUD Computing
   g. IF (Output is to be displayed)
   h. THEN (Output -> Display Module)
   i. ELSE IF (Output has to perform some action)
   j. THEN (Output -> Action/Alert Module)
   k. If (input = invalid), exit.

4. IF (connection = FALSE)
5. GOTO Step 2.

The proposed architecture can be used to develop applications for any of the IoT domains like agriculture, healthcare, insurance, transportation, environment monitoring, traffic management, smart manufacturing, business applications, etc. This architecture suits all the requirements which are needed by any IoT application.

4 Verification of Architecture with Respect to Various IoT Application Areas

The proposed architecture is aimed at providing a generic architecture; hence, it should be able to handle various applications of IoT with ease. We start comparing the various IoT applications against this architecture to validate and verify its usage.

4.1 Case 1: Healthcare

We test the healthcare applications for this architecture. The various types of input for this application will be blood pressure, pulse rate, heartbeat, body fat per cent, calories burnt, calorie intake and sugar levels. Each of these inputs has a permissible range, above or below that should be an alert/alarm to the user. This data is also stored for monitoring and long-term evaluation. This data is then displayed to the user, and if anything is alarming, then the action is taken accordingly.

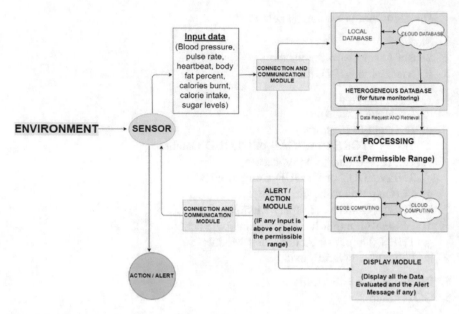

Fig. 4 Healthcare application in proposed architecture

As shown in Fig. 4, this architecture perfectly satisfies the healthcare applications wherein the developer needs to define the type of input, its permissible range and the action to be performed in case of an alert. The architecture is very flexible and accepts changes very easily, but simultaneously it is robust and provides a strong baseline for the application to work.

4.2 Case 2: Transportation

To test the transport applications, we take a look at the inputs which are amount of light, visibility in metres, rain level, snow level, temperature, humidity, congestion per square area and road conditions. These inputs need to be then evaluated against some criteria which in this case are the permissible upper limit or lower limit, i.e. alert should be triggered if visibility is less than a permissible value, rain/snow is greater than a permissible value, congestion is greater than a permissible value, amount of sunlight triggering the street lights, etc. These comparisons by processing of data lead to triggering some action or alert informing the people about the present conditions and handle them appropriately. This data can also be shared on cloud database in order to monitor a specific area and alert distant users about the condition (Fig. 5).

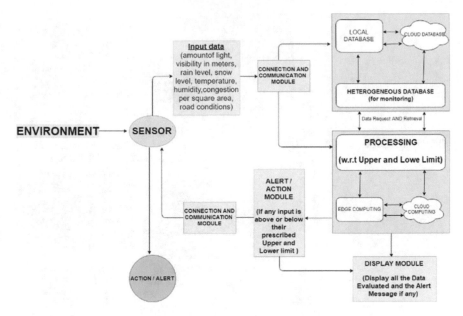

Fig. 5 Transportation application in proposed architecture

4.3 Case 3: Environment

The proposed architecture handles environment applications with ease. The inputs w.r.t environment like rain levels, snow levels, water levels, water contamination (particles level), air particles level, radiation level, ground vibrations, UV radiation level and groundwater levels can be quantized and measured. After their measurement, they can be processed and checked against their respective permissible levels and if they exceed it or go below it, then it should be an alarming condition, and these data should be stored for further research and increased availability for the purpose of monitoring. For example, if the radiation levels or the UV levels are higher than their permissible limit, then users and authorities should be alarmed. If the groundwater levels are below the threshold, then it should be informed, and if the rate of water level of the river and rain level is increasing rapidly, then it can be used to alarm and inform the people about the possible upcoming threats of flood, etc. (Fig. 6).

4.4 Case 4: Agriculture

Agriculture applications of IoT are also fulfilled completely by our suggested architecture as shown in Fig. 5. The various inputs like soil moisture content,

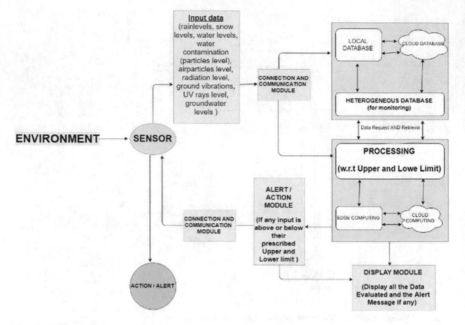

Fig. 6 Environment application in proposed architecture

soil nutrients content, fertilizer levels, irrigation water monitoring, groundwater level, etc. can be considered and processed against the minimum or maximum permissible levels, and actions can be taken correspondingly.

Agriculture can also benefit from data received by other applications such as weather forecast, environment monitoring which can be accessed from the cloud database and processed together with the information gathered locally, and a full report can be produced which helps the users.

For example, if the soil moisture levels are above or below a certain threshold, then it should be notified to the user, and if the soil nutrients level are below a certain requirement or a particularly high or low suited for one type of crop, then it should be known by the user, monitoring of irrigation water to avoid overirrigation and proper channelling of water to all the areas, etc.

Table 1 depicts a quick comparison between some of the heterogeneous input and the various criteria defined for their comparison and processing which can be further used to perform some action or alert (Fig. 7; Table 2).

Table 1 Modules of the proposed architecture

Modules	Functionality
Connection and communication module	Responsible for establishing connection between application and sensor and the communication between them
Heterogeneous database	Responsible for maintaining a local database and connecting to a cloud database in order to store data, keep data in sync and access it for later use
Processing	Responsible for edge/cloud computing. To perform some computing on local data or remote data and produce output
Alert/action module	Responsible for alerting and making the sensor perform some task or send the alert to display module as required
Display module	Responsible for display of monitoring data, alerts, user requested data, etc.

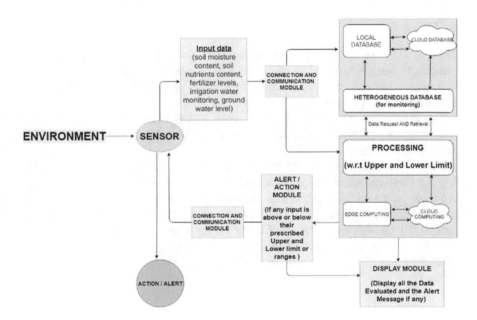

Fig. 7 Agricultural application in proposed architecture

5   Future Prospects and Conclusion

IoT is revolutionizing the very concept of being connected. With more and more extensive use of IoT across every aspect of our lives and surroundings, we are moving towards an era of being highly connected and feeling the presence of control and knowledge ubiquitously. With the advent of multiple platforms for the usage of IoT in their respective specific areas, a need will arise of a common

Table 2 Comparison of heterogeneous input and corresponding processing criteria

Types of I/P	Processing criteria				Actions performed
	Pattern	Upper limit	Lower limit	Ranges	
Light		✓	✓	✓	On, Off, Dim
Pulse rate		✓	✓	✓	Display, Alert
Pm10		✓			Display, Alert
Temperature		✓	✓	✓	Display, Alert
Humidity		✓	✓	✓	Display
Gesture	✓				Perform
Body temperature		✓	✓	✓	Display, Alert
Water level		✓			Display, Alert
Fertilizer content		✓	✓	✓	Display
Rain level		✓	✓	✓	Display, Alert
Ground vibrations		✓			Display, Alert

platform for developers and users to access all this data, process it and extract knowledge from it in order to produce better, versatile, timely and accurate predictions, monitoring and knowledge of personal and public beings in order to increase the ease of living.

Our proposed architecture is a step towards that future and we aim to provide a common framework to develop applications with limited efforts but multiple versatile functionalities.

Future works are planned to test this architecture further and then develop a single software solution which can act as a tool for every IoT application area being as simple yet powerful as possible along with introducing the concept of connected data, i.e. finding and using the connections between this heterogeneous data from various IoT devices to perform better analysis and produce more accurate results. We also aim by using less sensors and more data with a combination of edge computing and cloud computing to produce better, detailed analysis quickly and efficiently.

References

1. https://iot-analytics.com/10-internet-of-things-applications/
2. https://hackernoon.com/the-role-of-internet-of-things-in-the-healthcare-industry-759b2a1abe5
3. Yang, G., Xie, L., Mäntysalo, M., Zhou, X., Pang, Z., Da Xu, L., Kao-Walter, S., Chen, Q., Zheng, L.-R.: A health-IoT platform based on the integration of intelligent packaging, unobtrusive bio-sensor, and intelligent medicine box. IEEE Trans. Ind. Inform. **10**(4), 2180–2191 (2014)
4. Almari, A.: Monitoring system for patients using multimedia for smart healthcare. IEEE Access **6**, 23271–23276 (2018)
5. Bui, N., Zorzi. M: Health care application: a solution based on the internet of things (2011)

6. Widyantara, I.M.O., Sastra, N.P.: Internet of things for intelligent traffic monitoring system: a case study in Denpasar. Art. Int. J. Emerg. Trends Technol. Comput. Sci. (2015)
7. Nugra, H., Abad, A., Fuertes, W., Galárraga, F., Aules, H., Villacís, C., Toulkeridis, T.: A low-cost IoT application for the urban traffic of vehicles, based on wireless sensors using GSM technology. In: 2016 IEEE/ACM 20th International Symposium on Distributed Simulation and Real Time Applications (2016)
8. Huang, Y., Wang, L., Hou, Y., Zhang, W., Zhang, Y.: A prototype IOT based wireless sensor network for traffic information monitoring. Int. J. Pavement Res. Technol. **11**(2), 146–152 (2018)
9. Sastra, N.P., Wiharta, D.M.: Environmental Monitoring as an IoT Application in Building Smart Campus of Universitas Udayana
10. http://iotworm.com/internet-of-things-will-change-transportation/
11. Ibrahim, M., Elgamri, A., Babiker, S., Mohamed, A.: Internet of Things based Smart Environmental Monitoring using the Raspberry-Pi Computer
12. https://www.mendix.com/examples-of-iot-applications/
13. https://www.sciencedirect.com/science/article/pii/S1996681416302619

Performance Evaluation of Spark SQL for Batch Processing

K. Anusha and K. Usha Rani

Abstract Now-a-days, large amount of data is being generated at various organizations. In many organizations, there is an inefficiency of handling Big Data with higher volumes, velocity, and variety. Though data is a huge resource, organizing Big Data is a huge challenge in present days. Currently, number of companies adopted different types of NoSQL databases like Cassandra, MongoDB, HBase, etc., which can handle number of requests at a time. To process the Big Data, Apache Spark, one of the most powerful processing engines, has a number of benefits. The main programming notion in Apache Spark is Resilient Distributed Datasets (RDDs), which handles only procedural processing. However, the most regular data processing paradigms are relational queries which cannot be handled by RDD. To overcome this, there is a need to use several higher-level libraries on Apache Spark. Spark SQL is one of the novel components in Apache Spark Framework that integrates relational processing through Apache Spark's functional programming API. It allows Apache Spark programmers to use the benefits of relational processing. It also provides an integration of relational processing and procedural processing using a declarative Data Frame API. Hence, in this study, Spark SQL Data Frames are experimented to enhance the processing of weather data stored in Cassandra database. Further, the study has proved that the Spark SQL Data Frames have outperformed performance than Spark Core RDD which we have experimented earlier.

Keywords Apache Spark-SQL · Data frames · Apache Cassandra

K. Anusha (✉)
Research Scholar, Department of Computer Science,
Sri Padmavati Mahila Visvavidyalayam, Tirupati, India
e-mail: siri.bachina@gmail.com

K. Usha Rani
Professor, Department of Computer Science,
Sri Padmavati Mahila Visvavidyalayam, Tirupati, India
e-mail: usharanikuruba@yahoo.co.in

© Springer Nature Singapore Pte Ltd. 2020
P. Venkata Krishna and M. S. Obaidat (eds.), *Emerging Research in Data Engineering Systems and Computer Communications*, Advances in Intelligent Systems and Computing 1054, https://doi.org/10.1007/978-981-15-0135-7_13

1 Introduction

Big Data is a group of massive set of structured, semi-structured, and unstructured data that has the potential to be mined for information. Many organizations have Big Data, and they need to harness that data and extract value from it. Big Data Analytics is the major solution to extract useful data from large datasets. It is essentially used to uncover hidden patterns, correlations, etc., from large datasets. The new benefits that Big Data Analytics brings, however, are speed and efficiency. Apache Hadoop MapReduce, which is the general framework designed to store and process Big Data, is time consuming. One of the important frameworks for performing general Big Data Analytics is Apache Spark, and it has a much faster and more general data processing platform [1]. Apache Spark executes all programs up to 100 xs faster and quicker in memory, or 10 xs faster on disk, than Hadoop. The Apache Spark Core contains a set of higher-level libraries which are very powerful and can be effortlessly used in the same application. These libraries at present consist of Apache Spark SQL, Apache Spark Streaming, MLlib (for machine learning), and GraphX. The natural successor and complement to Hadoop are Apache Spark, which continues the Big Data development. Apache Spark provides a simple and flexible API to execute large number of jobs distributed across network for data analytics. Spark runs the jobs faster than other forms of analytics because much can be finished in-memory [2]. Apache Spark gives the total control of Big Data to developers on the way to provide real-time data analytics. The Spark ecosystem architecture and the Data Frames in Spark stack are shown in the following Fig. 1.

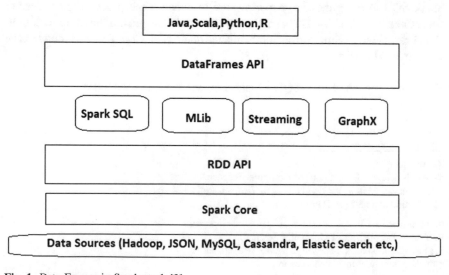

Fig. 1 Data Frames in Spark stack [3]

One of the important frameworks for general Big Data Analytics on distributed computing cluster like Hadoop is Apache Spark. It has the capability of an in-memory computation which greatly increases the processing speed compared to Hadoop MapReduce. Central concept in Apache Spark is Resilient Distributed Datasets (RDD). RDD is the heart of Apache Spark. It is the fundamental data structure of Apache Spark. In our previous study, we experimented Apache Spark and Cassandra Integration on weather data, and the results were proved that it has better performance than Hadoop MapReduce [4].

Even though RDD in Apache Spark increases the efficiency of work and makes it useful for faster computations, there are some limitations in Apache Spark RDD. Some of the limitations of RDD are related to inbuilt optimization, storage and processing speed, etc.

In this study, we considered Cassandra database for experimentation because the traditional databases are not able to handle large datasets because of their limited capacity. NoSQL databases are one of the solutions to store Big Data, which overcome the problem of traditional relational databases. One of the NoSQL databases, Apache Cassandra is an open-source distributed database organization and management systems. Cassandra NoSQL database is a scalable non-relational database that offers high availability, performance, etc. The design of Cassandra database is capable to scale, execute, and offer continuous uptime. The variety of key features and benefits provided by Cassandra makes it as the crucial database for modern online applications [4, 5].

To overcome the limitations in Apache Spark RDD, Data Frames are evolved which also reduces the processing time. Data Frames are the core concept of Spark SQL, which is one of the higher-level libraries of Apache Spark. Spark SQL is a flexible model and a powerful API provided by Apache Spark. Spark SQL allows running SQL queries easily. Hence, in this study, Spark SQL and Cassandra Integration are experimented to further enhance the processing speed.

The remaining sections of the paper are organized as follows: Sect. 2 describes literature review; Sect. 3 represents the methodology; Sect. 4 represents the proposed system; Sect. 5 represents the experimental analysis; Sect. 6 represents the conclusion.

2 Literature Review

Extensive studies have been carried out on Big Data Analytics using Apache Spark. In this section, few studies related to Big Data, Apache Spark, Apache Spark SQL, and Apache Cassandra are presented.

Zaharia et al. [6] proposed that Apache Spark is a unified engine for processing large datasets which handles both batch and stream processing. They represent that it is a simple programming model can capture streaming, batch, and interactive workloads and enable new applications that combine them.

Armbrust et al. [7] provide an introduction about relational data processing in Spark and optimizations in Spark SQL. They explained the features of Spark SQL and represented the pipeline performance of Data Frame API.

Bhutkar et al. [8] give a clarification that although customary databases are valuable for performing complex logical queries in wide variety of uses, they have a few issues while dealing with the gigantic measure of information. NoSQL advancements like HBase, Cassandra, and Mongo DB have picked up importance throughout the years in view of their capacity to deal with huge information in appropriated condition. These advancements are for the most part open source and give methods for taking care of an altogether vast volume of information with lower cost and simpler administration than customary RDBMS.

Bondiombouy et al. [9] give an outline of query processing in multi-store frameworks. The principle inspiration of this structure of another age of frameworks is multi-store frameworks which is hard to get to and incorporate information.

Xin et al. [10] clarify about running substantial scale Spark outstanding tasks at hand. They disclosed that because of the out separated size of information which surpasses the abilities of single machines, there is a need to utilize new machines to scale out calculations to various hubs.

Bhattacharya et al. [11] provide a brief review on the importance of Apache Spark in processing of Big Data. They explained about various tools for Big Data Analytics and also represented the statistics of Big Data. They also explained about various security issues of Big Data.

Miller [12] suggested that how the information parallel paradigm can be stretched out to the distributed case, utilizing Spark all through which is a quick, in-memory disseminated accumulations structure written in Scala.

Parab [13] gave an introduction about overview of Spark SQL. He also explained about Data Frames and various operations supported by Data Frames.

Chakraborty [14] gave an introduction about Spark SQL overview, Data Frames and their usage and about advantages and disadvantages of Spark SQL and its architecture.

3 Methodology

Brief description of Apache Spark Core RDD, Apache Spark SQL, and Cassandra is presented in this section.

3.1 Apache Spark-Cassandra Implementation

One of the base engines for extensive parallel and distributed data processing is Apache Spark Core. The main responsibility of Apache Spark is a memory management, debugging, spreading, and monitoring of disk storage. The intrinsic data separation concept (RDD) was introduced by Apache Spark. RDD is a set of objects that can be distributed to immutable errors that can work in parallel. RDD is an important data structure of Apache Spark, which has a collection of read only. It allows developers to compute large clusters with a method of refusing patience. RDD can load any type of object and is created by loading external data or distributing driver software [1]. Two types of base operations are supported by RDD. They are Transformations and Actions. Transformations contain a few tasks such as map, filter, join, and union that are performed on an RDD and provide a new RDD that contains the outcome [2, 11]. Actions are likewise some tasks such as reduce, count, first, etc., that return a value after running a calculation on an RDD. Transformations of an Apache Spark Framework are "lazy," i.e., they do not register their outcome immediately. They can just barely "recollect" the task to be performed and the dataset to which the activity is to be performed [15, 16]. These transformations in reality figured when an action is called and the outcome is come back to the driver program. This structure empowers Apache Spark to run all the more proficiently and successfully. For the most part, every transformed RDD might be recomputed each time an action keeps running on it [1]. The architecture for Spark-Cassandra connector is shown in [4].

3.2 Apache Spark SQL Implementation

Spark SQL is one of the Apache Spark parts that help querying of information by means of SQL or by means of the Hive Query Language. Spark SQL, alongside offering help for different information sources, it additionally makes it conceivable to interface SQL queries with code changes which results in a useful tool. The center of Spark SQL is known as Data Frame. A Data Frame just holds information as a gathering of lines and every segment in the column is named. Data Frames permit to effectively choose, plot and separating of information. In **Spark SQL, it** can automate the process and make it much easier on reading and writing data to do analysis. In Data Frame API, the information is sorted out into named segments as like a table in relational database. It is a changeless disseminated assembling of data. Data Frame in Apache Spark Framework enables engineers to force a structure onto a circulated accumulation of information, permitting larger amount deliberation. Data Frames use heap memory for serialization and have an inbuilt optimization which reduces the overhead. The architecture for Spark SQL is depicted in the following Fig. 2.

The query processor gets to the Apache Spark engine through the Apache Spark Java interface, as it gets to outside data sources like RDBMS or a key-value store utilizing Spark SQL common interface supported by JDBC drivers. It additionally incorporates two principle parts: the Data Frame API and the Catalyst query optimizer. The Data Frame API offers a rich joining among relational and

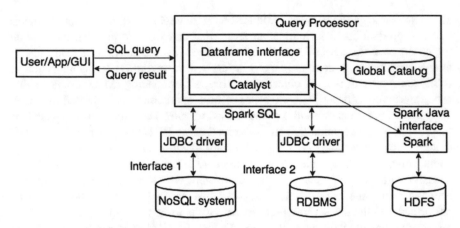

Fig. 2 Apache Spark SQL architecture [10]

procedural processing that enables relational operations to be performed on both outside data sources and Apache Spark's RDDs. It is fused into Apache Spark's supported programming dialects like Java, Scala, and Python and supports a simple inline depiction of client characterized capacities. Data Frames maintain a strategic distance from the enrollment process found in other database frameworks. Subsequently, the Data Frame API lets designers easily mix relational and procedural programming, to perform advanced analytics on gigantic information accumulations.

Apache Spark Catalyst Optimizer—Catalyst optimizer is the optimizer used in Apache Spark SQL, and all the queries written by Apache Spark SQL and Data Frame DSL are optimized by this tool. This optimizer is much better than the RDD and hence the performance of the system is increased [17, 18].

Databases

1. NoSQL Database—Cassandra: This database gives a strategy to capacity and recovery of information which is not quite the same as the forbidden relations utilized in customary relational databases. The use of these NOSQL databases in Big Data and constant Web applications is expanding. Apache Cassandra, a NOSQL database is best for its rapid and online value-based information [8]. It has an ace less ring design which is progressively effective, simple to set and fundamental tain [5].
2. Input Dataset: In this study, the climate informational index of size 512 MB is gathered from National Climatic Data Center (NCDC) site [19, 20]. NCDC contains information consistently with every day and hourly premise. The climate dataset contains fields like date, area id and observations for each climate parameters like temperature, mugginess, weight, and so forth.

4 Proposed System

To gain the advantages of Spark SQL API along with Cassandra, an integrated method is proposed to reduce the processing time. A brief description of Data Frames and the proposed algorithm are given in this section.

4.1 Data Frame

Data Frame is the main notion of Spark SQL's API. These Data Frames are optimized and compiled to Byte code through Spark SQL. Data Frame offers a rich set of coordination inside Apache Spark programs. These are accumulations of organized records that can be controlled utilizing Apache Spark's procedural API or utilizing new relational APIs that allow richer optimizations [16]. Apache Spark's built-in disseminated collections of items produce these Data Frames, specifically, empowering relational handling in existing Apache Spark. Data Frames are more suit-capable and more effective than Apache Spark's procedural API. They effectively work out numerous aggregates in a single pass utilizing SQL statement, which is unpredictable to express in regular functional APIs. Data Frames monitor their schema and support various relational operations that lead to more advanced execution which is unimaginable with RDD's. They infer the schema utilizing reflection. Apache Spark Data Frames are lazy in that each Data Frame object speaks to a sensible arrangement to register a dataset, yet no execution happens until the client calls a special "output operation" such as save. This empowers rich enhancement over all operations. Data Frames upgrade the Apache Spark's RDD show, making it a lot quicker and less demanding for Apache Spark engineers to work with organized information by giving essential techniques for filtering, aggregating, and projecting over vast datasets. Data Frames are accessible in Apache Spark's Java, Scala, and Python API [3, 21].

4.2 Proposed Algorithm

1. Download batch weather data collection from weather stations.
2. Start Cassandra database Service.
3. Load downloaded weather data collection into Cassandra shell.
4. Run Apache Spark shell.
5. Run the Apache Spark-Cassandra Connector.
6. Start Spark SQL shell.
7. Extract weather data from Cassandra through Apache Spark Data Frame API's.
8. Evaluate the time taken for processing whole data from Cassandra database.

Fig. 3 Spark-Cassandra
connector versus Spark SQL

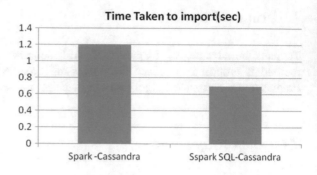

Table 1 Comparison of
processing time of integrated
approaches

Integrated approach	Time taken to import (s)	
Spark-Cassandra	1.2	
Spark SQL-Cassandra	0.7	

5 Experiment Analysis

The proposed system is experimented to process the weather data stored in
Cassandra database using Apache Spark SQL Data Frames. In this experiment, it is
observed that processing data using proposed method consumes less processing
time compared with Apache Spark Core RDD's [4]. The following Fig. 3 shows the
graphical representation of analysis. The results are tabulated in the following
Table 1.

6 Conclusion

A new component in Apache Spark, i.e., Apache Spark SQL has a rich interaction with
the relational processing. Spark SQL has a declarative Data Frame API which allows
relational processing and also offers benefits like automatic optimization. It also
allows users to write down complex pipelines that combine relational and complex
analytics. An integration of Spark SQL Data Frames and Cassandra is experimented
on weather data, and the result is compared with Spark Core and Cassandra
Integration. It is proved that the present study enhances the processing speed.

References

1. https://www.sas.com/en_in/insights/analytics/big-data-analytics.html
2. https://www.researchgate.net/publication/310613994_Apache_Spark_A_unified_engine_for_
   big_data_processing

3. https://insidebigdata.com/2015/11/30/an-overview-of-ApacheSpark-sql/
4. https://dzone.com/articles/analytics-with-apache-spark-tutorial-part-2-ApacheSpark
5. Apache Cassandra [Online]. Available: https://www.datastax.com/wp-content/uploads/2012/09/WPDataStax-HDFSvsCFS.pdf
6. Anusha, K., UshaRani, K.: Big data techniques for efficient storage and processing of weather data. Int. J. Res. Appl. Sci. Eng. Technol. (IJRASET) 5(VII) 2017. ISSN: 2321-9653
7. https://www.toptal.com/ApacheSpark/introduction-to-apache-spark
8. Bhutkar, B.: Data Management using Apache Cassandra. SAS Research and Development (India) Pvt. Ltd
9. https://www.researchgate.net/publication/304850049_Query_processing_in_multistore_systems_an_overview
10. Xin, R., Zaharia, M.: Lessons from running large-scale Spark workloads. http://tinyurl.com/largescale-spark
11. Bhattacharya, A., Bhatnagar, S.: Big data and apache spark: a review. Int. J. Eng. Res. Sci. 2(5), 206–210 (2016)
12. https://es.coursera.org/lecture/scala-spark-big-data/spark-sql-NlNqx
13. https://www.youtube.com/watch?v=4noellXBRA8
14. https://www.youtube.com/watch?v=S6jtHLr6UNs
15. https://www.semanticscholar.org/paper/Spark-SQL%3A-Relational-Data-Processing-in-Spark-Armbrust-Xin/080ed793c12d97436ae29851b5e34c54c07e3816
16. https://data-flair.training/blogs/apache-spark-rdd-vs-dataframe-vs-dataset/
17. https://intellipaat.com/blog/what-is-Spark-sql/
18. https://www.simplilearn.com/running-sql-queries-using-spark-sql-tutorial-video
19. https://www.ncdc.noaa.gov/data-access/land-based-station-data/land-baseddatasets/quality-controlledlocal-climatological-data-qclcd
20. NCDC weather data [online]. Available: https://www.ncdc.noaa.gov/orders/qclcd/
21. https://www.slideshare.net/databricks/large-scaleApacheSparktalk
22. Anusha, K., Usha Rani, K., Lakshmi, C.: A survey on big data techniques. Special Issue on Computational Science, Mathematics and Biology IJCSME- SCSMB-16March-2016, ISSN-23498439
23. https://www.youtube.com/watch?v=Mxw6QZk1CMY

IoT-Based Traffic Management

K. Lalitha and M. Pounambal

Abstract Many cities in the world face congestion problems in road traffic, particularly in rural areas. To alleviate the problems in transportation and service the safe journey objectives, normal cities have to be transformed into "smart cities." The information communication technologies and IoT delve into the window to convert the normal traffic into IoT-based traffic. The role of IoT and the platform are very important role in coordinating with the vehicles on the road and promote unique identification and safety. This paper proposes an IoT-based traffic management solution for smart cities and enables coordinate with the drivers of emergency vehicles to find the signal status and traffic density and choose the path, thereby the traffic are controlled dynamically with a minimum violations. The concept proposes without loss of generality enabling the emergency vehicles pass in the traffic reaching their destinations hassle free.

Keywords IoT based traffic congestion · Mobile node · Traffic junction sensors

1 Introduction

Among the trending computer architectures, the IoT holds a humble role gaining attention in the next-generation wireless telecommunications, providing vast number of objects, things connected with radio communication which includes sensors, sensor tags, actuators and mobile devices in interaction [1–3]. A network posing futuristic foreground for large multimedia applications even supporting video on demand (VoD), IPTV and voice over (VoIP) and growing. The current era demands security as a very critical element with importance of digital data,

K. Lalitha (✉)
SCOPE, Vellore Institute of Technology (Vellore Campus), Vellore, India
e-mail: lalithahpc@gmail.com

M. Pounambal
Vellore Institute of Technology (Vellore Campus), Vellore, India
e-mail: mpounambal@vit.ac.in

© Springer Nature Singapore Pte Ltd. 2020
P. Venkata Krishna and M. S. Obaidat (eds.), *Emerging Research in Data Engineering Systems and Computer Communications*, Advances in Intelligent Systems and Computing 1054, https://doi.org/10.1007/978-981-15-0135-7_14

particularly in multimedia applications of IoT. IoT architectures run unverified and user-customized applications from unknown users, which is an abundant resource of raising threats in IoT.

The vulnerable situation of such traffic applications with sensors is they are exploited by hackers, and IoT is accessed by users of malice to probe in launching of vicious service attacks. Use of shared multimedia data excessively on network resources might always prone to the interruption in services of other legitimate users [2–4]. At onset of the typical traffic scenario, the existing IoT has not really had the capabilities of security mechanism to deal with the threats mentioned hitherto. The importance of deploying the security strategy reasons to protect the security in critical multimedia applications that stream on IoT.

2 Existing Scenario

It is difficult to identify the traffic violators. Road safety and ascertained priority on roads are the major subjects within the transport policy of the road transport authorities [3, 4]. In-vehicle activities, fatigue, unresponsive and careless attitude of drivers causes many traffic accidents, which leads to vehicles of emergency such as ambulance and fire distracted from regular course of route. A framework and a planned setup are, thereby, required to prevent such accidents and enhance the driving with driving assistance system, which reports the status of vehicle in the traffic, traffic violations and faults [5, 6].

Manually monitoring the vehicles in the traffic is a very hectic task, the increasing number of road users raises difficulty in vehicle identification; unique identification of the vehicle is very essential in such applications. The limitations and non-availability of resources lack in performance of unique identification of vehicles in the increasing traffic. The authorities at traffic monitoring shall have to adapt new methods in intelligent monitoring of vehicles uniquely and thus overcoming complexities.

The paper now is aimed to propose a concept to automate the process of unique identification of the vehicle. In a typical running traffic, many users violate rules without any hesitation; such incidents cause trouble to other vehicles. No traffic lanes are exempt from junctions and crossings of road, though arranged by the traffic lights to alert the pedestrians and vehicle commuters. The traffic management cannot be effectively achieved without intervention of a traffic person (police), though organized by sophisticated infrastructures of automated traffic management. The ever increasing flow of vehicles on the road and the number of users cannot compete with limited resources available to traffic management system in monitoring the vehicle uniquely.

3 Proposed System

A server centered cloud supported approach is recommended for the application of traffic monitoring and management system. A vehicular data cloud as a service shall be a great support for IoT-based traffic monitoring and management system. Vehicular networks have their roots wireless technology, the idea encamps the roadside infrastructure and the radio-equipped vehicles communicating with wireless networks. The movement of the vehicles in the tracked by the networks and lot of data is thus populated. Various types of vehicular services are combined and used to implement the conceptual mapping, encapsulation, aggregation and composition, and the vehicles are allowed to interaction with various hosted services. Multilayer approaches and service-oriented architectures are supportive of the IoT-managed traffic monitoring system.

3.1 Traffic Computation

Certain sink nodes or mobile agents can be setup to process the traffic computations. The sequence that is followed by the mobile agents to visit the vehicles or the select nodes can have impact on the traffic computation. Thus, mobile agents compute traffic statically, dynamically and/or both. The job of the mobile agent is to detect source node and the movement in the current network in the destined path and suggesting the options for rerouting. Traffic computation shall be taken with three categories of vehicles, passenger vehicles, private vehicles and service vehicles. The priorities for the categories of vehicles are assigned, and traffic computation is processed for generating rerouting options. High alert vehicles are emergency or convoy vehicles. Traffic management shall also have to address these vehicles including in the database when they are in itineraries or long commutes. IoT-based traffic management: An IoT-based traffic management shall consist of mobile agent, traffic lights, graph of route map and a database of vehicles in the cloud. The vehicles are addressed by the identifiers that are moving by their category, source and destination and any other important alerts.

3.2 Traffic Decision Support System

A system is supported by big data analytics and traffic management system with IoT. The system is typically consisting of uniformly functioning modules that work autonomously and as well as integrate themselves into a combined networks. The modules work in departments of the traffic management system; the communication between the departments is by integrating the modules through an interconnection system. The solution provided in this system collects all crucial information using sensors, actuators and facilitates fetching of real-time data related to monitoring and control. The data obtained in the solution is also useful in prediction of traffic trends and allows simulations. The system assists decision makers of the typical traffic management systems with reliable information with scientific base.

3.3 Parking Issues and Traffic Congestion

The increase of population and buying more than number of sufficient vehicles per se strains transportation system and health of the cities. Traffic management systems in cities can make use of big data and big data analytics to gather the access to the city planners and identify the causes of congestion in the traffic. Typical city planners use data analytics to find out the vacant parking slots mostly preferred by the drivers. Smart devices, sensors and actuators connected in the IoT-based traffic management system shall be introduced to define the ratings of the parking lots based on the frequency of usage, density and personalization of parking lots.

3.4 Proposed System Architecture

The proposed system architecture contains a console managed by Raspberry Pi which is programmed to control the signals using WiFi cameras. WiFi cameras are installed in all the traffic junctions in order to identify the emergency vehicles and pave them a separate way and mitigate the traffic congestion. The managed WiFi camera reads the vehicle type by the patterns of the vehicle images and catches the importance of the vehicle and thus the console unit raises an alarming signal to the signal lights. The WiFi station connects to the nearby traffic signal light points and communicates the alert and allows the emergency vehicle get on route immediately. This kind of architectures has been imposed by several traffic management systems, where the image identification and alarming the importance of vehicles are failed, due to large networked systems. Instead, an autonomous system with a minimal WiFi network can be setup by using the frugal hardware to enact immediately causing favor to the emergency vehicles (Fig. 1).

4 Performance of the Proposed System

The proposed system is simulated using Raspberry Pi and WiFi with mobile nodes as vehicles built on a prototype traffic plan. 20 mobile nodes move in the traffic plan haphazardly taking trips to and fro for a period of 1 h. The mobile nodes are categorized as passenger vehicles, ambulance vehicles, fire vehicles, public trans-port vehicles, convoy of ministries and other emergency vehicles. Each mobile

Fig. 1 Prototype model of
proposed system architecture
with Raspberry Pi, WiFi and
mobile nodes as vehicles

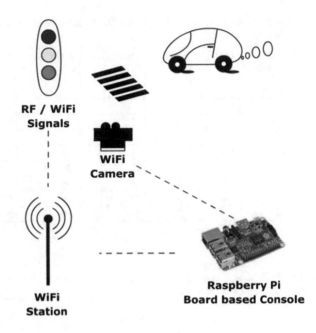

RF / WiFi
Signals

WiFi
Camera

WiFi
Station

Raspberry Pi
Board based Console

node is assigned priority to cross the traffic signals. The roads in the prototype are laid with lanes that provide facility to move passenger vehicles and transport vehicles without any hassles at the traffic junctions. The overall performance is defined with metrics for feasibility to allow the vehicles in priority and without causing any delay in their journeys. The performance of the system is evaluated using a Repeater Operating Curve (ROC) which shows ascending true positive rates with respect to the observational and predictive rates of success in passing the traffic junction without hassles. The following Table 1 shows the number of mobile nodes participating at an instance of one hour period and the hassle-free pass at the traffic junction. The experiment arrives at observational and predictive probabilities calculated based on the priority and the movement times at the traffic junction.

Table 1 Data obtained in the experimentation for ROC representation and the nature of AUC

Number of vehicles	No. of hassle-free passes at traffic junction	Observational probability at traffic junctions	Predictive probability at traffic junctions
2	15	1	0.94676
6	14	0.9218	0.88302
10	12	0.78276	0.73684
12	11	0.439824	0.49615
14	8	0.195658	0.24903
18	6	0.1093	0.09248
20	3	0.90242	0.94571

Table 2 Analysis of predictive probability at traffic junctions and their false positive and true positive rates

Predictive probability at traffic junctions	Failure	Success	Cumulative failure	Cumulative success	FPR	TPR	AUC
			0	0	1	1	0.090535
0.09248	22	5	22	5	0.909465	0.964029	0.102451
0.24903	24	8	46	13	0.8107	0.906475	0.113495
0.49615	25	16	71	29	0.707819	0.791367	0.166405
0.73684	32	21	103	50	0.576132	0.640288	0.237805
0.88302	37	19	140	69	0.423868	0.503597	0.34321
0.94571	42	28	182	97	0.251029	0.302158	0.830786
0.94676	61	42	243	139	0	0	

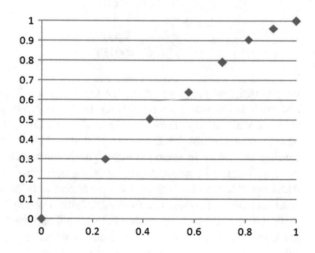

Fig. 2 ROC representing the ascension and the nature of AUC

The frequency of the mobile nodes at the traffic junction moving hassle free is charted in the following table, and the repeater operative curve shows the ascension indicating the success of experiment (Table 2; Fig. 2).

5 Conclusion

The overgrowing population, over purchasing of number of vehicles and congestion in moving in the city arise a high alert need of automation of traffic management and monitoring system. The traffic management system controls and applies decision making for the instant solutions. Whereas, traffic monitoring system logs in the information about the commutations and learns the behaviors and patterns. An IoT-based system is very ideal to track and log the traffic in busy commute areas of the city.

References

1. Misbahuddin, S., et al.: IoT based dynamic road traffic management for smart cities. In: 2015 12th International Conference on High-Capacity Optical Networks and Enabling/Emerging Technologies (HONET). IEEE (2015)
2. Zanella, A., et al.: Internet of things for smart cities. IEEE Internet Things J. **1**(1), 22–32 (2014)
3. Gubbi, J., et al.: Internet of things (IoT): a vision, architectural elements, and future directions. Future Gener. Comput. Syst. **29**(7), 1645–1660 (2013)
4. Rathore, M.M., et al.: Urban planning and building smart cities based on the internet of things using big data analytics. Comput. Netw. **101**, 63–80 (2016)
5. Prakash, B., et al.: An Iot Based Traffic Signal Monitoring and Controlling System Using Density Measure of Vehicles (2018)
6. Javaid, S., et al.: Smart traffic management system using Internet of Things. In: 2018 20th International Conference on Advanced Communication Technology (ICACT). IEEE (2018)

Detection of Autism Spectrum Disorder Effectively Using Modified Regression Algorithm

T. Lakshmi Praveena and N. V. Muthu Lakshmi

Abstract Person with impairments in social communication, abnormal behavior, and sensory activities are considered as suffering with neurodevelopmental syndrome and this syndrome is termed as autism spectrum disorder (ASD). Diagnosis process of ASD is based on observation of frequent movements, social communication skills. and eye contact of person. In some cases, standard questionnaires are used to assess the person. The objective of this paper is to automate the diagnosis process to generate accurate results, which are used to detect ASD. Diagnosing and predicting autism at early age help to take better treatment. Early detection of ASD in children can reduce the symptoms of ASD and they can mingle with normal children. In recent years, more research work has been done on ASD to find methodologies for ASD prediction. Machine-learning methods are efficient to analyze ASD as it generates accurate results with less computational power compared to other methods. An efficient regression-based algorithm is proposed to predict ASD with less computation time makes detection process faster. The proposed algorithm is applied over dataset collected from UCI machine-learning repository. Dataset consists of around 2000 people's information of varying age groups like adolescent, adult, child, and toddlers. The results obtained from this dataset are analyzed and present efficiency of algorithm to predict ASD at an early age. Performance analysis on proposed and existing algorithms is compared and analyzed.

Keywords Neurodevelopmental syndrome · Autism spectrum disorder · Machine learning · Regression method

T. Lakshmi Praveena (✉) · N. V. Muthu Lakshmi
Sri Padmavati Mahila Visvavidyalayam, Tirupati, AP, India
e-mail: praveenalaxmi1@gmail.com

N. V. Muthu Lakshmi
e-mail: nvmuthulakshmi@gmail.com

© Springer Nature Singapore Pte Ltd. 2020
P. Venkata Krishna and M. S. Obaidat (eds.), *Emerging Research in Data Engineering Systems and Computer Communications*, Advances in Intelligent Systems and Computing 1054, https://doi.org/10.1007/978-981-15-0135-7_15

163

1 Introduction

Autism spectrum disorder (ASD) is a neurodevelopmental disorder. Neurodevelopmental disorders are disabilities associated with brain and neurological system. The example of neurodevelopmental disorders is hyperactivity disorder (ADHD), autism, learning disorder, intellectual disorder, conduct disorder, and disability in hearing and vision [1]. Children with neurodevelopmental disorders face several problems with language, speech, motor skills, learning, memory, behavior, and other neurological functions. Autism spectrum disorder is combination of one or more neurodevelopmental disorders [2]. ASD caused by many factors, which may be complications in pregnancy or in delivery, consumption of medicines during pregnancy, dosage of vaccines or from ASD ancestors or siblings [3]. The diagnosis process of ASD is difficult and time taking process because it is a multi-disciplinary action done by many experts who are psychiatrist, therapist, special educators, counselors, and neurologist [4]. These experts may take 2–3 days to assess child in all aspects by interviewing the parents with standard questionnaires [5]. In similar way, treatment also takes long time to give better life to growing child. From this, we can say that qualitative and fast diagnosis is required to provide treatment to the ASD affected child for better results.

The objective of this paper is to find effective method to detect ASD with more accurate results and with less time computation process. Early detection of ASD in children can reduce the symptoms of ASD, which makes the ASD children to mingle with normal children [6]. If the ASD children are diagnosed effectively, quickly, and accurately in order to detect this syndrome then discrimination from normal child can be avoided. In recent years, more research work has been done on finding methodologies in diagnosing ASD [7]. Among the other methodologies, machine-learning methods are efficient to analyze ASD as it generate accurate results with less computational power and also make identification process faster [8]. The supervised learning algorithms of machine learning are used for prediction and classification [9]. The proposed methodology based on regression technique is used to detect ASD with less computation time. Datasets are collected from UCI machine-learning repository [10].

The dataset is shared by Dr. Fadi Tabtah. Dataset consists of around 2000 people's information of varying age groups like adolescent, adult, child, and toddlers. This paper presents how the proposed algorithm effectively helps to predict ASD at early stage or at least after recognizing the problem and this makes the people live better. Performance analysis on proposed and existing algorithms is compared and analyzed in this paper.

2   Autism Spectrum Disorder

A person having impairments in social communication, different behavior, and also with abnormal sensory activities is considered as neurodevelopmental syndrome and this syndrome is termed as autism spectrum disorder (ASD) [11]. ASD is associated with brain and neurological structure of children. Actually, ASD starts by birth but most of the people recognize only when it explicitly shows abnormal behavior in many times [12]. ASD was initially identified in 1940 and was conceptualized in the year of 1994 by publishing DSM-IV. In 2013, it is named as autism spectrum disorder with the publishing of DSM-V [13]. ASD children have learning disability, shows less interest in social communication, deficiency of motor skills, and also suffer with sensory activities deficiency, speech deficiency, listening deficiency, and behavioral deficiencies [14]. There are many theories and researches proposed by researchers on causes of autism but they are not proved to be as main cause of autism. Some important causes proposed by different researchers in their research of ASD are given [15]. Genetically, ASD can be affected to children from the parents with ASD or from the siblings with ASD. Prenatal environment can cause ASD n some cases like rubella infection in mother, pesticides, consumption of folic acid, and ultrasound waves [13]. Prenatal environment is risk factors found at the time of delivery like low birth weight, gestation duration, and hypoxia during birth [16]. Postnatal environment caused by MMR vaccine, leaky gut syndrome, viral infection, oxidative stress, and vitamin D deficiency. ASD is categorized into multiple disorders [17].

2.1   Importance of Early Detection of ASD

Early detection is very helpful to ASD children to live a normal life and behave like other children in society. In general, ASD can be identified at an early age of 18 months by observing child behavior and intelligence [18]. Importance of early diagnosis, intervention, and detection plays an important role in improving cognition language, adaptive behavior, daily activities, and social behavior. Researchers surveyed 1420 parents and ASD children of developmentally delayed group to find the impact of early detection of ASD [18]. The late detection of ASD, delays speech, social behavior, and increases parental stress. Research suggests that the early detection improves quality of life of ASD children [19]. Diagnosing at an age of 12–48 months improves behavioral changes in ASD children. Different screening methods and checklist are used for toddlers, child, young, and adult based on autism diagnostic observational schedule (ADOS) [20]. At present, screening is available from the age 4–5 years. Early detection of ASD, needs more research with advanced technical- and non-technical methods. It helps to live a better life and improved life by ASD children. It also helpful for therapists to propose better therapies. Doctors can suggest better treatments. Caregivers and special education trainers get help with this to improve ASD children intelligence and join them with regular children [21]. The following section discusses various methods existing for diagnosis process to detect ASD accurately.

2.2 Existing Methods for Detecting ASD

- **Parental Interviews**: Parents are the first observers who find the difference in their child's behavior. The standard questionnaires used for screening are American Society for Quality (ASQ-3), Modified Checklist for Autism in Toddlers, Revised with Follow-Up (M-CHAT-R/F), and Social Communication Questionnaire SCQ [22].
- **Behavior Observation**: Behavior observation is another assessment method used by psychiatrists to assess ASD children. Behavior observation scale (BOS) has 67 objectively specified behaviors helped to assess ASD children [23]. BOS has different set of objective behaviors for different ages and for different types of ASD.
- **Electroencephalogram (EEG) Analysis**: EEG analysis is a clinical method used for early detection of ASD at an early age of <24 months. EEG analysis is used to analyze brain structure and neural connectivity. Childs brain is fully developed under 12 months [24].
- **Magnetic Resonance Imaging (MRI) Processing**: MRI analysis is used to identify problems raised in the brain. Structural MRI helps to identify structural changes in brain [25].

3 Machine-Learning Methods to Detect ASD

Machine learning is one of the components of artificial intelligence (AI), uses the principles of data mining and it also generates automatic correlations, and machine is learned from existing data, to apply on new data or algorithms [26]. Machine learning can also be used for applications where human interaction is not required like IoT and robotics. Machine learning discovers new model from the past data or experience. Machine-learning algorithms are classified as supervised and unsupervised algorithms. Supervised algorithm trains a machine based on map of input to an output [27]. Supervised learning further classified into classification and regression. The significant algorithms are decision trees, support vector machine (SVM), naïve Bayes, random forest, linear regression, polynomial regression, and logistic regression. Unsupervised learning algorithms train the machine using information that is neither classified nor labeled and allowing the algorithm to act on that popularly used techniques are categorized into clustering, association rule mining, neural networks, and anomaly detection [28]. Machine learning has potential use in medical sciences and it gives accurate results in predictive analytics. It helps to create efficient algorithms for diagnosing ASD using data from manual diagnosis process and from diagnostic instruments. Machine learning makes diagnosis process simple and fast. Efficient and accurate results are generated with machine-learning algorithms.

3.1 Regression Methods

Predicting the values of numeric attributes and continuous attributes is known as regression method. Regression method is an interdisciplinary analytical method of statistical and machine learning. Machine learns from numerical or continuous values in different applications like IoT, robotics, and sports abilities. Most of the machine-learning applications work with classification methods. Regression method is different from classification; predicted result differed if the attributes are numerical or continuous. Regression methods are named as functional prediction, continuous class learning, and real value prediction. Regression methods are classified as linear regression, logistic regression, weighted regression, rule-based regression, and instance-based regression [14]. Regression methods are used in analyzing medical data. Predicted value helps to find important factors that effects disease. In this paper, linear regression methods are used to predict values. Linear regression finds the linear relationship between independent and dependent attributes. Linear regression can be used with single variable or with multiple variables. Linear regression generates a straight line with equation $Y = b_0 + b_1X$ where b_0 is y-intersect and b_1 is slope of line. Multi-variable regression method evaluated based on equation $Y = b_0 + b_1X_1 + b_2X_2 + \cdots + b_nX_n$ where $b1, b2, b3,\ldots, b_n$ are slope value of variables $X_1, X_2,\ldots, X_n$ [29]. Linear regression used to predict important factors of ASD and an improvised version of slope calculations is proposed. The proposed method reduces the computation complexity.

4 Proposed Methodology

The proposed methodology of machine learning uses regression methods to train the machine for fast detection of ASD. Regression method generates a straight-line equation based on the existing data. The beta coefficients B_0 and B_1 are calculated based on dependent variable Y and independent variables X_i. The computation is reduced by minimizing the computation process of beta values. The analytical results compare existing method and proposed method (Fig. 1).

Fig. 1 Overview of proposed methodology

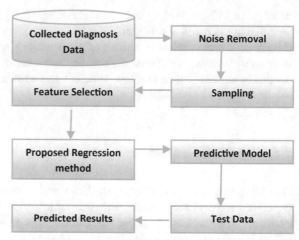

4.1 Algorithm for Proposed Regression Method

Regression methods are used for predictive analytics. Regression methods are classified into linear regression and logistic regression. Regression method generates straight line, which shows relationship between attributes. The important part of regression methods is calculating beta values. Beta values calculation process is simplified in proposed method. Beta values are computed by using closed form solution.

$$\beta = \left(x^{\bar{1}}x\right)^{-1}x^{\bar{1}}y$$

Algorithm regression ()
{
Input: ASD diagnosis dataset.
Output: Beta values B0, B1…..Bn.

1. Prepare dataset by converting all attributes to numerical or binary.
2. Identify correlation between attributes using correlation coefficient for each Xi and Yi.

$$\text{Correlation Coefficient } (r) = \frac{1}{n-1}\sum_{i=1}^{n}\left(\frac{x_i - \bar{x}}{S_x}\right)\left(\frac{y_i - \bar{y}}{S_y}\right)$$

3. Develop regression model with existing values to predict the future values.
4. Compute coefficients b0, b1, b2…bn using

$$b_i = \sum_{i=1}^{n} \frac{(x_i - \bar{x})(y_i - \bar{y})}{(x_i - \bar{x})^2}$$

$$b_0 = \bar{y} - b_i \bar{x}$$

$$\hat{y}_i = b_0 + b_i x_i$$

5. Compute residuals, the difference between existing and predicted values.
6. Calculate standard error (SE) for beta values.

$$b_0 = S\sqrt{\frac{\sum x^2}{n \sum x^2 - (\sum x)^2}} \quad b_1 = S\sqrt{\frac{n}{n \sum (x)^2 - (\sum x)^2}}$$

7. Compute mean square error (MSE) between existing and predicted class variable values.

$$a^2 = S^2 = \frac{\sum_{i=1}^{n}(y_i - \hat{y}_i)^2}{n - 2}$$

8. Compute standard squared error (SSE)

$$\sum_{i=1}^{n} ri^2 = \sum_{i=1}^{n}(y_i - \hat{y}_i)^2$$

9. Apply regression model on test data. Compute performance metrics. Compare performance metrics of training and test data.
10. Apply closed form solution to compute beta values. Repeat steps from 5 to 9.

$$\beta = \left(x^{\bar{1}}x\right)^{-1} x^{\bar{1}}y$$

11. Apply proposed model for test data. Compute performance metrics SE, MSE, and SSE.
12. Compare performance metrics of proposed model and existing model.

}

5   Implementation

ASD dataset is collected from UCI machine-learning repository. Dataset is provided
by Thabtah [17] for ASD classification. F. Thabtah proposed DSM-5 model for
ASD screening using classification algorithms of machine learning. F. Thabtah
recently released updated version dataset for four different age groups. Dataset is
provided for adult, adolescent, child, and toddler. Adult dataset has information for
the age group >18 years. Adolescent dataset has information for the age group
12–16 years. Child dataset has information for the get group of 4–11 years. Toddler
dataset has information for the age group 0.5–4 years. Attribute description and
possible values are given in Table 1 for all attributes of adult, adolescent, child, and
toddler datasets.

Table 1 ASD dataset attribute description

S. no.	Attribute name	Description	Possible values
1	A1 to A10	Diagnosis questionnaire result	Yes-1, No-0
2	Age	Age of the diagnostic patient. Continuous values	Minimum value: 1 year and maximum value: 80 years
3	Sex	Gender of the patient	M-Male, F-Female
4	Ethnicity	Is a category of people who identify with each other based on similarities such as common ancestry, language, history, society?	White, Black, Hispanic, Aboriginal, Latino Middle Eastern, Asian, South Asia, Others
5	Jaundice	Jaundice effected immediately after delivery or not	Yes-1, No-0
6	Family ASD	ASD existence in family hierarchy	Yes-1, No-0
7	Residence	Country or state of residence	Different states and countries in Asia, South Asia, and others
8	Used app before	Whether patient assessment is first time or not	Yes-1, No-0
9	Score	Sum of values of A1 to A10 attributes	Minimum value: 0 and Maximum value: 10
10	Screening type	Age group of patients selected for screening	1–3, 4–11, 12–16, 17, and above
11	Language	Language used for communication	English, Russian, Farsi, Spanish, Arabic, and French
12	User	Person who bought patient for screening	Self, parents, others, and relative
13	ASD	Class variable. Classified under ASD or not	Yes and No

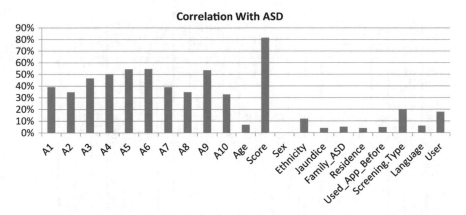

Fig. 2 Correlation between independent and dependent variables

6 Result Analysis

The correlation graph in Fig. 2 shows the relation between attributes with correlation coefficient. A1 to A10 attributes shows 40–60% of impact in predicting ASD. Score attribute has 80% impact in deciding ASD. Screening type has 20% impact in deciding ASD. Remaining attributes has very less impact in predicting ASD.

Linear regression with straight-line equation is used for computing coefficients of independent variables to predict dependent variables. Table 2 shows linear regression with closed form equation is used for computing coefficients of independent variables to predict dependent variables and also presents the coefficients of independent attributes calculated with proposed regression method using closed form equation for coefficients computation.

Table 3 summarizes the statistical measures of existing linear regression method and proposed linear regression method. Measures are improved in case of closed form-based linear regression method compared to straight-line-based linear regression method (Table 4).

7 Conclusion and Future Work

ASD is a neurodevelopmental disorder and is a lifelong disorder. The prediction process of ASD is a long process. For better treatment and improvement in behavior, detection process must be fast, efficient and accurate. Machine learning is one of the emerging methodology, which takes less time to predict new results. This paper proposes a methodology for early detection of ASD using linear regression method of machine-learning algorithms, which helps in predicting ASD. This work

Table 2 Linear regression results of existing and proposed method

Results of linear regression with straight-line equation				Result of linear regression with closed form equation			
Attributes	Estimated coefficients	Standard error	T value	Attributes	Expected coefficients	Standard error	T value
(Intercept)	0.82	0.102	8.071	Int	1.05	0.10206	8.071
A1	0.26	0.023	11.447	A1	0.31	0.023	11.447
A2	0.22	0.021	10.392	A2	0.24	0.021	10.392
A3	0.22	0.023	9.099	A3	0.28	0.023	9.099
A4	0.29	0.023	12.277	A4	0.3	0.023	12.277
A5	0.34	0.023	14.301	A5	0.37	0.023	14.301
A6	0.40	0.024	15.996	A6	0.34	0.024	15.996
A7	0.30	0.022	13.513	A7	0.24	0.022	13.513
A8	0.22	0.021	10.219	A8	0.25	0.021	10.219
A9	0.39	0.023	16.198	A9	0.35	0.023	16.198
A10	0.22	0.021	10.162	A10	0.24	0.021	10.162
Age	0.001	0.001	1.331	Age	0.001	0.001	1.331
Sex	-0.033	0.021	-1.573	Sex	-0.029	0.021	-1.573
Ethnicity	0.0007	0.001	0.409	Ethnicity	0.0007	0.001	0.409
Jaundice	0.030	0.026	1.163	Jaundice	0.035	0.026	1.163
Family_ASD	-0.033	0.027	-1.23	Family_ASD	-0.029	0.027	-1.23
Residence	0.0002	0.00028	0.869	Residence	0.0002	0.00028	0.869
Used_App_Before	-0.04	0.079	-0.545	Used_App_Before	-0.039	0.079	-0.545
Screening type	0.067	0.016	3.838	Screening type	0.071	0.016	3.838
Language	0.021	0.008	2.629	Language	0.028	0.008	2.629
User	0.020	0.005	3.421	User	0.0252	0.005	3.421

Table 3 Statistical and accuracy measures of existing and proposed regression methods

Statistical measures	Straight-line-based linear regression	Closed form-based linear regression	Accuracy measures	Straight-line-based linear regression	Closed form-based linear regression
correlation coefficient	−0.6996035	−0.719960	Accuracy	0.5688	0.6688
Residual standard error	0.5431	0.5031	Kappa	0.3986	0.4986
Multiple R-squared	0.6945	0.6945	Sensitivity	0.66564	0.68564
Adjusted R-squared	0.6924	0.6924	Specificity	0.97311	0.99311
F-statistic	330.5	330.5	Positive predicted value	0.7561	0.7761
p-value	<2.2e−16	<2.2e−16	Negative predicted value	0.95874	0.91874
Standard error	0.5431	0.4931	Prevalence	0.1113	0.1513
Mean square error	0.293	0.253	Detection rate	0.07409	0.07909
Standard squared error	857.71	852.71	Detection prevalence	0.09799	0.09999

Table 4 Estimation of proposed regression model accuracy

Methods to estimate model accuracy	Dataset/ sample size		RMSE (Repeated mean square error)	R-squared	MAE (Mean square error)
Bootstrap	2629		0.5473248	0.6884137	0.457611
K-fold cross validation	2636	10	0.5457446	0.689682	0.456921
	2511	7	0.5450491	0.6902811	0.456042
	2343	5	0.5464835	0.6889417	0.457516
	1953	3	0.5465098	0.689225	0.457048
Repeated cross validation for 10-fold three times	2629		0.5451202	0.6901611	0.456606
Leave one out cross validation	2628		0.5453541	0.6897356	0.456614

can extend to work with different types of ASD diagnosis data like MRI scan data, EEG data, and gene sequences data.

Acknowledgements Original data was collected by Fadi Fayez Thabtah who has developed mobile application for detecting autism spectrum disorder. Fadi Fayez Thabtah placed datasets in UCI machine-learning repository with open access to use by the researchers. Thankful to Fadi Fayez Thabtah for supporting research on ASD.

References

1. Daniels, A.M., Rosenberg, R.E., Law, J.K., Lord, C., Kaufmann, W.E., Law, P.A.: Stability of initial autism spectrum disorder diagnoses in community settings. J. Autism Dev. Disord. **41**(1), 110–121 (2011)
2. Becerra, T.A., Massillon, M.L., Yau, V.M., Owen-Smith, A.A., Lynch, F.L., Crawford, P.M., Pearson, K.A., Pomichowski, M.E., Quinn, V.P., Yoshida, C.K., Crone, L.A.: A Survey of Parents with Children on the Autism Spectrum: Experience with Services and Treatments. https://doi.org/10.7812/TPP/16-009
3. Whyatt, C.P., Torres, E.B.: Autism Research: An Objective Quantitative Review of Progress and Focus Between 1994 and 2015. https://doi.org/10.3389/fpsyg.2018.01526
4. Thabtah, F.: Machine learning in autistic spectrum disorder behavioral research: a review and ways forward. Inform. Health Soc. Care 1–20 (2018)
5. Volkmar, F.R., Reichow, B., McPartland, J.: Classification of autism and related conditions: progress, challenges, and opportunities. Dialogues Clin. Neurosci. **14**, 229–237 (2012)
6. Elder, J.H., Kreider, C.M., Brasher, S.N., Ansell, M.: Clinical impact of early diagnosis of autism on the prognosis and parent–child relationships. Published in Dove Press journal Psychology research and behavior management on 24 August 2017
7. Angra, S., Ahuja, S.: Machine learning and its, applications: a review. In: 2017 International Conference on Big Data Analytics and Computational Intelligence (ICBDAC). https://doi.org/10.1109/icbdaci.2017.8070809
8. Rasmussen, T.E.: Technology as a tool in autism spectrum disorder (ASD): an overview. Department of Psychology, UiT—Arctic University of Norway
9. Geetha Ramani, R., Sivaselvi, K.: Autism spectrum disorder identification using data mining techniques. Int. J. Pure Appl. Math. **117**(16), 427–436 (2017). ISSN: 1311-8080 (printed version); ISSN: 1314-3395
10. Dua, D., Graff, C.: UCI Machine Learning Repository [http://archive.ics.uci.edu/ml]. University of California, School of Information and Computer Science, Irvine, CA (2019)
11. Gök, M.: A novel machine learning model to predict autism spectrum disorders risk gene. In: Neural Computing and Applications, pp. 1–7. Springer
12. Maenner, M.J., Yeargin-Allsopp, M., Braun, K.V.N., Christensen, D.L., Schieve, L.A.: Development of a Machine Learning Algorithm for the Surveillance of Autism Spectrum Disorder. Published: 21 Dec 2016. https://doi.org/10.1371/journal.pone.0168224
13. Takara, K., Kondo, T.: Autism spectrum disorder among first-visit depressed adult patients: diagnostic clues from backgrounds and past history. Gen. Hosp. Psychiatry **36**, 737–742 (2014). Elsevier
14. Crippa, A., Salvatore, C., Perego, P., Forti, S., Nobile, M., Molteni, M., Castiglioni, I.: Use of machine learning to identify children with autism and their motor abnormalities. J. Autism Dev. Disord. **45**, 2146–2156 (2015). https://doi.org/10.1007/s10803-015-2379-8
15. Salvatore, C., Cerasa, A., Castiglioni, I., Gallivanone, F., Augimeri, A., Lopez, M., et al.: Machine learning on brain MRI data for differential diagnosis of Parkinson's disease and progressive supranuclear palsy. J. Neurosci. Methods **222**, 230–237 (2013)

16. Spaina, D., Sinc, J., Lindera, K.B., McMahond, J., Happéa, F.: Social anxiety in autism spectrum disorder: a systematic review. Res. Autism Spectr. Disord. **52**, 51–68 (2018)
17. Thabtah, F.: Autism Spectrum Disorder Screening: Machine Learning. CMHI'17, 20–22 May 2017, Taichung City, Taiwan © 2017 Association for Computing Machinery. ACM ISBN 978-1-4503- 5224- /17/05…$15.00. http://dx.doi.org/10.1145/3107514.3107515 Adaptation and DSM-5 Fulfillment, Nelson Marlborough Institute of Technology
18. Crippa, A., Salvatore, S., Perego, P., Forti, S., Nobile, M., Molteni, M., Castiglioni, I.: Use of Machine Learning to Identify Children with Autism and Their Motor Abnormalities. https://doi.org/10.1007/s10803-015-2379-8
19. Auyeung, B., Baron-Cohen, S., Wheelwright, S., Allison, C.: The autism spectrum quotient: childrens version (aq-child). J. Autism Dev. Disord. **38**(7), 1230–1240 (2008)
20. Mythili, M.S., Mohamed Shanavas, A.R.: A study on autism spectrum disorders using classification techniques. Int. J. Soft Comput. Eng. (IJSCE) **4**(5) (2014). ISSN: 2231-2307
21. Schneider, A., Hommel, G., Blettner, M.: Linear Regression Analysis, Part 14 of a Series on Evaluation of Scientific Publications
22. Freeman, B., Ritvo, E.R., Guthrie, D., Schroth, P., Ball, J.: The Behavior Observation Scale for Autism
23. Bosl, W.J., Tager-Flusberg, H., Nelson, C.A.: EEG Analytics for Early Detection of Autism Spectrum Disorder: A Data-Driven Approach. Published online 01 May 2018 by Scientific Reports
24. Chen, R., Jiao, Y., Herskovits, E.H.: Structural MRI in Autism Spectrum Disorder. Published in final edited form as Pediatr Res. **69**(5 Pt 2), 63R–68R (2011). https://doi.org/10.1203/pdr.0b013e318212c2b3
25. Bone, D., Bishop, S.L., Black, M.P., Goodwin, M.S., Lord, C., Narayanan, S.S.: Use of machine learning to improve autism screening and diagnostic instruments: effectiveness, efficiency, and multi-instrument fusion. J. Child Psychol. Psychiatry **57**(8), 927–937 (2016)
26. Geetha Ramani, R., Sivaselvi, K.: Autism spectrum disorder identification using data mining techniques. Int. J. Pure Appl. Math. **117**(16), 427–436 (2017). Issn: 1311-8080 (Printed Version); Issn: 1314-3395 (On-Line Version)
27. Angra, S., Ahuja, S.: Machine learning and its, applications: s review. In: Proceedings of International Conference on Big Data Analytics and Computational Intelligence (ICBDAC-2017). https://doi.org/10.1109/icbdaci.2017.8070809
28. Alarifi, H.S., Young, G.S.: Using multiple machine learning algorithms to predict autism in children. In Proceeding of International Conference Artificial Intelligence, ICAI'18. ISBN: 1-60132-480-4, CSREA Press
29. Uysal, I., Guè Venir, H.A.: An overview of regression techniques for knowledge discovery. Knowl. Eng. Rev. **14**(4), 319–340, Printed in the United Kingdom Copyright #1999, Cambridge University Press (1999)

Gene Sequence Analysis of Breast Cancer Using Genetic Algorithm

Peyakunta Bhargavi, Kanchi Lohitha Lakshmi and Singaraju Jyothi

Abstract This paper describes the use of genetic algorithm (GA) implementation on tissue of breast cancer tumor and normal breast gene data sequences to analyze and discriminate among these data sequences. This discrimination is done between the breast cancer and non-breast cancer genetic factor data sequences based on optimal values generated after implementation of genetic algorithm in every single generation. Genetic algorithm is population-based evolutionary algorithm of soft computing techniques. Genetic algorithm utilizes arbitrary investigation of problem joined with evolutionary procedures like mutation (transformation) and crossover to improve the chance of predicting optimal solutions. GA provides different approaches to expand the chance of solving real-world genetic-related problems which enables to upgrade the execution of calculation. The main technique of GA is to produce possible guesses on provided information. The values calculated in the present work provide a way for progressive approach to design a framework to discriminate cancer and non-cancer data sequences.

Keywords Genetic algorithm (GA) · Soft computing (SC) · Breast cancer diagnosis

1 Introduction

According to George Mendel (basic genetic researcher) it is determined that, to inherit a particular trait from parental to adolescent one part of a parent gene is distributed on to its successive descendant generations [1]. Genetic algorithm

P. Bhargavi (✉) · K. Lohitha Lakshmi (✉) · S. Jyothi (✉)
Department of Computer Science, SPMVV, Tirupati, India
e-mail: bhargavi18@yahoo.co.in

K. Lohitha Lakshmi
e-mail: lohita.kanchi@gmail.com

S. Jyothi
e-mail: jyothi.spmvv@gmail.com

© Springer Nature Singapore Pte Ltd. 2020
P. Venkata Krishna and M. S. Obaidat (eds.), *Emerging Research in Data Engineering Systems and Computer Communications*, Advances in Intelligent Systems and Computing 1054, https://doi.org/10.1007/978-981-15-0135-7_16

provides best approach to solve computational problems by using evolutionary techniques and to find globally accepted optimal solutions. The methods or procedures used in GA's are highly randomized. Computations performed in GA's produce progressive generations representing best parent solution for individual iteration. Best parent population will be chosen based on fitness value generated for individual iteration. Individuals generated with best fitness value will be treated as best parent for performing subsequent results [2]. The work of GA on natural biological approaches such as certain selection, crossover and mutation will be treated as the best parent in order to develop a new population based entirely on a fitness function and realign the subsequent sequences to produce a new sequence. To find similarities between homologous sequences, alignment of sequences will be actually performed on a sequence pair or more than a sequence pair. Multiple sequence alignment [MSA] is a powerful tool for naturally finding sequence patterns that might be hidden or scattered in the gene sequence database [3].

The produced resultant aligned sequences of MSA will be passed as input to GA to find best optimal solution based on fitness value generated after evaluation of each generation. The proposed evolutionary approach, i.e., combination of MSAGA is applied on the gene sequence database to obtain alignment on pairwise sequences and to get better fitness accuracy rates compared to other existing methods. Breast cancer is one of the most critical genetic diseases particularly in women. The early diagnosis is the only solution to survive. Identification in the advanced stage can lead to death in maximum cases. The present work also discovers the prospect of applying MSAGA on gene sequence database for not only breast cancer but also for other hereditary diseases by analyzing the gene sequences.

2 Multiple Sequence Alignment for Gene Sequence Analysis

Multiple sequence alignment [MSA] is used on the search space efficiently to discover a good solution by applying alignment on homologous sequences to find similar properties. In genomic analysis, MSA is used as the identification preserved sequence motifs, the prediction of the phenotypic differences between the sequences and the implication of the historical relationships of the genetic traits in different generations in a conventional set of gene sequences, an MSA provides a hidden treasure of specific information about physical and functional relationships. MSA plays a significant part in other genetic-related operations such as analyzing protein structures, genotypes (genetic changes) and functionality along with gene sequence analysis [4]. MSAs have even successively turned out to be gradually more important because of their relevance to novel test methods like high-throughput trials or next-generation sequencing (NGS). Consequently, MSA tools are as of now fundamental to the improvement of a few bioinformatics research studies. The only complexity is to find appropriate alignment tool from the already existing huge aligner tools which is suitable for current biological experiment [5]. The major difficulty to apply any computing technique is generation of test data. Solving this problem, several research works are performed to generate test data using several different MSA techniques. The importance of MSA is to find ideal alignment sequence from a group or pair of sequences [6].

3 Genetic Algorithm for Gene Sequence Analysis

Initially, we can summarize the genetic algorithm as defining initial population as first step, classify the population based on specific condition as second step, select best population for mating to produce offspring's and finally replace bad characters with best ones to produce optimal results. This process will be repeated until specific condition satisfied with desired result. GA is optimized technique with some input values (population), best values can be obtained as output or results [7]. In GA, each individual population is a part of optimal solution. This group of individuals underwent on different operations to reproduce new individual population. Those operations include selection process based on fitness value next crossover in which portions of chromosomes are exchanged, and next operation is mutation in which some bits are randomly exchanged between chromosomes. Fitness function is applied on possible solution. Population with best fitness value is best fittest for next generation. The main mechanism in genetic algorithms to generate output optimal solution initiates with the process of identifying suitable input data by applying software technique which satisfies given testing condition [8].

4 MSAGA for Gene Sequence Analysis

In this present implementation, MSA technique is used to analyze gene sequences, to find out relationship between sequences, to study evolution process of sequences and to find out similarity in structural relationship. In this present evolutionary methodology, two processes are adapted to generate initial population [9]. The first mechanism is MSA which is used to generate initial population by implementing alignment on pair-wise sequences from multiple sequences belong to same homogenous categories of cancer, noncancerous and one more category of one sequence from cancer sequences and one from cancer suppressor gene sequences. The result of alignment is formed as a separate sequence and translated into supported format to pass as input to second mechanism, i.e., GA in present research. GA is used to find best parent and best fitness values after performing certain fundamental operations after specific number of iterations till it meets a specific termination condition to produce solutions with optimal results. The experimental results pro-duced with this hybridized evolutionary approach are giving clear perspective of efficiency of algorithm. The results show that rather than implementing simple genetic algorithm use of MSAGA maximizes dimension of similarity. This was shown that use of PRRN software in MSA to generate initially aligned sequences to improve outcome with reduced complexity. A general evolutionary framework is technologically advanced using MSAGA rather than implementing simple GA makes extending objective functions to improve output [10]. Evolutionary methodologies may not compete with the progressive techniques in terms of speed and iterative approach, but these are capable of aligning misaligned sequences initially by using different software implementation techniques which are not supported by progressive methods [11]. A last consideration compared to simple GA and MSAGA is that MSAGA is more sensible with the changes in initial population which effects the final fitness values, similarity scores and final arrangement.

5 Implementation

In the recommended approach, GA is applied by Python language. Python language is expressive due to its clean syntax and semantics [12]. Many libraries and third-party tool kits are helpful to extend the functionality of the language into biological domains like sequence analysis, annotation, phylogenomics and performing computations on real-time stochastic data related to genomics to find ideal solutions [13]. Performance of any language usually decides by its computational efficiency and libraries. Scripts written in Python act as an interface between programs performing sequence analysis. These scripts are helpful in extracting information from large files usually FASTA files for extracting gene sequence information to enhance optimal results of production. Python is one of the more suitable languages to quickly implement due to its automatic memory organization and vast free libraries [14, 15]. The present methodology is applied on several data sets, and some of the experimental results are presented for demonstrating sample results.

- In present implementation, different iteration methods we support of software packages such as PRRN/PRRP which support pairwise alignment and iterative refinement by implementing hill climbing algorithm to get optimal alignment sequence corresponding with optimal alignment score [16].
- These optimal sequences or solutions generated by MSA will be passed as input to GA. In initial step of GA implementation, gene character set and MSA generated output sequence are initialized as target set.
- Calculate start time and generate initial best population and initial fitness value in the first iteration.
- Calculate and store the current generation and parent length and actual and expected fitness value.
- Return sum if actual fitness value is equal to expected fitness value and store best fitness value and best parent of the generation.
- Apply selection, crossover and mutation operations on the generated fitness score and best parent sequence until termination condition is satisfied.
- Display best parent and best fitness value of current generation.

In the present phenomenon, MSAGA is implemented on gene data sequences. Based on the deviation in generated fitness values determine the characteristic difference of different categories of sequences, i.e., diseased and normal gene sequences.

5.1 Flow Chart for MSAGA Algorithm Implementation in Python

Breast cancer gene data input sequences are collected from different government authorized websites, i.e., NCBI, UCI, NIH, etc. The details of input sequences are listed below with their accession numbers and description. In the present work, data is collected from a particular genetic disease, i.e., breast cancer database. Input data is classified into normal gene data sequences and suppressor gene data sequences of the same organism of diseased and healthy tissue. MSAGA algorithm implementation is presented in Fig. 1.

(i) **List of Breast Cancer Gene data Input Sequences with Accession No's and Description**

See Table 1.

(ii) **List of Breast Cancer Suppressor Gene data Input Sequences with Accession No's and Description**

See Table 2.

5.2 MSA Algorithm Steps

1. Start the process with the required input sequences.
2. Align first sequence with second sequence which belongs to same category of diseased or healthy sequences.
3. Distance matrix will be created for each pair of sequences.
4. Optimal value will be generated based on alignment score by using hill climbing algorithm.
5. Sequence generated relevant to optimal score will be considered as optimized sequence with this algorithm and is used as input for GA.

5.3 GA Algorithm Steps

1. Start GA with randomly generated output sequences of MSA as input to GA.
2. Initialize gene set with different character sets of nucleotide and MSA generated aligned output sequence as target character set.

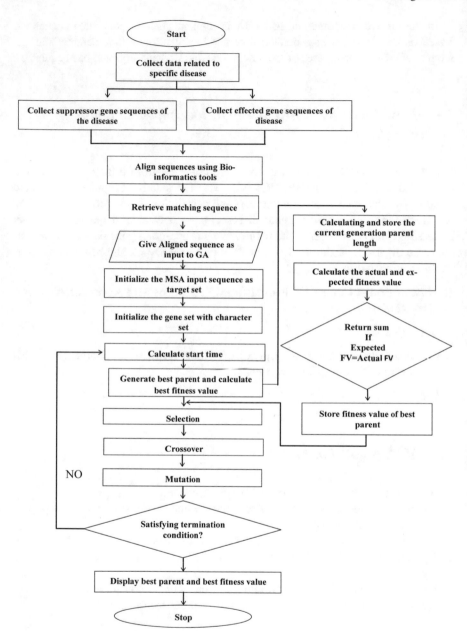

Fig. 1 Flow chart for MSAGA

Table 1 List of breast cancer gene data input sequences

S. no.	Accession no. and description
SEQ 1	AF507075.1 Homo sapiens IRCHS11A non-functional breast and ovarian cancer susceptibility protein (BRCA1) gene, exon 20 and partial sequence
SEQ 2	AY093484.1 Homo sapiens isolate IRCHS8A breast and ovarian cancer susceptibility protein (BRCA1) gene, partial cds
SEQ 3	AF507076.1 Homo sapiens IRCHS11B breast and ovarian cancer susceptibility protein (BRCA1) gene, exon 20 and partial cds
SEQ 4	AF507077.1 Homo sapiens IRCHS6A breast and ovarian cancer susceptibility protein (BRCA1) gene, exon 20 and partial cds
SEQ 5	AF507078.1 Homo sapiens IRCHS6B breast and ovarian cancer susceptibility protein (BRCA1) gene, exon 20 and partial cds
SEQ 6	AY093484.1 Homo sapiens isolate IRCHS8A breast and ovarian cancer susceptibility protein (BRCA1) gene, partial cds
SEQ 7	AY093486.1 Homo sapiens isolate IRCHS7A breast and ovarian cancer susceptibility protein (BRCA1) gene, partial cds
SEQ 8	AY093487.1 Homo sapiens isolate IRCHS7B breast and ovarian cancer susceptibility protein (BRCA1) gene, partial cds
SEQ 9	AY093488.1 Homo sapiens isolate IRCHS16A breast and ovarian cancer susceptibility-like protein (BRCA1) gene, partial sequence
SEQ 10	AY093489.1 Homo sapiens isolate IRCHS16B breast and ovarian cancer susceptibility protein (BRCA1) gene, partial cds
SEQ 11	AY093492.1 Homo sapiens isolate IRCHS4A breast and ovarian cancer susceptibility protein (BRCA1) gene, partial cds
SEQ 12	AY093493.1 Homo sapiens isolate IRCHS4B breast and ovarian cancer susceptibility protein (BRCA1) gene, partial cds
SEQ 13	AY093490.1 Homo sapiens isolate IRCHF4A breast and ovarian cancer susceptibility protein (BRCA1) gene, partial cds
SEQ 14	AY093491.1 Homo sapiens isolate IRCHF4B breast and ovarian cancer susceptibility protein (BRCA1) gene, partial cds
SEQ 15	AF288938.2 Homo sapiens breast cancer 2, early onset (BRCA2) gene, partial sequence
SEQ 16	AF317283.1 Homo sapiens mutant BRCA2 gene, partial sequence
SEQ 17	AF284812.1 Homo sapiens BRCAI (BRCA1) gene, exon 20 and partial cds
SEQ 18	AF309413.1 Homo sapiens BRCA2 protein (BRCA2) gene, partial cds

3. Initialize start time and compute execution time.
4. Evaluate fitness function and generate best parent and calculate best fitness value.
5. Calculate and store the current generation parent length.
6. Calculate and store the actual and expected fitness value.
7. Return sum if expected fitness value equals to actual fitness value.

Table 2 List of breast cancer suppressor gene data input sequences

S. no.	Accession no. and description
SEQ 1	AF209138.1 Homo sapiens tumor suppressor p53 (TP53) gene, partial cds
SEQ 2	AF066082.1 Homo sapiens mutant p53 transformation suppressor gene, exon 6 and partial cds
SEQ 3	AB700556.1 Homo sapiens gene for p53 protein, partial cds, exon 3, intron 3
SEQ 4	AF209128.1 Homo sapiens tumor suppressor p53 (TP53) gene, partial cds
SEQ 5	AF209129.1 Homo sapiens tumor suppressor p53 (TP53) gene, partial cds
SEQ 6	AF209130.1 Homo sapiens tumor suppressor p53 (TP53) gene, partial cds
SEQ 7	AB118156.1 Homo sapiens p53 gene for P53, exon 5, partial cds
SEQ 8	AB700556.1 Homo sapiens gene for p53 protein, partial cds, exon 3, intron 3
SEQ 9	AB699689.2 Homo sapiens gene for p53 protein, partial cds
SEQ 10	AB699689.2 Homo sapiens gene for p53 protein, partial cds
SEQ 11	AF209136.1 Homo sapiens cell-line HN5 tumor suppressor p53 (TP53) gene, partial cds
SEQ 12	AB699689.2 Homo sapiens gene for p53 protein, partial cds
SEQ 14	AF209139.1 Homo sapiens tumor suppressor p53 (TP53) gene, partial cds
SEQ 15	AB699004.1 Homo sapiens gene for P53 protein, partial cds and exon
SEQ 20	AB118156.1 Homo sapiens p53 gene for P53, exon 5, partial cds
SEQ 21	AF209132.1 Homo sapiens cell-line Molt4 tumor suppressor p53 (TP53) gene, partial cds
SEQ 22	AF209137.1 Homo sapiens tumor suppressor p53 (TP53) gene, partial cds
SEQ 23	AF209140.1 Homo sapiens tumor suppressor p53 (TP53) gene, partial cds
SEQ 24	EF178470.1 Homo sapiens tumor protein p53 (TP53) gene, exon 8 and partial cds

8. Store fitness value of best parent.
9. Apply selection, crossover and mutation operations on new offspring of each location of the gene.
10. Accept the newly generated offspring if termination condition is satisfied.
11. If termination condition is not satisfied return the steps from 3 to 10.
12. If the termination condition is satisfied, stop and return the best solution and best fitness value of current population.

6 Experimental Result Analysis

Genetic algorithm is a population-based stochastic search method which produces optimal results for a problem of natural evolution. GA becomes more powerful due to its optimal development and used in performing computations of natural evolution [17]. The resultant fitness function of GA accurately reflects the data input, and this leads to effect on performance of final solution [18]. Due to these constraints to improve the quality of output or experimental solution of the problem [19], MSA applies on genetic data input sequences of particular disease. MSA is used to find fundamental relationship and common characteristics of protein or nucleotide sequences by applying alignment and capable to exhibit great time-based and space complexity [20].

Table 3 represents fitness scores of ten iterations and similarity scores of each pair of MSA generated breast cancer gene data input sequences. When these similarity scores are observed, range of values lies 50 and above 50 percentages in case of breast cancer gene data input sequences. Similarity score is calculated built

Table 3 Result of MSAGA on cancer gene sequences

GA implementation on MSA aligned cancer gene sequences											
S.NO	MSA sequence pairs	MSAGA RESULT FOR 10 ITERATIONS										Similarity Score
1	AF507075.1, AY093484.1	5	28	28	26	27	32	25	27	25	31	60%
2	AF507076.1, AF507077.1	26	22	21	24	26	23	23	24	19	27	60%
3	AF507078.1, AY093484.1	23	19	21	19	22	22	21	26	24	22	60%
4	AY093486.1, AY093487.1	27	26	31	28	28	23	31	23	24	25	60%
5	AY093488.1, AY093489.1	25	28	28	24	25	31	24	28	24	26	80%
6	AY093492.1, AY093493.1	29	27	29	25	21	28	28	29	23	26	50%
7	AY093490.1, AY093491.1	28	21	20	24	25	23	21	20	24	25	70%
8	AF284812.1, AF309413.1	57	72	59	57	70	66	74	84	72	84	60%
9	AF284812.1, AF309413.1	16	25	20	16	20	24	30	27	28	20	50%

on the percentage of similar fitness values out of existing fitness values in ten iterations. MSA is applied on pairwise sequences belong to same category of breast cancer sequences.

6.1 Representation of Tabular Values with Similarity Scores

(i) **Best fitness scores after implementation of MSAGA on breast cancer gene data input sequences**

Below-mentioned Table 4 represents fitness values of breast cancer suppressor or healthy gene MSA aligned sequences of ten iterations and similarity score. Resultant similarity scores lie 50% and above 50% like as breast cancer gene data

Table 4 Result of MSAGA on cancer suppressor gene sequences

S.NO	MSA Sequence Pairs	BEST FITNESS SCORES FOR 10 ITERATIONS										Similarity Score
		GA implementation on MSA aligned cancer suppressor gene sequences										
1	AF209138.1, AF066082.1	10	11	7	10	11	10	16	8	10	16	50%
2	AB700556.1, AF209128.1	11	15	17	6	14	11	10	11	15	11	60%
3	AF209140.1, AB118156.1	17	17	16	16	17	17	16	18	15	12	70%
4	AF209140.1, AF209140.1	14	8	12	14	14	11	12	5	8	10	70%
5	AF209134.1 ,AF209140.1	14	15	13	7	13	20	14	17	14	14	50%
6	AF209136.1, AF209139.1	12	9	14	15	7	8	7	9	12	14	80%
7	AB699689.2, AB699004.1	27	31	25	29	29	34	19	27	25	27	70%
8	AB118156.1, AF209137.1	13	11	8	16	11	13	8	13	11	11	90%
9	AF209132.1, AF209137.1	10	9	9	14	10	10	15	6	13	15	70%

input sequences. MSA is applied on pairwise sequences belong to same category of sequences in each case.

(ii) **Best fitness scores after implementation of MSAGA on breast cancer suppressor gene data input sequences**

In Table 5, MSA is applied on one breast cancer and one breast cancer suppressor gene sequence. GA is applied on resultant paired aligned sequence. If the fitness values are observed ten iterations, the similarity among fitness values is less compared to previous cases, i.e., in Tables 3 and 4. Due to less similarity in fitness values, similarity score is moreover reduced for these different categories of sequences. Similarity scores lay less than 50% in this case.

This phenomenon works as a fundamental platform to generate deterministic and refined good quality of data which has no progressive improvement after certain

Table 5 Result of MSAGA on cancer and cancer suppressor gene sequences

S.NO	MSA Sequence Pairs	BEST FITNESS SCORES FOR 10 ITERATIONS										Similarity Score
1	KP255416.1 AB699689.2	29	21	25	32	23	21	33	30	28	26	20%
2	KP255415.1 AB699689.2	27	31	19	34	30	36	27	34	35	27	40%
3	KP255414.1 AF066082.1	6	13	9	13	9	19	15	12	11	14	40%
4	KP255412.1 AF209128.1	15	21	19	21	17	17	13	17	14	11	40%
5	KP255410.1 AF209139.1	10	11	11	10	13	7	9	14	10	6	40%
6	KP255409.1 AF209140.1	16	7	12	17	13	12	11	19	10	11	40%
7	KP255408.1 AB118156.1	22	15	8	18	12	24	26	26	20	14	20%
8	KP255407.1 AB700556.1	12	10	15	6	13	16	12	9	16	8	40%
9	KP255411.1 AF209137.1	10	8	10	11	9	17	11	13	7	15	40%

Table header spanning title: **GA on MSA aligned cancer &non cancer sequences**

iterations. This sort of data is passed as input for many complex techniques such as genetic algorithm, particle swarm optimization [PSO] and differential evaluation [DE] which is useful in gene sequence analysis, prediction of genetic traits through inheritance (gene annotation), construction of phylogenetic trees and producing optimal solutions to some complex gene-related problems.

(iii) **Best fitness scores after implementation of MSAGA on breast cancerand suppressor gene data input sequences**

In below mentioned line chart, comparison of similarity scores is mentioned in graphical format. Each line represents similarity score of every category of sequences. Blue line represents the similarity score of cancer sequences. Red line represents similarity score of cancer suppressor sequences. Green line indicates similarity score of cancer and non-cancer sequences. The graph clearly represents that similar category sequences related to both cancer and non-cancer sequences show high similarity scores than merged align sequences of different categories of cancer and noncancerous sequences.

In my previous work, sequence analysis is applied on PSO which is one of the population-based optimization soft computing techniques to determine the difference between breast cancer and breast cancer suppressor gene data sequences with the generated optimal values [21]. In the proposed evolutionary methodology, instead of implementing simple GA on gene analysis, the proposed hybridized method MSA with GA is getting more accurate results compared to previous experimental results. Fitness score will be generated for individual iteration. After generating fitness scores for approximately ten iterations based on the repetition of similar values, similar score is calculated. MSA sequence pairs, MSAGA implementation for ten iterations and similarity scores are represented in tabular forms for each category of sequences.

Graphical representation of comparison of fitness values of MSAGA on different categories of gene sequences

See Fig. 2.

7   Conclusion and Change in the Future

In the proposed evolutionary methodology, when the similarity scores are compared after implementation of MSAGA algorithm on different categories of cancer sequences, cancer suppressor sequences and cancer and noncancerous merged sequences, the similarity scores of cancer and cancer suppressor gene sequences are high, and similarity scores of merged category of cancer and noncancerous sequences are very less. The proposed framework may be helpful for further diagnosis in gene sequence analysis of other hereditary diseases like diabetes, leukemia, filariasis, color-blindness and autism by applying computational

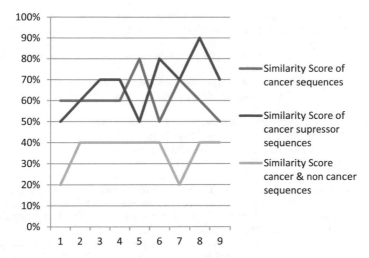

Fig. 2 MSAGA fitness values on different categories of sequences

techniques based on different fitness values. In my future work, such type of methodology will be applied with other hybrid soft computing algorithms for gene sequence analysis to get optimal results to compare and determine best approach.

References

1. Radenbaugh, A.J.: Applications of genetic algorithms in bioinformatics, UMI Microform 1458165 Copyright 2008 by ProQuest LLC. All rights reserved. This microform edition is protected against unauthorized copying under Title 17, United States Code, San Jose State University
2. Parsons, R., Forrest, S., Burks, C.: Genetic Algorithms for DNA Sequence Assembly, From: ISMB-93 Proceedings. Copyright © 1993, AAAI (www.aaai.org). All rights reserved
3. Gupta, R., Agarwal, P., Soni, A.K.: Genetic algorithm based approach for obtaining alignment of multiple sequences. Int. J. Adv. Comput. Sci. Appl. (IJACSA) **3**(12) (2012)
4. Lakshmi, N.J., Gavarraju, P., Jeevana, J.K., Karteeka, P.: A literature survey on multiple sequence alignment algorithms. Int. J. Adv. Res. Comput. Sci. **6**(3) (2016)
5. Ortuno, F., Valenzuela, O., Pomares, H., Rojas, I.: Determining the most suitable multiple sequence alignment methodology by using a set of heterogeneous biological features. In: IWBBIO 2013. Proceedings Granada, 18–20 Mar 2013
6. Lohitha Lakshmi, K., Rajesh, P.: An evolutionary optimization for multiple sequence alignment. IJCSN Int. J. Comput. Sci. Netw **3**(4) (2014). ISSN (Online): 2277-5420 Impact Factor: 0.274, www.IJCSN.org
7. Jain, S.: Introduction to genetic algorithm & their application in data science, 31 July 2017
8. Aljahdali, S.H., Ghiduk, A.S., El-Telbany, M.: The Limitation of Genetic Algorithms in Software Testing, June 2010. https://doi.org/10.1109/aiccsa.2010.5586984. Source IEEE Xplore

9. Naznin, F., Sarker, R., Essam, D.: Vertical decomposition with genetic algorithm for multiple sequence alignment. BMC Bioinform. **12**(1), 353 (2011). https://doi.org/10.1186/1471-2105-12-353

10. Gondro, C., Kinghorn, B.P.: A simple genetic algorithm for multiple sequence alignment. Published 2007 in ISMB (2007)

11. Gondro, C., Kinghorn, B.P.: A simple genetic algorithm for multiple sequence alignment. Genet. Mol. Res. **6**(4), 964–982 (2007). Received 3 Aug 2007, Accepted 25 Sept 2007, Published 5 Oct 2007

12. Ekmekci, B.: An Introduction to Programming for Bio scientists: A Python-Based Primer (2016)—Cited by 12—Related articles 7 June 2016

13. Multiple sequence alignment, From Wikipedia, the free encyclopaedia

14. Holland, J.: Adaptation in Natural and Artificial Systems. The University of Michigan (1975)

15. El-Mihoub, T.A., Hopgood, A.A., Nolle, L., Battersby, A.: Hybrid genetic algorithms: a review. Eng. Lett. **13**(2), EL_13_2_11 (Advance online publication: 4 August 2006

16. Fourment, M., Gillings, M.R.: A comparison of common programming languages used in bioinformatics. BMC Bioinform. **9**, 82 (2008). Published online 2008 Feb 5. https://doi.org/10.1186/1471-2105-9-82. PMCID: PMC2267699PMID: 18251993

17. Beasley, D., Bull, D.R., Martin, R.: An overview of genetic algorithms: part 1, fundamentals. Univ. Comput. **15**, 58–69, 1993(3) (PDF) Hybrid Genetic Algorithms: A Review. Available from: [accessed Feb 15 2019]

18. Ahmed, Z.H.: An experimental study of a hybrid genetic algorithm for the maximum traveling salesman problem (2013). https://doi.org/10.1186/2251-7456-7-10

19. Abdesslem, L., Soham, M., Mohamed, B.: Multiple sequence alignment by quantum genetic algorithm. 1-4244-0054-6/06/$20.00 ©2006 IEE

20. Lohitha Lakshmi, K., Bhargavi, P., Jyothi, S.: An analysis of breast cancer dna sequences using particle swam optimization, Copyright © 2018 Authors This is an open access article distributed under the Creative Commons Attribution License, which permits unrestricted use, distribution, and reproduction in any medium, provided the original work is properly cited. Int. J. Eng. Technol. **7**(4.7), 335–338 (2018)

Impact of Single Server Queue Having Machine Repair with Catastrophes Using Probability Generating Function Method

S. Anand Gnana Selvam, B. Chandrasekar, D. Moganraj
and S. Janci Rani

Abstract We consider the effect of Single Server Queue Having Machine repair with catastrophes utilizing likelihood creating capacities that is inferred. Here, the likelihood producing capacities for the quantity of units in the line when the server is occupied and is on machine repair are numerically inferred. The normal quantities of units in the line when the server is occupied and is on machine repair are gotten. Numerical examination has improved the situation investigation of different execution measures for different estimations of parameters.

Keywords Single server queue · Catastrophes · Machine repair · Probability generating functions

1 Introduction

The investigation of queueing models with administration breakdowns or some other sort of intrusions goes back to 1950s. Among some early papers around there, a Catastrophe can be seen as a general breakdown which makes every one of the occupations in the framework to be lost. For instance, if an open phone separates, every one of the clients in the holding up line leaves the pay phone. This sort of model has numerous applications in day-to-day life. Queueing models with server breakdowns are more reasonable in PC and correspondence exchanging

S. Anand Gnana Selvam (✉)
Department of Mathematics, A.E.T College, Attur, Salem, Tamil Nadu, India
e-mail: anandjuslin@gmail.com

B. Chandrasekar
Department of Mathematics, Sacred Heart College, Tirupattur, Tamil Nadu, India

D. Moganraj
Department of Mathematics, Arunai Engineering College, Tiruvannamalai,
Tamil Nadu, India

S. Janci Rani
Department of Mathematics, ACAR, Hosur, Tamil Nadu, India

© Springer Nature Singapore Pte Ltd. 2020
P. Venkata Krishna and M. S. Obaidat (eds.), *Emerging Research in Data Engineering Systems and Computer Communications*, Advances in Intelligent Systems and Computing 1054, https://doi.org/10.1007/978-981-15-0135-7_17

frameworks since the disappointment and repair of processors majorly affect the stream of occupations that must be taken care of by those processors Towsley and Tripathi. As of late, there has been extensive consideration paid to the investigation of M/G/1 compose queueing frameworks with two periods of administration under Bernoulli getaway calendar and distinctive getaway arrangements, for example observe [4–7] in which after two progressive periods of administration, the server may go for a Bernoulli getaway. The inspiration for these kinds of models originates from some PC systems and media transmission frameworks, where messages are handled in two phases by a solitary server. As current media transmission frameworks turn out to be more minds boggling and the handling intensity of microprocessors turns out to be more costly, the benefits of more complex planning turn out to be more evident. The requirement for planning the allotment of assets among at least two heterogeneous sorts of errands emerges additionally in numerous different applications. In the vast majority of the past examinations, it is expected that the server is accessible to the administration station on a lasting premise, and the administration station never fizzles. Be that as it may, these presumptions are for all intents and purposes far-fetched. By and by, Hassin [10] we regularly meet situations where benefit stations come up short and are repaired. Comparative marvels dependably happen in the zone of PC correspondence systems, adaptable assembling frameworks, and so forth. Since [11] the execution of such frameworks might be intensely influenced by the administration station breakdown and postponement in repair, these frameworks with a repairable administration station are well worth exploring from the queueing hypothesis perspective, and in addition, from the dependability perspective. Consequently, Fiems [9] considered the unwavering quality examination of a model under Bernoulli excursion plan with the suspicion that the server is liable to breakdowns and repairs. In the present investigation, we have considered an effect of Single Server Queue Having Machine repair with Catastrophes utilizing likelihood creating capacities. Consistent state probabilities have been discovered utilizing likelihood creating system. Whatever remains of the paper is composed as pursues: In area 2, we depict suppositions of the model. At that point, the depiction of the model is displayed. In area 3, numerical investigation has been done, and different execution measures have been ascertained.

1.1   Description of the Model

Consider a solitary server Markovian line in which the units touch base in clusters as indicated by Poisson process with rate λ. Give a_ξ a chance to be the main request likelihood that 'ξ' units touch base in a group amid a little time interim under the condition that $0 \leq a\_\xi \leq 1$, $0 \leq a_\xi \leq 1$ and $\sum_{\xi=1}^{n} a_\xi = 1$. The arriving units are presented with settled group measure, and the administration time pursues exponential circulation with parameter μ. In this framework, the idea of disaster and the server's machine repair are used to play out the queueing model. These two ideas pursue exponential circulation with parameters separately. It is accepted that the framework at first contains k units when the server goes into the framework and begins to benefit promptly in a bunch of size k. After fruition of an administration, if the server finds not as much as k units in the line, at that point the server goes for a different machine repair of length α. In the event, if there are more than k units in the line, at that point the primary k units will be chosen from the line and administration will be given as a clump. One may dissect the likelihood of fiasco, or, in other words calamity happens in the framework; every one of the units accessible in the framework will be totally annihilated, the framework winds up the void, and the server goes for a different machine repair. On the off chance that there are not as much as k units in the line on the server's arrival from machine repair, the server instantly leaves for another machine repair until the point that the server at long last discovers k or more units in the line.

1.2 Formulation of Basic Equations

Consider a Markov process $\{N(t), C(t)\}$. Here, $N(t) = n$ is the number of units in the queue at time t, and $C(t) = 1, 2$ represents the status of the server mentioned as busy and on machine repair, respectively, at time t.

$\frac{dP_{n,1}(t)}{dt}$ is the probability that there are n units in the queue at time t when the server is busy.

$\frac{dP_{n,2}(t)}{dt}$ is the probability that there are n units in the queue at time t when the server is on machine repair. Based on the above assumptions, the differential-difference equations under transient conditions are framed.

$$\frac{dP'_{0,1}(t)}{dt} = -(\lambda + v + \mu)P_{0,1}(t) + \mu P_{k,1}(t) + \alpha P_{k,2}(t) \tag{1}$$

$$\frac{dP'_{n,1}(t)}{dt} = -(\lambda + v + \mu)P_{n,1}(t) + \lambda \sum_{\xi=1}^{n} a_\xi P_{n-\xi,1}(t) + \mu P_{n+k,1}(t)$$
$$+ \alpha P_{n+k,2}(t) \quad \text{for } n = 1, 2, 3 \tag{2}$$

$$\frac{dP'_{0,2}(t)}{dt} = -(\lambda + v)P_{0,2}(t) + \mu P_{0,1}(t) + v \tag{3}$$

$$\frac{dP'_{n,2}(t)}{dt} = -(\lambda + \nu)P_{n,2}(t)$$

$$+ \lambda \sum_{\xi=1}^{n} a_\xi P_{n-\xi,2}(t) + \mu P_{n,1}(t) \quad \text{for } n = 1, 2, 3, \ldots, k-1 \tag{4}$$

$$\frac{dP'_{n,2}(t)}{dt} = -(\lambda + \nu + \alpha)P_{n,2}(t) + \lambda \sum_{\xi=1}^{n} a_\xi P_{n-\xi,2}(t) \quad \text{for } n \geq k \tag{5}$$

The transient state, Eqs. (1)–(5), is transformed into steady-state equations by assuming

$$\lim_{t \to \infty} P_{n,i}(t) = P_{n,i} \text{ and } \lim_{t \to \infty} P'_{n,i}(t) \text{ and is got as follows:}$$
$$(\lambda + \nu + \mu)P_{0,1} = \mu P_{k,1}(t) + \alpha P_{k,2}(t) \quad \text{for } n = 0 \tag{6}$$

$$(\lambda + \nu + \mu)P_{n,1} = \lambda \sum_{\xi=1}^{n} a_\xi P_{n-\xi,1} + \mu P_{n+k,1} + \alpha P_{n+k,2} \quad \text{for } n = 1, 2, 3, \ldots \tag{7}$$

$$(\lambda + \nu)P_{0,2} = \mu P_{0,1} + \nu \tag{8}$$

$$(\lambda + \nu)P_{n,2} = \lambda \sum_{\xi=1}^{n} a_\xi P_{n-\xi,2} + \mu P_{n,1} \quad \text{for } n = 1, 2, 3, \ldots, k-1 \tag{9}$$

$$(\lambda + \nu + \alpha)P_{n,2} = \lambda \sum_{\xi=1}^{n} a_\xi P_{n-\xi,2} \quad \text{for } n \geq k \tag{10}$$

1.3 Probability Generating Function Method

Now, define probability generating functions when the server is busy and on machine repair. The respective functions are denoted as $G(z)$ and $H(z)$. Let

$$G(z) = \sum_{m=0}^{\infty} P_{m,1}z^m \text{ and } H(z) = \sum_{m=0}^{\infty} P_{m,2}z^m \tag{11}$$

Multiply the Eqs. (6) and (7) by the proper powers of z and use the probability generating functions given in (1) which gives

$$\left[(\lambda+v+\mu)z^k - \lambda A(z)z^k - \mu\right]G(z) = \alpha H(z) - \mu\sum_{m=0}^{k-1}P_{m,1}z^m - \alpha\sum_{m=0}^{k-1}P_{m,2}z^m \quad (12)$$

Similarly, follow the same process in the Eqs. (8), (9), (10) and (1) which provide

$$[\lambda+\alpha+v - \lambda A(z)]H(z) = \alpha\sum_{m=0}^{k-1}P_{m,2}z^m + \mu\sum_{m=0}^{k-1}P_{m,1}z^m + v \quad (13)$$

Equation (13) is rewritten as

$$H(z) = \frac{B(z)+v}{\lambda+\alpha+v - \lambda A(z)} \quad (14)$$

where $B(z) = \alpha\sum_{m=0}^{k-1}P_{m,2}z^m + \mu\sum_{m=0}^{k-1}P_{m,1}z^m$.

In this stage, substitute the expression (14) in Eq. (12) and get

$$G(z) = \frac{(\lambda+v - \lambda A(z))B(z) - \alpha v}{\{\lambda+\alpha+v - \lambda A(z)\}\{\lambda A(z)z^k + \mu - (\lambda+v+\mu)z^k\}} \quad (15)$$

The expressions (14) and (15) are called the probability generating functions when the server is on machine repair and the server is busy, respectively.

1.4 Factorial Moments

On differentiating $G(z)$ and $H(z)$ and letting $z \to 1$, the expressions become an in determinant form. Hence, apply L'Hospital's rule and get the required results as follows:

$$G'(1) = \frac{k(\alpha+v)(v+\mu)B(1) + \lambda v A'(1)B(1) - v(\alpha+v)B'(1)}{v(\alpha+v)^2} \quad (16)$$

This is the mean queue size (first factorial moment) when the server is busy.

$$H'(1) = \frac{(\alpha+v)B'(1) + \lambda A'(1)\{B(1) - v\}}{(\alpha+v)^2} \quad (17)$$

This is the mean queue size when the server is on machine repair. Similarly, the second factorial moments are obtained as

$$G''(1) = \frac{M}{N} \tag{18}$$

where $M = v^2(\alpha+v)^2 \left[v(\alpha+v)\{2\lambda A'(1)B'(1) - vB''(1) + \lambda B(1)A''(1)\} + \lambda v\xi_2 A' \right.$
$(1) + (\alpha+v)\xi_1\xi_2 + (\alpha v - vB(1))[-\lambda\xi_1 A'(1) + \lambda vA''(1) + (\alpha+v)\{\lambda(k(k-1)l+2$
$kA'(1) + A''(1)) - k(k-1)(\lambda+v+\mu)\} \quad -\lambda\xi_1 A'(1)] + \{\lambda vA'(1) + (\alpha+v)\xi_1\}\xi_2]$
$+ 2(\alpha+v)v^2 \{\lambda vA'(1) + (\alpha+v)\xi_1\}[-(\alpha+v)\xi_2 + (\alpha - B(1)\{vA'(1) + (\alpha+v)\xi_1\}].$
And

$$N = v^4(\alpha+v)^4$$

where $\xi_1 = \lambda A'(1) - (v+\mu)k$

$$\xi_2 = \lambda B'(1) - \lambda B(1)A'(1)$$

$$H''(1) = (\alpha+v)^{-2}[(\alpha+v)B''(1) + \lambda A''(1)(B(1)+v)] \\ + 2\lambda(\alpha+v)^{-3}A'(1)\left\{ (\alpha+v)B'(1) + \lambda A'(1)(B(1)+v) \right\} \tag{19}$$

1.5 Numerical Illustration

The numerical values are computed for the mean number of units available in the queue when the server is busy and that the server is on machine repair. For this purpose, the expressions (16) and (17) are used after fixing the unknowns such as

$\alpha = 0.1$, $v = 0.2$, $k = 10$, $\mu = 5$, $B(1) = 0.04$, $B'(1) = 0.02$, $A'(1) = 0.2$.

The required mean number of units is computed corresponding to the different values of λ and presented in Table 1. According to [5], the table shows that the mean number of units increases as λ increases for both cases under which the server is busy as well as the server is on machine repair (Fig. 1).

2 Conclusion

The idea of numerous machine repair and fiasco is utilized. The mean number of units in the queue and factorial minutes when the server is occupied and that on machine repair are determined, and numerical outcomes demonstrated that the quantity of units in the line increments when the landing rate increments under various places of the server. The determined articulations are exceptionally valuable to talk about some uncommon cases.

Table 1 Mean queue size

λ	Mean number of units in the queue	
	Server is busy	Server is on machine repair
1	34.6888	0.4667
2	34.7778	0.8667
3	34.8667	1.2667
4	34.9556	1.6667
5	35.0444	2.0667
6	35.1333	2.4667
7	35.2222	2.8666
8	35.3111	3.2666
9	35.4000	3.6666
10	35.4888	4.0666

Fig. 1 Mean queue size estimate

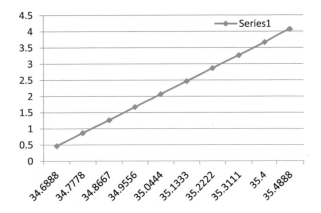

References

1. Ancker Jr., C.J., Gafarian, A.V.: Some queuing problems with balking and reneging, I. Oper. Res. **11**(1), 88–100 (1963)
2. Ancker, C., Gafarian, A.: Queueing with impatient customers who leave at random. J. Ind. Eng. **XIII**(84–90), 171–172 (1962)
3. Chaudhry, M.L., Templeton, J.G.C.: A First Course in Bulk Queues. Wiley and Sons, New York (1983)
4. Choudhury, G., Madan, K.C.: A two phases batch arrival queueing system with a vacation time under Bernoulli schedule. Appl. Math. Comput. **149**, 337–349 (2004)
5. Choudhury, G., Madan, K.C.: A two-stage batch arrival queueing system with a modified Bernoulli schedule vacation under N-policy. Math. Comput. Model. **42**, 71–85 (2005)
6. Choudhury, G., Paul, M.: A two phase queueing system with Bernoulli vacation schedule under multiple vacation policy. Stat. Methodol. **3**, 174–185 (2006)
7. Choudhury, G., Tadj, L., Paul, M.: Steady-state analysis of an M/G/1 queue with two phases of serviceand Bernoulli vacation schedule under multiple vacation policy. Appl. Math. Model. **31**, 1079–1091 (2007)

8. Choudhury, G., Deka, M.: A single server queueing system with two phases of service subject to server breakdown and Bernoulli vacation. Appl. Math. Model. **36**, 6050–6060 (2012)
9. Fiems, D., Maertens, T., Bruneel, H.: Queueing systems with different types of server interruptions. Eur. J. Oper. Res. **188**(3), 838–845 (2008)
10. Hassin, R., Haviv, M.: Equilibrium strategies for the queue with impatient customers. J. Oper. Res. **34**, 191–198 (2006)
11. Jain, M.: Transient analysis of machining systems with service interruption, mixed standbys and priority. Int. J. Math. Oper. Res. **5**(5), 604–625 (2013)
12. Takagi, H.: Queueing Analysis—A Foundation of Performance Evaluation, vol. I. Elsevier, Amsterdam (1991)

Utilization of Blockchain Technology to Overthrow the Challenges in Healthcare Industry

K. Hemalatha, K. Hema and V. Deepika

Abstract Information is exposing to cyber threats if a site's encryption is weak in distributed systems. In this situation, information hacking becomes effortless. To overcome this, information should be packed into blocks. Blockchain technology is such a new enabling technology which links the blocks of information to form a chain with other blocks of similar information. A blockchain is a decentralized computation and information sharing platform which enables multiple authoritative domains, which do not trust each other, to cooperate, coordinate and collaborate in a rational decision-making process. Information on a blockchain existed as a shared and continuously reconciled database. The blockchain database do not store in single location. No centralized version of the information exists which becomes difficult to hack. It allows the digital information to be shared in a distributed environment but not to copy in a single location. It created a new backbone of a new type of Internet. Today, blockchain is finding applications in every field like finance, healthcare, economics and legal, etc. Blockchain provides benefits such as reimbursement of healthcare services, exchange of health data, clinical trials and supply chains. Blockchain has the ability to use the increasing amounts of available health data from traditional sources, electronic health records and other digital health applications as well as non-traditional sources. However, there are certain challenges that are preventing the large-scale deployment of this technology in healthcare. Blockchain has the potential to overcome challenges that exist in the healthcare industry such as interoperability, security, cost of maintenance, integrity, traceability and universal access. In this study, few challenges faced by the healthcare industry are reviewed, and the potentiality of blockchain to overcome the challenges is also studied.

K. Hemalatha (✉) · K. Hema · V. Deepika
Sri Venkateswara College of Engineering (SVCE), Tirupati, Andhra Pradesh, India
e-mail: hemalathakulala@gmail.com

K. Hema
e-mail: goldenhema@gmail.com

V. Deepika
e-mail: deepika.v1@svcolleges.edu.in

© Springer Nature Singapore Pte Ltd. 2020   199
P. Venkata Krishna and M. S. Obaidat (eds.), *Emerging Research in Data Engineering Systems and Computer Communications*, Advances in Intelligent Systems and Computing 1054, https://doi.org/10.1007/978-981-15-0135-7_18

Keywords Blockchain · Healthcare · Interoperability · Security · Cost of
maintenance · Traceability · Universal access

1 Introduction

The term "blockchain" refers to the way of storing transaction data in "blocks" that
are linked together to form a "chain". The chain increases as the number of
transactions increases as every entry is stored as a block on a chain. In a blockchain
system, there is no central authority, and all the cells are linked to one another.
Transaction records are stored and distributed across all network participants for
visible also. The decentralized, open and cryptographic nature of blockchain allows
storing and sharing data in a highly secure manner [1].

The consensus mechanism in blockchain promises a common, unambiguous
order of transactions in blocks maintains the integrity and consistency of blockchain
across the distributed nodes [2]. Therefore, any alteration in the transaction data
should be reflected in all the nodes of the network which is practically impossible.
This leads to tamper-proof characteristic of blockchain.

Beyond cryptocurrencies, blockchain has a wide range of applications. Among
them, blockchain provides potential benefits for healthcare field. Healthcare is a
significant domain which deals with very large collection of data. Therefore, the
healthcare field is facing challenges such as data security, integrity and privacy [3].
These challenges can be solved using blockchain with its strong features such as
data integrity and distributed data storage.

2 Blockchain

Blockchain is related to the distributed technologies family which can maintain the
transactions record and share data between multiple servers in a decentralized
manner [4]. In the initial stages, blockchain is used only for online financial
transactions of the digital currency Bitcoin widely. Later by understanding the
potential of blockchain technology, they are deployed in diverse areas such as
electronic voting, smart system and certificate authorities [5].

Blockchain technology uses distributed ledger to process the shared data among
multiple servers or organizations. Distributed ledger is a type of database that is
stored in different locations [6]. Each transaction shared among the network will be
verified according to the consensus mechanism. The recorded transactions cannot
be deleted, but they can be recreated at any time.

2.1 Blockchain Working

In blockchain, transactions are grouped into blocks, and these blocks are linked with previous blocks. The chain of blocks will be sorted sequentially. The validity of the transaction will be verified by the process called mining [7]. In a typical architecture of a blockchain, every individual node in the network contains local copy of global data of the network. If any node wants to enter some information to the blockchain, it will get updated to the local copy of blockchain of every node. So, that blockchain architecture maintains a strong consistency among nodes in the network. The local copy of information is called as public ledger which is an important term in blockchain technology. Public ledger works like a database and stores the historical information that can be used for future computations.

The data stored in blockchain is made secure using the concept cryptography. Each block may consist of multiple transactions which are in encrypted form that will be verified by the peers. The detailed structure of a block is represented in Fig. 1.

A block is a container data structure that consists of a series of transactions. Average size of a block is around 1 MB. The size of the block may grow up to 8 megabyte which can help to process a large number of transactions. As per the context of Bitcoin, the structure of block consists of two components such as block header and list of transactions. Block header consists of metadata about block such as previous block hash, mining statistics and Merkle tree root. Previous block hash is used to create the new block hash as represented in Fig. 1 [8]. If all the transactions in a block are arranged in a tree form, then it is called as Merkle tree. In Merkle tree, every leaf node consists of the hash of the transactions, and the intermediate nodes consist of the combined hash values. The structure of Merkle tree is presented in Fig. 2.

At the root of the Merkle, tree contains the combined hash of the right tree and left tree. If anyone wants to change the second transaction, the hash will change and

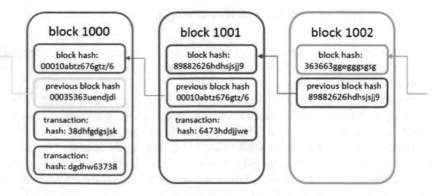

Fig. 1 Structure of a block [8]

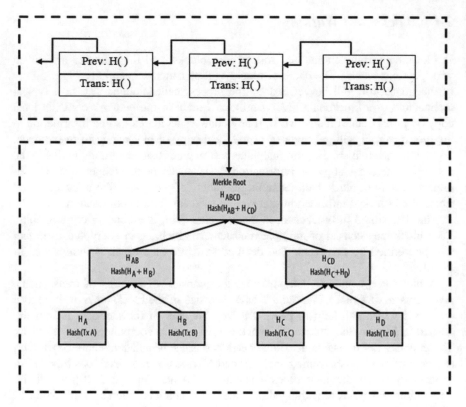

Fig. 2 Merkle tree in blockchain

at the same time the root hash will also change. If the root hash changes, the hash of the linked blocks will also change. This leads to tamper-proof architecture [9].

2.2 Features of Blockchain Technology

Transparency: Blockchain is a distributed ledger in which all the participants in the network share the same information. Hence, transactions become more transparent.

Security: The transactions in blockchain are encrypted and linked with the previous transaction. Therefore, it becomes difficult to hack so that the fraud and unauthorized access of critical information can be prevented [10].

Immutable: The term immutable refers to something which cannot be modified or deleted. Creating immutable ledgers is one of the significant features of blockchain technology. In blockchain, if the log of transactions created by the consensus among the participants is verified, then it cannot be replaced or edited [11].

Decentralized system: In the decentralized system of blockchain technology, few coordinators will be there instead of single coordinator. The few coordinators cooperate with each connected node. If particular node or multiple nodes failed at that time, the coordinators can connect to other individual node and share information. Hence, no damage will occur to the data even though failures occur in the network [12].

Because of the above-mentioned features, blockchain technology is receiving ever-increasing attention and it is adopted in different application domains. Among those applications in this study, we considered healthcare domain.

3 Healthcare Industry and Challenges

Healthcare industry is one of the significant domains which deal with huge data. In healthcare, large amount of data will be created when patient has undergone for various tests such computerized tomography and MRI scans. The test reports should be distributed among radiographers, physician and doctor for the diagnosis of illness. The same data may be accessed by another physician in the process of patient treatment. For effective treatment, the patient's complete information is necessary. Incomplete information may lead to repeated tests and may affect the treatment also. Hence, in the healthcare industry patient, electronic health record (EHR) management plays a significant role [3].

Centralized health record system is necessary to maintain the patient's complete health records which are very sensitive. In the current situations, patient's data is stored across multiple organizations. The sharing and accessing of patient's health records among multiple organizations without cyberattack are a crucial task [13]. Patient health record management consists of several issues. Some of those are presented in this section.

3.1 Interoperability

As per the healthcare industry, interoperability is the exchange, usage and communication of health records among different software applications and information technology systems. Efficient interoperability systems provide several benefits such as improved operational efficiency reduces the overall health system cost, reduces duplication. The ultimate goal of interoperability in healthcare is cost-effective and complete clinical care for the patients [14].

An individual patient's health data is available among numerous systems, and no institution contains complete control over the individual patient data. Significant collaboration, data sharing agreement and goverance are some of the major issues while exchanging the information [15]. In addition to the above-mentioned issues, interoperability deals with some technical barriers.

3.2 Security

Patient's health record consists of personal information and sensitive medical data that may attract cybercriminals. Cybercriminals want to get benefit by selling the theft patient's data to the third party. The data should be protected not only from cybercriminals but also from unauthorized access. The leakage of data or editing is legally liable and for that the health organizations should pay penalty. Hence, security is another important challenge of healthcare industry [16].

3.3 Cost of Maintenance

Patient health record maintenance system involves various operations such as data storage, retrieval and recovery. The maintenance cost plays an important role in above-mentioned operators. All the medical data transactions of the patient have associated with costs such as for verification, security, privacy, institutional agreement and interoperability. High maintenance cost may affect the clinical results and treatment of a patient. Therefore, cost-effective patient health records' maintenance system is necessary for healthcare industry [17].

4 Blockchain—An Enabler of Healthcare Challenges

Healthcare is one of the significant applications of blockchain in which the healthcare industry can get more benefits by adopting the blockchain technology [18]. Blockchain technology helps to reduce the most of the challenges faced by the healthcare industry. The components of blockchain enable efficient interoperability and can handle larger volumes of data effectively. As it consists of potential architecture with fault tolerance, data encryption and disaster recovery, it can handle the challenges faced by the healthcare field efficiently [19].

4.1 Blockchain Towards Interoperability

Interoperability issue in Healthcare becomes more patient-centric, and blockchain facilitates the data exchange and provides great control over their data. To overcome the interoperability issue, blockchain enables a centralized distributed mechanism for the management of authentication.

Gordon et al. [20] suggested a method for the interoperability by establishing an API connection to the systems from which the data should be used so that the patients can establish connections with that API and can collect the data properly.

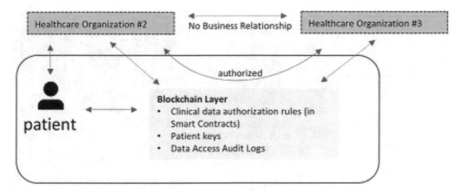

Fig. 3 Blockchain framework for the interoperability of patient's data [20]

They suggested another method using patient identity. As per that method, if two systems want to exchange the patient's data, the patient identity should be resolved initially. To verify the patient identity, an individual public key should be assigned using blockchain's public key infrastructure (PKI). That public key can be used to link the patient's data across institutions for the matching of patient's data. The diagrammatical representation of the methods is suggested for interoperability in Fig. 3.

4.2 Blockchain-Based Models for Security

Esposito et al. [3] described the conceptual blockchain-based EMR/EHR/PHR ecosystem. As per the proposed ecosystem when a new patient data is created, a new block will be created and distributed to the peers in the network. Once the majority of the patients in the network approved the created block, the block will be inserted in chain. If the block is not approved, the created block will be defined as an orphan that does not include in the main chain. The proposed design provides security by the decentralized consensus and consistency and it is represented in Fig. 4.

Linn et al. suggested a blockchain model for healthcare that include solutions for security and privacy [21].

As per the model suggested in Fig. 5, a mobile dashboard application is used to assign a set of access permissions to the users for the access. Once the healthcare provider granted the access, they can utilize the user's health data. Biometric identity systems are suggested for enhanced security over password and token-based methods. This model provides a singular control over the user's individual data and the power of granting access to specific health providers and entities for the disease diagnosis.

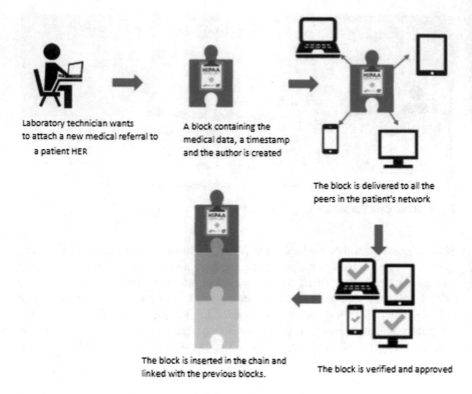

Fig. 4 Conceptual blockchain-based EMR/EHR/PHR ecosystem [3]

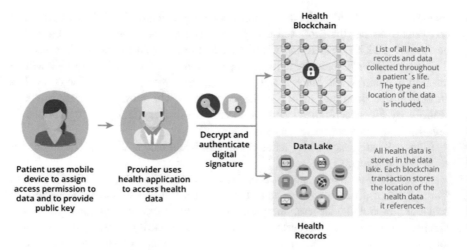

Fig. 5 Blockchain model for healthcare [21]

4.3 Prediction of Blockchain-Based System's Maintenance Cost

Even though the blockchain technologies perform quick and real-time transactions, prediction of such systems cost is still a challenge. Resources for continuous troubleshoot, parameter's updation, backup and recovery are required for the maintenance of the blockchain-based systems. Blockchain distributed nature, open-source technology and various properties can reduce the above-mentioned resource's cost significantly. However, the transparent natures of blockchain lead to more data exchange transactions and integration which is a time-consuming activity [22].

Significant computing process is necessary for blockchain-based applications. The cost of computing power is based on the volume and size of the transactions involved in that application. However, it is difficult to predict the cost of operating a blockchain at enterprise level. Therefore, to predict the cost of a fully scaled system based on blockchain technology customer need, target experiments and common guidelines are necessary.

5 Conclusions

Blockchain technology is a disruptive technology that has huge potential. The adoption of blockchain technology will impact the way of business which is done. Healthcare industry is one of the significant domains that deal with sensitive and huge volume of medical data. The challenges of healthcare such as interoperability, security and cost of maintenance are studied and the blockchain-based models proposed and suggested by the researchers to overcome the above-mentioned challenges which are also reviewed in this study. Various frameworks and methodologies are developed using blockchain technology to provide solutions to the challenges, interoperability and security. However, work related to the issue maintenance cost founded is less. Less effort has been taken for this issue when compared with remaining issues. Still, it is remained as a challenge for researcher to reduce the cost of blockchain-based applications.

References

1. https://medium.com/cossette-health/blockchain-in-health-care-bbc543fb58e. Retrieved on 28 Dec 2018
2. Wang, S., et al.: Blockchain-powered parallel healthcare systems based on the ACP approach. IEEE Trans. Comput. Soc. Syst. 5(4), 942–950 (2018)
3. Esposito, C., De Santis, A., Tortora, G., Chang, H., Choo, K.R.: Blockchain: a panacea for healthcare cloud-based data security and privacy? IEEE Cloud Comput. 5(1), 31–37 (2018)

4. Aste, T., Tasca, P., Di Matteo, T.: Blockchain technologies: the foreseeable impact on society and industry. Computer **50**(9), 18–28 (2017)
5. Henry, R., Herzberg, A., Kate, A.: Blockchain access privacy: challenges and directions. IEEE Secur. Priv. **16**(4), 38–45 (2018)
6. Fiaidhi, J., Mohammed, S., Mohammed, S.: EDI with blockchain as an enabler for extreme automation. IT Prof. **20**(4), 66–72 (2018)
7. Gatteschi, V., Lamberti, F., Demartini, C., Pranteda, C., Santamaría, V.: To blockchain or not to blockchain: that is the question. IT Prof. **20**(2), 62–74 (2018)
8. https://ico.conda.online/the-crypto-guide-for-beginners-%E2%80%93-what-is-blockchain/. Retrieved on 29 Dec 2018
9. https://nptel.ac.in/courses/106105184/3/lec3.pdf. Retrieved on 30 Dec 2018
10. https://www.ibm.com/blogs/blockchain/2018/02/top-five-blockchain-benefits-transforming-your-industry/. Retrieved on 30 Dec 2018
11. https://www.quora.com/What-makes-a-blockchain-network-immutable. Retrieved on 31 Dec 2018
12. https://nptel.ac.in/courses/106105184/3/lec1.pdf. Retrieved on 31 Dec 2018
13. Hölbl, M., Kompara, M., Kamišalić, A., NemecZlatolas, L.: A systematic review of the use of blockchain in healthcare. Symmetry **10**, 470 (2018). https://doi.org/10.3390/sym10100470
14. Zhou, Y., Ancker, J.S., Upadhye, M., McGeorge, N.M., Guarrera, T.K., Hegde, S., et al.: The impact of interoperability of electronic health records on ambulatory physician practices: a discrete-event simulation study. Inform Prim Care (2013)
15. Downing, N.L., Adler-Milstein, J., Palma, J.P., Lane, S., Eisenberg, M., Sharp, C., et al.: Health information exchange policies of 11 diverse health systems and the associated impact on volume of exchange. J. Am. Med. Inform. Assoc. (2017)
16. Zhang, J., Xue, N., Huang, X.: A secure system for pervasive social networkbased healthcare. IEEE Access **4**, 9239–9250 (2016)
17. Just, B.H., Marc, D., Munns, M., Sandcfer, R.: Why patient matching is a challenge: research on master patient index (MPI) data discrepancies in key identifying fields. Perspect. Health Inf. Manage. 1e, 1324:113–22 (2016)
18. European Coordination Committee of the Radiological: Blockchain in Healthcare; Technical report; European Coordination Committee of the Radiological: Brussels, Belgium (2017)
19. Patient-Centered Health on the Blockchain with Chelse Barabas. Precision Medicine Initiative Cohort Program. Retrieved 7 2016, from National Institutes of Health. https://www.nih.gov/precision-medicine-initiative-cohort-program
20. Gordon, W.J., Catalini, C.: Blockchain technology for healthcare: facilitating the transition to patient-driven interoperability. Comput. Struct. Biotechnol. J. **16**, 224–230 (2018)
21. https://www.healthit.gov/sites/default/files/11-74-blockchainforhealthcare.pdf. Retrieved on 2 Jan 2019
22. https://www2.deloitte.com/content/dam/Deloitte/us/Documents/public-sector/us-blockchain-opportunities-for-health-care.pdf

Monitor and Abolish the Wildfire Using Internet of Things and Cloud Computing

P. A. Ramesh, Y. A. Siva Prasad and G. Chidananda

Abstract The term IoT: Internet of Things paradigm provides an environment which facilitates the implementation of new technology in all social environments. Forests are one of the important aspects in our social life, which provides vegetation, place to wild animals, and ecosystem services that are critical to human welfare. But every year, they are attacked by wildfire; due to this, we lose huge vegetation, animal life and create air pollution. In this paper, we proposed an IoT-based model to provide a solution to the wildfire. The paper describes an approach to detect and reduce the effect of wildfire. We place smoke and temperature sensor poles in specific locations in the forest and allow the collection of data from these sensors continuously to the cloud. In the cloud, we define API to analyze the data and if any suspicious data is found, the API automatically sends necessary orders to drones, which can be deployed in that suspicious location and spread fire suppression gas to reduce the effect of wildfire, and also an instruction sends to Central Monitoring System of Wildfire.

Keywords Internet of Things · Cloud · Sensors · Wildfire · Social environment

1 Introduction

A wildfire or wildland fire is a fire which occurs in incendiary vegetation, especially that occurs in the countryside or outback area and it depends on the type of vegetation where it occurs [1]. Wildfires occur when the necessary fire elements come together in a susceptible area: a fire source is brought into contact with a

P. A. Ramesh (✉) · Y. A. Siva Prasad · G. Chidananda
Sri Venkateswara College of Engineering, Tirupati, Andhra Pradesh, India
e-mail: Ramesh.p@svcolleges.edu.in

Y. A. Siva Prasad
e-mail: sivaprasady@gmail.com

G. Chidananda
e-mail: chidananda.g1@svcolleges.edu.in

© Springer Nature Singapore Pte Ltd. 2020
P. Venkata Krishna and M. S. Obaidat (eds.), *Emerging Research in Data Engineering Systems and Computer Communications*, Advances in Intelligent Systems and Computing 1054, https://doi.org/10.1007/978-981-15-0135-7_19

flammable material such as vegetation that leads to sufficient heat and sufficient amount of oxygen from the ambient air. Wildfire spread with a high forward rate of spread (FROS), when burning through impenetrable forest and contain flammable elements; they can move as fast as 10.8 km per hour in forests and 22 km per hour in grasslands, which take away houses, animals lose their lives and also affect air pollution [2]. For this, we proposed an IoT for wildfire-based paper which reduces the effect of wildfire; we deploy the poles in different locations in a forest. The poles which contain smoke and heat sensors and by using IoT technology we read the data continuously by through sensors. If fire accident occurs, these sensors find that data and send to central monitor location; we also place the drones in surrounding locations of the forest after reading the wildfire data; we force the drones to spread carbon dioxide gas on that incident location. Our aim is to insure these drones have to do action automatically when wildfire occurs so that we can reduce the effect of wildfire before fire department takes action.

There are different articles that have been published on smoke detection; we identify the smoke by analyzing the thickness of smoke dark channel strength using motion history image [3]. The researchers improved the smoke detection sensitivity using homogenous semiconductor sensors [4]. Reference [5] designs a new type of smoke sensor.

Internet of Things is a technology, which allows the communication between livable and non-livable objects through internet. IoT can be explained as the framework used to collect information from the perceive devices such as sensors and RFID devices. After collecting the information, it forwards through network to cloud. Here, it analyzes the data and sends instructions to application layer. The IoT devices interact through internet and these objects can be operated and monitored by humans or other objects [6]. IoT is providing smart facilities in all aspects of society lifestyle, such as health care, smart cities, agriculture, saving energy, home automation, intelligent traffic system, and more [7].

Cloud is one of the layers in IoT technology. Cloud computing can be defined as a computing environment where computing needs of one party can be outsourced to another party and when the computing power or resources like database or emails are needed, they can be accessed them via internet [8]. There are three important services that are provided by cloud environment, i.e., (1) Platform as service, (2) Infrastructure as service, and (3) Software as a service. From cloud computing, we store the data and process that data for analyses and decision making. The cloud acts as middleware layer between perceive layer and application layer.

The rest of this paper is organized as follows: Sect. 2, provides the literature review on the technologies which are needed to IoT for wildfire including drones. Section 3 details the architecture of IoT for wildfire and also gives a case study which explains in detail. Section 4 illustrates obstacles we face in implementation of IoT for wildfire in real time, and Sect. 5 concludes the paper.

2 Literature Review

In present days, technology provides huge facilities which are safe, secure, and reliable. One of the important characteristics of fire is that it produces much smoke. The smoke detector detects smoke and toxic gasses when a fire occurred in a location. The other parameters of wildfire are temperature and humidity. We use an integrated temperature, humidity sensor with MQ2 smoke sensor to collect data, and have GPRS module to transfer data to cloud [9]. The MQ2 is a metal oxide semiconductor type gas sensor, which detects the smoke that changes in resistance of it's sensing material, and having features like high sensitivity for smoke, low cost and long service life with high stability. We use microprocessor CC2530 to control the sensor signals and convert analog to digital. We found a sensor module VT460 [10], which has characters like smoke sensitivity 0.5–2 db/m, measured temperature range −10 to 85 dc, and humidity range 0–95%.

The drones have transformed from being playing toys of children to a full-on cultural phenomenon. The Internet of Drones (IoD) is a layered network control architecture designed mainly for coordinating the access of unmanned aerial vehicles to controlled airspace and providing navigation services between locations referred to as nodes [11]. The drones provide services like package delivery, traffic surveillance, and also us in reuse operations. In present days, we have auto-piloted, directed, and monitored drones are available in market [12].

The architecture ETSI's M2M is a standard IoT architecture that has many advantages in terms of components, classification, and layers [13]. This architecture consists of two layers. First one is gateway that allows us to gather information from the perceive devices and the second layer is the network management layer, which includes retrieving information securely, routing, store, and analyze the information.

3 IoT Architecture for Wildfire

3.1 Case Study

The wildfire ignites in a natural area such as a forest, grassland, or prairie. It is often caused by humans or unfortunate lightning. They spread very quickly and increases the risk within periods of little rain and high winds. In this section, we provide a case study to illustrate how the IoT technology helps to reduce the effect of wildfire. First, we will place the smoke sensors at an elevation using cement poles randomly in a forest and place the CO gas equipped drones in the surroundings of the forest. The sensors and drones use solar power for energy (Fig. 1).

All these sensors and drones equipped with GPRS module to transfer and receive the data to cloud. Whenever an unplanned fire exists in a forest, the smoke sensors catch the signals and pass the data to cloud through GPRS module through Zigbee network.

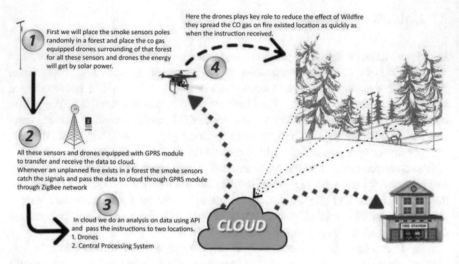

Fig. 1 A prototype IoT wildfire model

In cloud, we do an analysis on data using API and pass the instructions to two locations; one is to the drones. Here, the drones play a key role to reduce the effect of wildfire, they spread the CO gas on location as quickly as when the instructions are received. And second instruction passes to fire department. From this, we reduce the effect on wildfire so that we can able to save as much as wildlife and grassland in a forest.

There are lots of research articles that have been published on detecting smoke and fire exist. Using IoT, we integrate wireless smoke detection system to cloud environment for better results. For this, we propose an IoT architecture for wildfire which gives a smart solution to the wildfires.

3.2 IoT for Wildfire Architecture

IoT is the technology that provides facilities for us to do our work conveniently. Through internet, we can be able to communicate with huge number of devices, connect which some are connected devices and some are perceived. All these devices have their own IPv6 capabilities as well as the ability to measure data. The IoT for wildfire architecture to support the case study is shown in Fig. 2.

In the above architecture, we see four layers

1. Perceive Layer
2. Network Layer
3. Cloud Environment
4. Application layer.

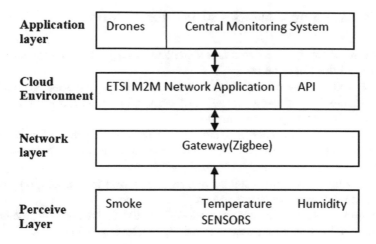

Application layer	Drones	Central Monitoring System	
Cloud Environment	ETSI M2M Network Application		API
Network layer	Gateway(Zigbee)		
Perceive Layer	Smoke	Temperature SENSORS	Humidity

Fig. 2 Architecture of IoT for wildfire

1. Perceive Layer

In perceive layer, we have a set of smoke sensors, which collect the data continuously. Whenever an unfortunate wildfire occurs, the sensors collect smoke and temperature data. The GPRS module that passes data to the cloud through Zigbee network.

2. Second Layer

The gateway of the network is the second layer in IoT architecture; The Zigbee-based ad hoc sensor network module allows a gateway to the sensor data and cloud. We use data clusters, which are responsible for data aggregation and data transmission to router.

3. Cloud Environment

Through internet, the collected data from sensors reach the cloud environment. Here, the cloud services analyze the data and take the necessary actions based on data. We implement predictive analytics in data mining techniques on perceived data and pass the instructions to drones and central monitoring system.

4. Application Layer

After analyzing data if any fire symptoms are found, then automatically the Software API passes instructions to application layer. Here, we have two different applications that are IoT for controlling wildfire. One is drones, which automatically employ on fire-affected location and spread CO gas, so that we may reduce the effect of wildfire before fire department people arrive. Second is to pass instructions to Central Monitoring System of wildfire, so that they will quickly take the necessary action.

4 Discussion on Obstacles of IoT for Wildfire

1. Wildfire occurs in forest, and it is a challenging work to place the sensors and drones in that deep forest.
2. There will be intercommunication between deployed sensors and maintains interoperability. All these sensors must work with a high level of functionality and consistency to give best results.
3. All these sensors in outer area, since the sensors are located in the forest area, the functioning of the sensors has to be monitored always so that always there will be monitoring on the functionality of them.
4. There will be technical problems in data transferring from nodes to cloud, like nodes to gateway, gateway to cloud.
5. We should always provide maintenance to the technology, equipment like sensors, GPRS modules, drones to the highest level possible.

5 Conclusion

The technology IoT is a heterogeneous technology that allows us enhance the quality of our society through this paper. We proposed IoT for wildfire architecture, which allows us to monitor fire exist in forest, so that if any unfortunate situations occur, we can able to detect and reduce the effect of wildfire; from this, we can save huge grasslands, animals, and environment.

References

1. https://en.wikipedia.org/wiki/Wildfire
2. https://nhmu.utah.edu/sites/default/files/attachments/All%20About%20Wildfires.pdf
3. Miao, L.G., Chen, Y.J., Wang, A.Z.: Video smoke detection algorithm using dark channel priori. In: IEEE Chinese Control Conference Nanjing (2015)
4. Kohl, D., Eberheim, A., Schieberle, P.: Detection mechanism of smoke compounds on homogeneous semiconductor sensor films. Thin Solid Films **490**, 1–6 (2005)
5. Liu, B.J., Alvarezossa, D., Kherani, N.P., Zukotynski, S., Chen, K.P.: Gamma-free smoke and particle detector using tritiated foils. IEEE Sens. J. **7**, 917–918 (2007)
6. Jia, X., Feng, Q., Fan, T., Lei, Q.: RFID technology and its applications in internet of things (IoT). In: IEEE Conference (2012)
7. Internet of Things-IOT: Definition, Characteristics, Architecture, Enabling Technologies, Application & Future Challenges
8. Jadeja, Y., Modi, K.: Cloud computing-concepts, architecture and challenges. In: ICCEET Conference (2012)
9. He, Z., Fang, Y., Sun, N., Liang, X.: Wireless communication-based smoke detection system design for forest fire monitoring. In: IEEE Conference (2016)
10. https://vutlan.atlassian.net/wiki/spaces/DEN/pages/43810826/VT460

11. Gharibi, M., Boutaba, R., Waslander, S.L.: Internet of drones. https://ieeexplore.ieee.org/document/7423671
12. https://www.dronethusiast.com/heavy-lift-drones
13. Ikram, M.A., Alshehri, M.D., Hussain, F.K.: Architecture of an IoT-based system for football supervision (IoT Football). IEEE (2015)

Critical Review on Privacy and Security Issues in Data Mining

Pasupuleti Nagendra Babu and S. Ramakrishna

Abstract In the present day, research of data mining security and privacy issues plays an important role in protecting the data. Data mining is the procedure that derives, categorize, and evaluates the effectiveness and suitability of the data from vast datasets given by numerous sources leading to privacy and security issues. This paper reviews various data mining methods used to identify the privacy, security threats and investigates their usage and limitations. The critical review on problems faced in denial of service, distributed denial of service, malware, botnet, spyware, probing, and ransomware has been thoroughly discussed and analyzed using data mining techniques.

Keywords Data mining · Security · Privacy · Algorithm · Dataset

1 Introduction

The dawn of information processing and retrieval in various fields of science and technology has led the researchers to work on data mining to more extent. The preprocessed data, which is very bulky amount of data, is stored in different formats like archives, forms, pictures, voice recording, videos, technical data, and various new data formats [1–3].

In the present day, communications, which is, rapidly developing with 4G+ capability in transmitting and receiving data, and there is always risk of privacy and security threat in terms of destruction of network, unidentified access to data and data modification. The loopholes in the network cause hackers to break into the network and steal the data. The attackers create proxy switch in the network and can steal money from banks by altering the transactions and customers' personal data.

P. Nagendra Babu (✉)
Department of Computer Science, Rayalaseema University, Kurnool, India
e-mail: pnagendrababu77@gmail.com

S. Ramakrishna
Department of Computer Science, S.V. University, Tirupati, India

© Springer Nature Singapore Pte Ltd. 2020
P. Venkata Krishna and M. S. Obaidat (eds.), *Emerging Research in Data Engineering Systems and Computer Communications*, Advances in Intelligent Systems and Computing 1054, https://doi.org/10.1007/978-981-15-0135-7_21

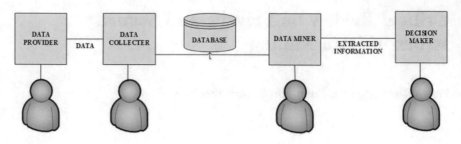

Fig. 1 Flow of data mining

The hackers can send falsified information over network and can destruct the services by various attacks which lead to network breakdown [4–6].

The flow of data mining starts with data provider providing the required data and forwarded to data collector, which is stored in database. Using different data mining algorithms, the required data is pruned and forwarded to decision maker. The resulted data is structured and useful as shown in Fig. 1. For combating the various attacks and securing the data, data mining has developed various algorithms. There are a lot of researchers who worked on data security to solve the problems such as data breach, intruder attacks, malware, adware, DoS attacks, and DDoS attacks. In this paper, we gave a critical review on explaining data mining and its role on security. The various aspects of security attacks and its solutions using data mining algorithms are deeply reviewed.

The continuation of this paper follows with review on data mining techniques and explains their pros and cons in Sect. 2. Section 3 starts with literature review on data mining applications in various types of attacks and critical review on the research done on the said attacks followed by conclusions.

2 Architecture of Data Mining

In data mining, the techniques used for contending security are based on predictive and descriptive mode. In predictive mode, the data is used for forecasting such as classification rules, clustering rules, and link analysis rules, and in descriptive mode, the data is categorized into relevant information such as association rules, clustering, sequence discovery, and summarization. Based on the extent of security provision, the techniques are arranged. These techniques help in providing real-time protection in the network and also provide assistance as shown in Fig. 2 [7–13].

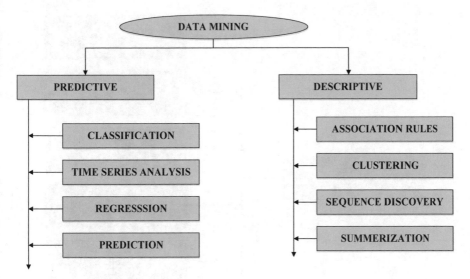

Fig. 2 Data mining techniques

3 Pros and Cons of Data Mining

3.1 Pros

1. The results in data mining provide accurate standards of data in marketing and retail industry.
2. In finance industry, data mining provides data about good and bad loans by checking the history of account holders.
3. In large-scale datasets, it can check and provide required patterns of useful data.

3.2 Cons

1. Privacy and security are big issues in data mining; for example, like stealing the customer records from companies and unauthorized logging into user profiles.

So, certain algorithms have to be developed for mitigating the security attacks in data mining.

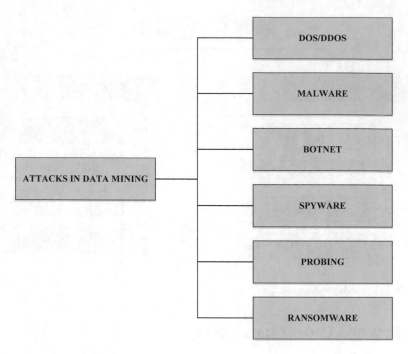

Fig. 3 Data mining attacks

4 Survey on Privacy and Security Issues in Data Mining

According to the survey of privacyrights.org, from 2005 to 2018, nearly 1157 crores of individual records have been compromised of various companies in the USA itself by the hackers [14]. Different companies are losing crores of rupees per year due to various attacks in data mining worldwide. This is due to vulnerabilities in the data mining procedure. We explain different types of attacks and present solutions in data mining to mitigate the attacks. In data mining, the main and popular types of attacks are DoS, DDoS, probing, botnet, malware, spyware, adware, and ransomware as shown in Fig. 3.

4.1 DoS/DDoS

DOS stands for denial of service. It is a type of attack used by hackers to attack the network, thereby stopping the services to the users [15]. DDoS stands for distributed denial of service. It is a type of attack where hackers try to attack the network server from multiple domains, thereby creating havoc to the network resulting in collapse of the network [16].

Zolotukhin et al. [17] have proposed an algorithm using SSL/TLS protocol for detection of denial of service attacks. The protocol will encrypt the dataflow in the network so that no intruder can attack the network. They used several data mining techniques such as k-means, k-nearest neighbor, and SVM algorithm for performance evaluation of DDoS attacks. The simulation results showed that the rate of detection using DBSCAN technique fared better than other techniques. The main drawback of this research is they never described size of the packet header which has impact on security, variation in packet size, and there is a chance of attack in DDOS.

Dongre et al. [18] have proposed a heuristic-based technique for checking the denial of service attack. They used information gain, gain ratio, and correlation aspects to limit the features of the vector. Information gain processes the randomness of each element; if the value is higher then it will be eliminated from dataset. The gain ratio is used for reduction further using number and size of the dataset branches. Correlation uses a dataset attribute on features basis by treating each feature as sign. For experiment setup, they used WEKA tool for calculations. In the results, they state that out of three features ranking such as information gain, gain ratio, and correlation, information gain fared well in the accuracy in detecting the DOS attack. The main drawback of this research is using heuristic-based approach which will give possible solution among all but cannot guarantee the best solution in the process.

Keshri et al. [19] have proposed a technique using firewall and IDS with various data mining techniques like Naïve Bayes, K-means, SVM for DoS detection. They used NSL–KDD dataset for testing; the results show that Naïve Bayes has less performance rate and SVM has better performance rate. The highest accuracy is found by Random Forest and Hybrid approach for DoS detection. The main drawback of this research is that the authors never discussed which ISO layer the results are obtained since each layer has its own impact on the DoS attack

Alkasassbeh et al. [20] have proposed a data mining technique for detecting DDoS attacks. For simulation purpose, they used dataset namely SIDDOS, HTTP flood, and UDP flood. The network traffic is generated by NS2 simulator. The results are compared with algorithms such as multilayer perceptron (MLP), Naïve Bayes, and random forest. According to the results shown, MLP succeeded with top accuracy rate of 98.63% in detection of DDoS attacks. The main drawback of this research is NS2, NS2 has limitations such as high computation time and scalability issues, and already there is newer version NS3 available which is advanced and highly accurate.

Nijim et al. [21] have proposed a framework named fast detect for DDOS detection and prevention. The technique is based on clustering all the incoming request from the attack device and testing it based on the history of resource usage of central processing unit time, memory allocation size, and network bandwidth. The main drawback of this research is there is no real-time or experimental implementation of the proposed framework which poses a question whether it can handle DDOS or not.

4.2 Malware

Malware stands for any type of software code designed to damage the computer which can be a server or a network. It can be introduced in the form of executable code and scripts [22].

Nourouzi et al. [23] have proposed classification technique for detecting malware for data mining. They created a XML file containing malware and converted into non-parse matrix. The application they used processes the non-parse matrix into WEKA input dataset. For training purpose, they used algorithm like Naïve Bayes, Bayes net, J48, and regression algorithms. According to experimental results, regression classification method fared well for classification of malware detection. Authors also proposed that they will try to develop a real behavior antivirus using classification technique. The main drawback of this paper is WEKA tool, as it runs on Java which uses more memory; for large datasets, it is unable to load and takes plenty of time in WEKA Explorer GUI. As WEKA is open source for commercial application, it needs to get license by purchasing from corporate entities which is costly.

Fraley et al. [24] proposed topological feature extraction for polymorphic malware detection for conducting the experiment they obtained samples from ClamAV, Virusshare. They used WEKA tool for validation. According to the results obtained for small set of sample files, the true positive rate is high in detecting the malware files. For larger datasets, the rate of detection is low. The main drawback of this research is that the work is still in preliminary stage due to sample parameters not fully explored like packet data for analysis.

Ahuja and Amandeep [25] reviewed the research on malicious detection technique for web security. According to them, malware constitutes worms, Trojan, rootkits, etc. The detection is based on signature-based technique and behavior-based techniques. The main drawback of this research is that the authors never discussed which technique is highly efficient in detecting the malware.

Edem et al. [26] proposed an automated malware investigation framework to mitigate malware attack. They used clustering technique and malware analysis in their framework. The suspicious files are grouped so that their behavior is observed and further processed for investigation. They used K-means algorithm with feature extraction technique. The main drawback of this paper is that present day malware intrusion can bypass API sandbox which requires more encryption techniques for mitigating malware attacks.

4.3 Botnet

Botnet is a bunch of computers connected in a similar fashion and controlled by a master computer. Their purpose is to send the required information to the master computer as instructed [27].

Bhaiyya et al. [28] proposed a framework for mitigating the botnet attack using K-means, clustering, and machine learning classifier C5.0. According to their framework, they created weblogs using KDD cup for botnets. C5.0 algorithm was used to find the botnets and K-means is used further to check in-depth finding of botnets. The main drawback of this research is authors only proposed the framework and never implemented for experimental purpose.

Haque et al. [29] proposed a data mining framework using meta-analysis for botnet detection. The framework helps to prevent a few features of botnet intruding the user machine. They have used optimized apriori algorithm for dataset analysis. They tested the botnet detection in Linux platform, Cisco IOS, and Windows Server. The main drawback of this research is that they do not describe how the confidence values are generated using association rules in their framework and with what parameters it can check the botnet detection.

Kannan et al. [30] proposed a framework for detecting the botnet. They used clustering technique for detection of botnet. They further used SVM classifier to check the traffic whether the traffic is attack or normal. The main drawback of this research is how the said framework has encryption if the botnet has encryption and attacking the target device.

Bijalwan et al. [31] proposed ensemble of classification algorithm for detecting botnet. They used ISCX dataset for their framework. They divided the datasets into normal traffic and botnet traffic and applied ensemble of classifier algorithm to test the response. The experimental results show 96.4% accuracy. The main drawback of this research is concept drifting of ensemble of classifier algorithm which requires a lot of training to train each individual algorithm and is very complex.

4.4 Spyware

A piece of software intrudes into user device and steals the information like passwords and its functioning from target devices over a network [32].

Border et al. [33] proposed a framework called webtap for detecting spyware. The framework tries to detect the inbound traffic via HTTP connections in firewall. It tries to check the inbound requests of size below 20 kb on daily basis. They used two methods to counter spyware: one is to check whether any site tries to send requests to access user several times. It will monitor further about the requested site and the second method is calculating the ratio of standard deviation of bandwidth with usage of mean bandwidth. The main drawback of this paper is if the hacker site sends a request more than 60 kb, then the proposed mechanism does not work to combat spyware.

Shahzad et al. [34] proposed a technique for combating spyware by mining executable files. They try to extract the binary features from executable files. They extracted features from the binary files and they applied feature reduction method to reduce complexity of dataset and convert it into attribute relation file format and execute it to find the spyware. They used WEKA tool to test the detection of spyware. The main drawback of this paper is if the executable file is encrypted and the proposed framework does not have such mechanism to combat the spyware.

Pandey et al. [35] proposed a framework for spyware detection using data mining. According to them, spyware is defined as any software which collects information from a system without consent of its system admin. For this framework, they collected datasets namely benign files meaning clean files and spyware files from internet sources. From the files, they created binary sources. Using Java programming, they converted byte array to hexadecimal dumps, and further, they convert into n-gram generations. Later, they extract the features by using FBFE frequency-based feature extraction and then feature reduction follows where the features with high frequency are included for training classifier and low frequency are not included. They used Bayesian reasoning for decision making to predict the datasets. The main drawback of this research is Java programming; for larger datasets, it is not possible to load from the GUI as it swaps more memory.

Divya et al. [36] proposed a framework using data mining algorithm such as association, classification, and regression to test the malicious code. In their experiment, they used UNIX testbed for generating hexadecimal dumps of binary files using XXD. They used Naïve Bayes classifier to check whether dataset file is spyware or normal file. The main drawback of this research is not testing the parameters of spyware which has lot of attributes where authors never mentioned about it in detecting the spyware.

4.5   Probing

Probing attack is defined as any intruder tries to enter the target device and monitor the computer and get required information [37].

Malik et al. [38] proposed a framework using multiobjective particle swarm optimization and random forest for probe attack detection. In the framework, they have used a hybrid attribute selection method that has a subset of attributes is selected and processed with random forest classifier for testing of the subset. Further, the random forest classifier gives feedback to particle swarm optimization so that it uses for accurate subsets selection of detection in probing attacks. The main drawback of this research is particle swarm optimization as it requires time-consuming operation for large and complex dataset due to repeated fitness function testing.

Vasilomanolakis et al. [39] proposed a framework named probe response attack framework (PREPARE) for detecting probe attacks in collaborative intrusion detection system. For programming, the framework used Python and C. They used DShield for simulation which detects the probe scanning attacks in the network. The main drawback of this research is using Python which is very slow and very weak language for computing and it is not very good choice for high memory tasks. There are many limitations for accessing database with Python coding.

4.6 Ransomware

Ransomware is defined as any software designed to encrypt the target computer and asks for money for getting decrypted. This is the most dangerous attack present day community is facing [40].

Kharaz et al. [41] proposed a framework named UNVEIL to detect ransomware namely file lockers and screen lockers. Their framework used a technique to control and access the system files of the target device. For the detection they use to take time to time screenshots of the infected computer before and after the ransomware attack using dissimilarity scores. Their framework is built based on Cuckoo sandbox. Out of 148,223 datasets of files, the proposed framework detected 13,647 files as malicious ransomware files. The main drawback of this research is they never discussed encryption and decryption standards of their framework, as ransomware is based on encryption and decryption method.

Homayoun et al. [42] proposed a framework for detecting ransomware using frequent pattern mining. In their framework, they used sequential pattern mining for feature identification for samples collected from various internet sources and find maximal frequent patterns of the datasets. For classification analysis, they used algorithms such as J48, random forest, bagging, and MLP algorithms. From the experimental results from the given samples, they achieved 99% rate of detection for ransomware and 96.5% rate in detecting the family of ransomware samples. The main drawback of this research is that the proposed framework is only applicable for detecting the ransomware to some extent and not mitigating the ransomware attack where encryption and decryption take place.

Chen et al. [43] proposed a technique for validating only WannaCry ransomware attack. According to then, WannaCry ransomware is one of the dangerous attacks which encrypts the victim's computer and asks for money. It is considered as most advanced cyberattack in present day. WannaCry ransomware created havoc worldwide by shutting down UK National Health Service and Japan's Honda motor company. For experiments, they used Cuckoo sandbox for analysis of WannaCry ransomware. They used four methods for validation such as extracting features of the WannaCry ransomware, differentiating the normal file from malicious file, finding the ransomware from huge datasets, a method to create patterns that can bypass 63 antivirus detection. The main drawback of this research is that the proposed framework cannot help in real-time protection if there is a WannaCry ransomware attack on target device.

The critical review on privacy and security issues in data mining is tabulated below for highlights from various researchers as shown in Table 1.

Table 1 Critical review on privacy and security attacks in data mining

S. No.	Attacks	Proposed by Authors and Year	Remarks	Limitations
1	Dos/DDos	Zolotukhin et al., 2015 [17]	Authors proposed an algorithm using SSL/TLS protocol for detection of denial of service attacks	The main drawback of this research is they never described size of the packet header which has impact on security, variation in packet size, and there is a chance of attack in DDOS
		Dongre et al., 2018 [18]	Authors proposed a heuristic = based technique for checking the denial of service attack	The main drawback of this research is using heuristic-based approach which will give possible solution among all but cannot guarantee the best solution in the process
		Keshri et al., 2016 [19]	Authors proposed a technique using firewall and IDS with various data mining techniques like Naïve Bayes, K-means, SVM for DoS detection	The main drawback of this research is that the authors never discussed which ISO layer the results are obtained since each layer has its own impact on the DoS attack
		Nijim et al., 2016 [20]	Authors proposed a data mining technique for detecting DDoS attacks. For simulation purpose, they used dataset namely SIDDOS, HTTP flood, UDP flood	The main drawback of this research is NS2, NS2 has limitations such as high computation time and scalability issues, and already there is newer version NS3 available which is advanced and highly accurate
		Nijim et al., 2017 [21]	Authors proposed a framework named fast detect for DDOS detection and prevention	Drawback of this research is there is no real-time or experimental implementation of the proposed framework which poses a question whether it can handle DDOS or not
2	Malware	Nourouzi et a.l, 2016 [23]	Authors proposed classification technique for detecting malware for data mining. They created a XML file containing malware and converted into non-parse matrix	The main drawback of this paper is WEKA tool, as it runs on Java which uses more memory; for large datasets; it is unable to load and takes plenty of time in WEKA Explorer GUI
		Fraley et al., 2016 [24]	Authors proposed topological feature extraction for polymorphic malware detection for conducting the experiment they obtained samples from ClamAV, Virusshare	The main drawback of this research is that the work is still in preliminary stage due to sample parameters not fully explored like packet data for analysis
		Ahuja and Amandeep 2016 [25]	Authors reviewed the research on malicious detection technique for web security. According to them, malware constitutes worms, Trojan, rootkits, etc.	The main drawback of this research is that the authors never discussed which technique is highly efficient in detecting the malware
		Edem et al., 2014 [26]	Authors proposed an automated malware investigation framework to mitigate malware attack. They used clustering technique and malware analysis in their framework	The main drawback of this paper is that present day malware intrusion can bypass API sandbox, which requires more encryption techniques for mitigating malware attacks

(continued)

Table 1 (continued)

S. No.	Attacks	Proposed by Authors and Year	Remarks	Limitations
3	Botnet	Bhaiyya et al., 2015 [27]	Authors proposed a framework for mitigating the botnet attack using K-means, clustering, and machine learning classifier C5.0	The main drawback of this research is authors only proposed the framework and never implemented for experimental purpose
		Haque et al., 2018 [29]	Authors proposed a data mining framework using meta-analysis for botnet detection. The framework helps to prevent a few features of botnet intruding the user machine	The main drawback of this research is that they do not describe how the confidence values are generated using association rules in their framework and with what parameters it can check the botnet detection
		Kannan et al., 2017 [30]	Authors proposed a framework for detecting the botnet. They used clustering technique for detection of botnet	The main drawback of this research is how the said framework has encryption if the botnet has encryption and attacking the target device
		Bijalwan et al., 2016 [31]	Authors proposed ensemble of classification algorithm for detecting botnet. They used ISCX dataset for their framework	The main drawback of this research is concept drifting of ensemble of classifier algorithm which requires a lot of training to train each individual algorithm and is very complex
4	Spyware	Border et al., 2004 [33]	Authors proposed a framework called webtap for detecting spyware. The framework tries to detect the inbound traffic via HTTP connections in firewall	The main drawback of this paper is if the hacker site sends a request more than 60 kb then the proposed mechanism does not work to combat spyware
		Shahzad et al., 2010 [34]	Authors proposed a technique for combating spyware by mining executable files. They try to extract the binary features from executable files	The main drawback of this paper is if the executable file is encrypted and the proposed framework does not have such mechanism to combat the spyware
		Pandey et al., 2015 [35]	Authors proposed a framework for spyware detection using data mining. They used Bayesian reasoning for decision making to predict the datasets	The main drawback of this research is Java programming; for larger datasets, it is not possible to load from the GUI as it swaps additional memory
		Divya et al., 2012 [36]	Authors proposed a framework using data mining algorithm such as association, classification and regression to test the malicious code	The main drawback of this research is not testing the parameters of spyware which has lot of attributes where authors never mentioned about it in detecting the spyware
5	Probing	Malik et al., 2013 [38]	Authors proposed a framework using multiobjective particle swarm optimization and random forest for probe attack detection	The main drawback of this research is particle swarm optimization as it requires time-consuming operation for large and complex dataset due to repeated fitness function testing
		Vasilomanolakis et al., 2016 [39]	Authors proposed a framework named PREPARE (Probe response attack framework) for detecting probe attacks in collaborative intrusion detection system	Python is very slow and very weak language for computing and it is not very good option for high memory tasks. There are many limitations for accessing database with Python coding

(continued)

Table 1 (continued)

S. No.	Attacks	Proposed by Authors and Year	Remarks	Limitations
6	Ransomware	Kharaz et al., 2016 [41]	Authors proposed a framework named UNVEIL to detect ransomware namely file lockers and screen lockers	The main drawback of this research is they never discussed encryption and decryption standards of their framework, as ransomware is based on encryption and decryption method
		Homayoun et al., 2017 [42]	Authors proposed a framework for detecting ransomware using frequent pattern mining	The main drawback of this research is the proposed framework is only applicable for detecting the ransomware to some extent and not mitigating the ransomware attack where encryption and decryption takes place
		Chen et al., 2017 [43]	Authors proposed a technique for validating only WannaCry ransomware attack	The main drawback of this research is that the proposed framework cannot help in real-time protection if there is a WannaCry ransomware attack on target device

5   Conclusions

In this paper, we reviewed the flow of data mining and described the architecture of data mining algorithms. We further discussed the pros and cons of data mining. Further, we explained different types of attacks in data mining and applications of data mining algorithms and techniques on attacks detection. We reviewed all the attacks namely DOS/DDOS, malware, botnet, spyware, probing, and ransomware in data mining and did a critical review highlighting the research points and explained the drawbacks of the proposed techniques and frameworks of the researchers.

References

1. Alcalá, R., Gacto, M.J., Alcalá-Fdez, J.: Evolutionary data mining and applications: a revision on the most cited papers from the last 10 years (2007–2017). Wiley Interdisc. Rev. Data Min. Knowl. Discovery **8**(2), e1239 (2017)
2. Yamini, O., Ramakrishna, S.: A study on advantages of data mining classification techniques. Int. J. Eng. Res. Technol. **4** (2015)
3. Hassani, H., Huang, X., Silva, E.S., Ghodsi, M.: A review of data mining applications in crime. Stat. Anal. Data Min. ASA Data Sci. J. **9**(3), 139–154 (2016)
4. Madni, H.A., Anwar, Z., Shah, M.A.: Data mining techniques and applications—a decade review. In: 2017 23rd International Conference on Automation and Computing (ICAC), pp. 123–135 (2017)

5. Wang, R., Ji, W., Liu, M., Wang, X., Weng, J., Deng, S., Yuan, C.: Review on mining data from multiple data sources. Pattern Recogn. Lett. **109**, 120–128 (2018)
6. Zanin, M., Papo, D., Sousa, P.A., Menasalvas, E., Nicchi, A., Kubik, E., Boccaletti, S.: Combining complex networks and data mining: why and how. Phys. Rep. **635**, 1–44 (2016)
7. Kesavaraj, G., Sukumaran, S.: A study on classification techniques in data mining. In: 2013 Fourth International Conference on Computing, Communications and Networking Technologies (ICCCNT), Tiruchengode, India, 4–6 July 2013
8. Fu, T.: A review on time series data mining. Eng. Appl. Artif. Intell. Elsevier B.V **24**(1), 164–181 (2011)
9. Dong, G., Taslimitehrani, V.: Pattern-aided regression modeling and prediction model analysis. IEEE Trans. Knowl. Data Eng. **27**(9), 2452–2465 (2015)
10. Chi, X., Fang, Z.W.: Review of association rule mining algorithm in data mining. In: 2011 IEEE 3rd International Conference on Communication Software and Networks, 27–29 May, 2011, Xian, China (2011)
11. Jayakameswaraiah, M., Ramakrishna, S.: Development of data mining system to analyze cars using TkNN clustering algorithm. Int. J. Adv. Res. Comput. Eng. Technol. **3**(7), 2365–2373 (2014). ISSN: 2278-1323
12. Chung-Ching, Yu., Chen, Yen-Liang: Mining sequential patterns from multidimensional sequence data. IEEE Trans. Knowl. Data Eng. **17**(1), 136–140 (2005)
13. Bafna, K., Toshniwal, D.: Feature based summarization of customers' reviews of online products. Procedia Comput. Sci. **22**, 142–151 (2013)
14. Data breach record of USA from 2005–2018. URL: https://www.privacyrights.org/data-breaches. Last accessed 9 Mar 2019
15. Kim, M., Jung, S., Park, M.: A distributed self-organizing map for DoS attack detection. In: Seventh International Conference on Ubiquitous and Future Networks, Sapporo, Japan, pp. 19–22, 7–10 July 2015
16. Kaur, P., Kumar, M., Bhandari, A.: A review of detection approaches for distributed denial of service attacks. Syst. Sci. Control Eng. **5**(1), 301–320 (2017)
17. Zolotukhin, M., Hämäläinen, T., Kokkonen, T., Niemelä, A., Siltanen, J.: Data mining approach for detection of DDoS attacks utilizing SSL/TLS protocol. In: Balandin, S., Andreev, S., Koucheryavy, Y. (eds.) Internet of Things, Smart Spaces, and Next Generation Networks and Systems. ruSMART 2015, NEW2AN 2015. Lecture Notes in Computer Science, vol. 9247. Springer, Cham, (2015)
18. Dongre, S., Chawla, M.: Analysis of feature selection techniques for denial of service (DoS) attacks. In: 2018 4th International Conference on Recent Advances in Information Technology (RAIT), Dhanbad, India, 15–17 Mar 2018
19. Keshri, A., Singh, S., Agarwal, M., Nandiy. S.K.: DoS attacks prevention using IDS and data mining. In: 2016 International Conference on Accessibility to Digital World (ICADW), Guwahati, India, 16–18 Dec 2016
20. Alkasassbeh, M., Hassanat, A.B., Al-Naymat, G., Almseidin, M.: Detecting distributed denial of service attacks using data mining techniques. Int. J. Adv. Comput. Sci. Appl. (IJACSA) **7**(1), 436–445 (2016)
21. Nijim, M., Albataineh, H., Khan, M., Rao, D.: FastDetict: a data mining engine for predicting and preventing DDoS attacks. In: 2017 IEEE International Symposium on Technologies for Homeland Security (HST), Massachusetts, USA, pp. 1–5 (2017)
22. Razak, M.F.A., Anuar, N.B., Salleh, R., Firdaus, A.: The rise of "malware": bibliometric analysis of malware study. J. Netw. Comput. Appl. **75**, 58–76 (2016)
23. Norouzi, M., Souri, A., Samad Zamini, M.: A data mining classification approach for behavioral malware detection. J. Comput. Netw. Commun. pp. 1–9 (2016). Hindawi Publishing Corporation
24. Fraley, J.B., Figueroa, M.: Polymorphic malware detection using topological feature extraction with data mining. In: South east Conference, Norfolk, VA, USA, 30 Mar–3 Apr 2016

25. Ahuja, K., Amandeep: A Survey on malicious detection technique using data mining and analyzing in web security. Int. J. Eng. Dev. Res. (IJEDR) **4**(2), 319–322 (2016)
26. Edem, E.I., Benzaid, C., Al-Nemrat, A., Watters, P.: Analysis of malware behaviour: using data mining clustering techniques to support forensics investigation. In: 2014 Fifth Cybercrime and Trustworthy Computing Conference, Auckland, New Zealand, pp. 54–63, 24–25 Nov 2014
27. Feily, M., Shahrestani, A., Ramadass, S.: A survey of botnet and botnet detection. In: 2009 Third International Conference on Emerging Security Information, Systems and Technologies, Athens, Glyfada, Greece, pp. 268–273, 18–23 June 2009
28. Bhaiyya, A., Bodkhe, S., Pimpalkar, A.: Botnet identification system using clustering and machine learning C5.0. Int. J. Comput. Sci. Mob. Comput. **4**(4), 313–316 (2015)
29. Haque, A., Ayyar, A.V., Singh, S.: A meta data mining framework for botnet analysis. Int. J. Comput. Appl. 1–8 (2018)
30. Kannan, R.: A reliable & scalable frame work for HTTP BotNet detection. Int. J. Comput. Trends Technol. (IJCTT) **54**(1), 19–23 (2017)
31. Bijalwan, A., Chand, N., Pilli, E.S., Rama Krishna, C.: Botnet analysis using ensemble classifier. Perspect. Sci. **8**, 502–504 (2016)
32. Wang, T.-Y., Horng, S.-J., Su, M.-Y., Wu, C.-H., Wang,P.-C., Su, W.-Z.: A surveillance spyware detection system based on data mining methods. In: 2006 IEEE Congress on Evolutionary Computation Sheraton Vancouver Wall Centre Hotel, Vancouver, BC, Canada, pp. 3236–3241, 16–21 July 2006
33. Borders, K., Prakash, A.: Web tap: detecting the covert web traffic. In: Proceedings of the 11th ACM Conference on Computer and Communications Security, Washington, DC, USA, 25–29 Oct 2004
34. Shahzad, R.K., Haider, S.I., Lavesson, N.: Detection of spyware by mining executable files. In: 2010 International Conference on Availability, Reliability and Security, Krakow, Poland, pp. 295–302 (2010)
35. Pandey, K., Naik, M., Qamar, J., Patil, M.: Spyware detection using data mining. Int. J. Res. Appl. Sci. Eng. Technol. (IJRASET) **3**(III), 488–492 (2015)
36. Divya, P., Rajalakshmi, S.: Classifying spyware files using data mining algorithms and hexadecimal representation. Int. J. Sci. Eng. Res. IJSER. June, India (2012)
37. Paliwal, S., Gupta, R.: Denial-of-service, probing & remote to user (R2L) attack detection using genetic algorithm. Int. J. Compute. Appl. **60**(19), 57–62 (2012)
38. Malik, A.J., Khan, F.A.: A hybrid technique using multi-objective particle swarm optimization and random forests for PROBE attacks detection in a network. In: IEEE International Conference on Systems, Man, and Cybernetics, Manchester, UK, pp. 2473–2478, 13–16 Oct 2013
39. Vasilomanolakis, E., Stahn, M., Cordero, C.G., Mühlhäuser, M.: On probe-response attacks in collaborative intrusion detection systems. In: IEEE Conference on Communications and Network Security (CNS), Philadelphia, USA, 17–19 Oct 2016
40. Silva, J.A.H., Hernandez-Alvarez, M.: Large scale ransomware detection by cognitive security. In: IEEE Second Ecuador Technical Chapters Meeting (ETCM), Salinas, Ecuador, 16–20 Oct 2017
41. Kharaz, A., Arshad, S., Mulliner, C.: UNVEIL: a large-scale, automated approach to detecting ransomware. In: 25th Usenix Security Symposium, Austin, Texas, USA, pp. 757–772, 10–12 Aug 2016
42. Homayoun, S., Dehghantanha, A., Ahmadzadeh, M., Hashemi, S., Khayami, R.: Know abnormal, find evil: frequent pattern mining for ransomware threat hunting and intelligence. IEEE Trans. Emerg. Top. Comput. 1–11 (2017)
43. Chen, Q., Bridges, R.A.: Automated behavioral analysis of malware a case study of wannacry ransomware. In: 2017 16th IEEE International Conference on Machine Learning and Applications, Cancun, Mexico, pp. 454–460, 18–21 Dec 2017

A Review of Semantic Annotation Models for Analysis of Healthcare Data Based on Data Mining Techniques

M. Manonmani and Sarojini Balakrishnan

Abstract The evolution of medial data has made interaction and communication between devices very important and need of the hour to address the problems and requirements of the people in the medical domain. It seems to be an important task to overcome the compatibility issues between the communicating devices in the healthcare sector and provide meaningful solutions to the users of medical data. In the present era, interconnection of various sensors and devices in the medical domain is possible to a great extent with the advent of Internet. In the field of medicine, the major problems, viz. interoperability of heterogeneous medical devices, security of the patient information, and personalized visualization of the processed data seems to be very vital. This paper presents an analysis of the different data mining techniques applied in healthcare domain and semantic annotation of healthcare data for overcoming the issue of interoperability. The paper also discusses about the proposed research work which includes creating a semantic model for handling heterogeneous healthcare data and the application of feature selection algorithms and classification algorithms for medical diagnosis.

Keywords Medical domain · Internet · Interoperability · Sensors · Semantic model · Data mining techniques

1 Introduction

The Internet has made possible the logical association of number devices on a large network where these devices negotiate with each other and are potential enough to move with one another. This in fact is driving the automation to a next level

M. Manonmani · S. Balakrishnan (✉)
Department of Computer Science, Avinashilingam Institute for Home
Science and Higher Education for Women, Coimbatore, India
e-mail: dr.b.sarojini@gmail.com

M. Manonmani
e-mail: manonmaniatcbe@gmail.com

© Springer Nature Singapore Pte Ltd. 2020
P. Venkata Krishna and M. S. Obaidat (eds.), *Emerging Research in Data
Engineering Systems and Computer Communications*, Advances in Intelligent
Systems and Computing 1054, https://doi.org/10.1007/978-981-15-0135-7_22

wherever the devices can exchange valuable information with one another and build selections on their own without any human intervention.

In the medical domain, the various devices are controlled and accessed remotely. This has resulted in direct connection and integration of the real world to the computer-based systems using different sensors and net. The interconnection of those multiple sensor devices is leading to computerization in almost all fields and conjointly sanctioning applications which apply new technologies. This has led to improved accuracy, potency, and economic profit with reduced human intervention. Semantic annotation of healthcare data aids in processing the keywords attached to the semantic features and deriving the relationship of each feature with the other and how the features are used in the classification of diseases. In the medical sector, huge amount of information, viz. patient details, diagnostic procedures, treatment plans, insurance coverage, and formal procedures are present and there arises a need to understand the interrelationships between these data. The complexity of the medical data hinders communication between the patient and the physician which can be simplified with semantic annotation by providing meaningful abstraction of the features in diagnosis of chronic diseases. Semantic annotation process provides a clear classification scheme of the features of that is relevant in interpretation of medical data.

Internet of medical things (IoMT) is another emerging term to be used in connection with the healthcare services environment where machine-to-machine interaction and online data processing will be involved between infinite ranges of medical services. The adoption of IoT in healthcare is, in fact, an added advantage to the healthcare sector with AI and robotics already making remarkable foot in the medical field.

2 Semantic Annotation Process

In the present times, the services provided by the healthcare sector are expensive and since the number of people in the aging group is higher, the range of population affected by chronic diseases is on an increase. The expenses incurred toward diagnostic procedures more compared to other expenses in the health sector. The right diagnosing will reduce the necessity of hospitalization. A new paradigm to the already existing system would be to semantically annotate the incoming healthcare data and further select the subset of features that aid in classification of the diseases. Semantic annotation models describe the resource provided by the user by the process of annotation of the data.

2.1 Overcoming the Issue of Interoperability in Healthcare Infrastructure with the Help of Semantization Annotation Models

In the medical domain, the issue of interoperability involves the need to interconnect various heterogeneous devices that exchange the information received from the sensors to the Internet healthcare cloud. Due to the lack of worldwide satisfactory standards, the problem of interoperability among the various devices continues at the hardware level. But semantic models can be built to overcome the issue of interoperability in heterogeneous infrastructure by annotating the data from the electronic health records. In the healthcare domain, the patients and physicians need to communicate with each other in case of remote health monitoring system. The sensors will collect the required information from the sensors and transfer it to the healthcare cloud server either immediately or the information may be stored in EHRs. When needed, the information stored in the health cloud server has to be made available in a format that is understood by both the patient and the physician. For this purpose, the healthcare data needs to be semantically annotated to provide meaningful communication between the patient and the physician. Semantic models provide meaningful annotation of the healthcare data collected from heterogeneous devices. Thus, semantic models prove to be justifiable in overcoming the problem of interoperability in the healthcare sector and at large even to the healthcare data collected from IoT devices too.

Semantic annotations can enable interoperability to refine or add new features to the collected healthcare data with the help of tags. These tags are annotated with the help of RDFs that can provide meaningful relationships between the features that describe the various disease-related data. The RDF is a semantic web framework that combines the medical knowledge and context information with the collected sensor data and makes it easier for machines to understand and further process the annotated data. Thus, semantic interoperability among the various heterogeneous healthcare devices can be achieved with the help of data semantization using RDFs.

The third section provides a review of the related research work done in providing semantic interoperability in the healthcare domain and the fourth section describes the proposed research work that aims at providing semantic annotation to the healthcare data, in turn aiming at providing feasible solutions in diagnosing chronic diseases.

3 Literature Review

Ringsquandl et al. [1] have presented a features selection method based on semantic annotation method for handling high dimensional data in the industrial automation sector. High dimensional data may be of great concern in the field of industry. In order to handle the high dimensional data that is present in the automation system, the authors have used semantic analysis for feature selection. In this model, the authors have developed two-feature selection approaches for semantic analysis and this feature section approach is evaluated against different data applications in the industry field. In order to eliminate the highly correlated features and to increase the processing time, the authors have presented the semantic-guided feature selection model.

Mahdavinejad et al. [2] have discussed different machine learning methods with smart city implementation challenges and issues. The main issues of machine learning and use of IoT in data analytics is focused in this paper. The data that is being studied in this paper is Asrhus Smart City traffic data and the data mining algorithm applied is support vector machine (SVM).

Jabbar et al. [3] have proposed a semantic framework for providing semantic interoperability in IoT. This model was proposed to handle heterogeneous data that is highly present in an IoT system when huge data are transferred between sensor devices in the medical domain. The current health status of the patients is monitored and communicates to them with the help of the sensor devices that operate remotely. These data are recorded and communicated to the physicians as well. The data that is collected through the sensors is annotated semantically and this information is shared in a significant manner between that physician as well as the patient. The authors have put forward a proposal to create a light-weight semantic annotation model that is based on heterogeneous devices in IoT. In order to use the IoT data semantically meaningful, Resource Description Framework (RDF) is employed which relates things using triples. The information thus exchanged is analyzed and then semantically annotated. Then the data that is annotated is passed to the intelligent health cloud in order to match the medicines that are prescribed form pharmaceutical corporations. This information, in turn, is sent to the patient IoT. SPARQL query then analyzes the triples and provides information about the patient's disease using the RDF graph. The major benefit of this semantic interoperability model is that the devices are remotely available at any time and the information about the patient's current health status can be queried by the physicians with the help of the IoT device at any time and moreover remotely.

Joshi et al. [4] have focused on delivering a combined proposal that includes the features of data mining and Internet of things. Based on the results of the combined approach, decisions, and plan of actions could be implemented effectively and quickly. The authors have conducted the research in Hungarian database for diagnosing heart disease. The research work includes a case study based on evaluation of various data mining algorithms, viz. naïve bayes, decision tree, and K means clustering. Rapid Miner software is used in analysis of the results. Datasets are taken from "UCI machine learning repository." "Exang" values for the datasets are calculated and experimental results indicated that "exang" value more than five is indicative of high survival rate compared to "exang" values less than five. The "exang" value evaluation is based on the implementation of decision tree algorithm. Based on the features, K means clustering divides the dataset into two different clusters. These clusters are indicative of the presence or absence of the chronic disease. On the other hand, naïve bayes algorithm provides classification results with high accuracy for classifying the patients into different classes. Thus, this case study has evaluated the advantage of combined approach of data mining and IoT.

Gia et al. [5] have presented an enhanced health observation system based on fog computing. The authors have used electrocardiogram (ECG) feature extraction in their case study as ECG reports are inevitable in diagnosing cardiac diseases. Light-weight wavelet transform mechanism is employed to examining the ECG signals. The results that are evaluated based on the P wave and T wave values indicate that fog computing has achiever more bandwidth efficiency of 90% with low-latency real-time response. In this paper, the authors have built a health monitoring system based on fog computing which includes advanced features like interoperability, distributed database, real-time notification mechanism, location awareness, and graphical user interface with access management. The authors have also provided a light-weight template for extracting ECG feature for prediction of cardiac diseases.

Ma et al. [6] have presented an application based on the healthcare data obtained from IoT and big data applications. This paper discusses the big health system applications and the techniques of using IoT in big data applications. The whole system is divided into different layers with includes health perception layer, transport layer, and huge health cloud service layer. Big health cloud layer is divided into two sub-layers, cloud service support sub-layer and cloud service application sub-layer are the two divisions of the big health cloud layer and the functions of each layer are discussed in detail. The key technologies, viz. wearable devices, mobile communication technology, cognitive computing, healthcare cloud robot, health cloud, and health big data, and typical applications of big health system are discussed in detail.

Chui et al. [7] have presented the topic of disease diagnosis in smart healthcare. This paper gives a summary of the recent optimization algorithms and machine learning algorithms. Evolutionary optimization, stochastic optimization, and combinatorial optimization are discussed. Owning to the very fact that there are lots of applications in healthcare, four applications within the field of diseases identification (which conjointly list within the top ten causes of global death in 2015), namely cardiovascular diseases, diabetes mellitus, Alzheimer's disease and other forms of dementia, and tuberculosis, are considered for discussion in this paper. In addition, challenges in the deployment of disease diagnosis in healthcare have also been discussed.

Antunes et al. [8] have developed an unsupervised model that can learn word categories spontaneously. In this paper, the authors have discussed the current storage limitations and analytical solutions in data mining and IoT. They have aimed to provide a solution for semantic annotation where the dataset taken for evaluation includes that semantic IoT dataset and the Miller-Charles dataset. Among human classification, the correlation achieved was 0.63. The reference dataset, i.e., the Miller-Charles dataset was used to find the semantic similarity. In this paper, 38 human subjects were used to rate 30 work-pairs and a rating scale from 0 to 4 was used to rank word-pairs. A rating scale was used to rate no similarity and perfect synonymy. A value of 0 in the rating scaled indicates "no similarity" and a value of 4 in the rating scale indicated "perfect synonymy." The correlation result was 0.8 for human classification. Unsupervised training methods were used by the author to mark the groups and also to improve the accuracy. The model was evaluated using mean squared error (MSE). In order to take into consideration multiple word categories, the model was extended further to accomplish the multiple words and a novel method based on unsupervised learning was developed. The issues like noisy dimension from distributional profiles and sense-conflation were taken into consideration and in order to curb this, dimensional reduction filters and clustering were employed in this model. By this method, the accuracy can be increased and also this model can be made used in a more potential way. A correlation of 0.63 was achieved after evaluating the results against the Miller-Charles dataset and an IoT semantic dataset.

Sharma et al. [9] have presented a method that evaluates the stemming and stop word techniques for problems based on text classification. They have summarized how the feature selection methods are influenced by the stop word and stemming words. The experiment was conducted with 64 documents having 9998 unique terms. The experiments have been conducted using nine documents frequency threshold values (sparsity value in %), namely, 10, 20, 30, 40, 50, 60, 70, 80, and 90. The threshold is that the proportion value instead of the sparsity value. Experimental results show that the removal of stop words decrease the size of feature set. They have found the maximum decrement in feature set at sparsity value 0.9 as 90%. Results indicate that the stemming process significantly affects the size of feature set with different sparsity value. As they increased the sparsity value the size of feature set also increased. Only for sparsity value 0.9 the feature set decrease from 9793 to 936. Experimental results reveal an important fact that stemming, even though is very important is not making only very negligible difference based on the number of terms selected. From the experimental results, it could be seen that preprocessing has a huge impact on performances of classification. The goal of preprocessing is to cut back the amount of features that was successfully met by the chosen techniques.

4 Proposed Research Work

The proposed research work aims to collect healthcare data from the sensors, wearable devices or other benchmark dataset and perform semantic annotation of the collected data. The main aim of the research is to develop a semantic model and perform feature selection process for medical diagnosis and, in turn, evaluate the accuracy of the classification algorithms. Semantic annotation of the healthcare data is created using protégé tool. Protégé tool is an ontology editor and provides frame works for building intelligent systems. The protégé tool supports the Web Ontology

Language (OWL). Description logic is used in OWL and it supports automated reasoning which proves to be applicable in creating semantic models where inter-operability and security becomes a major issue that needs to be addressed when we process medical data. The proposed research methodology is depicted in Fig. 1.

The ontology-based sematic annotation model as depicted in Fig. 1 performs semantic tagging to create tags for identification of the features derived from healthcare dataset. After creating data tags, the semantically annotated features will well-describe their relationships with the main classes and other sub-classes created in the semantic tagging section performed in the back end of the protégé tool. The contribution of the research work lies in creating relationships between the semantically annotated features based on fuzzy rule mining. The fuzzy rule mining technique helps us to determine whether the associated classes in the dataset dispense accurate classification of the patients based on the disease while maintaining the integrity of information conveyed by the numerical attributes. The results of annotation produce a context-aware data that adds meaningful information to healthcare data and in turn provides a well-defined communication between the user interface and other stakeholders of healthcare sector.

After creating a semantic model, the context-aware data is set to undergo feature selection process with optimization algorithms to select the subset of the most relevant features from the original set of features. Once the relevant features subset is arrived, data mining algorithms can be applied to evaluate and classify the patients as having the disease or not having the disease and the accuracy of the classifiers will be analyzed for providing better classification results.

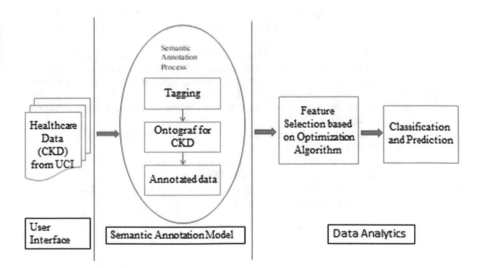

Fig. 1 Proposed research methodology

5   Conclusion

This paper has provided a study of the different data mining techniques that can be employed to process the medical information that is gathered from sensors and other IoT and non-IoT devices. The paper also discusses about adding semantics to the healthcare data and points out the importance of creating a semantic model that includes semantic annotation and deals with the issues of interoperability among various devices used in the hospital. The paper discusses about the steps to be followed in the proposed research work and how the data mining techniques prove to be viable in analyzing the healthcare data from UCI and making it meaningful to the user interface for making important decisions.

Thus, the proposed semantic model will be very useful for the patients and doctors to interact with each other and exchange information via the Internet for treatment of diseases and regular follow up regarding their treatment and recovery.

References

1. Ringsqunadi, M., et al.: Semantic-guided feature selection for industrial automation systems. In: International Semantic Web Conference, Springer, LNCS 9367, pp. 225–240 (2015)
2. Mahdavinejad, M.S., Rezvan, M., et al.: Machine learning for internet of things data analysis: a survey. Digit. Commun. Netw. **4**, 161–175 (2018)
3. Jabbar, S., Ullah, F., Khalid, S., Khan, M., Han, K.: Semantic interoperability in heterogeneous IoT infrastructure for healthcare. Wireless Communications and Mobile Computing, 10 pages (2017)
4. Joshi, M., et al.: An application of IoT on Hungarian database using Data mining Techniques: a collaborative approach. In: 2017 3rd International Conference on Advances in Computing, Communication & Automation (ICACCA), pp. 1–6. IEEE (2017)
5. Gia, T.N., et al.: Fog computing in healthcare internet of things: a case study on ECG feature extraction. In: 2015 IEEE International Conference on Computer and Information Technology; Ubiquitous Computing and Communications; Dependable, Autonomic and Secure Computing; Pervasive Intelligence and Computing, pp. 356–363 (2015)
6. Ma, Y., et al.: Big health application system based on health internet of things and big data. IEEE Access **5**, 7885–7897 (2017)
7. Chui, K.T., et al.: Disease diagnosis in smart healthcare: innovation, technologies and applications. Sustainability **9**, 2–23 (2017)
8. Antunes, M., Gomes, D., Aguiar, R.: Towards IoT data classification through semantic features. Future Gener. Comput. Syst. 20 pages (2017)
9. Sharma, D., et al.: Evaluation of stemming and stop word techniques on text classification problem. Int. J. Sci. Res. Comput. Sci. Eng. **3**(2), 1–4 (2015)

Review of Optimization-Based Feature Selection Algorithms on Healthcare Dataset

M. Manonmani and Sarojini Balakrishnan

Abstract Feature selection is an important function in classification and prediction technique, especially in medical data mining. It is embedded with the task of selecting a subset of relevant features that can be used in constructing a model. Optimization algorithms play a significant role in medical data mining especially in diagnosing chronic disease because it offers good efficiency in less computational cost and time. Also, the classification algorithms yield better results when the feature selection algorithm is based on an objective function. This paper aims to provide a review of optimized feature selection algorithms and an overview of the proposed improved teaching learning based optimization algorithm (ITLBO) for classification and prediction of chronic kidney disease (CKD). The proposed algorithm aims to reduce the number of features required for diagnosing the CKD with an objective function that optimizes the features and selects them based on the updated weight of the fitness function.

Keywords Optimization algorithm · Objective function · Chronic kidney disease (CKD) · Improved teacher–learner-based optimization algorithm (ITLBO) · Fitness function

1 Introduction

Medical data mining deals with the creation and manipulation of medical knowledge base for clinical decision support. The knowledge base is generated by discovering the hidden patterns in the stored clinical data. The generated knowledge and interpretations of decisions will help the physician to know the new

M. Manonmani · S. Balakrishnan (✉)
Department of Computer Science, Avinashilingam Institute for Home Science
and Higher Education for Women, Coimbatore, India
e-mail: dr.b.sarojini@gmail.com

M. Manonmani
e-mail: manonmaniatcbe@gmail.com

© Springer Nature Singapore Pte Ltd. 2020   239
P. Venkata Krishna and M. S. Obaidat (eds.), *Emerging Research in Data
Engineering Systems and Computer Communications*, Advances in Intelligent
Systems and Computing 1054, https://doi.org/10.1007/978-981-15-0135-7_23

interrelations and regularities in the features that could not be seen in an explicit form. However, there are many challenges and issues in mining the medical data, viz. high dimensionality, heterogeneity, imprecise data, and data interpretation becoming difficult. Due to the high dimensionality in healthcare data, feature selection methods are being applied wherein the features are mapped into lower-dimensional space (feature extraction) or the features are selected (feature selection) in order to reduce the dimensionality. This, besides computational benefits, yields better classification accuracy. Feature selection methods are preferred to feature extraction methods for medical data as the original features and their values are maintained. Feature selection based on optimization techniques help to find the most discriminatory feature subset that are relevant in medical diagnosis with less computational time and effort.

Feature selection, as a process of selecting a subset of original features according to certain criteria, is important and frequently used as a reduction technique for data mining. Feature selection has been an active research area for decades in fields such as machine learning and data mining [1].

In recent research undertaking, feature selection methods are based on the efficiency of optimization algorithms for selecting the subset of relevant features for getting better classification results. Optimized feature selection algorithms are based on the process of optimizing the parameters of the feature selection algorithm for a specific problem. When the parameters in the feature selection algorithm are not optimized, the computational effort involved in the processing may be either increased or there might be a problem of local optima. There arises a need to find the optimal function in the solution space of the algorithm in addition to the tuning of other control parameters that further enhances the efficiency of the algorithm. As a general solution to the problem of tuning algorithm-specific parameters, data mining analysts have opted for the optimized algorithm. Hence, optimized feature selection forms a better solution for enhancing the results of an algorithm. Feature selection with parameter optimization forms the basis of most machine learning systems as these systems provide sensible results for any problem at hand.

The second section provides a review of optimization-based feature selection algorithms in the healthcare domain and the third section describes the proposed research work that aims at optimized feature selection based on ITLBO algorithm and the fourth section deals with conclusion.

2 Literature Survey

Kelwade et al. [2] have carried out experiments on the heart rate time series to predict normal and abnormal rhythms using particle swarm optimization in radial basis function neural network. A combined model of particle swarm optimization (PSO) and radial basis function neural network (RBFN) has been proposed by the authors to improve the accuracy of cardiac arrhythmia prediction. It was found that a set of features both linear and nonlinear were retrieved from the RR interval time series datasets and PSO was used to optimize the spread parameter in order to improve the performance of RBFN. The datasets were extracted from the MIT-BIH arrhythmia database. In order to predict the eight types of cardiac signals that indicate normal and abnormal rhythm, the heart rate (RR interval) time series data were put under experiment. Results derived from the experiments indicate that PSO-RBFN model proves to be more efficient in terms of prediction accuracy for cardiac disorder diagnosis. The overall prediction accuracy achieved by the hybrid PSO-RBFN was 96.3%.

Rusdi et al. [3] have proposed the implementation of Artificial Bee Colony (ABC) algorithm to medical image dataset to overcome the problem of curve fitting. Data of computed tomography (CT) images from two different patients were collected for this study. The research work has been undertaken in four phases. Firstly the digital images have been converted to binary images; secondly, boundary and corner point has been detected; thirdly, parameterization has been done; and finally, curve reconstruction has been done using ABC algorithm. The distance of the fitted cubic Bezier curve was evaluated using Sum Square Error (SSE) formula. After the tuning of parameters based on ABC algorithm, the error rate of both skulls was found to be smallest. The proposed method has proved to be efficient in finding the best-fitted Bezier curve that resembled the original medical images. The Douglas–Peucker algorithm was used to improve the performance of the proposed system with minimal computation time. It is evident that the author has proposed an alternative method to redesign the medical image as it results only in a minimal error.

Celik and Yurtay [4] have undertaken a research to evaluate the classification accuracy of the ant colony optimization algorithm for the diagnosis of primary headaches using a Web site questionnaire expert system that was completed by patients. The study was conducted among 850 patients who were suffering from headache from three cities in Turkey. Detailed study was conducted from the questionnaire filled by the patients. The diagnostic results were compared with the classification algorithm based on ant colony optimization (ACO) algorithm. The overall accuracy obtained from ACO-based classification algorithm was 96.9412%. The classification accuracy of patients diagnosed with migraine, tension-type, and cluster headaches were 98.2, 92.4, and 98.2%, respectively. The study indicates that the Web site-based algorithm with ant colony optimization parameter tuning will be beneficial to the neurologists for gathering quick results and for tracing patients for headache symptoms. At large, medication and treatment follow-up can be addressed by the neurologists by using the electronic health records from the Internet.

Jaddi et al. [5] have presented an optimization algorithm based on the kidney process in the human body. The kidney-inspired algorithm is a population-based algorithm. The objective function in this algorithm is calculated based on the mean of all the solutions in the current population. By this way, the solutions are filtered based on the mean of the objective function that is calculated at each iteration. Thus, the filtered solutions form the better solutions and are moved to the filtered blood and the rest of the solutions are considered as worse solutions and are transferred to the waste. This process of filtering the better solutions and eliminating the worse solutions is similar to the glomerular filtration process in the kidney.

This process of filtration and reabsorption is repeated at each iteration till the best solution is arrived and the waste and the filtered blood are combined to become the new population and then the filtration rate is updated to the new rate. The algorithm is evaluated on eight well-known benchmark test functions and the results are compared with other algorithms. The performance of the proposed algorithm is found to be better on seven among eight test functions when compared to the recent researches.

Long et al. [6] have suggested interval type-2 fuzzy logic system (IT2FLS) for diagnosing heart disease. The IT2FLS system is based on the application of rough sets for attribute reduction. The integrated approach of rough sets-based attribute reduction and IT2FLS has the advantage of manipulating medical datasets with high dimension. The proposed algorithm is based on hybrid learning process which comprised the fuzzy c-means clustering algorithm and chaos firefly algorithm for optimization. The optimal reduction is calculated using chaos firefly algorithm combined with the attribute reduction results of rough sets. The chaos firefly algorithm increases the efficiency of the IT2FLS and in turn reduces the computational burden of the proposed system. The proposed rough sets-based attribute reductions using the discrete chaos firefly algorithm was evaluated against the heart disease dataset and SPECTF dataset. The performance of a few classification algorithms was tested. It was found that the accuracy of Naïve Bayes was improved by approximately 2% from 83.3 to 85.2%. The accuracy of SVM rose from 75.9 to 81.5%. The accuracy of ANN and new approach increased from 78.8 to 81.5% and 86 to 88.3%, respectively. For the SPECTF dataset, the accuracy of ANN and new approach improved from 73.3 to 77% and 79.1 to 87.3%, respectively. Experimental results clearly indicated a significant improvement in the accuracy of the proposed system compared with other machine learning algorithms like Naïve Bayers, SVM, and ANN. The proposed system can provide fruitful results in diagnosing heart disease.

Doreswamy and Salma [7] have proposed a hybrid model based on binary bat algorithm (BBA) and feedforward neural network (FNN). In this model, the advantages of BBA and effectiveness of FNN are utilized for classifying the three benchmark breast cancer datasets as either malignant or benign. For training the network, the BBA is used and a fitness function is used for error minimization. Experiments and results indicate that FNNBBA-based classification produces an accuracy of 92.61% for training data and 89.95% for testing data.

Cai et al. [8] have proposed a RF-BFO-SVM framework to differentiate Parkinson's disease (PD) patients from healthy controls by combining the feature selection with bacterial foraging optimization (BFO)-based SVM. The vocal data of PD patients were normalized by scaling to the range $[-1, 1]$ and relief method was used to remove the redundant and non-relevant features. BFO algorithm is used to optimize the parameters of SVM. The main aim in parameter optimization is to evaluate the performance of each set of candidate parameters using BFO algorithm. In this study, the voice recordings of 31 subjects, including 23 patients with PD (16 males and 7 females) and 8 healthy controls (3 males and 5 females) were used. Results indicate that RF-BFO-SVM presented best five features in terms of performance metrics. The classification accuracy was further improved by the application of relief feature selection. After the application of feature selection, the ACC, sensitivity, and specificity were improved further by 0.53, 0.54, and 0.67%, respectively.

Nalluri [9] have presented a metaheuristic feature selection algorithm based on artificial fish swarm optimization (AFSO) for diagnosing medical data. The algorithm is inspired by the behavior of fish under water. The AFSO algorithm was evaluated against nine medical datasets having binary and multiple classes and the classification accuracy of SVM was implemented. Results indicate an efficient increase in the accuracy of SVM classifier with wrapper-based AFSO algorithm for feature selection.

Allam and Nandhini [10] have produced a new wrapper-based feature selection method called binary teaching–learning-based optimization (FSBTLBO). The proposed algorithm is based on common controlling parameters like population size and a number of generations in order to get a subset of optimal features from the dataset. The authors have to implement the objective function using different classifiers in order to evaluate the fitness of individuals for finding the efficiency of the proposed system. The FS_BTLBO algorithm was evaluated against the Winconsin diagnosis breast cancer (WDBC) dataset. Experimental results have indicated that the proposed algorithm produces higher accuracy with a minimal number of features from the breast cancer diagnosis medical datasets. The algorithm was used to classify the dataset as malignant and benign tumors. The classification accuracy achieved was 98.43% which was found to be more that other compared technique.

3 Proposed Feature Selection Method

In the proposed research work, we aim to reduce the features in the original dataset that are relevant in diagnosing chronic kidney disease (CKD) by applying a variation in calculating the fitness function in the original teaching learning based optimization (TLBO) algorithm. The TLBO is based on swarm intelligence optimization algorithms based on the teaching—learning process in a classroom and requires only common control parameters like population size, number of generations, etc. The TLBO algorithm has the advantage that it does not require any algorithm-specific parameters in implementing the algorithm. The TLBO algorithm consists of two phases, the teacher phase and the learner phase. The learners interact with each other and the best learner becomes the teacher based on the "fitness" value of the optimization algorithm. The best solution arrived in the entire population is considered as the teacher. The parameters involved in the objective function are the design variables used in the optimization problem and the best value of the objective function is the best solution arrived at the ith iteration.

The decision about the selection of a particular feature that fits the given objective function is based on the weight assigned to the objective function.

In the original TLBO algorithm, the weight of the objective function is calculated on the basis of the Euclidean distance vector.

In the proposed improved teaching learning based optimization (ITLBO) algorithm, the weight of the objective function is calculated based on Chebyshev distance vector which calculates the distance between the two vectors based on the greatest difference along any coordinate dimension. In contrast to the original TLBO algorithm, the proposed work consists of computing the new learner value based on the previous value decided by a weight factor w. A general overview of the proposed feature selection process is depicted in Fig. 1.

The general description of the proposed feature selection process represented in Fig. 1 is as follows: The chronic kidney disease (CKD) dataset is taken from the UCI and ITLBO algorithm is applied to the dataset. The ITLBO aims to find the best features that help in early prediction of the chronic kidney disease in patients suffering from chronic kidney illness. The main objective is to obtain an overall reduction in the number of features present in the original dataset. The reduced feature subset is then fed to the classifier models for evaluating the performance metrics. The accuracy of the classifier models can be further studied on the basis of before feature selection and after feature selection.

The advantage of ITLBO algorithm is that instead of calculating the mean of the distance vector in the solution space, the best value of the fitness function is calculated by updating the weight of the objective function at each iteration using the Chebyshev distance to find the best solution at the ith iteration.

4 Conclusion

Feature selection with optimization is found to produce better classification results particularly in medical data mining. Feature selection thus forms an important and inevitable process in medical diagnosis. The proposed ITLBO will be evaluated against the chronic kidney disease (CKD) dataset taken from the UCI repository and the behavior of the classification algorithm before and after application of ITLBO is aimed to be implemented. This paper has provided a review of optimization-based feature selection algorithms and has discussed the proposed ITLBO algorithm for improving the feature selection process in medical data mining.

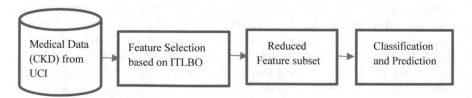

Fig. 1 A general overview of the proposed feature selection process

References

1. Harb, H.M., Desuky, A.S.: Feature selection on classification of medical datasets based on particle swarm optimization. Int. J. Comput. Appl. **104**(5), 14–17 (2014)
2. Kelwade, J.P., Salankar, S.S.: Prediction of heart abnormalities using particle swarm optimization in radial basis function neural network. In: 2016 International Conference on Automatic Control and Dynamic Optimization Techniques (ICACDOT), pp. 793–797. IEEE, Pune, India (2016)
3. Rusdi, N.A., et al.: Reconstruction of medical images using artificial bee colony algorithm. Math. Prob. Eng., 7 pages (2018)
4. Celik, U., Yurtay, N.: An ant colony optimization algorithm-based classification for the diagnosis of primary headaches using a website questionnaire expert system. Turk. J. Electr. Eng. Comput. Sci. **25**, 4200–4210 (2017)
5. Jaddi, N.S., et al.: Kidney-inspired algorithm for optimization problems. Commun. Nonlinear Sci. Numer. Simul. **42**, 358–369 (2017)
6. Long, N.C., et al.: A highly accurate firefly based algorithm for heart disease prediction. Expert Syst. Appl. **42**, 8221–8231 (2015)
7. Doreswamy, Salma, M.U.: A binary bat inspired algorithm for the classification of breast cancer data. Int. J. Soft Comput. Artif. Intell. Appl. (IJSCAI), **5**(2/3), 1–21(2016)
8. Cai, Z., et al.: A new hybrid intelligent framework for predicting Parkinson's disease. IEEE Access (Spec. Section Health Inform. Developing World) **5**, 17188–17200 (2017)
9. Nalluri, M.S.R., et al.: An efficient feature selection using artificial fish swarm optimization and SVM classifier. In: International Conference on Networks & Advances in Computational Technologies (NetACT), pp. 407–411. Thiruvananthapuram (2017)
10. Allam, M., Nandhini, M.: Optimal feature selection using binary teaching learning based optimization algorithm. J. King Saud Univ.—Comput. Inf. Sci., 1–13 (2018)

Survey: A Comparative Study of Different Security Issues in Big Data

Ravinder Nellutla⬦ and Moulana Mohammed⬦

Abstract Data is starting at now a champion among the most basic assets for associations in each field. The consistent improvement in the essentialness and to manage and process large amount of data has made another issue: it cannot be dealt with by standard examination strategies and to store such kind of data to use different file system. This issue was, thusly, comprehended through the formation of another worldview: big data. Be that as it may, big data started new problem related to the large quantity data exclusively or the assortment of the information, yet additionally to information security and protection and another side of big data security in terms of file system. To manage such type of data, we are using different file systems starting from Google. This file system works on distributed and parallel systems. We study the different load balancing algorithm and found some of the differences. In this paper, we present what is big data, then how the big data needed security and different types of security issues in big data focused major issues. The main motto of this paper is to provide a survey report on the latest trends in big data and latest security mechanisms that in form of different levels of security like data and infrastructure security.

Keywords Big data · Data security · Network security · Hadoop file system

1 Introduction

These days, numerous individuals get associated with each other in one virtual world known as "Digital Society" rather than physically associated. So that is the reason vast data is produced by the many people. Majorly, the data is produced from the different social networking sites and large organization, and this type of data is known as big data.

R. Nellutla (✉) · M. Mohammed
CSE Department, Koneru Lakshmaiah Education Foundation, Vaddeswaram, Guntur,
Andhra Pradesh, India
e-mail: ravindernellutla@gmail.com

© Springer Nature Singapore Pte Ltd. 2020
P. Venkata Krishna and M. S. Obaidat (eds.), *Emerging Research in Data
Engineering Systems and Computer Communications*, Advances in Intelligent
Systems and Computing 1054, https://doi.org/10.1007/978-981-15-0135-7_24

Coming to the, big data, it is an extensive measure of information that is unable to store and unable to maintain, and this is known as big data. Ordinarily, we chip away at information of size MB (Worddoc, Excel) or greatest GB (Movies, Codes), yet information in petabytes, i.e., 10^{15} bytes measure is called big data. It is expressed that just about 90% of the present information has been produced in the past few years, and day by day the data size is increasing.

The term huge information has come into utilization as of late to allude to the regularly expanding measure of data that associations are putting away, handling and examining, attributable to the developing number of data sources being used. 90% of the information on the web has been made since 2016, as indicated by an IBM Marketing Cloud contemplate. Individuals, organizations, and gadgets have all progressed toward becoming information industrial facilities that are drawing out extraordinary measures of data to the web every day. In the year of 2014, there were 2.4 billion of web clients only. That number developed to 3.4 billion by the end of the year 2016, and in 2017, there were 300 million of web clients were included—making a sum of 3.8 billion web clients in the year of 2017 (as of April 2017). This is a 42% expansion in individuals utilizing the web in only three years.

The data is producing from the following sources that are

- Social networking sites are Facebook, Twitter, and many other communication-related software.
- Next one is the E-commerce sites like Amazon, Flipkart, Myntra, and so on.
- Satellites are also producing large amount of data like weather forecasting, in the weather forecasting that data which are stored and manipulated.
- Share market is also producing large amount of data.

2 Literature Survey

According to the life cycle of data processing, the big data involves different kinds of like structural data, un-structural data and semi-structural data. Then, it will perform the different operations on our data like cleaning of the data, conversion, and integration. The data collects from different sources. It will use some of the key techniques like batch computing, iterative computing, and from the different stages, we need different security mechanism like for the data acquisition, there is a data privacy, for data storage and management, there is a data management security, for the computing model, there is an infrastructure security, for data analysis, there is a security that is reactive security.

In this paper, we mainly focus on big data privacy and security issues which exist in all stages of big data processing and discussed different securities category wise that are infrastructure security and data privacy [2].

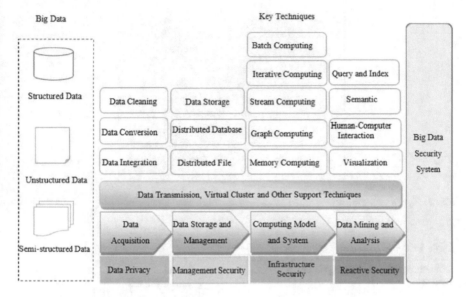

Fig. 1 Framework of big data

2.1 Framework of Big Data

The framework of the big data consists of what type of data is processing that is structured or semi-structured or unstructured, this is we are considering as big data and how the different operations can be taken place in the data, before processing the what type of security mechanisms are following and specifies different security mechanisms [1]. The following diagram shows the framework of big data in terms of storing, processing, and security (Fig. 1).

Our aim is, how to provide the security for the big data that is whatever the data we are getting from the different sites already we discussed earlier.

2.2 Types of Security

Major security challenges will be categorized into the following categories (Fig. 2).

Here, the categorization of the security, there are four types, and from these types, again we can divide as different subtypes. Table 1 shows different security mechanisms and the subtypes of each aspect of the security.

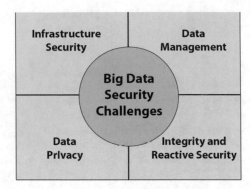

Fig. 2 Types of security

Table 1 Types of security

Infrastructure security	Data privacy	Data management	Integrity and reactive security
Security for Hadoop	Cryptography	Security at collection or storage	Integrity
Availability	Access control	Polices, laws, or government	Attack detection
Architecture security	Confidentiality	Sharing algorithms	Recovery
Authentication	Privacy preserving queries		
Communication security	Anonymization privacy at social networks differential privacy		

3 Literature Review of Data Privacy

In the present review, I have selected multiple papers from the different conferences in the area of big data and the security mechanisms in the big data. All the articles present in this paper review from the past eight years that are 2009–2017. The characterizations are been imperative for tremendous data, security, or assurance separately.

For the purpose of surveying, we gathered majority of the papers from different conferences. We have used papers from top gatherings, situated from highly reputed conferences. Inside and out, twelve appropriate social events have been picked, including each one of the three of best-situated security and insurance gatherings. There, moreover, exists a couple of new, promising gatherings in colossal data.

Here, the total papers will be divided as two segments that are stream one and stream two. Stream one spotlights on identifying the papers related to security or

Table 2 Categories of articles

Category	Three V's	Security and privacy
Confidentiality		Yes
Data analysis	Value	
Data format	Variety, volume	
Data integrity	Veracity	Yes
Privacy		Yes

insurance in humongous data gatherings. This request is intentionally worked to get a broad assortment of security and assurance papers, including essential papers that have ignored "gigantic data" from the title. In addition, stream two is expected to find tremendous data papers in any of the social affairs, not at all like stream one. The inspiration to manner join Stream B is first to get gigantic data papers in security and assurance gatherings. Stream two will have the ability to find colossal data papers in interchange gatherings, which allow to get security or assurance papers that were not starting at now got by stream one.

By then, each paper is requested into not less than one of the arrangements. These characterizations were picked in perspective of the characteristics of the big data with additional security feature and assurance classes included in the set. Along these lines, the orders get both the natural qualities of immense data, and what's more security and insurance (Table 2).

From total articles, only papers (82). In the wake of isolating without endpapers and playing out the quality evaluation, papers (82) remain. Coming to stream one majority of the papers, results are matched, and the results of the stream two papers not provided that much of information. In the going, table, the amount of papers from each social affair is showed up for stream one and stream two independently.

Presently, we see the diverse classes of security like specified in Table 3.

Table 3 Different streams of articles

Conference name	Stream one	Stream two
ICDE	22	0
ICDM	4	0
SIGMOD	21	1
VLDB	25	1
ICML	5	0
NIPS	0	
S&P		1
CCS		0
Total	78	4

Table 4 Confidentiality

Author	C	DA	DF	DI	P
Liu et al. [3]	Yes	Yes			
Chu et al. [4]	Yes	Yes	Yes		
Bender et al. [5]	Yes				
Meacham and Shasha [6]	Yes	Yes			

3.1 Confidentiality

It is a key credit to ensure when precarious information is overseen, particularly since being able to store and process information while ensuring riddle could be an inducing capacity to spur endorsement to gather information. Through 23 papers are arranged as protection papers. Some of the papers (5) use homomorphic encryption, which is a system that empowers certain math exercises to be performed on mixed data. Among some papers, one uses totally homomorphic encryption which supports any number juggling movement [15], while the rest uses fragmentary homomorphic encryption which reinforces given calculating errands (Table 4).

Another point secured by a couple of papers is to get the opportunity to control. Through, four papers discuss get to control. Distinctive focuses secured were secure multiparty count, a thought where various components play out a computation while keeping each substance's information private, imprudent trade, where a sender may trade a scrap of information to the recipient anonymous piece is sent, and assorted mixed documents used for improving look for time efficiency. Out and out of, papers (3) use secure multiparty figuring, papers (2) use careless trade and papers (2) use mixed records.

3.2 Data Integrity

Information trustworthiness is the authenticity and nature of data. It is along these lines decidedly connected with veracity, one of the characteristics of the big data. All together papers (5) secured data dependability. Since there is just a little approach of information dependability papers, no reasonable subject tendency was spotted. In any case, one paper shows a strike on uprightness, two papers are on mess up change and information purifying, and two papers utilize settled rigging to ensure dependability of the information.

Xiao et al. [7] demonstrate that it is sufficient to harm 5% of the preparation esteems, an informational index utilized exclusively to prepare a machine learning calculation. Arasu et al. [8] executed a SQL database called cipher base that spotlights on secrecy of information and honesty (Table 5).

Table 5 Data integrity

Author	C	DA	DF	DI	P
Xiao et al. [7]		Yes		Yes	
Arasu et al. [8]	Yes		Yes	Yes	
Lallali et al. [9]	Yes			Yes	

Table 6 Privacy

Author	C	DA	DF	DI	P
Wang and Zheng [10]		Yes		Yes	
Jurczyl et al. [11]				Yes	
Cao and Karras [12]				Yes	
Acs et al. [13]				Yes	
Cormode et al. [14]				Yes	

3.3 Privacy

A basic idea is insurance for tremendous data since it can contain delicate data about individuals. To ease the security issue, data can be de-perceived by ousting qualities that would recognize a man. This is an approach that works, if done successfully, both when data are regulated and when released.

A couple of security models, for instance, k-anonymity, l-diversity, t-closeness, and differential insurance, can be used to anonymize data. Out of total of 61 papers, one paper [10] livelihoods.

Algorithm k-lack of definition, and another paper [11] uses l-arranged assortment and t-closeness yet also differential insurance to anonymize data. Additionally, Cao and Karras [12] familiarize a successor with t-closeness, called β-likeness which they affirm is more valuable and clearer. In connection, a generous section, 46 papers, of the security arranged papers focuses just on differential assurance as their insurance illustrates. Among these are differentially private histograms [13] and unmistakable data structures for differentially private multidimensional data [14]. An interesting discernment by Hu et al. is that differential security can large influence precision of the result (Table 6).

4 Literature Review of Infrastructure Security

Apart from the data security, I investigated about the infrastructure security. Under the infrastructure security, we have different categories and that are also investigated. Coming to the infrastructure security, security for Hadoop, availability, architecture security, authentication, and communication security.

The general architecture of Hadoop is as follows: (Fig. 3).

HDFS Architecture

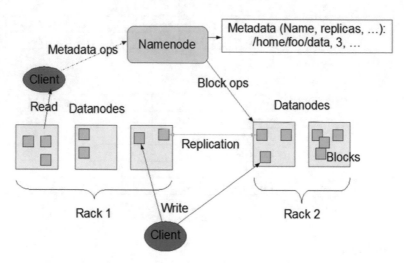

Fig. 3 Hadoop architecture

4.1 Security for Hadoop

Hadoop is an open-source framework given by apache software foundation for storing and processing the huge data set with cluster of commodity hardware. This framework is used by the plain programming models and which handle the distributed parallel programming models. Hadoop handles the size of the data sets that is petabytes and exabytes across clusters of computers so that cluster of Hadoops can easily scale out. Hadoop is made up of two components those are Hadoop distributed file system and map-reduce. The overview of the map and reduce phases is as follows.

Map Phase
See Fig. 4.

Reduce Phase
Nowadays, this big data is having vast popularity in each and every field, and also, there are millions of opportunities by using this. But at the same time, we need to more focus on the security issues to protect our data in the network (Fig. 5).

Coming to, the literature survey of map reducing feature in Hadoop (load balancing and skew). The map reduces, when an user has to implement large varieties of possibilities to implement user-defined function, and it leads to longer execution time. In this, the challenge is to distribute workload among all clusters. When I was focusing on the load balancing, some applications produce the imbalance work on the cluster during execution.

Map Phase

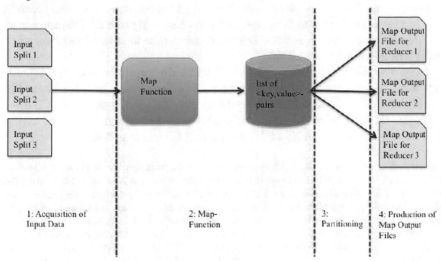

Fig. 4 Architecture of map phase

Reduce Phase

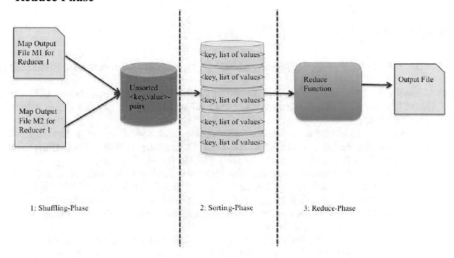

Fig. 5 Architecture of reduce phase

So, in the year 2013, the team of Munich University and the team of Bozen-Bolzano University combinedly worked on the Hadoop file system security, they implemented the algorithm called as "Top Cluster Algorithm," [16] and tested. This algorithm was given better results to distribute the load to all reducers eventually and avoid the skew also.

In the year 2015, the authors "Chi-yi Lin and Ying-Chin in" proposed a different algorithm which gets better results compared with the existing algorithms like Chord ring [17]. The algorithm designed by the "Chi-yi Lin and Ying-Chin in" influences the load balancing state, and a new balance node was introduced to help in matching the heavy loaded and light loaded of the data nodes.

Acknowledgements While this audit examines security and protection for enormous information, it does not cover entire papers accessible inside the subject, since it is not possible to complete physically survey them all. Instead, the focal point of this survey is to investigate late papers and to give both a subjective examination, so as to make a preview of the present best in class. There are a few fascinating thoughts for tending to security and protection issues inside the setting of huge information. In this paper, 208 late papers have been gathered from A∗ meetings, to give a review of the present best in class. In the end, 82 papers were ordered in the wake of passing the sifting and quality appraisal arrange, collected different papers for infrastructure security, and clearly mentioned the work of different authors also. At last, we find that the security and protection for enormous information, in view of the explored papers, are not quite the same as security furthermore, security look into by and large.

References

1. Kim, H.S.: Privacy preserving security framework for cognitive radio networks. IETE Tech. Rev. **30**(2), 142–148 (2014)
2. Manyika, J., Chui, M., Brown, B., Bughin, J., Dobbs, R., Roxburgh, C., Byers, A.H.: Big Data: The Next Frontier for Innovation, Competition, and Productivity. McKinsey Global Institute (2011)
3. Liu, A., et al.: Efficient secure similarity computation on encrypted trajectory data. In: 2015 IEEE 31st International Conference on Data Engineering (ICDE). 2015 IEEE 31st International Conference on Data Engineering (ICDE), pp. 66–77 (2015)
4. Chu, Y.-W., et al.: Privacy-preserving SimRank over distributed information network. In: 2012 IEEE 12th International Conference on Data Mining (ICDM), pp. 840–845 (2012)
5. Bender, G., et al.: Explainable security for relational databases. In: Proceedings of the 2014 ACM SIGMOD International Conference on Management of Data. SIGMOD '14, pp. 1411–1422. ACM, New York, USA 2014
6. Meacham, A., Shasha, D.: JustMyFriends: full SQL, full transactional amenities, and access privacy. In: Proceedings of the 2012 ACM SIGMOD International Conference on Management of Data. SIGMOD '12. ACM, New York, USA, pp. 633–636 (2012)
7. Xiao, H., et al.: Is feature selection secure against training data poisoning? In: Proceedings of the 32nd International Conference on Machine Learning (ICML- 15), pp. 1689–1698 (2015)
8. Arasu, A., et al.: Secure database-as-a-service with cipherbase. In: Proceedings of the 2013 ACM SIGMOD International Conference on Management of Data. SIGMOD '13, pp. 1033–1036. ACM, New York, USA (2013)

9. Lallali, S., et al.: A secure search engine for the personal cloud. In: Proceedings of the 2015 ACM SIGMOD International Conference on Management of Data. SIGMOD '15. ACM, New York, USA, pp. 1445–1450 (2015)
10. Wang, Y., Zheng, B.: Preserving privacy in social networks against connection fingerprint attacks. In: 2015 IEEE 31st International Conference on Data Engineering (ICDE), pp. 54–65 (2015)
11. Jurczyk, P., et al.: DObjects+: enabling privacy- preserving data federation services. In: 2012 IEEE 28th International Conference on Data Engineering (ICDE), pp. 1325–1328 (2012)
12. Cao, J., Karras, P.: Publishing microdata with a robust privacy guarantee. Proc. VLDB Endow. 5(11), 1388–1399 (2012)
13. Acs, G., et al.: Differentially private histogram publishing through lossy compression. In: 2012 IEEE 12th International Conference on Data Mining (ICDM), pp. 1–10 (2012)
14. Cormode, G., et al.: Differentially private spatial decompositions. In: 2012 IEEE 28th International Conference on Data Engineering (ICDE), pp. 20–31 (2012)
15. The Centre for Democracy & Technology.: Building a Strong Privacy and Security Policy Framework for Personal Health Records, The Centre for Democracy & Technology (CDT) Washington, July 2010 (www.cdt.org)
16. A Load Balancing Algorithm implementation in Hadoop by "Thomas Watter Brenner" dated August Ist 2013
17. A Load Balancing Algorithm for Hadoop Distributed File System by "Chi-yi Lin and Ying-Chin in" Tamkang University dated by 2015

Mineral Identification Using Unsupervised Classification from Hyperspectral Data

Priyanka Gupta and M. Venkatesan

Abstract Hyperspectral imagery is one of the research areas in the field of remote sensing. Hyperspectral sensors record reflectance of object or material or region across the electromagnetic spectrum. Mineral identification is an urban application in the field of remote sensing of Hyperspectral data. Challenges with the hyperspectral data include high dimensionality and size of the hyperspectral data. Principle component analysis (PCA) is used to reduce the dimension of data by band selection approach. Unsupervised classification technique is one of the hot research topics. Due to the unavailability of ground truth data, unsupervised algorithm is used to classify the minerals present in the remotely sensed hyperspectral data. K-means is unsupervised clustering algorithm used to classify the mineral and then further SVM is used to check the classification accuracy. K-means is applied to end member data only. SVM used k-means result as a labelled data and classify another set of dataset.

Keywords Hyperspectral imagery · Unsupervised classification · K-means · PCA · SVM

1 Introduction

Hyperspectral sensors are used to target detection, classification, pattern recognition and discrimination. These sensors collect images of earth surface in the form of narrow, continuous and discrete spectral bands. These spectral bands form a complete, continuous spectral pattern of each pixel. Most of the study [1], using

P. Gupta (✉) · M. Venkatesan (✉)
Department of Computer Science Engineering,
National Institute of Technology, Surathkal, Karnataka, India
e-mail: prigupta9875@gmail.com

M. Venkatesan
e-mail: venkisakthi77@gmail.com

© Springer Nature Singapore Pte Ltd. 2020
P. Venkata Krishna and M. S. Obaidat (eds.), *Emerging Research in Data Engineering Systems and Computer Communications*, Advances in Intelligent Systems and Computing 1054, https://doi.org/10.1007/978-981-15-0135-7_25

259

hyperspectral data for geological applications, have so far addressed in the different regions of climates. A mineral or combinations of minerals are source of material, known by a combination of different minerals that is in many different forms like solid, organic and inorganic. Each pixel contains a mixture of different spectra due to the multiple components available in the surface that form the ground surface. This complexity results in incorrect identification and/or misclassification of surface materials. Therefore, the classification of materials and minerals from different areas of earth's surface is one of the most important research topics in remote sensing of hyperspectral data.

Hyperspectral image classification can be of three types—supervised, semi-supervised and unsupervised classification based on ground truth data availability. The semi-supervised classification [2] is where we used some labelled data and based on that classifying the unlabelled data. However, in unsupervised classifier [3], a remote-sensing image is divided into a number of groups of similar characteristic of the image values and then classified into classes, without any knowledge of ground truth data. Two unsupervised classification algorithms—k-means and its variant, and the iterative self-organizing data analysis (ISODATA) technique—are the most commonly used classifiers. Both give the same set of clusters when a number of clusters are same. In k-means number of cluster knowledge is priori while in ISODATA no need of prior knowledge of number of clusters. In this work, we focus on some unsupervised classification technique and combine with some supervised technique. Here, we used the k-means and SVM classifier with sigmoid kernels to get better classification accuracy.

Hyperspectral remote-sensing datasets are represented as a 3D data cube with spatial and spectral information such that X–Y plane contains spatial information and Z-direction contains spectral information. The hyperspectral datasets have more than 200 narrow and contiguous wavelength bands at bandwidths of about 5–10 nm. Dataset used for this study is obtained from EO-1 Hyperion satellite which does not has the ground truth data.

2 Background

Mineral mapping is used to map different types of minerals with their contents and characteristics. It is one of the important applications in high resolution of remote sensing data by hyperspectral technique. So in the application of remote sensing technology and hyperspectral technology, the analysis of rocks and minerals has superiority to traditional ways. Many research have been presented by analysing mineral spectra of visible and near-infrared (VNIR) bands which give a promising identification model of minerals with acceptable accuracy to acquire the mineral category and content in rock images taken from hyperspectral sensor. Satellite and airborne images are used to classify crops, examine their health, finding disease, etc.

Classification of hyperspectral image are based on both spectral and spatial features. In reference to [4], their proposed work is based on low spatial resolution because spatial features can be changed within metres that affect different factors of the images, such as imperfect imaging and atmospheric scattering while detecting reflectance. Other factors which degrade image quality are sensor noise and secondary illumination effect, spatial resolutions helps to remove these effects to improve in quality of hyperspectral images.

In earlier study, many of the works are done in the field of remote sensing and mineral exploration in different areas. According to [5], work is done on spectral analysis and mapping of different minerals in part of Latehar and Gumla District, Jharkhand. In this study, used EO-1 Hyperion data for AL + OH mineral from rocks first compensate the atmospherical effect from data and MNF transformation is used to reduce the data noise from it. To find out the end member as the target member apply singular angular mapping and matched filtering in it.

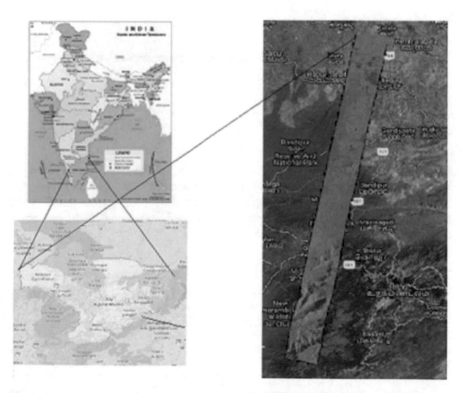

Fig. 1 Satellite view of Nilgiris hill

3 Study Area and Data

Study area [6] is located in the state of Tamil Nadu, southern India at latitude 11 08′ to 11 37′ N and longitude of 76 27′ E to 77 4′ E shown in Fig. 1. Minerals present in Nilgiris district of Tamil Nadu are iron ore, bauxite and clay utilize in different industries. There is an excellence of different types of minerals distributed in different regions with different compositions. Iron ore and bauxite are found near Kotagiri. Magnetize is found close to Thengumaranada in Moyar valley and Quartz is in Devala and Clay in the constitution of China clay arises in Cherambadi of Nilgiris district.

Hyperion EO-1 is a US satellite which is used to collect the data from space. This type of data has 242 narrow continuous spectral bands with spectral variety between 0.4 and 2.5 m at 10 nm interval calibrated with 16 bit resolution and 30 m spatial resolution.

4 Methodology

In this section, we described a pixel-based classification technique in unsupervised way.

4.1 Data Pre-processing

Pre-processing in high dimensional data is one of the important challenges. Pre-processing makes the data more accurate and noise-free; so that, it can give better result. Pre-processing will remove all the dead pixels, water band, noisy pixels, etc. In hyperspectral data, pre-processing will remove all bad band, noisy band and zero band instead of pixels.

Zero Band Removal
The bands which do not have any pixel information in hyperspectral data is called zero band. By using ENVI software (used for visualizing a geological data), we visualize that some set of bands are zero band which we need to remove it. In EO-1 Hyperion dataset's, zero bands are listed in Table 1.

Table 1 List of zero band

S. No.	Zero bands	Reason
1.	1–7	Zero bands
2.	58–78	Overlap region
3.	120–132, 165–182, 218–224	Water vapour absorption
4.	184–186, 225–242	Bad bands

Destriping of Band

Vertical stripe may occur in the region where brightness of pixel varies relatives to nearby pixels. These stripes make the image unclear and contain pixels with wrong information that will give a negative impact on further processing algorithms. Using local destriping algorithm, we can remove this type of strips to some extent. Used equation by the algorithm is

$$\sum_{j=1}^{n} [(x_i - 1, j, k) + (x_i + 1, j, k)]/2n \tag{1}$$

Atmospheric Corrections

The reflected solar energy travels through the atmosphere. Based on the amount of atmospheric reflection, types of particles and gases available, atmospheric absorption and atmospheric scattering, light interacts with atmosphere and materials and reflected energy store in the form of spectrum. Atmospheric correction is compulsory to remove all these unwanted effects. FLAASH stands for fast line-of-sight atmospheric analysis of hypercube and able to process wavelengths in VNIR and SWIR region up to 3 μm. In this study, we used ENVI for atmospheric correction. FLAASH will also be able to remove adjacency consequence, cirrus and opaque cloud map and also compute a scene-average visibility. This removes all water vapour windows from the images. After this step, all the bands present in the dataset are noise-free. FLAASH will also be able to remove adjacency consequence, cirrus and opaque cloud map and also compute a scene-average visibility. This removes all water vapour windows from the images. After this step, all the bands present in the dataset are noise-free.

4.2 Dimensionality Reduction

Principle component analysis (PCA) is a dimensionality reduction technique. It is used as feature extraction, and in hyperspectral data, feature extraction can be done by band selection. This process is defined as reducing the number of random variables under consideration, by obtaining a set of principal component, i.e. selection of band. In high dimensional data, feature extraction can minimize execution time for hyperspectral data. PCA is based on eigenvalue decomposition of covariance matrix. Let us consider hyperspectral image is of $M \times N \times B$ size. Pixel vector is calculated using all bands as in stack as shown in Fig. 2. In [7], show pixel vector in hyperspectral images.

$$X_i = [x_1, x_2, x_3, \ldots, x_n]$$

where B is number of band and M and N are number of rows and columns, respectively, $i = 1, 2, 3, \ldots, M_1$ and

$$M_1 = M \times N$$

Mean will be calculated by

$$m = \frac{1}{M_1} \sum_{i=1}^{M_1} X_i = ([x_1, x_2, x_3, \ldots, x_n])^{\mathrm{T}}$$

Covariance matrix will be calculated by

$$C_x = \frac{1}{M_1} \sum_{i=1}^{M_1} (X_i - M)(X_i - M)^{\mathrm{T}}$$

For eigenvalue decomposition of the covariance matrix

$$C_x = \mathrm{ADA}^{\mathrm{T}}$$

where $D = \mathrm{diag}(\lambda_1, \lambda_2, \lambda_3, \ldots, \lambda_N)$ is the diagonal matrix composed of the eigenvalues $\lambda_1, \lambda_2, \lambda_3, \ldots, \lambda_N$ of the covariance matrix C_x and A is the orthonormal matrix composed of the corresponding N dimension eigenvectors $a_k(k = 1, 2, \ldots, n)$ as follows:

$$A = (a_1, a_2, \ldots, a_n)$$

The linear transformation is

$$Y_i = A^{\mathrm{T}} X_i (i = 1, 2, 3, \ldots, M_1)$$

is the PCA pixel vector, and all these pixel vectors form the PCA bands of the original images.

Fig. 2 Pixel vector

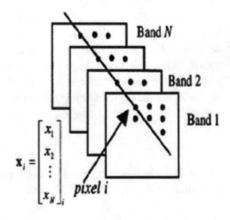

4.3 Classification Approach

Unsupervised classification can be defined as the identification of classes by grouping the pixels of similar type within one class. Based only on their statistics without using previous knowledge about the spectral classes presented in the image, pixels are clustered into classes. K-means is one of the commonly used methods in unsupervised classification. But we need prior knowledge of number of classes is prerequisite in K-means. To estimate the number of cluster, we used elbow method.

Elbow Method
To find out the k (number of cluster) value, use elbow method. According to [8], it examines the percentage of variance as a function of cluster k. It runs k-means on the dataset for the range of value k, e.g. 1–10 and for each value of k, it calculates the sum of squared error (SSE). And then, it plots a graph of SSE for each value of k. After that observe the graph and check that at which value of k graph go flatten that would be an optimal number of cluster. In Fig. 3, red-dotted circle shows that after 4, the graph start flatten and after 6, it becomes flat. So, the optimal number of classes possible is six, i.e. k will be equal to six.

K-means Clustering
After finding the number of classes, next, we run k-means on the extracted end member. The main idea behind k-means is to initialize one value to each cluster which is called centroid or mean. These centroid values are assigned to each clusters and then pick each pixel value and assign it to those cluster which nearest to centroid. Next task is to recompute the centroid for each cluster and repeat the process until convergence criteria are met.

We applied k-means only on extracted end members which give high probability of correct classification. It classifies the image into four classes such that each index of the image is corresponding to one of the class. Because we do not have ground

Fig. 3 Elbow graph

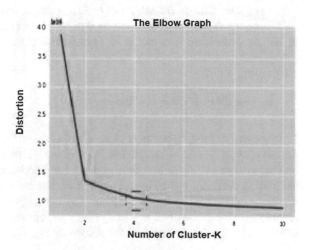

Fig. 4 Ground truth
generated by k-means

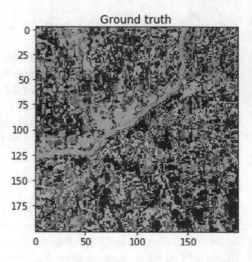

truth data we consider this as a ground truth data. Figure 4 shows the obtained ground truth data. Because ground truth data is not available to check the accuracy of clustering method is challenging task. Still, there are some methods used to check that how better is clustering result. Davies-Bouldin index is one of those methods that can be used to evaluate the model. Where, a lower DB index relates to a model with better separation between the clusters and vice versa. In this k-means algorithm DB index score is equal to 0.316.

SVM Classifier

In this work, support vector machine considers for multiclass problem of hyperspectral imagery. SVM is supervised technique, so k-means result is considered as a ground truth labelling. According to [9], SVMs perform better than other classification techniques and also in pattern recognition and provide higher classification accuracy. Furthermore, SVMs also give good accuracies even in the presence of heterogeneous classes for which only few training samples are available. In this, different kernels can be used based on dataset and set of problems. Here used kernel in tangent-based function called as sigmoid function.

5 Experiment and Result

In proposed method, hyperspectral image is of size $M\times \leftarrow N \times B$ size. Where B is the number of bands in image and $M \times N$ is size of each band. Total number of bands in dataset is 224 after removal of all bad bands and noise correction left band are 138 out of 224. Next PCA is applied on left bands. And out of 138 bands, 12 bands are selected as an end member. On these end members, we used k-means clustering which classifies the image into six classes. Now we train our original sample with these labelled data (as shown in Fig. 4), using SVM classifier.

Fig. 5 Predicted result using SVM

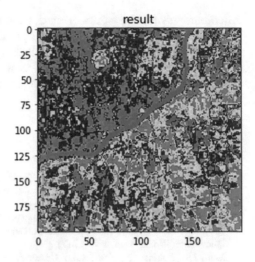

Fig. 6 Spectra profiles of minerals

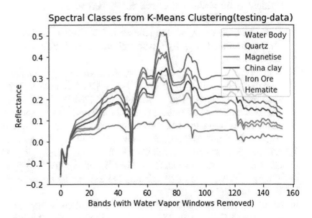

We used some part of dataset as a test data on which we predict result using SVMs as shown in Fig. 5. This is able to classify into four classes successfully and SVM gives accuracy score as 76.03%. Spectra profile of these six classes is shown in Fig. 6. In which, green spectra profile shows water body and rest of the types of minerals. The range of reflectance is in between 0.4 and 2.5, therefore in graph, spectra profile exists from 0.3 to up to 3 μm.

6 Conclusion

Mineral classification from the hyperspectral data is unsupervised classification because labelled data is not available. K-means is used here to classify the data into desired classes. Hyperion data with 242 bands used for classifying minerals.

After removing all noise from the data, band selection performed using PCA. Then, K-means applied on selected band and got the labelled data which is further used for SVM classifier. SVM classifier gives approximately 76.03% accuracy. This classifier also depends on the dataset as well as on different parameters used in SVMs. For the future work, we can work on classifier to improve the accuracy.

References

1. Zhu, X.X., Tuia, D., Mou, L., Xia, G.-S., Zhang, L., Xu, F., Fraundorfer, F.: Deep learning in remote sensing: a comprehensive review and list of resources. IEEE Geosci. Remote Sens. Mag. **5**(4), 8–36 (2017)
2. Sawant, S.S., Prabukumar, M.: Semi-supervised techniques based hyper-spectral image classification: a survey. In: Power and Advanced Computing Technologies (i-PACT), 2017 Innovations in IEEE, pp. 1–8 (2017)
3. Mou, L., Ghamisi, P., Zhu, X.X.: Unsupervised spectral–spatial feature learning via deep residual conv–deconv network for hyperspectral image classification. IEEE Trans. Geosci. Remote Sens. **56**(1), 391–406 (2018)
4. Villa, A., Chanussot, J., Benediktsson, J.A., Jutten, C., Dambreville, R.: Unsupervised methods for the classification of hyperspectral images with low spatial resolution. Pattern Recogn. **46**(6), 1556–1568 (2013)
5. Satpathy, R., Singh, V.K., Parveen, R., Jeyaseelan, A.T.: Spectral analysis of hyperion data for mapping the spatial variation of AL + OH minerals in a part of Latehar & Gumla district, Jharkhand. J. Geogr. Inf. Syst. **2**(4), 210 (2010)
6. Vigneshkumar, M., Yarakkula, K.: Nontronite mineral identification in Nilgiri hills of Tamil Nadu using hyperspectral remote sensing. In: IOP Conference Series: Materials Science and Engineering, vol. 263, no. 3, p. 032001. IOP Publishing (2017)
7. Ranjan, S., Nayak, D., Satish Kumar, K., Dash, R., Majhi, B.: Hyperspectral Image Classification: A k-means Clustering Based Approach, pp. 1–7 (2017)
8. Kingrani, S.K., Levene, M., Zhang, D.: Estimating the number of clusters using diversity. Artif. Intell. Res. **7**(1), 15 (2017)
9. Moughal, T.: Hyperspectral image classification using support vector machine. In: J. Phys.: Conference Series, vol. 439, no. 1, p. 012042. IOP Publishing, (2013)

Accountable Communication
in Ubiquitous Computing

I. S. N. Pradeep, K. Athmaram and K. Mritymjaya Rao

Abstract This is a cognitive computing era with Internet of things (IoT) revolution, and ubiquitous computing concept is going to come alive soon. Today, we depend on our smart devices almost for every reason. Most of the gadgets designed to habituate private information sharing mechanisms to fellow users via social media. However, dependency and security are two sides of the coin. With reference to existing literature, various means of security recommendations in server/hosting device/cloud point view like private set insertion (PSI) protocol, private information retrieval (PIR) protocol, etc. but inadequate details present on how to control the sharing at end-user's end. The research is motivated by practical usage of certain gadgets, while they started sharing the statistical reports in social media, along with important private information, where we could not control the flow even after the uninstallation/disabling the usage of the app. Thus, the purpose of study is to portray that privacy matters are not seriously considered by citizens as gadgets give fun and dynamic adoptability for distinct services. Many a times, the information is being shared is irrespective of the actual needs of what to share. Thus, the research tried to provide a client-level ergonomic solution for privacy control. Our contribution towards this research is to build a simple user interface (UI)-based accountable information sharing method with the help of multimedia elements (text, images, audio and video) and integration of notifications in regional language by using Unicode to interact with all types of end-users in order to build an acceptable pervasive (ubiquitous) computing systems with reliability to serve all kinds of business or domestic deployable purposes with minimized social security anomalies.

Keywords Ubiquitous computing · Privacy and security · Interactive · Multimedia · Accountable communication · Open source

I. S. N. Pradeep (✉) · K. Athmaram
Department of Computer Science, Dravidian University, Kuppam,
Andhra Pradesh, India
e-mail: isnpradeep@gmail.com

K. Mritymjaya Rao
Department of Computer Science, Kakatiya University, Warangal,
Telangana, India

© Springer Nature Singapore Pte Ltd. 2020
P. Venkata Krishna and M. S. Obaidat (eds.), *Emerging Research in Data
Engineering Systems and Computer Communications*, Advances in Intelligent
Systems and Computing 1054, https://doi.org/10.1007/978-981-15-0135-7_26

1 Introduction to Smart Communication

Evolution has changed from multi-user—single device to single device—single user at initial decades; but today it has even improved to single user—multi-device style with the introduction of smart computing era with the gadgets that act smart with the help of internet but not fully intelligent devices yet. The term 'smart' defines the functionality being adopted in life styles with all these latest gadgets like smartphones, computers, music devices and smart home appliances, etc. So according to the increased usage, the way users get connected to these devices is getting changed by means of simple homogeneous to huge range of heterogeneous users and powerful lightweight—autonomous devices connected via server-based or even server-less heterogeneous architectures via cloud computing paradigm. Such technologies aim to impart computers in the form of micro-devices via Internet of things (IoT) platform in our everyday life with respect to various domains like public services, education, health, etc. It could seem tremendous to enjoy info-sharing today, but once the ubiquitous computing vision becomes the way of life in near future, everything will be connected to everywhere for infinite reasons and information sharing from sensors might become out of control where big data plays a vital role since huge amounts of data needs to be shared among such micro-devices, with a danger of unnoticeable surveillance causing loss of personal privacy. Mobile devices with IoT produce some problems such as privacy violations, inappropriate location information and difficulty of controlling at end-user level. Preservation of privacy with several protocols and achieving efficiency plus scalability of systems became difficult in big data era. Thus, privacy preservation in recent years along with transmitting sensitive data via sharable network has got considerable attention [1], but emergence of these devices has created new security problems. So homogeneous security policy is not desirable.

Today's smart world opens up four main possible ways to interact with smart applications and devices: (1) various independent services in the form of mobile or desktop apps, smart gadgets connected to web that are developed and constantly available via Internet for minimizing the end-user's need to physically interact with service providers. (2) Various independent devices are interconnected in an IOT model with constantly available smart apps via Internet for minimizing the end-user's need to interact with computers as computers. (3) Interdependent service providing devices connected to each other to maintain accountability of traffic with respect to demand and supply, finance inflow and outflow and many other services that are evaluated in real time based upon the business needs. (4) Systems in ubiquitous computing are usually *distributed, deal with personnel data of users,* and *that are relative to context.* Middleware is also a very essential layer to be organized in this architecture [1], which bridges communication in heterogeneous environment.

Fig. 1 Reference image that shows the smart supermarket environment [2]

In Fig. 1, for example, while we enter a smart supermarket equipped with automatic features like floor sensors for foot traffic, smart shopping cart, cashier-less checkout, etc. features, everything is accountable in this context with respect to interconnected communication among all the devices and a central server that monitors every automatic feature and sends seamless responses to the smart devices across the network.

2 Privacy Policies and Risks Involved in Smart Communication

Due to the enthusiasm involved in buying new smart devices or opting for a free app service or sometimes using an app that comes as a package with a new device, majority of the users accept terms and conditions/privacy policies of sharing the device storage or personal information without even reading the policies because user's tendency is tuned to enjoy a newly purchased gadget or recently downloaded app as their right for the optimum usage, without any knowledge of inventions, intentions and background happenings. Sometimes though users know that they are going to share their personal information, they just accept the terms and conditions by the app provider in order enjoy the fun but who knows if it is safe or not?!

The below case study portrays a serious privacy issues to the research community.

2.1 Smart Wristbands—A Simple Case Study for Privacy Issues

Mi-Fit wristbands are considered here as simple example case study. Today, users monitor their walk cycles, sleep intensity and other health information of a person and to provide graphical-visual output to the user in an impressive and interesting manner by using multimedia, but these devices use smartphones as gate ways to share the private information mixed with daily reports to compare with fellow social media users, whether they are known or unknown. It analyses sleep quality, walk cycles and provides a statistical synopsis like: 'You are ahead of so and so number of users!', etc. to excite the users to perform more. Also, such any personal information will not only be open to social media once the permission–acceptance is given by the users by any chance, but also the manufacturer of such bands can have all the personal and non-personal data also from user's mobile, as the product's privacy policy says. Sharing such personal data may or may not harm in any way at this point of time, and also the research is not against the usage of any such health bands too; but the research focuses to provide an accountability mechanism in info-sharing methods online whether the user is aware or unaware of the situations so that the same can be extended to ubiquitous systems where number of devices act as computers that are connected everywhere around us for real-time interaction of devices with humans.

Joey Ranting describes in his blog when Norton's report [3] cautioned him about the privacy risks of the same Mi-Fit band and app by Xiaomi that he might have installed the app from Google play store and even after the uninstallation of the same app, Norton Mobile still cautioned about the risk, and then he read the privacy policy of the app, which states that Xiaomi collects both personal data to use to identify the user and non-personal data from mobile, including accessing contacts, emails, phone numbers of friends and camera as well to mention a few, the below pictures say it all (Figs. 2, 3 and 4).

Fig. 2 Health reports from Xiaomi Mi-Smart Band's sharable statistics via social media [3]

Fig. 3 Norton's report on Xiaomi Mi-Fit app of wristband from Joey Ranting's Blog [4]

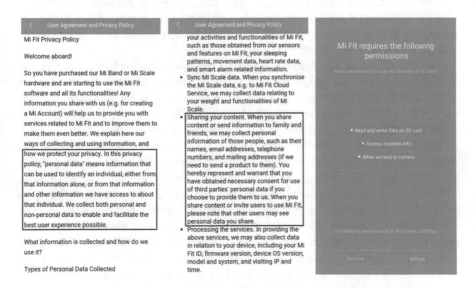

Fig. 4 Xiaomi Mi-Fit app's privacy policy of Mi-Fit band from Joey Ranting's Blog [4]

3 Ubiquitous Systems—Overview

Ubiquitous computing is a proposed fully functional futuristic environment in which infrastructure is set up in such a way that interfaces will usually be intuitive enough and all these micro-computers will be 'invisible', essentially blending into the periphery [5] to serve humans in real-time situations spontaneously based on pre-collected data or processed data time to time. With a simple smart assistant like wrist health band, if we can observe this level of data sharing model across social media, imagine what if we interact dynamically with micro-devices that possess all our data in large scale in UbiComp environment, while we move anywhere in the world. Number of peculiarities persist in UbiComp that varies from conventional computing (examples are listed in [6]).

4 Methodology and Techniques Used in Ubiquitous Scenario

The dream of 'ubiquitous computing' is to have 'interconnected computers—invisibly everywhere' through which we interact seamlessly with the world around us. Ubiquitous computing is a changeable environment based on the requirements of specific applications. These are categorized as context-aware applications, in which context means any information that characterizes an entity like: location/name/weight/health reports of an end-user. Creating a smart environment saturated with the maximum communication capability is the prime purpose of the ubiquitous computing [7]. In [7], various needs of UbiComp and essential characteristics of ubiquitous projects are mentioned clearly for study.

The innovation of this research on UbiComp model (using any device, in any location and in any format) is to refine the apps that share user's information via micro-devices to a point where the usage of such devices by end-users is made transparent with the help of multimedia elements and Unicode usage in UI designing in a simple manner. Basic design study was taken and understood from [8].

When it comes for methodology, based on multilateral security conceptual approach from [9–11]: 'each party is required to only minimally trust in the honesty of others, thus considering them as potential attackers. In a heterogeneous UbiComp environment, security and privacy conflicts are the norm and should be solved through a negotiation process. The concept implies what degree of security can be achieved against whom with respect to what kind or data of functionality, and what privacy setting is desired to be enforced against whom and in which context' [12]. Since 'privacy' is an important entity which is the resultant of negotiating and enforcing conditions like how, in which context, to what extent, which data of entity—is disclosed to whom, and when? [12].

Thus, we gave at most importance to privacy via an innovative UI design approach as an outcome of this research study. As it is observed that thousands of experiments have been already conducted on automation and imparting cognitive intelligence for devices globally; as a matter effect, where there is dependency, the need of perfect procedure and reliable protocols for zero communication gap. In this context, *transparent accessibility, loss of control* and also *self-governess* are a few problems from a number that may be arisen which may threaten the security and privacy of individuals in ubiquitous systems, apart from many of its obvious benefits [12].

5 Expected Services Versus Privacy Scenarios in Ubiquitous System

The UbiComp environment needs not be familiar to users where sensors collectively collect user's personal data, which is very sensitive. In this context, protecting the user's privacy has central importance. Users' relationship with devices in UbiComp environment dynamically changes so as the rights of accessing also change with the mechanisms of generation of data and many a times the type/group of users is not a predetermined criteria. Users of a correct set determine how a piece of information the system is associated with, while it is being produced [13]. A few upcoming intelligent services at a proposed Smart Home Model along with assumed sharable user's data is formatted in Table 1 as sample cases.

Not only the above-given domestic cases, but also intelligent medical/public services require all our details time to time that are very sensitive with respect to privacy.

Table 1 Smart Home Model with pre-assumed capacities and data sharing requirements

Intelligent device	Cognitive capability expected	User's data required to be shared for tasks
Intelligent chair	Understands your posture, sitting duration, reminds you about health precautions	Any neck- or backbone-related problems
Intelligent floor	Understands the moments of people and responds to various needs	Source and destination points, activities
Intelligent A/C	Understands the room occupation and senses the temperature outside and acts accordingly	Not required
Intelligent desk	Understands everything on your desk and gives the indications, reminders if required	Gadgets, work calendar, appointments and many details
Intelligent ceiling	Understands everything under the roof and responds to the real-time situation by giving instructions to the other electric devices	Not required
Intelligent mirror	Gives instructions when some changes occur on your face or body	Needs all your biometric details along with moles etc.
Intelligent touch screen	Interacts with all needy resources right from personal gadgets to social media navigations	Requires all the confidential users data

Data Discretion With the direct access and control over live information or recorded data being produced or available at 'real-world' situations, users need not have access in elsewhere situations [13]. In [13], a 'smart room' augmented model —case study was taken to test with a variety of sensors, and performance evaluation has been done to create a trust-worthy UbiComp environment.

6 Cognitive Hacking—An Assumed Threat in UbiComp Systems

'Cognitive hacking' is a newly proposed term out of this research with an assumption that what if the interests and disinterests of a user are captured by various means and shared, stored and monitored across social media for some destructive purposes?! The collected data may be misused by hackers or other organizations who can capture this chance as their opportunity. Such involuntary data extraction is assisting the user for health or pre-cautionary factors besides causing a backdoor facility to unauthorized members who contribute their efforts leads to damage rather supporting any kind of social constructive aspects, which may lead to downfall of humanity instead of helping our society. Such kind of data thefts are presumed with the name 'cognitive hacking' in this research, as one of the proposed futuristic risks.

For example, based on social media posts and reports from devices in UbiComp environment, if a user's or group's behaviour study is conducted illegally by some hacker(s) based on everyday's active information that is being captured, and if at all they are able to make a detailed list of people who suffer from depression or loneliness kind of scenarios, along with their health reports by exploiting the social media and the proposed intelligent systems to make out a new marketing strategy that can attract such specific user group and sell drugs in some other form to make them to get addicted to such products; imagine the results! This is some kind of unethical hacking scenario where the science can be used for downfall of humanity instead of helping our society. Such kind of data thefts are presumed with the name 'cognitive hacking' in this research, as one of the proposed futuristic problems to extend this research.

7 Techniques and Fact Findings from the Literature Study

To satisfy various security needs, a principle of distributing users' data among several set of servers with isolated user authenticated mechanisms for accessing only the required data stored in particular server (described as Nodes) will be randomly distributed across a virtual structure—without having all the data sets in any server. In this architecture, no single Node contains entire data or structure to store the data. A Node knows only its predecessor and successor. At least one certificate is to be issued to each user for the domain affiliation to access the server. A tree structure is followed in a randomized format to distribute the domains over the Nodes. User's private data will be on distinct set of servers with only necessary information to be accessed [1]. This is a great proposition of storing and accessing private users data via a distributed architectural manner, but this is only subject to the server-level security and privacy matter and client device that is directly accessed by user won't have any possibility of controlling what, when and to whom the data to be chosen to share.

8   Summary of Interpretations and Implications

1. Though the distributed model as described above in paragraph throws light on server-level privacy protection feature to secure private data of users, a client-level privacy control mechanism is to be designed in UbiComp or IoT models.
2. Users should be able to understand and control what kind of data is getting shared across the network and when to switch on and off the sharing, also the users should have total access to monitor and control which information is getting shared at live situation, also should be able to alter/update/delete the personal data when and then required, based up on one's own will and wish.
3. Context of every interaction in the environment should be notified to users.
4. While connecting the mobile devices to the UbiComp network, the screens of seeking permission in the interface should contain relevant icons, gif animations and other necessary multimedia elements to be easily understood.
5. User should be able to understand the clear instructions and information sharing details in the form of Unicode text of their own/preferable language, along with a voice description option as well on demand. This is to implement 'Think Global, Act Local model' for any area users across the globe.
6. User interface designs to instruct and interact with connected devices in the proposed ubiquitous computing model should be as simple as possible. Thus, simple technology platforms and easily understandable tools are to be used for any kind of users with clear and animated instructions help reduce redundancy of understanding the typical layouts while sharing and interacting.
7. Last but not least, open-source methods are required to be adopted in order to build human-centric computing to maintain transparency with developer community involvement, which helps to reduce autonomous control of devices over humans.

Though certifying such open tools would be a difficult task [13], proper guidelines/features are to be mentioned by ubiquitous environment developers so that to achieve great flexibility and specified features in open source itself, and it is useful for contributing great updates to develop and deploy—at par with standard certified software tools. It also helps to explore and bring an inclusive and acceptable green communication network platform as well—as an economical solution for countries under development since building UbiComp environment is usually a costly affair.

9 Accountability in Ubiquitous Computing

For achieving accountability in UbiComp systems, a few ideas are implemented in a sample mobile app for control of smart home user's privacy protection purpose and same thing can be applied in an IoT architecture too as provided in the screens below (Fig. 5).

Apart from health/public service areas, the Smart Home Model offers a great opportunity with ubiquitous computing that enables to augment lives of citizens with increased communications with powerful connectivity, better awareness and pervasive functionality [14]. Invisible computing with micro-computers connected everywhere around us with distributed and context-aware computing are the characteristics of ubiquitous computers, as described by Weiser [15]. There are seven challenges to check as per [16]. Connectivity across the globe with mobility is the key feature used to open gates for various opportunities along with various difficulties revealed in real-time scenarios. Thus, that indicates that these embedded devices are integrated in the distributing computing environment where every device is wirelessly connected and can process communication among themselves without any help of a server too.

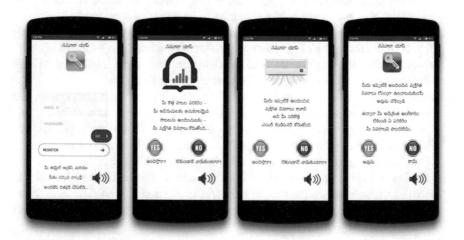

Fig. 5 Our solution for controlling private data of users to be shared via smart devices across the network in user's regional language with Unicode and multimedia elements support like text, images, audio and if required videos as well so that user can opt in or out of sharing one's own information at any time by understanding the situations well according to the need

10 Conclusion

So the need of the current hour was to understand how the interconnected devices communicate with each other as independent computers and to derive several important recommendations to build secured UI reference model via a user-friendly app with multimedia support integrated, while the private data sharing permissions are properly understood and controlled right from the user's level itself.

As per the objective of research is to not to lose the control of any private information of end-users by any means in this Omnipresent and Pervasive information processing system.

So with the above specifically derived points that shall help to provide accountability in UbiComp model, the app designers and code developers can ergonomically build the user-friendly interfaces with the help of multimedia elements in user's own language features included.

Cognitive hacking is a new term derived in the research so that new researchers can take privilege to develop solutions to avoid such attempts via social media in future. Privacy and data security are the two key aspects, while users are connected to and depend upon the ubiquitous computing network in future. Thus, the observations are presented based on several literature reviews to support researchers and user interface designers for deriving the future directions of their research to develop solutions that are possible with UI designing and development.

References

1. Yaici, M., Ameza, S., Houari, R., Hammachi, S.: Private Data Protection in Ubiquitous Computing
2. Image Source: Ossia Inc.
3. Images Source: Xiaomi Mi-Fit App
4. Joey Ranting's Blog
5. Rosenheck, L.: Learning with Ubiquitous Computing
6. Posland, S.: Ubiquitous Computing: Smart Devices, Environments and Interactions. Wiley Publishing (2009)
7. Mishra, A.K., Yadav, P., Singh, L.: Essential Characteristics for Ubiquitous Projects
8. Sharma, P., Goel, N.: Security Issues in Ubiquitous Computing: A Literature Review
9. Pfitzmann, A.: Multilateral Security in Communications, Chapter Technologies for Multilateral Security, pp. 85–91. Addison-Wesley-Longman (1999)
10. Rannenberg, K., Pfitzmann, A., M¨uller, G.: Multilateral Security in Communications, Chapter IT Security and Multilateral Security, pp. 21–29. Addison-Wesley-Longman (1999)
11. M¨uller, G., Rannenberg, K., (ed.). Multilateral Security in Communications. Addison-Wesley Longman (1999)
12. Gudymenko, I., Borcea-Pfitzmann, K.: Privacy in Ubiquitous Computing

13. Duan, Y., Canny, J.: Protecting User Data in Ubiquitous Computing: Towards Trustworthy Environments
14. Creese, S., Reed, G.M., Roscoe, A.W. Sanders, J.W.: Security and trust for ubiquitous communication. In: Keith Edwards, W., Grinter, R.E. (eds.) At Home with Ubiquitous Computing: Seven Challenges
15. Keith Edwards, W., Grinter, R.E.: At Home with Ubiquitous Computing: Seven Challenges
16. Weiser, Mark: The computer for the 21st century. Sci. Am. **265**(3), 66–75 (1991)

Prediction of Real Estate Price Using Clustering Techniques

C. Pradeepthi, V. J. Vijaya Geetha, Somula Ramasubbareddy
and K. Govinda

Abstract A standout among the most difficult assignments continuously condition is the expectation of house sales. Understanding the different components that impacts house sales is as much as critical on knowing a technique on the most proficient method to play out the forecast. Henceforth, it is a most extreme important to distinguish the relationship among various properties utilized in the dataset. These datasets can be received from different open databases and information repositories. Even though it has two distinct models (predictive and descriptive) for extracting the data, utilized grouping strategy for existing model, the support for utilizing such technique is clarified in this segment. With respect to the above expressed fundamental facts in this article, we play out the expectation.

Keywords Prediction · K-means · DBSCAN · Density-based clustering · Farthest first clustering

1 Introduction

The real estate sales estimation plays a vital role in day-to-day life. A person must be well knowledgeable about the terms and information about the selling conditions, and what are all the attributes or things that play the major role in buying a properly built house based on some legacy datasets in his fingertips. Real estate sales estimation is important as it provides the answer for various questions in one's mind that do we really want to buy the house, do we get all necessities in and

C. Pradeepthi · V. J. Vijaya Geetha
SOET, SPMVV University, Tirupathi, Andhra Pradesh, India

S. Ramasubbareddy (✉)
Information Techgnology, VNRVJIET, Vignana Jyothi Nagar, Nizampet Rd, Pragathi Nagar, Hyderabad, Telangana, India
e-mail: svramasubbareddy1219@gmail.com

K. Govinda
SCOPE School, VIT University, Vellore, Tamil Nadu, India

© Springer Nature Singapore Pte Ltd. 2020
P. Venkata Krishna and M. S. Obaidat (eds.), *Emerging Research in Data Engineering Systems and Computer Communications*, Advances in Intelligent Systems and Computing 1054, https://doi.org/10.1007/978-981-15-0135-7_27

around the house, does the house worth for the money we are spending, etc. There are a few essential data mining strategies which have been creating and utilizing in data mining ventures as of late including affiliation, arrangement, grouping, expectation, successive examples and choice tree. Clustering is a data mining procedure that makes a significant or helpful cluster of articles which have comparable attributes utilizing the programmed strategy. The clustering strategy characterizes the classes and places protests in each class. There are distinctive sorts of clustering techniques, for example, k-means, agglomerative hierarchical clustering and DBSCAN and so on. From many techniques, here we have performed five clustering algorithms, and finally at last, we have contrasted the outcomes with discover which calculation is more effective when contrasted with alternate calculations.

2 Background

Precious stone and Cambell, together with Lowengart, propose the meaning of business advancement [1], describing it as here and now incitements of procurement or offers of an items or administration. To include, Lowengart, after the investigation of numerous sources recommend more exhaustive definition, depicting it as various strategic (yet not vital) things created as a picce of vital promoting system, with the point of increasing the value of the results of administrations to accomplish a predefined advertising target. Albeit changed inside various gatherings and focused on business sectors, sales advancements are turned out to be a successful device in expanding the sales, notwithstanding the relative simplicity of estimating this viability [2]. It is additionally considered as a compelling strategy while focusing on sales advancements to a scope of the specific specialties, and furthermore while endeavoring to build the level of separation, so as to grow more exact crusade with the high shots for progress [1, 3, 4]. In any case, Uva and Lichtenstein pull out about the risks identified with the diminished apparent esteem if the business advancements are inappropriately focused on, and Uva likewise sees that the best execution of offers advancement should be possible giving the predominant item quality and esteem, yet which has an indication of low item mindfulness, inferring the way that it very well may be a decent instrument amid the presentation of another item. Kotler takes note of that business advancements can be utilized as an option for progressively expensive publicizing efforts, while Philip Jones takes note of that there are the perils of the costs advancements identified with the sensational diminishing in benefits as the aftereffect of inappropriately created strategies of offers advancements. To include, Low and Mohr claims that brands with higher spending on publicizing, contrasting with sales advancements, normally have more positive client demeanors, higher estimation of brand and higher piece of the overall industry. Kotler claims that while promoting is utilized as a long haul technique for improving the brand esteem [5], sales advancements are generally utilized keeping in mind the end goal to make a transient interest for the items, consequently sales advancements are effectively and progressively utilized by brand chiefs because of their adequate level in a portion of the types. Considering, it is an advantage for each advertiser to audit such an imperative instrument as sales advancements. As indicated by Srinivasan and Anderson [6], and furthermore d'Astous and Landreville [7], there are a few fundamental goals for executing sales advancements, which were uncovered in their topics about. A standout among the most apparent purposes behind sales promotions implementation is to expand the sales, yet it likewise utilized for expanding the consciousness of effectively existing item and empowering the sales amid the off-top time.

3 Methodology

The most ideal approach to clarify the proposed way is through the philosophy, here in this proposed demonstrate for sales anticipating we have utilized a few grouping procedures, for example, basic k-means, hierarchical clustering, density-based clustering, filtered clustering and farthest first grouping. This model not just gives the investigation results from the dataset utilizing different clustering procedures yet additionally thinks about the aftereffects of each grouping calculation and gives the most appropriate calculation among them. Before proceeding onward to the means on the most proficient method to play out the examination, we ought to comprehend why we have utilized clustering for this sales estimating. The most widely recognized type of unsupervised strategy is clustering which prepares for finding unlabeled information structure. Clustering regularly implies gathering in view of similitudes. As we are anticipating the sales for the up and coming day, it is difficult to have an objective property which continues as before all through the procedure. Sales normally implies causes the adjustments in condition all the more much of the time. So it is difficult to utilize order for sales estimating. The dataset [8] we have utilized comprises of a few properties which incorporates formatted date, summary, price, bedrooms, floors, bathrooms and day-by-day synopsis. This dataset does not contain an objective characteristic or class mark consequently it additionally one among the few explanations behind utilizing grouping in this model. The accompanying gives the essential steps to be done before continuing with the clustering methods which called as data preprocessing.

The definitions for different clustering methods are clarified underneath.

Simple K-means clustering
This involves partition of n observations into k clusters with the nearest mean valued cluster.

$$J(V) = \sum_{i=1}^{C} \sum_{j=1}^{C_i} \left(\|x_i - v_j\| \right)^2$$

where '$\|x_i - v_j\|$' is the Euclidean distance between x_i and v_j. 'C_i' is the number of data points in ith cluster.

'C' is the number of cluster centers.

Hierarchical Clustering

This method is used to build a hierarchy of clusters.

$$L(r, s) = \frac{1}{n_r n_s} \sum_{i=1}^{n_r} \sum_{j=1}^{n_s} D\left(x_{ri}, x_{sj}\right)$$

where L and S are different clusters.

Density-Based Clustering

This method does not require a number of clusters rather it builds clusters based on the data.

$$N_\varepsilon(p){:}\{q/d(p,q) \leq \varepsilon\}$$

Farthest First Clustering

This is a variant of k-means clustering.

$$\mathbf{Min}\,\{\mathbf{max}\,\mathrm{dist}(p_i,\,p_1), \mathbf{max}\,\mathrm{dist}(p_i,\,p_2)\ldots\}$$

4 Implementation

4.1 Data Processing

Data preprocessing is an information mining system that includes changing raw information into a reasonable arrangement. Certifiable information is regularly inadequate, conflicting, as well as in specific practices or inclines, and is probably going to contain numerous mistakes. Data preprocessing is a demonstrated strategy for settling such issues. Data preprocessing gets ready pure and raw information for further preparing. Information preprocessing is utilized database-driven applications, for example, client relationship administration and control-based applications (like neural systems). Information experiences a progression of ventures amid preprocessing.

4.2 Data Cleaning

Data is corrected down through procedures, for example, filling in missing qualities, smoothing the inappropriate information, or settling the irregularities in the information.

4.3 Data Integration

Data with various conditions is assembled and collision inside the data is settled.

4.4 Data Transformation

Data is standardized, accumulated and summed up.

4.5 Data Reduction

This progression plans to display a decreased view of the information in an information distribution center.

4.6 Data Discretization

Involves the decrease in various estimations of a constant quality by partitioning the scope of characteristic terms.

5 Result Analysis

The way Weka evaluates the clustering depends on the cluster mode selected. Four different cluster modes are available (as buttons in the cluster mode panel) .

Figure 1 shows instances in X-axis and price in Y-axis, and the model exhibits perfectly defined clusters which can be identified as that cluster 0 (red) and cluster 1 (blue) have separated effectively to its nearest mean value. This farthest first approach has outlined above the accuracy for such spaces are usually high. This is also the situation shown above, where the accuracy is 100%.

Figure 2 shows instances in X-axis and price in Y-axis, and the model exhibits a perfectly defined clusters which can be identified as that cluster 0 (red) and cluster 1 (blue) have separated effectively to its nearest mean value.

In K-means approach, the two clusters red and blue indicates the clusters 0 and clusters 1. The number of instances in X-axis and price in Y-axis. The red cluster rate is high when compared to farthest first clustering approach. The price ranges are even it will be in normal range and some cluster 1 are mixed.

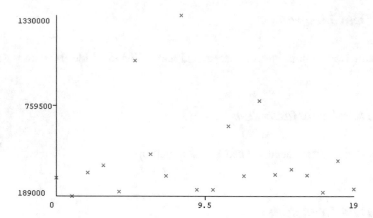

Fig. 1 Farthest first approach

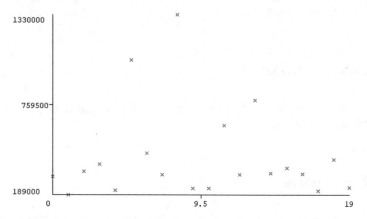

Fig. 2 Simple K-means

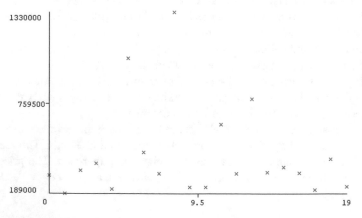

Fig. 3 Hierarchical clustering

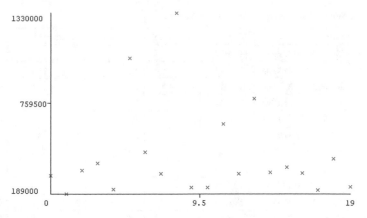

Fig. 4 Density-based clustering

Figure 3 shows instances in X-axis and price in Y-axis, and the model exhibits scattered clusters which can be identified as the cluster 1 (blue) has separated effectively to its nearest mean value.

Figure 4 shows instances in X-axis and price in Y-axis, and the model exhibits perfectly defined clusters which can be identified as that cluster 0 (red) and cluster 1 (blue) have separated effectively to its nearest mean value. The price range is in higher range, and the clusters are also combined in the normal range.

The above-represented graph shows the instance number in X-axis and price in Y-axis. Tables 1 and 2 show the comparison of analyzed results and graphs obtained from those results. As explained before, this model is not only on working with the datasets for sales information and providing the effective algorithm. From Table 1, neglect the Make density-based clustering algorithm because it produces the clusters based on the rather than based on the attributes and it consumes much time than other algorithms, the hierarchical clustering and filtered clustering are also neglected since we cannot define the required number of clusters in hierarchical clustering as defining the number of clusters plays a major role in unsupervised learning; even though it separates the cluster effectively from Fig. 2, the results can be justified. From farthest first clustering can be able to visually see from Fig. 4, the instances are not separated effectively.

6 Conclusion

Data mining has a vast application domain. By the use of the open-source tool Weka, we got a clear insight into how raw data is preprocessed and then based on the type of data it is being subject to either classification or clustering. Beyond the

Table 1 Sample dataset with parameters of weather

id	Date	Price	Bedrooms	Bathrooms	Floors	Grade	Zipcode
7.13E+09	20141013T000000	221,900	3	1	1	7	98,178
6.41E+09	20141209T000000	538,000	3	2.25	2	7	98,125
5.63E+09	20150225T000000	180,000	2	1	1	6	98,028
2.49E+09	20141209T000000	604,000	4	3	1	7	98,136
1.95E+09	20150218T000000	510,000	3	2	1	8	98,074
7.24E+09	20140512T000000	1.23E+06	4	4.5	1	11	98,053
1.32E+09	20140627T000000	257,500	3	2.25	2	7	98,003
2.01E+09	20150115T000000	291,850	3	1.5	1	7	98,198
2.41E+09	20150415T000000	229,500	3	1	1	7	98,146
3.79E+09	20150312T000000	323,000	3	2.5	2	7	98,038
1.74E+09	20150403T000000	662,500	3	2.5	1	8	98,007
9.21E+09	20140527T000000	468,000	2	1	1	7	98,115
1.14E+08	20140528T000000	310,000	3	1	1.5	7	98,028
6.05E+09	20141007T000000	400,000	3	1.75	1	7	98,074
1.18E+09	20150312T000000	530,000	5	2	1.5	7	98,107
9.3E+09	20150124T000000	650,000	4	3	2	9	98,126
1.88E+09	20140731T000000	395,000	3	2	2	7	98,019
6.87E+09	20140529T000000	485,000	4	1	1.5	7	98,103
16000397	20141205T000000	189,000	2	1	1	7	98,002

Table 2 Output comparison for all algorithms

Algorithm	Number of clusters	Time taken (s)	Number of iterations	Clustered instance
Simple K-means	2	0.01	9	0 (58%) 1 (42%)
Hierarchical clustering	2	0.02	–	0 (99%) 1 (1%)
Make density-based cluster	2	0.01	9	0 (58%) 1 (42%)
Filtered cluster	2	0.01	9	0 (58%) 1 (42%)
Farthest first	2	0	–	0 (89%) 1 (11%)

textual understanding of mining, the real-world experience of having worked with one is clearly more beneficial. Among the various types of algorithms present, the farthest first clustering produced the most accurate results for our topic which is product sale information.

References

1. Diamond, W., Campbell, L.: The framing of sales promotions: effects on reference price change. Adv. Consum. Res. Assoc. Consum. Res. **16**, 241–247 (1989)
2. Brookins, M.: Effectiveness of Sales Promotion. Accessed from: http://www.ehow.com/about_5463003_effectiveness-sales-promotion.html (2009). Accessed on 17 Dec. 2011
3. Chandon, P., Wansink, B., Laurent, G.: A benefit congruency framework of sales promotion effectiveness. J. Mark. **64**(4), 65–81 (2000)
4. Boulding, W., Lee, E., Staelin, R.: Mastering the mix: do advertising, promotion, and sales force activities lead to differentiation? J. Mark. Res. **31**(2), 159–172 (1994)
5. Cengage, G.: Sales Promotion. Accessed from: http://www.enotes.com/sales-promotionreference/sales-promotion-178752 (2002). Accessed on 17 Dec. 2011
6. Anderson, A.: Major Objectives of Sales Promotion. Accessed from: http://www.ehow.com/list_6707925_major-objectives-sales-promotion.html (2009). Accessed on 17 Dec. 2011
7. d'Astous, A., Landreville, V.: An experimental investigation of factors affecting consumers' perceptions of sales promotions. Eur. J. Mark. **37**(11/12), 1746–1761 (2003)
8. Abraham, M.M., Lodish, L.M.: Getting the most out of advertising and promotion. Harvard Bus. Rev. **68**(3), 50 (1990)

Study and Analysis of Matrix Operations in RLNC Using Various Computing

I. Jothinayagan, S. J. Sumitha, Kinnera Bharath Kumar Sai
and M. Rajasekhara Babu

Abstract Random linear network coding gained its importance in recent days with its greater potential to enhance the performance of the IoT systems. But the challenging issue is the matrix multiplications and inversions involved in it. Nowadays, with increase in multimedia streaming formats, IoT devices like smartphones will try to make full use of heterogeneous multicore architectures, which are drawing everyone's attention. The approach presented in this paper is the improvement of matrix operations through optimized operations on matrix blocks. We can schedule the operations on matrix blocks in the heterogeneous cores through directed acyclic graph (DAG). The utilization of computer technology to complete the task is known as computing. It is the process of using computer to complete a given goal-oriented task. Here, we make use of different types of computing in order to solve the problem of high computation of matrix operations. RLNC encoding and decoding achieved higher throughputs than already available approaches.

Keywords Random linear network coding · Matrix operations · Parallel computing · Directed acyclic graph

1 Introduction

Random linear network coding is a familiar coding approach for expeditiously transferring data in complicated or lossy networks, like wireless networks. It divides the original data into symbols, combines these symbols linearly based on random coefficients, and sends these coded symbols together with the coefficients [1]. In order to retrieve the original data, the receiver on the other end requires enough coded symbols to solve the linear system given by the random coefficients. But these encoding and decoding operations increase the computation complexity in the

I. Jothinayagan · S. J. Sumitha · K. Bharath Kumar Sai · M. Rajasekhara Babu (✉)
School of Computer Science and Engineering—SCOPE, Vellore Institute of Technology,
Vellore 632014, India
e-mail: mrajasekharababu@vit.ac.in

© Springer Nature Singapore Pte Ltd. 2020
P. Venkata Krishna and M. S. Obaidat (eds.), *Emerging Research in Data
Engineering Systems and Computer Communications*, Advances in Intelligent
Systems and Computing 1054, https://doi.org/10.1007/978-981-15-0135-7_28

IoT nodes. Now, laptops, smartphones, and so on will make use of more number of heterogeneous CPU cores. Here, we develop and measure parallelization methods for boosting up RLNC on heterogeneous multicore architectures. We optimize matrix operations through processing those matrix blocks and improvement of basic matrix operations. We will schedule the tasks involving the matrix block operations with DAG. The results show a significant improvement in sequential time versus parallel time of processing the matrix operations when compared to the coefficient matrix duplication approach [2].

2   Related Work

This portion tells a broad view on the various existing research perspectives on how the matrix operations are getting well optimized in RLNC. Wunderlich et al. [1] proposed the method to simplify the network coding by considering the Galois Field. This paper made a thorough comparison of Galois field approach with traditional RLNC approach. Koetter et al. [3] approach relied on GPU to handle the task efficiently, but for IoT nodes being low profile are not well compatible to work with GPU threads all the time. But we will make full use of multicore CPUs for parallelizing the matrix operations [4]. The blocks are assigned to various threads. Role division progressive decoding is found to solve the synchronization issue between master and worker threads, but still, it has dependencies between supervisor and worker threads whereas the coefficient matrix duplication approach provides a replica of the coefficient matrix to each thread and assigns it to each core [5]. Thus, each core (thread) handles the coefficient matrix in a parallelized fashion [6].

3   Comparative Study

3.1   Centralized Computing

Centralized computing is a sort of computing architecture where all processing/computing is performed on a centralized server. The central server, in turn, is liable for delivering application logic, process, and providing computing resources (both basic and complex) to the hooked-up client machines. Client computer systems are connected through the network to a central server that processes their computations. The central server is deployed with the first application, huge computing resources, storage, and different high-end computing-intensive options. All the client nodes are entirely hooked into the central server for any application access, computing, storage, web access, and security [4, 7].

3.2 Cloud Computing

Cloud computing is that the on-demand delivery of database storage, applications, and alternative IT resources through a cloud services platform via the net with pay-as-you-go rating. Advocates note that cloud computing permits firms to avoid or minimize up-front IT infrastructure prices. The provision of high-capacity networks at a comparatively lesser price and the storage devices have gained importance in recent days, because of the widespread adoption of hardware virtualization, service-oriented design and utility computing which collectively led to the growth in cloud computing [8].

Figure 1 explains how the back-end features are connected to the front-end using cloud and that allow the user to interact with cloud data securely. It is the responsibility of the back-end to produce the security of information for cloud users together with the control mechanism [9].

3.3 Edge Computing

Edge computing refers to the technologies that enable computation to be performed at the client's network [10]. It could be a non-trivial extension of cloud computing from the core network to the edge network [11, 3]. It is a distributed, open IT design that options localized process power, allowing mobile computing and Internet of Things (IoT) technologies. In edge computing, data is processed by the device itself or by a system or server, instead of being transmitted to an information center. Outage reduction is one among a significant reason to use edge computing. By pushing everything to the cloud, it ends up in ISP failures and cloud server period [12].

3.4 Fog Computing

Fog computing is a complicated technique to decrease latency and network congestion and supply economic gains for Internet of Things (IoT) networks. In IoT,

Fig. 1 Cloud architecture

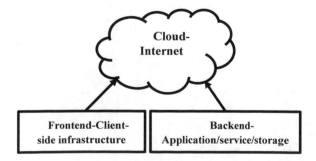

fog computing [9, 10, 6], coined by Cisco, is projected as a promising resolution. Fog computing solely pushes relevant information to the info centers and permits computing, decision-making, and action-taking to happen via multiple low-power computing devices. So, fog computing will give low-latency, fast-response, and placement awareness service [13].

3.5 Grid Computing

Grid computing is a processor design that incorporates computer resources from totally different domains to attain a main goal [6]. Grid computing is additionally referred to as distributed computing. It is a group of computers that is functioning along to perform various tasks. It shares the work with multiple systems and permits computers to contribute their individual resources to attain a typical goal [14]. Grid computing has many various scientific applications (Fig. 2).

3.6 Sequential Computing

In sequential computing, only one task is completed at a time. Consider an example of people standing in a queue to book a railway ticket where only one person can get the ticket at a time [15]. In this case, only one person can get a ticket a time.

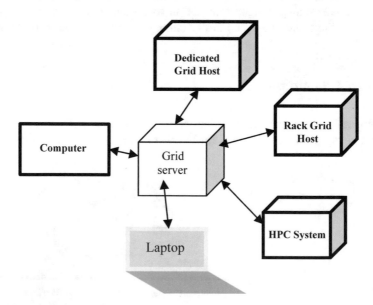

Fig. 2 Grid computing

Suppose there are two queues of people and one cashier is handling both the queues then one person can get the ticket at a time from both queues [16]. Similarly, in this type of computing, when there are list of tasks to be executed by a processor, each task will be addressed sequentially where the rest of the tasks will be on hold till the first one gets executed. This type of processing is also known as sequential processing. In serial processing, same tasks are completed at the same time but in parallel time may vary. The load is high in single core and follows bit by bit form in serial whereas in parallel data transfers in byte. Serial takes more time than parallel processor. Parallel computing is easier than the serial computing (Fig. 3).

3.7 Parallel Computing

In parallel computing, many executions are carried out in a simultaneous manner. In this computing, the task is divided into subtasks and whose results are combined at last [17]. The main aim of parallel computing is to increase speedup and decrease time complexity [18]. The computing can be defined using speedup as a result of parallelization given by Amdahl's law (Fig. 4).

$$\text{Speedup} = \frac{\text{Serial Execution Time}}{\text{Parallel Execution Time}} \qquad (1)$$

To parallelize the serial code, we can use the OpenMP (Open Multi-Processing) API which supports multiple platforms source code directives, functions, and environment variables [19]. The steps to do parallelization are as follows:

1. First, we need input array size.
2. We divide the problem into threads by using #pragma omp parallel.

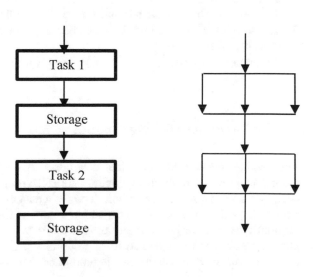

Fig. 3 Flow of sequential and parallel computing

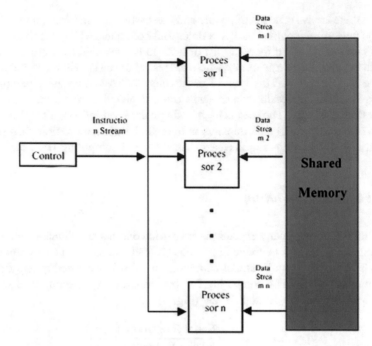

Fig. 4 Parallel processing system

3. We divide the cycle of "for loop" into threads.
4. Then we add "start_time" before starting the processes and add "end_time" at end of processes.
5. Find searching time by subtracting the end-time and start time by running the sequential code.
6. Then we run the two threads and four threads to find the searching time.
7. The speedup is found, and the efficiency is given as (Efficiency = Speed-up)/(Number of processors)
8. Performace Improvement = $(t_p - t_s)$/(Serial Time)

4 Basics of Networking Coding

Initially, the sender splits the initial data into n symbols, where each symbol has some length m. Then alternate symbols are still grouped into generations, where every generation contains g symbols. The main data of one generation can then be expressed as matrix M with g rows and m columns. Here, each row indicates one original symbol. In order to do encoding, we tend to assume r number of redundant packets that are to be generated for a group of g packets [20]. Now, a coefficient matrix C with random coefficients is multiplied with symbol matrix M

$$\alpha = MC \tag{2}$$

Now, the coded symbol β, which is the row of matrix α, is then transmitted as a coded packet along with its corresponding coefficient vector c of matrix C. The receiver on the other end creates a new coefficient matrix $\overline{C}$ on receiving at least g coded packets from sender and matrix $\overline{\alpha}$. In order to reconstruct the original symbol matrix M, the receiver is supposed to calculate:

$$M = (\overline{C})^{-1}\overline{\alpha} \tag{3}$$

5 Optimized Block Operations on Matrices for RLNC

5.1 Earlier Approach

The above-mentioned matrix operations are commonly found in many applications and thereby heavily used in the area of high-performance computing fields. Many libraries that are helpful for optimizing the matrix operations are considered into account for maximizing performance. The limitation is that these libraries operate on floating-point numbers. Few libraries convert the finite field elements to floating-point numbers and back. This conversion approach is good for large problem sizes. Also, the conversions itself add some overhead for small problem sizes.

5.2 Operations Based on Matrix Blocking

i. Partitioning matrix into blocks for parallel thread processing: Instead of allowing the parallel threads to work on complete rows and columns of the matrix, it is better to let the parallel threads work on square block partitions of matrices. Here, we consider the size to be 16×16 or 32×32 bytes for each block. It is usually chosen such that all the parallel threads are working parallel and all the operands can be easily fit into the $L1$ cache. For the sake of easy understanding, we focus on a single level of blocking in this paper and we introduce sub-blocking in the later stage of this paper. The only constraint of our blocking technique is that all coded packets need to be received before decoding step, so as a result of this constraint, it is not at all fit for the online decoding.
ii. Inversion operation on matrix

For this inversion operation, we introduced an LU factorization. There are three steps involved in this factorization.

a. Do the LU factoring on the given matrix X,
b. Now perform the inversion on matrix U,
c. Solve $X - 1L = U - 1$ to get $X - 1$.

Again internally, there are many matrix operations on the blocks. We will optimize it with the few single instruction multiple data (SIMD). We can represent the block operation as kernel operation which can be represented as a job with blocks containing input and blocks containing output in memory. Here, we divided the whole problem into eight-block operations. Later, we can optimize the block operations which are frequently used. Using SIMD operations, multiplication of matrices can be well optimized. To make the job easy, we make the multiplication of matrix into number of scalar-vector operations such that it can be readily optimized with SIMD instructions. These SIMD instructions are usually used within 32-bit instruction set architecture which is heavily used for IoT applications. We can improvize the optimization of matrix multiplication further by enhancing the cache efficiency through sub-blocking. It is to note that here, we are not sub-blocking in order to share the workload among the threads which is the reason for blocking in earlier section; rather, we always look forward to enhance the efficiency of the cache. In the below example, 64×64 blocks are represented and each block is again subdivided into 16 sub-blocks. If we perform the matrix kernel operations on the sub-blocks, then we can perform the multiplication of the matrix on the entire block. With the help of this sub-blocking method, we will be able to do the computations on the larger blocks, especially where input and output blocks cannot fit into the $L1$ cache of low-end IoT nodes. But larger sub-blocks which cannot be fitted into $L1$ cache will reduce the overall performance.

6 DAG Scheduling Graphs for Optimized Parallelization of RLNC

Principle involved in DAG scheduling—consider operation on each matrix block as an individual task with both inputs and outputs are done in memory. The primary idea is to first design an algorithm on how to make full use of data dependencies among the tasks in order to make the matrix inversion parallelized [3]. The various data operations performed on the blocks are recorded, and the data dependencies which are resolved are presented in directed acyclic graph. Nodes indicate the separate tasks, and edges indicate data dependencies within the nodes. So, to perform the inversion on a matrix, the scheduler divides the tasks till the DAG is fully processed. Here are the steps to record the DAG schedule i.e, firstly, create the scheduling and it records the transactions of each block operation [7]. Secondly, we identify the dependencies between the tasks and form a representation of the dependencies. Last step is that DAG schedule recording makes the task that does not have any dependencies into a "task queue" which is a linked list and tasks in this queue do not rely on the next tasks. DAG schedule execution—we initiate multiple threads, and then each thread picks up the task in the task queue. When the core processes the task, a traversal will be done on the tasks that are just completed [15]. Whenever the task is successfully processed, the counter set for dependency is decremented and the counter indicates the total dependencies that have to be analyzed before the task scheduling is done. Finally, when the counter for a particular task reaches zero, every dependency is cleared and the task is added to the task queue.

Preventing cache line conflicts—an important limitation in size selection of block is the cache line. On observing many IoT platforms, the size of $L1$ cache is 64 bytes. We prefer column-major order as the standard format to store the data into matrix. We can also go with row-major order, but it demands re-programming and re-design of the existing algorithms. The primary task is that scheduling the first task in the task queue for the next available core disregarding it is fast or slow core. Mostly, all the scheduling methods prefer fast cores to slow cores [10]. Also, they instruct cores which are slow to go to idle state and wake up the fast cores, when the slow core finds the fast core waiting for task that is newly created. Then, it schedules a task from the queue which has less dependencies on compared with the rest of the tasks and assign it to the appropriate core. Next step is to schedule the data locality by setting the priority for the tasks operating on the cache line based on its score. We need to prioritize the tasks on a particular block that is currently in the cache. If two tasks work on same blocks, then it gives a score to all the blocks provided that we have one input block and two output blocks. Finally, we could distinguish the fast and slow cores. Finally, we subtract the dependency count from the data locality score which tells the slow core such that we can assign the less priority task to it. But for the fast cores, we need to add the dependency count with the data locality score in order to get a total priority score which helps us to assign the high priority task [10, 18, 5]. Hence, the tasks with more dependencies are assigned with more priority to get the fast cores, and we always choose the task with the high score to process first.

Figure 5 shows the graph plotted on sequential time versus parallel computation time where there is a significant change in the execution times in parallel computation for various input matrix sizes.

As a result, we can get four as the theoretical speed-up. Figure 6 shows that how the practical speed-up is ramping up toward the theoretical speed-up.

Table 1 shows experimental results that are performed on different input matrix sizes during matrix-matrix multiplications.

Table 2 shows the experimental results that are observed when a more number of threads are being introduced into the matrix operations which results in significant change in speedup. Figure 7 represents the ramping of speed-up on allocating more threads to handle the loop regions.

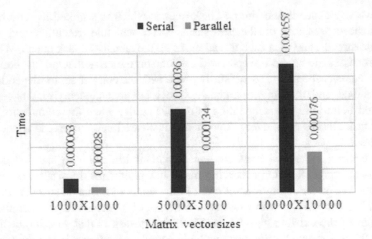

Fig. 5 Graph—serial time versus parallel time

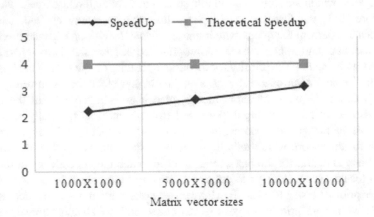

Fig. 6 Comparison of theoretical and practical speed-ups

Table 1 Comparison of execution times for different inputs

Matrix-vector size	1000 × 1000	5000 × 5000	10,000 × 10,000
Serial	0.000063	0.00036	0.000557
Parallel	0.000028	0.000134	0.000176
Speedup	2.25	2.686	3.164
Theoretical speedup	4	4	4

Table 2 Comparison of execution times while increasing the threads

Matrix-vector size 1000 × 1000			
No. of threads	Serial time	Parallel time	Speedup
1	0.000052	0.000069	0.753
2	0.0000646	0.000051	1.161
3	0.0000604	0.00005	1.208
4	0.0000726	0.000051	1.434

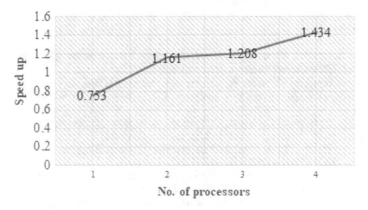

Fig. 7 Graph—number of processes versus speed-up

7 Conclusion

We have analyzed the serial execution time versus parallel execution time and found significant change in the speed-up thereby an indication that we can enhance the matrix computations involved in the RLNC. The proposed system of DAG and CD approaches are taken as benchmarks. Also, this could be used in another context of some other problems wherever huge computation task is going on.

References

1. Wunderlich, S., Cabrera, J., Fitzek, F.H.P., Reisslein, M.: Network coding in heterogeneous multicore IoT nodes with DAG scheduling of parallel matrix block operations. IEEE Internet Things J. 1–17 (2017)
2. Luszczek, P., Kurzak, J., Dongarra, J.: Looking back at dense linear algebra software. J. Parallel Distrib. Comput. **74**(7), 2548–2560 (2014)
3. Ho, T., Me´dard, M., Koetter, R., Karger, D.R., Effros, M., Shi, J., Leong, B.: A random linear network coding approach to multicast. IEEE Trans. Inf. Theor. **52**(10), 4413–4430 (2006)

4. Yi, S., Hao, Z., Qin, Z., Li, Q.: Fog computing: platform and applications. In: Proceedings of 3rd IEEE Workshop Hot Topics Web System Technology, pp. 73–78 (2015)
5. Zhuo, G., Jia, Q., Guo, L., Li, M., Li, P.: Privacy-preserving verifiable set operation in big data for cloud-assisted mobile crowdsourcing. IEEE Internet Things J. **4**(2), 572–582 (2017)
6. Johnston, S., Apetroaie-Cristea, M., Scott, M., Cox, S.: Applicability of commodity, low cost, single board computers for Internet of Things devices. In: Proceedings of World Forum on Internet of Things, pp. 1–6 (2016)
7. Osanaiye, O., et al.: From cloud to fog computing: a review and a conceptual live vm migration framework. IEEE Access **5**, 8284–8300 (2017)
8. Kamilaris, A., Pitsillides, A.: Mobile phone computing and the internet of things: a survey. IEEE Internet Things J. **3**(6), 885–898 (2016)
9. Bondi, L., Baroffio, L., Cesana, M., Redondi, A., Tagliasacchi, M.: EZ-VSN: an open-source and flexible framework for visual sensor networks. IEEE Internet Things J. **3**(5), 767–778 (2016)
10. Huang, X., Ansari, N.: Content caching and distribution in smart grid enabled wireless networks. IEEE Internet Things J. **4**(2), 513–520 (2017)
11. Ho, T., M´edard, M., Koetter, R., Karger, D.R., Effros, M., Shi, J., Leong, B.: A random linear network coding approach to multicast. IEEE Trans. Inf. Theor. **52**(10), 4413–4430 (2006)
12. Kolios, P., Panayiotou, C., Ellinas, G., Polycarpou, M.: Data-driven event triggering for IoT applications. IEEE Internet Things J. **3**(6), 1146–1158 (2016)
13. Chiang, M., Zhang, T.: Fog and IoT: an overview of research opportunities. IEEE Internet Things J. **3**(6), 854–864 (2016)
14. Sun, K., Zhang, H., Wu, D., Zhuang, H.: MPC-based delay-aware fountain codes for real-time video communication. IEEE Internet Things J. (in print) (2017)
15. Ahuja, S.P., Myers, J.R.: A survey on wireless grid computing. J. Supercomputer. **37**(1), 3–21 (2006)
16. Shi, W., Cao, J., Zhang, Q., Li, Y., Xu, L.: Edge computing: vision and challenges. IEEE Internet Things J. **3**(6), 637–646 (2016)
17. Yi, S., Hao, Z., Qin, Z., Li, Q.: Fog computing: platform and applications. In: Proceedings of IEEE Hot Web, pp. 73–78 (2015)
18. Hong, H.-J., Fan, C.-L., Lin, Y.-C., Hsu, C.-H.: Optimizing cloud- based video crowdsensing. IEEE Internet Things J. **3**(3), 299–313 (2016)
19. Satyanarayanan, M., Bahl, P., Caceres, R., Davies, N.: The case for VM-based cloudlets in mobile computing. IEEE Pervasive Comput. **8**(4), 14–23 (2009)
20. Conti, M., Kumar, M.: Opportunities in opportunistic computing. Computing **43**(1), 42–50 (2010)

Comprehensive Survey of IoT Communication Technologies

T. Ramathulasi and M. Rajasekhara Babu

Abstract The development of brilliant things associated with the Web is relatively in-wrinkling on the world. The IoT-network is empowering the association between the articles or devices and the backend frameworks by means of the Internet to impart. Existed correspondence advances gathers in-field information and after that transmit it to the data center where analytics are examined are connected to it, however, this is no longer survive capable by the development of IoT in which interchanges are more than the communication of human-to-human (H2H) and machine-to-machine (M2M). Effective correspondence innovation is regularly testing because of ultra-low power utilization, handling and capacity abilities, which have significant sway on the feasible throughput and parcel gathering proportion just as idleness. Therefore, every one of these variables makes it is hard to choose a suitable innovation to streamline correspondence execution, which dominatingly relies upon the given application. New methodologies and new advancements are required to change huge measures of gathered information into significant data and furthermore can meet the necessities for gigantic IoT in the 5G period. This paper provides a picture of the main wired medium, wireless communication technologies, and cellular communication technologies for massive IoT in the 5G era (eMTC and NB-IoT) with its features and technical characteristics that means to enable the user to pick a reasonable and satisfactory IoT correspondence innovation for claim requests in the colossal number of assortment. This review gives an exhaustive perspective on the correspondence technologies for IoT availability.

Keywords Communication technologies (CT's) · LPWA standards · Cellular standards · eMTC · NB-IoT

T. Ramathulasi · M. Rajasekhara Babu (✉)
School of Computer Science and Engineering, Vellore Institute of Technology,
Vellore, Tamil Nadu, India
e-mail: mrajasekharababu@vit.ac.in

T. Ramathulasi
e-mail: t.ramathulasi2018@vitstudent.ac.in

© Springer Nature Singapore Pte Ltd. 2020
P. Venkata Krishna and M. S. Obaidat (eds.), *Emerging Research in Data Engineering Systems and Computer Communications*, Advances in Intelligent Systems and Computing 1054, https://doi.org/10.1007/978-981-15-0135-7_29

1 Introduction

The expression "Internet of things (IoT)" was instituted by Kevin Ashton in 1999 [1]. A number which specifies the physical devices are being connected to internet increases at an uncommon rate understanding the position of IoT. One of the important key issues in IoT-based application development is CT by which the information shared to legitimate correspondence headway. Figure 1 illustrates the overall concept of the IoT market and shows the usage of low-power wide-area technologies like LoRa, Sigfox percentage is increased by 11%, whereas cellular IoT will increase more than 55% by 2030 [2].

Figure 2 outlines the commitments of this paper with respect to the ongoing writing in the field IoT CT's can be condensed. Whatever is left of this paper is organized as follows summary of the wired medium emerging CT's the overall short-range and long-range with cellular and cellular IoT wireless CT's, respectively. Finally, presents a summary of lessons learned and concludes this study.

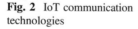

Fig. 1 IoT markets by access technology, 2017–2030

Fig. 2 IoT communication technologies

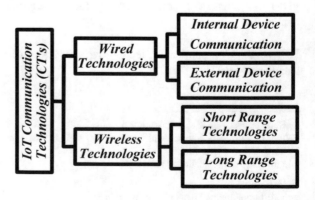

2 Wired Technologies

There are wired medium technologies for connecting devices which are connected through wired medium. It has two ways, one is communicate internal devices in a master device as internal communication, and another way is different devices to communicate as external communication [3].

2.1 Internal Communication

I2C "inter-integrated circuit (I2C)" is a bus was developed in 1980 by Philips Semiconductor. I2C devices used sensors, clocks, interface controls, and also have data interfaces. Types of I2C respective of its speeds like 100, 400 kbps, and 3.4 Mbps as slow, fast, and high speed.

SPI "serial peripheral interface bus (SPI)" is a specification for serial communication interface was developed in 1980 by Motorola. Secure Digital cards and liquid crystal displays are the typical application of SPI.

2.2 External Communication

Ethernet The framework is to connect two or more systems to form a local area network (LAN), with protocols for controlling the information passing through and simultaneous transmission to be avoided. In the mid of 1970s at the Xerox Palo Alto Research Centers (PARC), researchers have developed the Ethernet technology. A unique identifier as a MAC address is given to every Ethernet network interface card (NIC).

RS-232 Recommended Standard 232 is an interface standard basically used in serial ports of computer system. Electrical characteristics and signals of timings specify the standard number. TIA-232-F is the standard version issued in 1997.

RS-485 Recommended Standard 485 is an improvement of RS-285 by increasing the number of devices from 10 to 32. It is serial connection can be extended the conversation using cable up to feet of 4000. For industry application, which has noise immunity and multi-drop capability can choose this as option defines adequate signal voltages and electrical characteristics to the maximum extent.

UART Universal asynchronous receiver-transmitter (UART) is hardware device had a chip and contains the features of *I/O*. It can be used when high speed is not required when making a link between two devices, and also can get at low cost. Because of no clock signal transmitted so it is asynchronous.

USART Universal asynchronous and synchronous receiver-transmitter (USART), to transfer and get information in the mode of either synchronous or asynchronous by using the two types of pins called *I/O* pins. Both transmission and getting information can be done at the same time so it is called 'full-duplex.'

USB It could be an agent fringe interface. USB refers to "Universal serial bus." It gives a sequential transport standard to interfacing gadgets, for the most part to a PC, yet it additionally is being used on different gadgets, for example, set-top boxes, diversion consoles, and PDAs (Table 1).

3 Wireless Technologies

3.1 Short-Range Standards

Short-range standards for IoT's communication among different types of networks where the range is limited are very easy and convenient with low-power consumption [4]. Some of those standards are

6LoWPA An IP-based internetworking protocol, first most commonly used standard for IoT's communication is 6LoWPA, supports 2128 IP address. It was invented by the IETF group, IP communication over wireless networks. Without any gateways and proxies, it can connect with any other IP-based network. It supports different types of mesh and star topologies. As like Zigbee, it uses IEEE 802.15.4 at the respective layer of its architecture [4–6].

ZigBee One of low-power consumption, low-data rate, years of battery life, device security, and IEEE 802.15.4 wireless network standard is ZigBee, is discovered by ZigBee Alliance. It is suited for personal area networks, low-cost high-level application. It supports different types of topologies [4, 7].

IrDA (Infrared Communication) Developers uses the IR devices in which wireless personal area networks involves with laptops, computers, PDA's, and cellular phones which are spread over the range of 30 ft. [8].

RFID The systems having reader device and RF tag that is a small radio frequency transponder electronically programmed with information that has the characteristic of reading. High expensive and use high frequencies and have two techniques called active reader and passive reader. It can be used to measure the diagnostic of data directly [9, 10].

NFC Near-field communication (NFC) uses the similar technology principles are in RFID, not only for identification and elaborates the communication between the devices. It contains less amount of data for very nearby devices. Mostly used in industry applications, mobile phones, and payment systems.

Z-Wave Another low power, suitable for transfer small size messages with low-speed 100 kbps and 30 m long features of MAC protocol is Z-Wave developed by Zensys. In real time, the scenarios of light control, energy control, and healthcare control applications can be used. The mesh topology supported by Z-Wave [4, 10].

UWB Ultra-wide band (UWB) is as of now getting unique consideration and is a significant hotly discussed issue in industry and hypothetical. UWB as far as a transmission from a receiving wire for which the radiated flag data transmission surpasses the lesser of 500 MHz or 20% of the inside recurrence [11–13]. UWB transmissions transmit data by creating radio vitality at explicit time moments and possessing a vast transfer speed [13].

Li-Fi Li-Fi (light fidelity) is a technology using white light LED bulbs for transmission applying constant current and output at high speeds. This constant current used in this setup and operational procedure is simple if LED in state of 'on.' A controller loads that code data into LED's and achieved high-speed transmission of data at 100 MB/s [13].

BLE BLE is also known as Bluetooth smart which is a significant protocol for IoT application. It is designed and enhanced for short-range, low bandwidth, and low latency for IoT applications. The advantages of BLE classic Bluetooth include lower-power consumption, lower-setup time, and supporting star network topology with an unlimited number of node [4, 7].

Wi-Fi The architecture of IEEE 802.11 contains the several components to communicate that is connected and provides a local LAN without any medium that supports transparently, mobility to upper-level layers. Wireless fidelity (Wi-Fi) also includes in IEEE 802.11 architecture standards of 802.11/a/b/g/ for WLAN [8].

3.2 Long Range Standards

A "low-control wide-region network (LPWAN)" is a sort of remote media transmission innovation is viewed as a definitive arrangement, working in both the authorized and unlicensed range, and has risen to fill the crevice [14].

LPWAN Standards with Unlicensed Spectrum.

SIGFOX Uses ultra NB (narrow band), operates in ISM band and achieved operations with slower frequency width lower than 1 kHz in Sigfox technology. It has noise levels are low and works with very low power gives the throughput 100 bps. It uses binary phase-shift keying (BPSK) for data transfer modulation. It also supports stat type of network topology for downlink transmission.

LoRa Alliance Long Range (LoRa) partnership LPWAN arrangement depends on two noteworthy segments, to be specific LoRa and LoRa-WAN and works in sub-GHz ISM band.

INGENU The new version of On-Ramp Wireless is Ingenu is also operating in ISM band. It provides transmission power is high, gives high through put and having more capacity. The range of Ingenu is very short.

WEIGHTLESS One of open standard belongs to LPWA, operates in the sub-GHz spectrum which is unlicensed is Weightless. It works in three different open standards with different features, data ranges, and different power consumptions [15].

Cellular Standards or LPWAN Standards with Licensed Spectrum.

The 3GPP 3rd Generation Partnership Project (3GPP) in its releases provides three LPWA technologies using licensed spectrum for IoT application in very long distance. These solutions are (Extended Coverage GSM for Internet of Things) EC-GSM-IoT, (Long-Term Evolution Machine Type Communications Category M1) LTE MTC Cat M1, and (Narrowband IoT) NB-IoT which are discussed in the following [15].

LTE MTC Cat MI An IoT technology connects the devices to the network 4G without having batteries and any gateways, deployed in present LTE-base stations is LTE MTC Cat M1 or LTE-M. It provides long-life battery lifetime uses different modulation for downlink and uplink. Cellular IoT is one has solutions for optimizing transmission of data [16, 17].

NB-IoT The technology which supports many types of mobile equipment and with co-existence of each other is NB-IoT. It means no need to invest in extra towers and other requirements. It provides operators to use in very convenient to operate and work with this technology. In the forms of three operating modes, this NB-IoT technology is deployed [18].

EC-GSM IoT (Extended Coverage GSM IoT) The advancement of 3GPP releases proposed one of IoT technology provides extended coverage, more variable rates of data that meets the diversity is called EC-GSM IoT. The most peak data rate can be reached. It deployed in the GSM upgrade software and increases the features of security. In the case, to give variable information rates to chance the assorted variety of intelligence application (Table 2).

4 Conclusions

This present reality is embracing IoT in all respects quickly. The exponential development of the wearable gadgets, amazing correspondence advancements, and cloud-based information scientific strategies are giving another time of future continuously frameworks. It gives a sequential transport standard to interfacing gadgets, for the most part to a PC, yet it additionally is being used on different gadgets, for example, set-top boxes, diversion consoles, and PDAs.

After the investigation of the above areas, the outline of Wired CT's appeared in Table 1 and the sketch of Wireless CT's demonstrated in Table 2. Mapping the distinctive utilizations of the IoT in future partner with the ongoing system

Table 1 Wired medium technologies

IoT communication technologies (CT's)	Data rate	Range
I2C (Inter-integrated circuit)	100, 400 kbps, 1, 3.4 mbps varies depends on type of I2C version	Very low
SPI (Serial peripheral interface bus)	10 mbps to upto 100 mbps	Very low
RS-235 (Recommended standard 232)	1 kbps–20 kbps	15 m
RS-485 (Recommended standard 485)	100 kbps–35 mbps	1200 m
Standard ethernet (IEEE 802.3)	10 mbps	>100 m
Fast ethernet (IEEE 802.3u)	100 mbps	100 m
Giga ethernet (IEEE 802.3z)	1 Gbps	>25 m
10 Giga ethernet (IEEE 802.3ae)	10 Gbps	>10 km
UART (Universal asynchronous receiver and transmitter)	Bit by bit	Packets
USART (Universal synchronous and asynchronous data receiver and transmitter)	Bit by bit	Analog form
USB (Universal serial bus)	Bit by bit	Data with clock pulse

Table 2 Wireless technologies

IoT communication technologies	Frequency band	Data rate	Range
6LoWPA	868 MHz (EU) 915 MHz (USA) 2.4 GHz (Global)	250 Kbps	10–100 m
ZigBee	2.4 GHz	250 Kbps	10 m
IrDA (Infrared communication)	300–400 GHz	Serial infrared 300–115,200 bps Fast infrared 4 Mbps	About 10–30 m
RFID	125 kHz, 13.56 MHz, 902–928 MHz	128 Kbps	200 m
NFC (Near-field communication)	13.56 MHz	424 Kbps	<5 cm
Z-Wave	908.42 MHz	40–100 Kbps	<30 m
UWB (Ultra-wide band)	3.1–10.6 GHz	110 Mb/s	10 m
Li-Fi (Light fidelity)	100 THz	1 Gbps	10 m
Home plug GP	1.8 MHz–30 MHz	4–10 Mbps/20–200 m/bps	<100 m
BLE (Bluetooth LE)	2.4–2.5 GHz	305 Kbps	<50 m
Wi-Fi (wireless fidelity)	2.4, 5 GHz	54 Mbps	100 m
SIGFOX	Sub-GHz ISM	10–100 bps	Urban: 10 km Rural: 50 km

(continued)

Table 2 (continued)

IoT communication technologies		Frequency band	Data rate	Range
LoRa Alliance:		Sub-GHz ISM	0.3–50 Kbps	Urban: 5 km Rural: 15 km
INGENU		ISM 2.4 GHz	8 bps–8 Kbps	Urban: 3 km Rural: 10 km
WEIGHTLESS	Weightless-P	5 MHz	200–100 Kbps	Urban: 2 km
	Weightless-N	Ultra-narrow band (200 Hz)	30 Kbps–100 Kbps	Urban: 5 km
	Weightless-W	12.5 kHz	1 Kbps–10 Mbps	Urban: 5 km
LTE-M (Rel. 13)		700–900 MHz	375 kbps	<15 km
EC-GSM		800–900 MHz	70 kbps	<15 km
NB-IoT		700–900 MHz	20–65 kbps	<35 km

correspondence advancements is vital for investigating future research difficulties and headings. The future challenge is a requirement for new dependable and ultra-low inactivity conversation conventions alongside the development of their architecture.

References

1. Affairs, M.: Internet of Things in the Netherlands Applications Trends and Potential Impact on Radio Spectrum (2015)
2. Ropert, S., Leader, I.p.: IoT: a Market Keeping Its Promise of 10% Annual Growth Thanks to Fast-Growing Industrial Markets, (Online). Available: https://en.idate.org/marche-iot-2018
3. Munjal, Communication (Wired) Protocols in IOT (2017)
4. Bensky, A.: Short-Range Wireless Communication: Fundamentals of RF System Design and Application (2004)
5. Salman, T.: Internet of Things Protocols and Standards (2005)
6. Hossen, M., Kabir, A., Khan, R.H., Azfar, A.: Interconnection between 802.15. 4 devices and IPv6: implications and existing approaches (2010)
7. Lu, C.-W., Li, S.-C., Wu, Q.: Interconnecting ZigBee and 6LoWPAN wireless sensor networks for smart grid applications (2011)
8. Samie, F., Bauer, L., Henkel, J.: IoT technologies for embedded computing: a survey (2016)
9. Infrared Communication (IrDA), (Online). Available: https://iotpoint.wordpress.com/infrared-communication-irda/
10. Porkodi, R., Bhuvaneswari, V.: The Internet of Things (IoT) applications and communication enabling technology standards: an overview (2014)

11. Rahman, A.B.A., Jain, R.: Comparison of Internet of Things (IoT) Data Link Protocols (2015)
12. Yang, L., Giannakis, G.B.: Ultra-Wideband Communications: An Idea Whose Time Has Come, vol. 21, pp. 26–54 (2004)
13. Hirt, W.: Ultra-wideband radio technology: overview and future research. Comput. Commun. **26**(1), 46–52 (2003)
14. Su, Y., Shen, C., Lee, J.: A comparative study of wireless protocols: bluetooth, UWB, ZigBee, and Wi-Fi. In: IECON 2007—33rd Annual Conference of the IEEE Industrial Electronics Society, Taipei, 2007, pp. 46–51. doi: 2007.4460126 (2007)
15. Subhas, M., Noushin Poursafar, N., Alahi, M.E.E.: Long-range Wireless Technologies for IoT Applications: A Review (2017)
16. NOKIA., LTE evolution for IoT connectivity (2016)
17. Lescuyer, P., Lucidarme, T.: Evolved Packet System (EPS): The LTE and SAE Evolution of 3G UMTS. Wiley, New York (2008)
18. Wang, Y.-P.E., Lin, X., Adhikary, A., Grovlen, A., Sui, Y., Blankenship, Y.: A primer on 3g pp narrowband internet of things. IEEE Commun. Mag. **55**, 117–123 (2017)

Cost Effective Model for Using Different Cloud Services

**Chintalapati Jagadesesh Raju, M. Rajasekhara Babu
and M. NarayanaMoorthy**

Abstract The trend of cloud computing has started from the beginning of
Amazon's cloud services in 2006 despite its extravagant beginning in its starting
years now cloud computing is easily one of the trending concepts in computer
science engineering. Cloud computing has shown tremendous amount of results in
various sectors. This case study helps to find out the similarities and contrasts
between different cloud service providers in terms of their technical aspects and
range of services provided by them. This case study was used as a subjective
methodology for sampling. Since we all know the boom of cloud services that play
a vital role in our lives, we do not know the background process in the cloud world.
Cloud services are mostly offering renting access to enormous pool of computational power and accounting capacities to the resources. Generally, these are
cloud-based infrastructure which describes on characterizing the accounting
arrangements for obtaining resources. Most vitally cost-effective strategy is most
useful so that the cloud can be purchased at low price and can be used as proficiently as conceivable as possible. With this case study, the readers can get insight
of what cloud computing is, what are the various cloud services considered for this
paper and various cloud providing systems so that users can purchase which is best
and cheap related to the platform they are using.

Keywords Cloud service · Cloud service providers · Cloud cost

C. Jagadesesh Raju · M. Rajasekhara Babu (✉) · M. NarayanaMoorthy
School of Computer Science and Engineering, Vellore Institute of Technology,
Vellore, Tamil Nadu, India
e-mail: mrajasekharababu@vit.ac.in

C. Jagadesesh Raju
e-mail: jagadeeshraju.ch@gmail.com

M. NarayanaMoorthy
e-mail: mnarayanamoothy@vit.ac.in

© Springer Nature Singapore Pte Ltd. 2020
P. Venkata Krishna and M. S. Obaidat (eds.), *Emerging Research in Data
Engineering Systems and Computer Communications*, Advances in Intelligent
Systems and Computing 1054, https://doi.org/10.1007/978-981-15-0135-7_30

1 Introduction

We can outline distributed computing in method of representing various increasing advantages, on-demand arrangement for resources to typical pool of configurable figuring resources that can be immediately seen and can be released with immaterial administration with effort in organization interaction [1]. We realize that we are entering a period of "All-as-a-Service" where assets are shared at an unparalleled scale. Supposed processing the utility model, based on explaining the distributed system frameworks, winds up inescapable in the venture IT scene. The quick improvement of parallel and conveyed processing ideal models, driven by the expanding interest for registering force and system data transmission, invigorates the advancement of an assortment of enormously dispersed figuring stages such as grid, cluster, peer-to-peer computing rising scholarly including modern networks [2]. As of late, we saw the co-alteration of registering, stockpiling, system, and programming: just assets turned into the subject of business exchanges. Management of resources that involves efficiency in acquiring and increasing the status of computer networking and organizing assets is a standout among the most imperative research points in utility and distributed computing [3]. Cloud services are mostly offering renting access to enormous pool of computational power and accounting capacities to the resources. Generally, utility computing model is based on a cloud-based foundation and characterizes approaching for resources obtaining. It enables clients to get to a huge measure of PC resources. Venturing over period in between registering of utility and cloud, everyone can hope for new observational figuring standards executing "All-as-a-Service" within to be priced at minimal charge [4] essentially the executives systems are exceedingly alluring so resources used requiring little to no effort, so productively possible [5]. Again, the increasing adaptability furthermore presentation of heterogeneity colossal provocation and everyone need to adapt to changing supplier side asset accessibility and valuing, just as fluctuating client-side application arrangement and request [6].

2 Cloud Service Providing System

Since we all know the boom of cloud services that play a vital role in our lives, we do not know the background process in the cloud world. So, here, we need to know few of the terms CSB, CSP, CSC and check their roles in our cloud business deals [7]. Cloud service needs a standard portal to trade, so they evolve with the multiple gateway points like CSB (cloud service broker), CSP (cloud service provider), and CSC (content providers) [8].

2.1 CSB System

This system mainly helps in providing and managing the portal of CSB. It mainly provides CSP. A CSB ought to give CSP, the CSB portal to help CSPs to enlist them framework and programming administrations in the administration register-list. A CSB ought to oversee approval and validation of CSP and CSC. A CSB should know the status of accessible resources (e.g., server, memory, stockpiling, system, and administrations). Besides, a CSB ought to consult with countless CSPs to locate the reasonable CSP for the agreement which a CSC asked for his/her custom administration [9].

2.2 CSP System

First, they should be recognized part of portal CSB system then should provide business by doing their own service. When the user gets CSP logged in portal of CSB, the user can register and they are provided with various like (IaaS) (SaaS), the service for registering products. This system also helps in finding out if service level agreement mentioned in the portal of CSP will not be much considered [10, 11].

2.3 CSC System

In CSC, the providing of the content such as smartphone, smart TV etc., the providers has to find and they have to select the best cloud services. CSC requests its own platform and content to its provider by the following service such as pay-as-you-go. CSC manages services on its own. Passing into the portal of CSB this system should have the knowledge of various providers and their status [12, 13].

3 Pricing Models

The three diverse evaluating models for cloud are as per the following: [6] (1) On-Demand instance: This is the hourly rate with no long haul responsibility [14]. (2) Reserved instance: This is a buy of an agreement for each case you use with an altogether lower hourly use charge after you have paid for the booking [15, 16]. (3) Spot instance: This is a technique for offering on unused EC2 limit. This element offers a fundamentally lower cost [17].

3.1 Case Study #1: GOOGLE INSTANCE CLOUD

Google Cloud Platform is comprised of a variety of administrations and answers for use a similar programming and equipment foundation that Google utilizes for its very own items (like YouTube and Gmail). A portion of the primary GCP's advantages is that it is one of the biggest and most exceptional PC systems, and it gives you an entrance to the various apparatuses to enable you to concentrate on building your application. Stack driver monitoring, stack driver debugger, stack driver logging, and security scanner administration [4, 6, 18]. The modules included in Google cloud environment are as follows.

Google Compute Engine
Google Compute Engine enables clients to dispatch virtual machines on interest. This is one of the essential administrations for complete confinement and pro-grammed scaling from single occurrences to worldwide. This engine's boot quickly runs on dependable execution. In this system, various plans include various virtual servers with various sizes for explicit requirements [4, 6, 19].

GCP Deployment Manager
Google Cloud Deployment Manager enables you to indicate every one of the assets required for your application in a decisive arrangement utilizing YAML. This implies instead of carefully posting each progression that will be required for a sending, DevOps groups can disclose to Deployment Manager what a last orga-nization should look like and GCP will utilize the vital apparatuses and procedures for you. At the point when an ideal organization strategy is created, it is spared to be repeatable and versatile on interest. With Google Cloud Deployment Manager you can do send numerous assets at one time, in parallel, pass factors into your formats and get yield esteems back [4, 6, 19].

GCP Cloud Console
Cloud Console gives you a clear perspective on everything about your DevOps in the cloud. Web applications, information investigation, virtual machines, infor-mation store, databases, organizing, engineer administrations… Google Cloud Console encourages you to send, scale, and analyze generation issues in a straightforward online interface [4, 6, 19].

3.2 Case Study #2: MICROSOFT AZURE

Microsoft purplish blue is a distributed computing stage planned by Microsoft to effectively assemble sends and oversee applications and administrations through a worldwide system of information [20]. Microsoft Azure services are mainly divided into three platforms:

PAAS By using Microsoft Azure the developers can use it as a stage for developing, building, and conveying applications. They make the code with the platform given by azure and after that virtual machine execute the principles of the application utilizing windows server [4, 6, 21].

SAAS Like Cloud Services, Azure's mobile services give you the instruments to make and convey applications; however, clearly for this situation the applications are focused for cell phones. The data that gets gotten to by the application running on your gadget is put away in what is known as a back-end database [4, 6, 21].

IAAS In the discussion of VHD is the variety of virtual composition of hard drive on customary PC. It is the point of confinement unit on which the majority of the reports and applications are spared. Microsoft Azure offers resources to windows and Linux VHDs so it suits originators with limit in either. Also with this association moreover you just pay by how much time the VM is really running [4, 6, 21].

RESERVED INSTANCE: [22]

Provider	Instance type	V_CPU	Mem (GB)	1 year Linux	1 year windows	Linux price delta	Wintel price delta
GCP	n1-standard2	2	7.5	$582.60	$1283.40	14.88%	Low price
AZURE	D2 v3	2	8	$507.12	$1313.04	Low price	2.31%

Provider	Instance type	V_CPU	Mem (GB)	1 year Linux	1 year windows	Linux price delta	Wintel price delta
GCP	n1-standard4	4	15	$1165.08	$2566.68	15.01%	Low price
AZURE	D4 v3	4	16	$1013.04	$2624.88	Low price	2.27%

ON–DEMAND INSTANCE: [22]

Provider	Instance type	V_CPU	Mem (GB)	1 year Linux	1 year windows	Linux price delta	Wintel price delta
GCP	n1-standard2	2	7.5	$0.0950	$0.1750	Low price	Low price
AZURE	D2 v3	2	8	$0.0960	$0.1880	1.05%	7.43%

Provider	Instance type	V_CPU	Mem (GB)	1 year Linux	1 year windows	Linux price delta	Wintel price delta
GCP	n1-standard2	4	15	$0.1900	$0.3500	Low price	Low price
AZURE	D2 v3	4	16	$0.1920	$0.3760	1.05%	7.43%

4   Conclusion

As per, reserved instance pricing: Although Google has the most difficult pricing for discount valuing, there are trump cards that will unquestionably influence the estimating the Microsoft Azure that can be convincing with its firmly incorporated half and half cloud innovation and it would be difficult with Windows permitting with Azure Hybrid Use Benefit. Be keen about the amount and sort of RIs you are purchasing. It is tied in with deciding a suitable responsibility level, amplifying your discount and not focusing on resources you will never utilize. The estimating contrasts can be removed on the off chance that if your association is happy to go multi-cloud and at that point use it as an iron block to arrange progressively good pricing. On-Demand pricing: If you are hoping to make infant strides and need to begin with On-Demand examples and would prefer not to need to converse with anybody or arrange any arrangements Google is the best approach. You can not beat the sustained use discount in Microsoft's Azure in cost. Keep in mind your pricing will differ contingent upon the exceptional qualities of your applications, your helping cloud framework, and when you choose to hop into the market.

References

1. Mcir, R.: What are the Google cloud service platforms, 10 Feb. 2017 (Online). Available: https://hackernoon.com/what-are-the-google-cloud-platform-gcp-services-285f1988957a
2. Sosinsky, B.: Cloud Computing Bible, vol. 762, Wiley, New York (2019)
3. Hurwitz, J.S.: Cloud Computing for Dummies. Wiley, New York (2010)
4. Windows-azure-iaas-paas-saas-overview, March 2014 (Online). Available: http://robertgreiner.com/2014/03/windows-azure-iaas-paas-saas-overview/
5. AWSAzureGooglrpricematrix (Online). Available: https://www.heroix.com/download/AWSAzureGooglePriceMatrix.pdf
6. The NIST Definition of Cloud (Online). Available: https://nvlpubs.nist.gov/nistpubs/legacy/sp/nistspecialpublication800-145.pdf
7. Di Martino, B., Venticinque, S.,. Amato, A.: Cloud brokering as a service. In: Proceedings of P2P, Parallel, Grid, Cloud and Internet Computing, pp. 9–16 (2013)
8. Fastest way of revenue opportunities, 2019 (Online). Available: https://www.marketsandmarkets.com
9. Li, X., Zhao, H.: Resource Management in Utility and Cloud Computing (2013)
10. The first open cloud broker (Online). Available: www.compatibleone.org
11. Cloud brokerage system, 2019 (Online). Available: https://www.channelfutures.com/business-models/cloud-services-brokerage-company-list-and-faq
12. Badidi, E.: A cloud service broker for SLA-based SaaS provisioning. In: Proceedings of Information Society, pp. 61–66 (2013)
13. Cloudslang-alibaba-cloud-content, 2019 (Online)
14. Munteanu, V.I., Negru, V., Forti, T.F.: Towards a service friendly cloud ecosystem. In: ISPDC, pp. 172–179 (2012)
15. Chapman, C.: Towards a Federated Cloud Ecosystem. ICSE, p. 967 (2012)
16. Zhang, L.-J., Liu, D., Xie, Y.-F., Luo, L.-h., Hu, B.: A cloud oriented account service mechanism for SME SaaS ecosystem. In: Proceedings of SCC, pp. 3363–343 (2012)
17. Google cloud pricing (Online). Available: https://cloud.google.com/pricing

18. Pricing (Online). Available: https://www.alibabacloud.com/pricing
19. Azure pricing (Online). Available: https://azure.microsoft.com/en-in/pricing/
20. Amazons-ec2-generating-220 m-annually (Online). Available: http://cloudscaling.com/blog/cloud-computing/amazons-ec2-generating-220m-annually
21. Google cloud platform (Online). Available: https://searchcloudcomputing.techtarget.com/definition/Google-Cloud-Platform
22. Microsoft azure (Online). Available: https://azure.microsoft.com/en-in/

Sentiment Analysis on Movie Reviews

B. Lakshmi Devi, V. Varaswathi Bai, Somula Ramasubbareddy
and K. Govinda

Abstract Movie reviews help users decide if the movie is worth their time. A summary of all reviews for a movie can help users make this decision by not wasting their time reading all reviews. Movie-rating websites are often used by critics to post comments and rate movies which help viewers decide if the movie is worth watching. Sentiment analysis can determine the attitude of critics depending on their reviews. Sentiment analysis of a movie review can rate how positive or negative a movie review is and hence the overall rating for a movie. Therefore, the process of understanding if a review is positive or negative can be automated as the machine learns through training and testing the data. This project aims to rate reviews using two classifiers and compare which gives better and more accurate results. Classification is a data mining methodology that assigns classes to a collection of data in order to help in more accurate predictions and analysis. Naïve Bayes and decision tree classifications will be used and the results of sentiment analysis compared.

Keywords Prediction · Movie reviews · Naive Bayes · Decision tree · SLIQ

1 Introduction

Sentiment analysis of movie-rating sites can be of various applications like for other online movie-rating sites and for sites regarding opinions on books, products and various other things. Sentiment analysis or opinion mining is a method to

B. Lakshmi Devi · V. Varaswathi Bai
SOET, SPMVV University, Tirupati, Andhra Pradesh, India

S. Ramasubbareddy (✉)
Information Technology, VNRVJIET, Vignana Jyothi Nagar, Nizampet Rd,
Pragathi Nagar, Hyderabad, Telangana 500090, India
e-mail: svramasubbareddy1219@gmail.com

K. Govinda
SCOPE, VIT University, Vellore, Tamil Nadu, India

© Springer Nature Singapore Pte Ltd. 2020
P. Venkata Krishna and M. S. Obaidat (eds.), *Emerging Research in Data
Engineering Systems and Computer Communications*, Advances in Intelligent
Systems and Computing 1054, https://doi.org/10.1007/978-981-15-0135-7_31

systematically extract and identify affective state and the subjective information. In a bunch of text, we can use sentiment analysis for separating words which denote 'happy' emotion from words which denote 'sad' emotion, as a simple example. Opinion mining (also called emotion AI or sentiment analysis) is the utilization of computational linguistics, natural language processing, biometrics and text analysis to consistently establish, study, quantify and extract subjective data and emotional states. Sentiment analysis is extensively useful to voice of the client materials like survey rejoinders and reviews, online and social mass media, and healthcare materials for requests that vary from customer service to promoting to clinical medicine. Naive Bayes classifiers, in the field of machine learning, are a class of simple 'probabilistic classifiers' based on the application of Bayes' theorem with robust unconventional norms amongst the features. Naive Bayes has been examined broadly since 1950. The community involved with the retrieval of texts was introduced to it in the early 1960s and remained a preferred methodology for categorization of text, the issue of deciding texts as fitting into one class or the opposite with frequency of words as the characteristics. With applicable pre-processing, it is competitive in this area with additional cutting-edge methods as well as support vector machines. Naïve Bayes additionally finds use in automated clinical analysis. Decision tree-based learning makes use of a decision tree to travel from remarks concerning an item (denoted within the branches) to deductions concerning the element's goal worth (signified within the leaves). It is one amongst the predictive modelling tactics employed in machine learning, data mining and statistics. In classification trees, the tree model can take a distinct set of values for the target variable. Class labels are represented by leaves, and the combination of features which come to end with these class labels is represented by branches. Regression trees are those decision trees where the targeted variable can take uninterrupted values like in the case of real numbers. Decision trees are used to explicitly and visually signify decisions and higher cognitive process in decision analysis. A decision tree labels data in data mining, but the ensuing classification tree can be used for decision-making. Decision tree shapes regression or classification models in the kind of a tree assembly. It disrupts a data set into smaller subsets, whereas at a similar time a linked decision tree is established incrementally. The concluding outcome is a tree with leaf and decision nodes. Decision trees can manage numerical and categorical information. Advantages of such a system are that it permits for machine-controlled film scoring arrangement. Incorrect rating cannot be given because system computes on the basis of clients' remarks. It eliminates human blunders that normally arise throughout physical scrutiny. An unbiased result is provided by the system. Therefore, the system dismisses need of human labours and conserves resources and time. System disadvantages are that it should be provided with true inputs; else wrong outcomes are formed. Cloud hosting of the system needs to be done to process and receive results across the nation.

2 Background

Sentiment analysis could be a sort of language process for following the mood of the general public a couple of specific product or topic. Sentiment analysis, that is additionally referred to as opinion mining, involves in building a system to gather and examine opinions concerning the merchandise created in weblog posts, comments, reviews or tweets. Sentiment analysis is helpful in many ways that like in decision-making the success of a commercial campaign or new product launch confirms that versions of a product or service are standard and even determine that demographics like or dislike specific options. There are many challenges in sentiment analysis. The primary is associate degree opinion word that's thought-about to be positive in one state of affairs is also thought-about negative in another state of affairs. A second challenge is that individuals do not invariably express opinions in a same manner [2]. Each film review comment is going to be coordinated in contrast to the opinion terms in sentiment dictionary and classified into three categories: positive (+1), neutral (0) and negative (−1), reliant on the comparative number of negative and positive opinion terms in the remark, i.e., $|neg(c)|$ and $|pos(c)|$, respectively. Explicitly, the sentiment polarization of comment c is outlined as: $polarity(c) = sgn(|pos(c)| - |neg(c)|)$ [3]. To upsurge the precision of the classification, we should reject regular n-grams, i.e., n-grams that neither do not powerfully specify any objectivity nor show sentimentality of a sentence. In all data sets, such n-grams occur consistently [5]. Recommendation systems have become extremely important in the last decade with the rapid increase in the size and complexity of data provided over the World Wide Web. Internet users usually desire to be fed by simplified and customized procedures to access the information they require. A key step in the success of a satisfactory recommendation system is the effective prediction of the rating that will be potentially given by a user to a specific item. In this way, the system can recommend the users the items that they likely enjoy [1]. As we tend to all apprehend that the rating to a moving picture will mirror the favour degree of the user and also the same moving picture can receive totally different rates from individuals with different preferences. As a result of the user interest on movies that typically varies with the attributes of flicks, like director, actor and so on, the influence of user rating is often delineated by the attributes of the moving picture to some extent [4]. Despite its delusive unconventionality assumption, the Naive Bayes classifier is astonishingly effective in practice since its classification decision could usually be correct although its chance estimates are inaccurate [6]. Even though the discriminative logistical regression algorithmic rule features a lower asymptotic fault, the generative Naive Bayes classifier may additionally converge more quickly to its asymptotic fault. Thus, because the variety of training examples is exaggerated, one would expect generative Naive Bayes to at first do higher, except for discriminative logistical regression to eventually catch up to, and quite seemingly overtake, the performance of Naive Bayes [7]. Regardless of its straightforwardness, Naive Bayes can every so often beat more complex classification procedures. In a few fields, its functioning has been revealed to be analogous to that of decision tree learning and neural network [8]. The group of accounts accessible for evolving classification approaches are commonly disintegrated into two disconnected subsets, test set and training set. Training set is used for originating the classifier, while the test set is used to assess the accurateness of the classifier. SPRINT and SLIQ are decision tree classifiers which have been shown to achieve decent efficiency, accuracy and compactness for very bulky data sets [9]. A great quantity of programmes has been fashioned by the machine learning community to generate decision trees for classification. Quite prominent midst these for data classification include ScalParc, SPRINT, SLIQ, CHAID, CART, C5, C4.5, ID3 [10].

3 Proposed Method

Sentiment analysis of the review data set will be done by two methods: Naïve Bayes classifier and decision tree classifier. Data set used is a set of 500 IMDB movie reviews (half positive and half negative reviews). First, the data set will be read using pandas and data frame will be used for further processing. Pre-processing part involves cleaning of data, i.e., tokenization, removal of stop words, removal of special characters, stemming, etc. In this way, the data will be cleaned. Then, the data can be used for further classification and hence sentiment analysis. For classification, tf-idf matrix of review column is used and labelled sentiment column is used. Naive Bayes is a straightforward method for building classifiers: representations that allocate class tags to problem examples, depicted as trajectories of feature standards, wherever the category tags are obtained from some restricted set. There is no one formula to train such classifiers, however a category of procedures supported by a standard opinion: the whole lot of Naive Bayes classifiers adopt that the significance of a specific attribute is detached from the usefulness of the other element, provided the group variable. Naive Bayes classifiers are a family of simple probability-based classifiers based on applying Bayes' theorem with robust non-aligned conventions amongst the traits. Using Bayes' theorem, the conditional probability will be decomposed as

$$p(C_k/x) = \frac{p(C_k)p(x/C_k)}{p(x)}$$

To create the classifier model, the possibility of a given set of entries is found for altogether probable values of the category variable y and the output with highest likelihood is picked up. The conforming classifier, a Bayes classifier, is the operation that allocates a class label for $\hat{y} = C_k$ some k as the following:

$$\hat{y} = \arg\max_{k \in \{1,\dots,k\}} p(C_k) \prod_{i=1}^{n} p(x_i/C_k)$$

Feature vectors denote the occurrences with which specific outcomes have been produced by a multinomial distribution. This is the occurrence prototype characteristically used for article cataloguing.

Decision tree classifier
There are couple of algorithms to develop a decision tree.

- Classification and regression tree (CART) makes use of Gini index as metric.
- Iterative dichotomizer 3 (ID3) makes use of entropy function and information gain as metrics.

A decision tree is a tree-like structure in which each interior node characterizes a 'test' on an attribute, each branch signifies the consequence of the test, and each leaf node denotes a class label (decision taken subsequently figuring out all attributes). The paths from root to leaf signify classification guidelines.

Tree-generating processes is done using C5.0, C4.5 and ID3. The theory of entropy from information theory is the basis of information gain.

Entropy is expressed as the following:

$$H(T) = I_E(p_1, p_2, \dots, p_J) = -\sum_{i=1}^{J} p_i \log_2 p_i$$

where $p_1, p_2 \dots$ are fractions that sum up to 1 and denote the proportion of each class existing in the child node that outcomes from a fragment in the tree.

$$\overbrace{IG(T,a)}^{\text{InformationGain}} = \overbrace{H(T)}^{\text{Entropy(parent)}} - \overbrace{H(T/a)}^{\text{WeightedSumofEntropy(children)}}$$

$$= -\sum_{i=1}^{J} p_i \log_2 p_i - \sum_{a} p(a) \sum_{i=1}^{J} -pr(i/a) \log_2 pr(i/a)$$

Information gain is employed to determine that attribute to divide on at every stage in constructing the tree. Uncomplicatedness is preferable; therefore, tree should be kept tiny. To do so, at every phase we should always opt for the rift that ends up in the deepest daughter nodes. A normally used degree of transparency is called information that is deliberated in bits. For each node of the tree, the data worth represents the expected amount of information that may be required to specify whether or not a new instance ought to be classified yes or no, provided that the example reached that node.

The results of both the classifiers will be compared, and a conclusion will be made as to which model is a better fit for the data set in use.

Table 1 Accuracies of proposed approach

	Naïve Bayes	Decision tree
Precision	0.86	0.65
Recall	0.80	0.65
F-measure	0.80	0.65
AUC	0.97	0.65

4 Results

For the purpose of sentiment analysis in movie reviews, a database of 500 movie reviews has been used. Both Naïve Bayes classification and decision tree classification have been implemented on the data set. To compare both classifiers, the comparison parameters that have been evaluated are confusion matrix, precision, recall, f-measure, receiver operating characteristic curve (ROC curve) and the area under the curve (AUC) (AUC gives the accuracy of the classifier) (Table 1).

Below is tabulated the various parameters that have been evaluated for both classifiers:

The data set used is of 500 IMDB movie reviews, half positive and half negative reviews. From the actual table of reviews [1], reviews from row 24,750 to 25,250

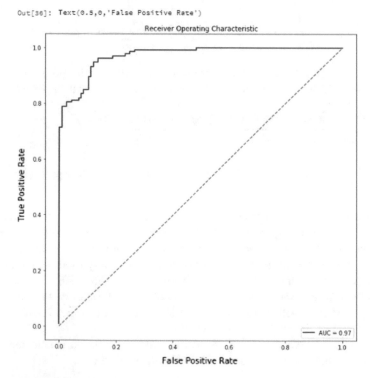

Fig. 1 Accuracy of Naïve Bayes

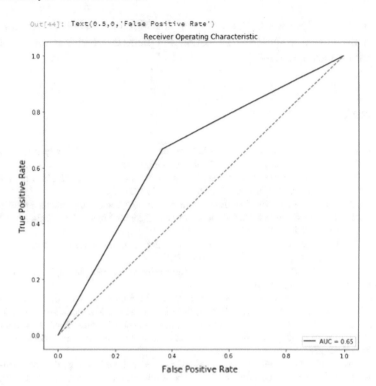

Fig. 2 Accuracy of decision tree

have been used, since the model of laptop used could not handle the memory load of using the previously intended 1 lakh reviews. From the data set, 50% reviews were randomly used as training data and the other 50% were randomly used as test data. After applying classification, the predicted values of sentiment for test data are, ROC obtained are [Y-axis: true positive rate (sensitivity), X-axis: false positive rate (1- specificity)].

Here, Fig. 1 shows the accuracy is said to be higher, since the curve is more inclined towards the top left corner of the graph.

Here, Fig. 2 shows the accuracy is said to be lower, since the curve is less inclined towards the top left corner of the graph.

5 Conclusion

Sentiment analysis can successfully rate the movie and deduce the emotion of the review almost accurately. After careful analysis of results obtained from both classifiers, I have concluded that when training set is used from the reviews itself, the accuracy of Naïve Bayes classifier (0.97) is more than that of decision tree classifier (0.65). Therefore, Naïve Bayes classifier is a better fit for the movie review data set used. Naïve Bayes classification predicts accurate sentiment analysis for the given movie reviews.

References

1. Ogul, H., Ekmekciler, E.: Two-way collaborative filtering on semantically enhanced movie ratings. In: The Proceedings of the ITI, pp. 361–366 (2012)
2. Vinodhini, G., Chandrasekaran, R.: Sentiment analysis and opinion mining: a survey. Int. J. Adv. Res. Comput. Sci. Softw. Eng. 2(6), 282–292 (2012)
3. Wang, J., Liu, T.: Improving sentiment rating of movie review comments for recommendation. In: The IEEE International Conference on Consumer Electronics—Taiwan, ICCE-TW 2017, pp. 433–434 (2017)
4. Li, J., Xu, W., Wan, W., Sun, J.: Movie recommendation based on bridging movie feature and user interest. J. Comput. Sci. 26, 128–134 (2018)
5. Pak, A., Paroubek, P.: Twitter as a Corpus for sentiment analysis and opinion mining in the computer, pp. 1320–1326 (2010)
6. Rish, I.: An empirical study of the Naive Bayes classifier. In: The Empirical Methods in Artificial Intelligence Workshop, IJCAI (2001) 22230, Jan. 2001, pp. 41–46 (2001)
7. Ng, A.Y., Jordan, M.I.: On discriminative versus generative classifiers: a comparison of logistic regression and Naive Bayes. In: Advances in Neural Information Processing Systems, pp. 841–848 (2002)
8. Islam, M., Wu, Q., Ahmadi, M., Sid-Ahmed, M.: Investigating the performance of Naïve-Bayes classifiers and K-Nearest neighbour classifiers. J. Converg. Inf. Technol. 5(5), 133–137 (2010)
9. Alsabti, K., Ranka, S., Singh V.: CLOUDS: tree classifier for large datasets. In: The Proceedings of the Fourth Knowledge Discovery and Data Mining Conference, pp. 2–8 (1998)
10. Lavanya, D., Usha Rani, K.: Ensemble decision tree classifier for breast cancer data 2(1), 17–24 (2012)

Privacy-Preserving Data Mining in Spatiotemporal Databases Based on Mining Negative Association Rules

K. S. Ranjith and A. Geetha Mary

Abstract In the real world, most of the entities are involved with space and time, from any starting point to the end point of the space. The conventional data mining process is extended to the mining knowledge of the spatiotemporal databases. The major knowledge is to mine the association rules in the spatiotemporal databases; the traditional approaches are not sufficient to do mining in the spatiotemporal databases. While mining the association rules, the privacy is the main concern. This paper proposed privacy preserved data mining technique for spatiotemporal databases based on the mining negative association rules and cryptography with low storage and communication cost. In the proposed approach first, the partial support for all the distributed sites is calculated, and then finally, the actual support was calculated to achieve privacy preserve data mining. The mathematical calculation was done and proved that this approach is best for mining association rules for spatiotemporal databases.

Keywords Data mining · Association rules · Privacy-preserving · Spatiotemporal databases · Distributed databases

1 Introduction

Data mining is a vast research area where it has applications on almost in every field like irrigation, finance, industry, etc., Data mining is to identify rules, patterns, or functions which occur from a large amount of databases [1]. It has applications like mining association rules, classification, production, and clustering. Among them, mining association rules are a trendy research area; specifically, users can extract

K. S. Ranjith (✉) · A. Geetha Mary
School of Computing Science and Engineering, VIT University,
Vellore, Tamil Nadu 632 014, India
e-mail: ksranjith2000@gmail.com

A. Geetha Mary
e-mail: geethamary@vit.ac.in

© Springer Nature Singapore Pte Ltd. 2020
P. Venkata Krishna and M. S. Obaidat (eds.), *Emerging Research in Data Engineering Systems and Computer Communications*, Advances in Intelligent Systems and Computing 1054, https://doi.org/10.1007/978-981-15-0135-7_32

the frequent patterns based on that can take effective decisions. Nowadays, due to the IoT nodes, the data is distributed [2]. Distributed databases may be horizontal, vertically, hybrid partitioned databases. Zhang et al. [3] distributed data mining of large databases may lead to problems like local mining may not be global [4]. Due to the connection between each and every object that surrounds us, create a massive amount of data. Mining that kind of distributed data is a major challenge. The data is related to spatiotemporal data. Spatiotemporal data is a data which is related to both space and time. Finding association rules which are interesting may disclose some patterns for selective marketing, forecasting the weather or financial or medical diagnosis, decision supports [5].

Spatiotemporal data has a large space of data that is stored on many computers. Some examples of that kind of data stocks exchanging, environmental data, medical data. Searching is similar patterns in a spatiotemporal database is a very essential in a data mining [6]. The existing distance data mining algorithms are not fully designed and not implemented for the recent trend of data-generating systems like IoT nodes [7]. Many databases are distributed in real time, and many of the algorithms will not suit due to processing power memory sizes [8].

While sharing the data, the privacy of two parties is very essential. So, privacy-preserving mining for distributed databases is very essential [9]. The privacy-preserving techniques may classify on the dimensions like data/rule hiding, data mining algorithm, data distribution, and privacy presentation [10]. Spatial databases consist of shape, size, distance, position, etc. [11]. In association rules, mining association rules were founded by satisfying the predefined minimum concept and support given by the database. Temporal data was obtained by monitoring processes and workflows and registering events the spatial data is obtained by robotics, CAD, GIS, mobile computing, computer vision, etc. [12]. The metrics distance and nonmetric shape, direction, etc. for spatial and the metrics before, after for temporal need to be taken for spatiotemporal databases [13].

The collection of facts, observations, raw data, etc., which is stored in an organized and systemized manner facilitates to extract the required data as per user requirement. An example of a database is an insurance database, client details, policy details, etc. Distributed database is a collection of databases which has its own processing unit and can be managed by the distributed database manager. In this environment, the database can be partitioned vertically, horizontally, and hybrid mode. In the spatiotemporal database, the data is related to the attributes space and time. Both the parameters used to describe the state of the object, change of the pattern, related patterns for any object. Extracting the useful information, knowledge, and database is called a data mining. The knowledge can be association rules, clusters, etc., and the major technique in data mining is classification and prediction.

1.1 Mining Association Rules

Let a, b be an item sets which are subset to the item set i_1, i_2, i_3 from distributed ddb_1, ddb_2,...., ddb_n. Association rules consist of two steps. In the first step, frequent patterns will be generated. In the second step, by using the minimum support and minimum confidence may mine the association rules.

$$\text{Support}(S) : (a \Rightarrow b) = \sum (aUb)/n$$
$$\text{Confidence}(C) : (a \Rightarrow b) = \sum (aUb)/\sum a$$

1.2 Mining Association Rules in Spatiotemporal Database

Spatiotemporal database introduces the concepts of task-relevant objects, reference objects, and concept hierarchy. The concept hierarchy is taxonomy of different concepts from high-level concepts to generalizations on low-level concepts based on the attributes that oriented. The example is shown in Fig. 1. Thus, the spatiotemporal object which is relevant to the task is called the task-relevant object. The main subject of the description is called reference object [11]. The spatiotemporal association rules may be in the form of $a \rightarrow b$ (t, $s\%$, $c \%$). Here, 'a' and 'b' are the set of predicates. $S\%$ and $c\%$ are support and the confidence and t is the time step. The negative support is denoted as $1 - P(A$ intersection $B)$ in a database, ddb_1, ddb_2, ..., ddb_n the negative confidence can be denoted as $1 - P(a/b)$ where 'a' is satisfied by the member of the ddb_1, when 'b' is satisfied by the same member of $ddb_{1,}$. The conjunction of 'm' single spatiotemporal predicates is called m-predicate, $m \geq 1$.

Each element of a spatiotemporal association rule is called a predicate. It is a conjunction of the spatial relations which includes the topological, direction, distance relations and the temporal relations like after, before, etc., and spatiotemporal

Fig. 1 Spatialtemporal states and process [3]

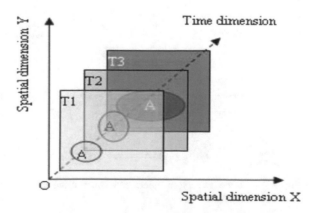

association rule may have a different start time but should have the same timestamp. The examples of the spatiotemporal association rule may be 'western specific version warm pool' and rainfall of the southeast of China. The town references object and water resources, tour a resource, traffic facilitates as task-relevant objects are shown in Fig. 2. Privacy-preserving is an issue that occurs both in centralized and distributed databases. In the centralized database, it will be in single place and multiple users are able to access the same database. Here, the privacy-preserving is done to mining the sensitive information from different users, whereas in the distributed database, the databases will be in different places, the privacy-preserving mining will be global mining where each and every individual site information data need to hide so that every site can be able to access the global results that are useful for the analysis of data based on the user requirement.

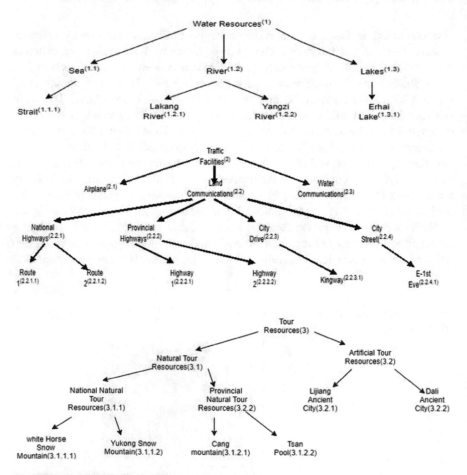

Fig. 2 The town references object and water resources, tour a resource, traffic facilitates as task-relevant objects

2 Related Work

In [1], presented algorithms that use to build polygons and polylines from a range of ultrasonic data, i.e., spatial databases. In [14], presents how to control the data from large databases of spatiotemporal by using the cultural algorithms with evolution programs. The paper [9] presents preserving mining of association rules on a horizontally partitioned data. In [15] survey of privacy-preserving, data mining was done. In Kantarcioglu et al. [16, 17] privacy-preserving association rules mining problem was investigated on horizontally distributed partitioned data. In [18], Kantarcioglu et al. proposed the enhanced Kantarcioglu and Clifton's scheme for privacy-preserving. Rakesh agarwal et al. [1] proposed a procedure for privacy-preserving, data mining. The objective of the procedure is to develop a model for aggregated data in data mining without accessing the private information of individual data in the records by using reconstruction decision tree procedure authors developed the model for aggregated data. In [2], Yehuda lindell preserves a protocol to secure multiparty databases by using the id 3 algorithm for providing privacy-preserving data mining In [3], j. Mennis demonstrates that how to mine the association rules from spatiotemporal database by considering the case study of US region on urban growth and the results have been produced that which shows the land cover and socioeconomic changes in the places Denver, USA (1970.1990).

3 Proposed Framework

Due to rapid growth of data generated notes, the demand for extracting knowledge in every field like industries, education, financial sectors, etc. is increasing. It is necessary to collect data for all generating nodes then need to store and provide the required patterns depending upon the user perspective. Storing data can be done by different an organization that maintains to store the data. Many data mining techniques are existed to extract the knowledge from different databases. The fundamental goals of data mining techniques are prediction or description [11, 16, 19]. It has some techniques like clustering, classification, association rules, etc.; among them, association rules have many applications which are used to generate the relationship between attributes from databases. Association rules can be mined by user-specified measures, minimum and minimum support (Figs. 3, 4 and 5).

To find the association rules, firstly, we find the frequent patterns from various transactions in a database; secondly, we find the association rules by using user-specified support and the confidence. Here, the authors considered the different distributed databases like ddb_1, ddb_2,..., ddb_n.

Fig. 3 Communication
among sites and distributed
mining [2]

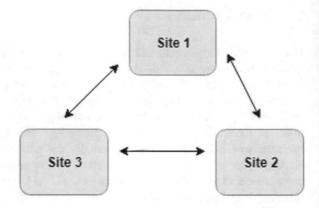

4 Proposed Algorithm

The process of the proposed algorithm is the item sets i_1, i_2, ..., i_n for each site ddb_1, ddb_2,..., ddb_n, the transactions t_1, t_2, ..., T_n. for site ddb_1, the transaction t_1, t_2, t for site ddb_2... are considered. From the transactions of each and every site, by using the distributed FP growth algorithm (DFPGA), calculate the frequent patterns. Due to DFPG algorithm, no candidate set was generated and it reduces consumption of memory, number of iterations, duplicates of data, etc. when compared to distributed mining of association rules (DMA). After generating the frequent patterns by using the negative support and confident may mine the negative association rules support at every site that monitors the remaining sites which is in the distributed environment assume site is coordinates, the negative association rules support partial support is sent to the coordinator site finally the coordinator site will find actual support and mine the global association rules for the distributed environment.

Algorithm

Step 1: START
Step 2: Consider the distributed databases ddb_1, ddb_2, ..., ddb_n.
Step 3: Select the sites from ddb_1, ddb_2, ..., ddb_n.
Step 4: Give unique number for every site and apply cryptography.
Step 5: Generate the frequent patterns by DFPGA.
Step 6: Calculate the negative support and confidence by $1 - P(A$ intersection $B)$, $1 - P(a/b)$.
Step 7: Arrange each site in ring architecture; each site has its own longitude and latitude with respect to the time (given an identification number for every site that indicates the space and time).
Step 8: Calculate the negative partial support.
Step 9: Send negative partial support to all other sites by $P_s = 1 - [x.\text{sup-}$ port $- (ddb_1*\text{minimum support}) +\text{encrypted identification number}]$.
Step 10: Now, all the sites send the value to the coordinator ($s1$)

Fig. 4 Flow chart for
proposed algorithm

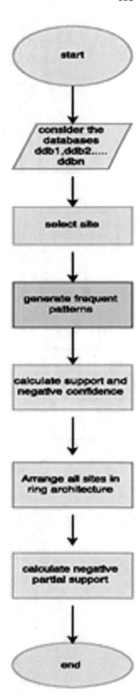

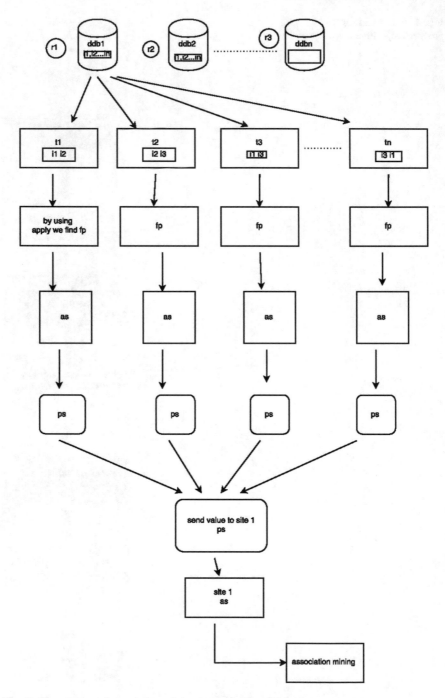

Fig. 5 The process of association mining in spatiotemporal databases

Step 11: Now, coordinator site 1 subtracts the cryptographic value and calculates the actual support by $A_s = 1 - (\sum_{i=1}^{n} \neg p)$

Step 12: Now, site 1 broadcasts the actual support to every site that presents in the distributed environment.

Step 13: END

5 Analytical Results

To perform the analysis, Nepal schools dataset was considered; it contains the spatiotemporal data with the different attributes but with the serial number and the spatiotemporal id (cental-1, East-2, etc.....). In this, the different spatiotemporal id is taken as the separate site. So for the dataset, Nepal has total six sites, i.e., six different space and the time ids. Threshold support 40% is considered to analyze the spatiotemporal database for all the sites where each site has a unique number. First, each site calculates the support count using the DFPGA, the negative partial support by using the defined formula, and the cryptographic id was added and then sent to all the other sites with same process in a ring manner to the coordinator site; finally, the actual support was calculated and extracted the association rules globally with privacy-preserving.

Site 1: Negative (−ve) Support

DB1 = 5.3, Encrypted Value = 1, Support Count = 4/75 = 0.05

Support = 1 − Support Count = 0.95

Partial Support $(P_s) = 1 - [x.\text{support} - (ddb_1 * \text{minimum support}) + \text{encrypted identification number}]$.

Partial Support $(P_s) = 1 - [0.95 - (5.3 * 0.4) + 1]$

Partial Support $(P_s) = 1 - (-0.17)$

Partial Support $(P_s) = 1.17$

Site 2: Negative (−ve) Support

DB1 = 25.3, Encrypted Value = 2, Support Count = 19/75 = 0.25

Support = 1 − Support Count = 0.75

Partial Support $(P_s) = 1 - [x.\text{support} - (ddb_1 * \text{minimum support}) + \text{encrypted identification number}]$.

Partial Support $(P_s) = 1 - [0.75 - (25.3 * 0.4) + 2]$

Partial Support $(P_s) = 1 - (-7.37)$

Partial Support $(P_s) = 8.37$

Site 3: Negative (−ve) Support

DB1 = 21.3, Encrypted Value = 3, Support Count = 16/75 = 0.21

Support = 1 − Support Count = 0.79

Partial Support $(P_s) = 1 - [x.\text{support} - (ddb_1 * \text{minimum support}) + \text{encrypted identification number}]$.

Partial Support $(P_s) = 1 - [0.79 - (21.3 * 0.4) + 3]$

Partial Support (P_s) = 1 − (−4.73)

Partial Support (P_s) = 5.73

Site4: Negative (−ve) Support

DB1 = 12, Encrypted Value = 4, Support Count = 9/75 = 0.12

Support = 1 − Support Count = 0.88

Partial Support (P_s) = 1 − [x.support − (ddb$_1$ * minimum support) + encrypted identification number].

Partial Support (P_s) = 1 − [0.88 − (12 * 0.4) + 4]

Partial Support (P_s) = 1 − 0.08

Partial Support (P_s) = 0.92

Site 5: Negative (−ve) Support

DB1 = 16, Encrypted Value = 5, Support Count = 12/75 = 0.16

Support = 1 − Support Count = 0.84

Partial Support (P_s) = 1 − [x.support − (ddb$_1$ * minimum support) +encrypted identification number].

Partial Support (P_s) = 1 − [0.84 − (16 * 0.4) + 5]

Partial Support (P_s) = 1 − (−0.56)

Partial Support (P_s) = 1.56

Site 6: Negative (−ve) Support

DB1 = 20, Encrypted Value = 6, Support Count = 15/75 = 0.2

Support = 1 − Support Count = 0.8

Partial Support (P_s) = 1 − [x.support − (ddb$_1$ * minimum support) + encrypted identification number].

Partial Support (P_s) = 1 − [0.8 − (20 * 0.4) + 6]

Partial Support (P_s) = 1 − (−1.2)

Partial Support (P_s) = 2.2

Actual Support (A_s) = 1 − ($\sum_{i=1}^{n} \neg p$)

A_s = 1 − (19.95)

Actual Support (A_s) = −18.95

6 Conclusion

This paper presents privacy-preserving association rule mining in the distributed environment. The cryptographic technique was used for encrypting the identification number of every distributed site for privacy purpose. Locally, all the sites calculate the negative support and confidence, partial support then partial support was sent to the coordinator site. So that coordinator site calculates the actual support and broadcasts to all sites and finds the association rules. The example for distributed environment is spatiotemporal databases. The proposed algorithm will best fit for the spatiotemporal databases to mine association rules.

References

1. Getta, J.R., McKerrow, L., McKerrow, P.J.: The application of database mining techniques to data fusion in spatial databases. In: Proceeding of 1st Australian Data Fusion Symposium, pp. 135–140. IEEE (1996)
2. Sahu, A.K., Kumar, R., Rahim, N.: Mining negative association rules in distributed environment. In: Proceedings International Conference on Computational Intelligence and Communication Networks (CICN), pp. 934–937. IEEE (2015)
3. Zhang, X., Su, F., Du, Y., Shi, Y.: Association rule mining on spatio-temporal processes. In: Proceedings 4th International Conference on Wireless Communications, Networking and Mobile Computing, pp. 1–4. IEEE (2008)
4. Cheung, D.W., Ng, V.T., Fu, A.W., Fu, Y.: Efficient mining of association rules in distributed databases. IEEE Trans. Knowl. Data Eng. **1**(6), 911–922 (1996)
5. Neerugatti, V., Reddy, R.M.: A survey on secure connectivity techniques for internet of things environment. Int. J. Eng. Res. Comput. Sci. Eng. (IJERCSE) **4**(3) (2017)
6. Cheung, D.W., Han, J., Ng, V.T., Fu, A.W., Fu, Y.: A fast distributed algorithm for mining association rules. In: Fourth International Conference on Parallel and Distributed Information Systems, pp. 31–42. IEEE (1996)
7. Chen, M.S., Han, J., Yu, P.S.: Data mining: an overview from a database perspective. IEEE Trans. Knowl. Data Eng. **8**(6), 866–883 (1996)
8. Abraham, T., Roddick, J.F.: Survey of spatio-temporal databases. GeoInformatica **3**(1), 61–99 (1999)
9. Wang, C., Huang, H., Li, H.: A fast distributed mining algorithm for association rules with item constraints. In: SMC 2000 Conference Proceedings. 2000 IEEE International Conference on Systems, Man and Cybernetics' Cybernetics Evolving to Systems, Humans, Organizations, and Their Complex Interactions, vol. **3**(1), pp. 1900–1905. IEEE (2000)
10. Verykios, V.S., Bertino, E., Fovino, I.N., Provenza, L.P., Saygin, Y., Theodoridis, Y.: State-of-the-art in privacy preserving data mining. ACM Sigmod Rec. **33**(1), 50–57 (2004)
11. Bertino, E., Fovino, I.N., Provenza, L.P.: A framework for evaluating privacy preserving data mining algorithms. Data Min. Knowl. Disc. **11**(2), 121–154 (2005)
12. Chang, C.C., Yeh, J.S., Li, Y.C.: Privacy-preserving mining of association rules on distributed databases (2006)
13. Kotsiantis, S., Kanellopoulos, D.: Association rules mining: A recent overview. GESTS Int. Trans. Comput. Sci. Eng. **32**(1), 71–82 (2006)
14. Neerugatti, V., Reddy, R.M.: An introduction, reference models, applications, open challenges in Internet of Things. Int. J. Mod. Sci. Eng. Technol. (IJMSET) (2017)
15. Andrienko, G., Malerba, D., May, M., Teisseire, M.: Mining spatio-temporal data. J. Intell. Inf. Syst. **27**(3), 187–190 (2006)
16. Wang, L., Xie, K., Chen, T., Ma, X.: Efficient discovery of multilevel spatial association rules using partitions. Inf. Softw. Technol. **47**(13), 829–840 (2005)
17. Wang, J., Luo, Y., Zhao, Y., Le, J.: A survey on privacy preserving data mining. In: Proceeding First International Workshop on Database Technology and Applications, pp. 111–114. IEEE (2009)
18. Gurevich, A., Gudes, E.: Privacy preserving data mining algorithms without the use of secure computation or perturbation. In: Proceeding 10th International Database Engineering and Applications Symposium (IDEAS'06), pp. 121–128. IEEE (2006)
19. Agrawal, R., Srikant, R.: Fast algorithms for mining association rules. In: Proceedings VLDB Conference, Santiago, pp. 487–499. IEEE (1994)

Spatial Data-Based Prediction Models for Crop Yield Analysis: A Systematic Review

Alkha Mohan and M. Venkatesan

Abstract Agriculture plays a vital role in the global economy. WHO states that there are three pillars of food security: availability, access, and usage. Among these three pillars, availability is the most important one. Ensuring food for the entire population of a country is achieved only through an increase in crop production. Accurate and timely forecasting of the weather can help to increase the yield production. Early prediction of crop yield has a vital role in food availability measure. Researchers monitor different parameters that affect the crop yield regularly. Yield prediction did through either statistical data or spatial data. Crop monitoring through remote sensing can cover a vast land area. Therefore, spatial data-based prediction is widespread in recent decades. Satellite images such as multispectral, hyperspectral, and radar images were used to calculate crop area, soil moisture, field greenness, etc. Among these imaging modalities, hyperspectral images give more accurate results, but its higher dimensionality is a challenging issue. Optimal band selection from hyperspectral images helps to reduce this curse of dimensionality problem. Crop area is one of the essential parameters for yield prediction. The exact crop area measure can be achieved only through the best crop discrimination methods. This paper provides a comprehensive review of crop yield prediction using hyperspectral images. Besides, we explore the research challenges and open issues in this area.

Keywords Crop yield prediction · Hyperspectral images · Optimal band selection · Vegetation indices · Classification

A. Mohan (✉) · M. Venkatesan
Department of CSE, National Institute of Technology Karnataka, Surathkal, India
e-mail: mohan.alkha@gmail.com

M. Venkatesan
e-mail: venkisakthi77@gmail.com

© Springer Nature Singapore Pte Ltd. 2020   341
P. Venkata Krishna and M. S. Obaidat (eds.), *Emerging Research in Data Engineering Systems and Computer Communications*, Advances in Intelligent Systems and Computing 1054, https://doi.org/10.1007/978-981-15-0135-7_33

1 Introduction

Food security is a significant problem faced by our world. Improvements in agri-cultural production ensure the availability of food for the whole population. Climatic changes due to global warming negatively affect food production. The effect of climatic change on agriculture is a trendy research area. Accurate weather forecasting helps farmers to overcome the crises in crop production due to climate change. Farmers can change the planting time based on these timely exact weather forecast. Crop yield depends on many parameters such as climatic factors, the water content in soil, mineral distribution in soil, sowing time, and harvesting methods. Researchers select different combinations of these parameters for predicting crop yield. In past decades, these parameters and crop area information were measured using ground survey. Use of satellite images helped researchers to overcome these difficulties. Satellite images such as multispectral and hyperspectral images used to find crop area, soil moisture, field greenness, etc. In most of the existing works, yield mapping performed only at the end of a season, however, variations in parameter values during the season depending on the final production. Therefore, continuous monitoring of the crop area and parameter evaluation is necessary for accurate yield prediction.

Yield prediction from remote sensing data became very popular in recent dec-ades. When the light hits the earth surface, it exhibits different phenomena such as scattering, reflection, emission, and absorption as shown in Fig. 1. The reflected energy from any material has a unique footprint. It is known as the spectral sig-nature of the material, which can be used to differentiate various types of materials on the earth's surface. The idea of spectral signature used for remote sensing data-based yield prediction [1]. Figure 2 shows the spectral signature of different materials in the urban environment.

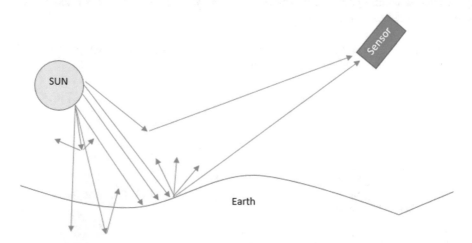

Fig. 1 Idea behind remote sensing imaging technique

Fig. 2 Spectral signature of common materials found in urban environment [3]

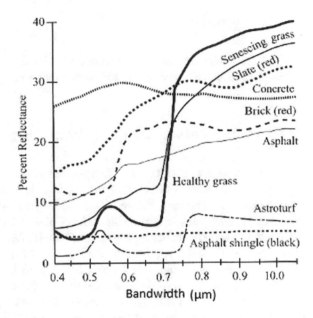

Higher number of bands in hyperspectral images helps to discriminate vegetation widely. However, a significant problem in hyperspectral image analysis is their computational complexity due to the higher number of bands. Thus, optimal band selection from hyperspectral images became a challenge. This paper familiarizes different existing methods used for optimal band selection. Calculating the crop area from hyperspectral images for yield prediction is another challenge. Since the spatial resolution of hyperspectral images is in meters, a single pixel may contain different vegetative parts named as mixed pixels. Both spatial and spectral features are necessary to classify different crop areas from a hyperspectral image.

Evolution of data science and machine learning introduces different prediction models. Based on the literature, prediction models classified into two namely: statistical models and machine learning models. The regression model is the most commonly used statistical model. Nonlinear methods give more accurate results in crop yield prediction because different parameters affect the yield differently. Decision support systems, rule-based systems, artificial neural networks, etc., are the major machine learning approaches in this field. Deep learning method in multispectral and hyperspectral satellite image gives a highly accurate crop yield prediction [2].

The remaining portion of the paper organized as follows: Sect. 2 focuses on various satellite imaging modalities and parameters relevant to crop yield prediction and its calculation from satellite images. Section 3 describes the literature of significant steps for yield prediction from hyperspectral images done so far, and Sect. 4 presents the challenges and open issues in this field. We conclude the review in Sect. 5.

2 Background

This section describes different background topics related to spatial data-based crop yield prediction such as imaging modalities and yield prediction parameters.

2.1 Multispectral and Hyperspectral Images

A spectral image worths thousand images. They have both spatial information and spectral information. Multispectral images captured beyond the visible spectrum and the spectral band may change from 3 to 15. Main spectral bands are blue, green, red, near infrared, mid-infrared, far infrared, thermal infrared, and radar. Landsat program is a series of earth-observing satellite with different spectral bands used for the higher differential application. Active Landsat programs are Landsat 7 and Landsat 8 with seven and nine spectral bands, respectively.

Hyperspectral images (HSIs) are contiguous band images. It contains spectral bands for all pixels in the image. Hyperspectral remote sensing originally developed for detecting and mapping minerals. In hyperspectral images, the bandwidth is very less and continuous; therefore, we can discriminate between the substrates in an image very clearly using their unique spectral signature. Increase in spectral capability increases the power of discrimination. Since hyperspectral images contain a large number of bands, its processing is complicated and time-consuming. Therefore, different feature selection algorithms are used to find the optimal band from hyperspectral images. Figure 3 shows the comparison of multispectral and hyperspectral images.

2.2 Vegetation Indices

Vegetation indices (VIs) are measured from remote sensing images using simple algorithms. Based on application, VIs used for both qualitative and quantitative measure of vegetation growth, cover, health, and so on. Beyond the growth and health of crop, vegetation indices are used to measure water content, pigmentation, carbohydrate, and protein content [3]. This section describes most commonly used VIs and their effect on vegetation analysis.

Normalized Difference Vegetation Index (NDVI)
Pioneer indices in the history of VIs are Ratio Vegetation Index (RVI) and Vegetation Index Number (VIN). RVI is sensitive to atmospheric effects and reduces the discriminating power. Thus, a new VI has been introduced named Normalized Difference Vegetation Index (NDVI) [4]. NDVI is the most commonly used vegetation index for imaging vegetation in multispectral images, and its value ranges from 0 to 1.

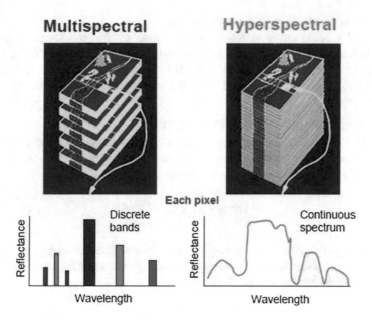

Fig. 3 Multispectral versus hyperspectral remote sensing

$$NDVI = \frac{(NIR - R)}{(NIR + R)} \qquad (1)$$

NDVI is calculated using the Eq. (3). This value also depends on many biophysical effects such as climatic effects, atmospheric effects, and clouds.

Vegetation Greenness (VG)
For better interpretation of field values in multispectral images, some authors calculate VG parameter from NDVI. VG indicates how green the pixel in an image is, i.e., the pixel is sparsely green or densely green. It helps to find the stress of the vegetation.

$$VG = \frac{NDVI}{(0.66 * I)} * 100 \qquad (2)$$

Here, NDVI measured for a given period (e.g., two weeks).

Leaf Area Index (LAI)
Total leaf area of a crop canopy is measured using LAI. It is commonly used to measure many biophysical phenomena such as evapotranspiration and photosynthesis. LAI is dimensionless and considered as a reliable indicator of crop productivity. LAI is the ratio of leaf area to per unit ground area. Scientists provide different LAIs based on disciplines and applications [5, 6].

Photosynthetically Active Radiation (PAR)

PAR is the spectral range of active radiation used by the crop for photosynthesis. This portion of the radiation is highly correlated to the net production of yield. It depends on actual sunlight hours, total daily radiation, and so on. A fraction of absorbed PAR (fPAR) calculated from the multispectral images can be used to identify crop production [7].

Normalized Difference Water Index (NDWI)

NDWI is used to find the change of water content in leaves. Near-infrared and shortwave infrared bands are used to identify the water content from spectral images. NDWI is calculated using the following equation

$$NDWI = \frac{X_{NIR} - X_{SWIR}}{X_{NIR} + X_{SWIR}} \tag{3}$$

Climatic parameters such as rainfall rate, temperature, humidity, precipitation rate, evapotranspiration rate, water stress, and global radiation rate are directly influencing the net production of yield.

3 Hyperspectral Image-Based Crop Yield Prediction

Measuring biophysical and biochemical attributes of crops for productivity calculation, stress monitoring, etc., through the land survey is a tedious task. In 2000, Thenkabail et al. [8] found a relationship between crop characteristics and vegetation indices using 512 band spectroradiometer. These narrowband spectrum-based models overshot the accuracy of all previous wideband image-based analyses. Later, researchers started using hyperspectral images for various agricultural applications. Thus, our review goes through various essential steps needed for hyperspectral image-based crop yield prediction.

3.1 Optimal Band Selection

The higher number of bands of HSIs helps in discriminating crops using clearly defined spectral signature. However, as the number of bands increases, the number of training data required to maintain minimum functionality and statistical confidence for classification purposes must be high. Since the availability of hyperspectral images is very less, the only solution for the classification problem is dimensionality reduction. Table 1 summarizes various research works done in optimal band selection from hyperspectral images in recent decades.

Table 1 Summary of research works done in optimal band selection of hyperspectral images

Approach used	Limitations
Spatially coherent locally linear embedding [9]	Accuracy depends on the size of image patch selection
Fractal-based dimensionality reduction [10]	Noise in image lead to misclassification
Wavelet-based noise removal + fractal dimension [11]	Proper choice of noise removal filter is a challenge
SVM + ICA [12]	Complex in multiclass classification
Information gain band + PCA [13]	Good result in small multimission satellite images
Fuzzy roughset model [14]	High computational time
Sparse graph-based discriminant analysis [15]	Construction of ll graph is a computational burden
Global mixture coordination factor analysis [16]	Estimation of expectation maximization is tedious
Semi-supervised double sparse graph [17]	Pseudo to a real label mapping error
Kernel PCA + DBSCAN [18]	Finding more suitable kernel
Semi-supervised local fisher discriminant analysis [19]	Fails in active learning

3.2 Crop Classification

Optimal band selection is the preprocessing step for hyperspectral image analysis. Use of these selected bands reduces the computational complexity of further processing. Each hyperspectral image covers a wide geographical area. Crop area is an essential measure in crop yield calculation. Therefore, the next step of crop yield prediction is the segmentation of different crop areas in HSI using classification techniques.

Classification of hyperspectral images can be accomplished using supervised, semi-supervised, or unsupervised learning techniques. The availability of hyperspectral images and its ground truth is comparatively less, hence supervised classification is accurate. Table 2 summarizes various research works done for classification of hyperspectral images in recent decades.

3.3 Parameter Selection

Next stage of crop yield prediction from the hyperspectral image is the calculation of various parameters that depend on crop yield. Different researchers use different combinations of parameters for crop yield prediction. It includes spatial features of images, climatic features, various vegetation indices extracted from spectral images, and survey data. Continuous monitoring of crop helps to predict the yield

Table 2 Summary of research works done in classification of hyperspectral images

Approach used	Limitations
Semi-supervised classification (EM + Gaussian mixture model) [20]	Useful for classes with Gaussian mixture distribution
Hyperspectral reflectance data [21]	Spectral similarity is observed between varieties of rice and sugarcane crops
Partial least square LDA on spectral information [22]	Classification error occurs due to variability of images taken in different time
Object-based classification [23]	High resolution image gives better accuracy
Spectral angle mapper [24]	Pixel-based approach takes more time
Kernel-based SVM [25]	Image segmentation accuracy depends heavily on hyperspectral data dimensionality reduction method
Convolutional neural network (CNN) [26]	Fails to classify categories with small number of samples
Deep belief network (DBN) [27]	Hyper segmentation depends on classification accuracy
Pixel pair feature + CNN [28]	Compromises computational complexity
3D CNN [29]	Can not classify unlabeled samples
DBN + texture feature enhancement [30]	Selection of best texture features

accurately. Classification helps to find the cultivated area of each crop, and the segmented area is used to find the vegetation indices for each crop separately.

Usage of hyperspectral images helps to calculate vegetation indices, stress parameters, water content, etc. However, climatic parameters are equally crucial for prediction of yield. Proper selection of parameters improves the prediction accuracy. Drummond et al. [31] used only the old statistical data for yield prediction. Lobell et al. [32] calculated NDVI and simple ratio index from the image and predicted the yield using fPAR, light-use efficiency (LUE), and harvest index (HI). Jiang et al. used NDVI, temperature, water index, APAR, and average yield for crop yield prediction [7]. Analysis of previous works implies that NDVI is the most commonly used vegetation indices for crop yield prediction from satellite data.

3.4 Prediction Models

The last stage of crop yield prediction is the prediction model design. Various vegetation indices, climatic factors, crop cultivated areas, and survey results are the inputs to the prediction model. Based on literature prediction, models are classified into two categories: statistical approaches and machine learning approach. Most commonly used statistical method is regression analysis.

Major drawbacks of the statistical model are:

- Generalization of parameters is very difficult because the parameters depend on the sites.
- Prior domain knowledge and design skills are needed for these models.

Evolution of machine learning and data science improves crop yield prediction accuracy. Mainly used machine learning approaches are decision-based systems, regression trees, random forest method, artificial neural network (ANN), deep learning techniques, and so on. Artificial neural networks are analogous to biological neurons. They collect input as stimulus, learn based on this stimulus and produce output.

4 Discussions

Even though there are wide varieties of methods available for yield prediction, prediction using hyperspectral images is relatively a new area and has a significant scope in future. Based on the literature study, here, we point out the significant challenges in crop yield prediction from hyperspectral images.

Hyperspectral images give more accurate prediction results than multispectral data and ground survey data. The main problems in hyperspectral image analysis are the availability of images and its high dimension. Identifying optimal bands from HSI is a viral research area. Hyperspectral images are nonlinear. Thus, linear dimensionality reduction techniques fail. Most nonlinear techniques compromise the classification accuracy, and some of them fail in noise handling. The availability of hyperspectral images is less compared to other satellite images. There is no accurate method available for finding the optimal bands in hyperspectral images. Therefore, a new optimal band selection method reduces computational complexity and improves classification accuracy that is necessary.

Classification of crop area from the HSI is another challenging issue. Since the spatial resolution of satellite images are in meters, a single pixel contains information about more than one crop. So, classification using spatial parameters is not an accurate method. Some methods classify the image by considering either spatial or spectral data. However, spatial-spectral features give efficient performance than others. Deep learning technique combined with different CNN classification methods outperformed all the other previous techniques. Designing a new deep neural network considering both spatial and spectral features to produce accurate hyperspectral image classification using a limited number of training data is an existing challenge.

Crop yield depends on multiple parameters such as climatic parameters, soil parameters, and vegetation indices. Multiple regression model is the most commonly used statistical method for crop yield prediction. However, the relationship between these parameters and yield varies for each field. Thus, creating a generic crop yield prediction model using a statistical method is challenging. Artificial neural network-based prediction method gives more accurate result than any other prediction methods. The accuracy of prediction depends on proper selection of prediction parameters. Improper crop area classification and noisy data negatively affect the result of the prediction.

5 Conclusions

This paper presented a structured review of various steps in the hyperspectral image-based crop yield prediction. Yield determining parameter extraction from the hyperspectral image is the first phase for prediction. Hyperspectral images cover a large geometric area so that a single image may contain different types of plants. Thus, parameter calculation for each crop needs crop classification from the image. Crop classification in hyperspectral images is tedious. To reduce the computational complexity of classification, optimal band selection from hyperspectral images became necessary. From these perceptions, our paper focused the significant research works done for optimal band selection, crop classification, and yield forecast using different prediction models. Even though plenty of works completed in this area, it still needs refinements.

References

1. Fonseca, L.M.G., Namikawa, L.M., Castejon, E.F.: Digital image processing in remote sensing. In: Tutorials of the XXII Brazilian Symposium on Computer Graphics and Image Processing, no. C, pp. 59–71 (2009)
2. Kuwata, K., Shibasaki, R.: Estimating corn yield in the United States with Modis Evi and machine learning methods. ISPRS Ann. Photogramm. Remote Sens. Spat. Inf. Sci. **III-8**, 131–136 (2016)
3. Batten, G.D.: Plant analysis using near infrared reflectance spectroscopy: the potential and the limitations. Aust. J. Exp. Agric. **38**(7), 697–706 (1998)
4. Rouse, J.W., Hass, R.H., Schell, J.A., Deering, D.W., Harlan, J.C.: Monitoring the Vernal Advancement and Retrogradation (Green Wave Effect) of Natural Vegetation. College Station, Texas (1974)
5. Zheng, G., Moskal, L.M.: Retrieving leaf area index (LAI) using remote sensing: theories, methods and sensors. Sensors **9**(4), 2719–2745 (2009)
6. Jensen, J.R.: Remote Sensing of the Environment: An Earth Resource Perspective, 2 edn. Prentice Hall, Upper Saddle (2007)
7. Jiang, D., Yang, X., Clinton, N., Wang, N.: An artificial neural network model for estimating crop yields using remotely sensed information. Int. J. Remote Sens. **25**(9), 1723–1732 (2004)
8. Thenkabail, P.S., Smith, R.B., De Pauw, E.: Hyperspectral vegetation indices and their relationships with agricultural crop characteristics. Remote Sens. Environ. **71**(2), 158–182 (2000)
9. Mohan, A., Sapiro, G., Bosch, E.: Spatially coherent nonlinear dimensionality reduction and segmentation of hyperspectral images. IEEE Geosci. Remote Sens. Lett. **4**(2), 206–210 (2007)

10. Ghosh, J.K., Somvanshi, A.: Fractal-based dimensionality reduction of hyperspectral images. J. Indian Soc. Remote Sens. **36**(3), 235–241 (2008)
11. S. Junying, S., Ning, S.: A dimensionality reduction algorithm of hyper spectral image based on fract analysis. Int. Arch. Photogramm. Remote Sens. Spat. Inf. Sci. **XXXVII**(Part B7), 297–302 (2008)
12. Moon, S., Qi, H.: Hybrid dimensionality reduction method based on support vector machine and independent component analysis. IEEE Trans. Neural Netw. Learn. Syst. **23**(5), 749–761 (2012)
13. Koonsanit, K., Jaruskulchai, C., Eiumnoh, A.: Band selection for dimension reduction in hyper spectral image using integrated information gain and principal components analysis technique. Int. J. Mach. Learn. Comput. **2**(3), 248–251 (2012)
14. Lodha, S.P., Kamlapur, S.M.: Dimensionality reduction techniques for hyperspectral images. Int. J. Appl. Innov. Eng. Manag. **3**(10), 1–5 (2014)
15. Ly, N.H., Du, Q., Fowler, J.E.: Sparse graph-based discriminant analysis for hyperspectral imagery. IEEE Trans. Geosci. Remote Sens. **52**(7), 3872–3884 (2014)
16. Wang, S., Wang, C.: Research on dimension reduction method for hyperspectral remote sensing image based on global mixture coordination factor analysis. ISPRS—Int. Arch. Photogram. Remote Sens. Spat. Inf. Sci. **XL-7/W4**, 159–167 (2015)
17. Chen, P., Jiao, L., Liu, F., Zhao, J., Zhao, Z., Liu, S.: Semi-supervised double sparse graphs based discriminant analysis for dimensionality reduction. Pattern Recognit. **61**, 361–378 (2017)
18. Datta, A., Ghosh, S., Ghosh, A.: Unsupervised band extraction for hyperspectral images using clustering and kernel principal component analysis. Int. J. Remote Sens. **38**(3), 850–873 (2017)
19. Wu, H., Prasad, S.: Semi-supervised dimensionality reduction of hyperspectral imagery using pseudo-labels. Pattern Recognit. **74**, 212–224 (2018)
20. Gomez-Chova, L., et al.: Semi-supervised classification method for hyperspectral remote sensing images. In: 2003 IEEE International Geoscience and Remote Sensing Symposium, vol. 3, pp. 1776–1778 (2003)
21. Rao, N.R.: Development of a crop—specific spectral library and discrimination of various agricultural crop varieties using hyperspectral imagery. Int. J. Remote Sens. **1161** (2010)
22. Hadoux, X., Gorretta, N., Rabatel, G.: Weeds-wheat discrimination using hyperspectral imagery. In: CIGR-Ageng 2012. International Conference on Agricultural Engineering, pp. 6 (2012)
23. Alganci, U., Sertel, E., Ozdogan, M., Ormeci, C.: Parcel-level identification of crop types using different classification algorithms and multi-resolution imagery in Southeastern Turkey. Photogramm. Eng. Remote Sens. **79**(11), 1053–1065 (2013)
24. Boitt, M., Ndegwa, C., Pellikka, P.: Using hyperspectral data to identify crops in a cultivated agricultural landscape—a case study of Taita Hills, Kenya. J. Earth Sci. Climatic Chang. **5**(9) (2014)
25. Liu, X., Bo, Y.: Object-based crop species classification based on the combination of airborne hyperspectral images and LiDAR data. Remote Sens. **7**, pp. 922–950 (2015)
26. Hu, W., Huang, Y., Wei, L., Zhang, F., Li, H.: Deep convolutional neural networks for hyperspectral image classification. J. Sens. (2015)
27. Mughees, A., Ali, A., Tao, L.: Hyperspectral image classification via shape-adaptive deep learning. In: 2017 IEEE International Conference on Image Processing (ICIP), pp. 375–379 (2017)
28. Li, W., Wu, G., Zhang, F., Du, Q.: Hyperspectral image classification using deep pixel-pair features. IEEE Trans. Geosci. Remote Sens. **55**(2) (2017)
29. Li, Y., Zhang, H., Shen, Q.: Spectral-spatial classification of hyperspectral imagery with 3D convolutional neural network. Remote Sens. **9**(1) (2017)
30. Li, J., Xi, B., Li, Y., Du, Q., Wang, K.: Hyperspectral classification based on texture feature enhancement and deep belief networks. Remote Sens. **10**(3) (2018)

31. Drummond, S.T., Sudduth, K.A., Joshi, A., Birrell, S.J., Kitchen, N.R.: Statistical and neural methods for site-specific yield prediction. Trans. ASAE **46**(1), 5–14 (2003)
32. Lobell, D.B., Asner, G.P.: Comparison of earth observing-1 ALI and Landsat ETM + for crop identification and yield prediction in Mexico. IEEE Trans. Geosci. Remote Sens. **41**(6), PART I, 1277–1282 (2003)

Review on Neural Network Algorithms for Air Pollution Analysis

Sumaya Sanober and K. Usha Rani

Abstract Neural network is a layer-based optimization technique to solve a real-time problem adjusting the weight values of the neuron based on its activation function. It aids to construct a model to compute optimum results in business analytical process, prediction analysis, financial forecasting, environmental analysis, etc. The environmental analysis are having two approaches namely determine the pollution or identifying the quality using environmental factors such as air, water, and land. The air pollution analysis and predication is to control the pollution. It is a challenging process due to its computational complexity. The environmental research community is working on air pollution factor analysis, pollution index computation, and predication. Present research addresses the findings of various artificial neural network algorithms and presented same. It is recognized that the obtained neural network models are providing sufficient reliable forecast that indicates an effective tool for analyzing and predicting the air pollution. Thus, the study aims to provide various ongoing research results of air pollution analysis and presented the usage of artificial neural network for analysis and prediction of air pollution.

Keywords Neural network models · Environmental mining · Optimization techniques · Air pollution analysis and prediction

S. Sanober (✉) · K. Usha Rani
Department of Computer Science, Sri Padmavati Mahila Visvavidyalayam,
Tirupati, Andhra Pradesh, India
e-mail: sumayacsdept@gmail.com

K. Usha Rani
e-mail: usharanikuruba@yahoo.co.in

© Springer Nature Singapore Pte Ltd. 2020
P. Venkata Krishna and M. S. Obaidat (eds.), *Emerging Research in Data Engineering Systems and Computer Communications*, Advances in Intelligent Systems and Computing 1054, https://doi.org/10.1007/978-981-15-0135-7_34

1 Introduction

The computational models are used to enhance the lifestyle of human being fore-casting the changes and recommending necessary preventive actions using envi-ronmental mining process. It includes the analysis of variations on environmental factor analysis such as air, soil, and water. The international community is viewing all the environmental factors and its impact closely to ensure the pollution-free earth for next generation [1]. The computation and predication of these environmental changes are having more complexity due its dynamism [2]. It required an opti-mization process in its computation of analysis and predictions because the analyses are representing chemical elements and its reactions which are more complex in computing its nature [3]. The study review ongoing research in this filed focused on optimization and prediction of variations in air pollution analysis and predications. It includes conceptual model, air components, algorithms, and techniques used for computation along with findings of various research works.

The artificial neural networks are a machine learning approaches that mimics human brain using artificial neurons. They have specific mathematical models, which are intended in accordance with the functions of biological neural networks [4], these models were used for specific purposes and predictions. The neural networks are programmed to obtain an effective solutions based on its training data set in accordance to the connection among them. Through the training process, neural network is capable to adopt complex functionality to compute output from the given input. After completion of training phase, neural network is capable to forecast the upcoming values which are constructed from the training. Through recent memory and processing influence, there is an incredible possibility for artificial neural network structural design such as convolutional and recurrent system that shows enhancements in deep learning [5]. The predefined design is a feedforward network also called as multi-layer perceptron (MLP). It consist of three layers namely input layer that is the initial layer connected to network, hidden layer or the middle layer that is specified by user and act as an intermediate node between input and output layer. The final layer is known as output layer to obtain the result in the network.

The most important factor of multi-layer perceptron is the association weights and biases. The output of every node is enumerated in the following blocks. Initially, the weighted summarizes of the input is enumerated based on calculation:

$$S_k = \sum_{i=1}^{n} W_{ij} I_i + \beta_j \qquad (1)$$

where I_i is the input, w_{ij} is the weight associated among I_i and hidden neuron x, β_j is bias. Future more, activation function is established to produce output neuron by summing up the weights value. They are basically six type of activation function logistic, hyperbolic, exponential, bi-polar, binary, and sigmoid which is used by multi-layer perceptron. Among them sigmoid function is used mostly [6].

Sigmoid function:

$$g(x) = 1/1 + e^{-Sk} \tag{2}$$

After enumeration, multi-layer perceptron enumerate final output.

$$Y_j = \sum_{i=1}^{m} w_{jk}g + \beta_j \tag{3}$$

The modern computation process is extracting the knowledge by the exiting set of data to enhance an application performance according to their domain nature. Since it is integrated with real-time applications [7], the computational complexities are high. The environmental data analysis used for identifying the variations from the previous observations on components on air. The analysis results accuracy are high depends on duration of the data and its volume [8]. The weather forecasting accuracy is achieved based on the selections of lengthy duration, which consist of huge data set [9] representing the changes of multiple locations at an instant of time [10]. The forecasting accuracy used for decision-making to plan integrated activities related on human life style. The environmental analysis represented in shows that computation complexities are high as well as it can produce accurate results while using appropriate algorithms. Therefore, this review summarized on algorithms that are used different researchers and its major purpose.

This paper Sect. 2 provides the review of air pollution index characters and the aforementioned computation along with ongoing researchers in AQI. Sections 2.1 and 2.2 provide the air pollution predication research area along with summary of other ongoing research consider for the review process. Section 3 provides conclusion of the paper and followed by the references.

2 Review of Air Pollution Index

The air pollution is computed by calculating air quality and rate of pollution. The computed air quality or the pollution level presented as air quality index (AQI). It is computed by considering (O_3), (CO), (SO_2), (N_2), and (PM) [11].

According to W. R. Ott "An AQI could be stated as an inclusive structure that converts the weighed values of distinct air pollution-related parameters into a single or set of numbers" [12]. AQI has six levels of health concern based on its range values according to Environment Protection Agency (EPA), USA [13] are presented in Table 1.

Table 1 AQI range values and its description

Standard range (AQI)	Air quality conditions	Color indicators	Description
R(0–50)	Good	Green	It is measured to be adequate and pointed to low or negligible risk
R(51–100)	Moderate	Yellow	It is considered to be conventional but for sensitive people to air pollution could be a cause of concern
R(101–150)	Unhealthy for sensitive groups	Orange	It could affect the people who are associated to of sensitive groups and other may experience discomfort
R(151–200)	Unhealthy	Red	This zone is to be considered unhealthy and people began to feel uneasy and sensitive group's needs serious attention as it could raise a serious health issues
R(201–300)	Very Unhealthy	Purple	This zone is measured to be very unhealthy: everybody may experience major health effects
R(301–500)	Hazardous	Maroon	It is the most dangerous health concern and most likely to be considered to be an emergency condition by targeting to the entire people

2.1 Air Pollution Computation

The research process is highly focusing to compute as well as predict the AQI to take necessary action to prevent the pollution. This section provided the details of predication of AQI using computing process.

The air pollution is computed using sub-indices formation and its aggregation. The foundation of sub-indices $(I_1, I_2, \ldots I_n)$ for n pollutant variables $(X_1, X_2, \ldots X_n)$ is enumerated using air quality measures and health effects. It is computed as:

$$SI = f(X_{SI}) \quad \text{where} \quad SI = 1, 2, 3 \ldots n \tag{4}$$

The sub-indices are calculated based on maximum values and its multiplication and represents the air quality factors as sub-indices. Subsequently, when the sub-indices are calculated they are combined. The aggregation function is summation of individual environmental factors and its weights for influence of air pollution as presented [14]. Mr. Green in 1966 [15] introduces the initial air pollution indices that consist of SO_2 and COH (coefficient of haze). Furthermore, the green index is calculated as the arithmetic mean of SO_2 and COH:

$$I = \text{Aggregated Key} = \sum w_i I_i \quad (\text{For } i = 1 \ldots n) \tag{5}$$

where

$\sum w_i = 1$
I_i = sub-index for pollutant i
n = number of pollutant variables
w_i = weights of the pollutant

$$GI = 0.5 * (ISO_2 + ICOH) \tag{6}$$

whereas

ISO_2 = sulfur dioxide sub-index
ICOH = coefficient of haze sub-index
X = observed pollutant concentration

Fenstock (1969) [16] offered a directory to evaluate the complexity of air pollution, In order to do so AQI of twenty-nine towns of USA were consider. This was termed to be the initial directory to evaluate air pollutant concentrations for the data on source emissions and atmospheric surroundings for individual town.

This index is mostly appropriate for square urban region.

$$\text{Air Quality Index} = W_i I_i \tag{7}$$

where

W_i= weightages for CO, TSP, and SO_2
I_i = expected values of sub-indices

2.2 Air Quality Index Research

See Table 2.

3 Conclusion

The air pollution focused to identify air quality factors, computation methods used for computing air quality index, analysis and compare the algorithms that are frequently used for pollution analysis and prediction.

The first part of the paper defines the air quality index and its measures to provide the decision for the community and related to its health status by indicating through colors [21]. If the values of air quality index are up-to 150 it is acceptable for the living condition according to Environmental Protection Agency, USA.

Table 2 Major ongoing research related to AQI

RF No	Algorithm or techniques	Pollutants	Main contribution
Air Pollution Analysis: The review of the research works done in the air quality index analysis is listed:			
[17]	MLP, radial basis function, Elman network, SVM and linear ARX model	Particulate matter (PM_{10})	Here, the wavelet transformation and ANN is used to predict the day by day regular concentration of particulate matter. The various trials to forecast the PM_{10} pollution in Warsaw present an acceptable outcome by providing an accuracy of prediction as far as the entire examined methods of quality
[18]	Autoregressive integrated moving average (ARIMA), Fuzzy time series (FTS) and artificial neural network (ANNs) methods	O_3, CO_2, SO_2, NO_2	These techniques were applied to forecast the API values for past ten years by using periodic data set of API situated in industrialized, domestic monitoring stations in Malaysia. The result shows for each technique gets compared with *root mean square error (RMSE)*. The outcome concludes, artificial neural network could be considered as a reliable approach compared to FTS and ARIMA
[4]	Neural network models. Spatial stochastic simulations	NO, NO_2, CO, PM_{10}	Initially, *neural network models* have been applied to produce short-term temporary forecasts for predicting pollution of air, weather forecasting data, and evaluation of forecast are examined in contrast to autonomous pervious data. Hence, by implementing method of spatio-temporal dispersion a prediction of pollutant can be easily found. Subsequently the analysis results, shows that the proposed technique offers a good substitute for the classification of metropolitan air quality

(continued)

Table 2 (continued)

RF No	Algorithm or techniques	Pollutants	Main contribution
[14]	Principal component analysis and artificial neural network	CO, O_3, PM_{10}, NO_2, CH_4, NmHC, THC, wind direction, humidity, and ambient temperature	The forecast strategy used here examine and proved that the replaced principal component scores (RPCs) remained the finest input factors in determining API. Among the four RPCs taken as input parameter, the substantial parameters to predict API were (carbon monoxide, ozone, particulate matter 10, nitrogen oxide, CH_4, NmHC, THC, wind direction, humidity, and ambient temperature). It also concludes that ANN could be considered to opt for better atmospheric management
[19]	Temporal and spatial NN model	O_3	Describes two types of prediction models as *temporal and spatial*. The temporal model focuses to estimate instant prediction of the pollutants content in the air for the nearby days and a spatial model estimates prediction of atmospheric pollution index anywhere in the city for determining API on the basis of neural networks. The obtained ANN model shows the adequate, consistent prediction that indicates they were efficient tools for forecasting and predicting the conduct of pollution in air for industrial zone

(continued)

Table 2 (continued)

RF No	Algorithm or techniques	Pollutants	Main contribution
[20]	Statistical forecasting techniques, artificial intelligence techniques, and numerical forecasting techniques	PM_{10}, NO_2, and SO_2	Here, research examines theory and implementation for the predicting methods by establishing the assessment on various prediction techniques and also listed the pros and cons of those techniques. It is concluded that the artificial intelligence approaches such as neural network method ensure the enhancement and compute the nonlinear data on the other hand the models are unbalanced and depends on the data set

Air pollution predictions: The review of the research works done in the air quality index prediction is listed:

RF No	Algorithm or techniques	Pollutants	Main contribution
[21]	Neural network and back-propagation model with genetic algorithm	PM_{10}	The model design consists of ANN methods used to predict air pollution analysis for both short- and long-term data. The result enumerates were accurately for shorter time periods when compared to long terms. As the huge variation of data in long-term gives adverse outcome on performance of network
[22]	LSTM recurrent neural network (RNN)	–	This model uses back-propagation neural network MLP for day-to-day meteorological prediction variables and their respective pollutant predictor takes as input and verified the proposed technique from September 2013 to October 2014 (almost 1 year). The outcomes show that the route creates geographic model and wavelet transformation can be an efficient implementation to enhance predicting accuracy of $PM_{2.5}$. It also concludes that the model could be opt by countries€™ for air quality predicting systems

(continued)

Table 2 (continued)

RF No	Algorithm or techniques	Pollutants	Main contribution
[23]	Recurrent network model Change Point detection model with RNM, sequential network construction model, and self organizing feature maps	VOC, NOX, CO, SO_2, PM_{10} $PM_{2.5}$, NH_3	This research is intent to discover ANN-based air quality predictors with various neural network models that are evaluated to implement air quality prediction the models. Subsequently these models are compared and implemented on short-term and long-term data. The outcomes enlighten that SOFM model implemented tremendously fine
[24]	Artificial neural network approach and ARIMAX model	NO_2	The research illustrates two model frameworks for air quality forecasting. The scope was to incorporate the techniques to existing traffic management support organization for an ecological flexibility of motorway for means of transportation in metropolitan zones. The research explains that the aptitude to predict exceedances of permissible pollution boundaries could be improved by implementing traffic management strategies and methods when the forecast concentration surpasses a minor threshold when compare to normal one
[25]	Artificial neural network with genetic algorithm (ANN) model	PM_{10}	This paper presented two modeling technique that is artificial neural network and conventional models to forecast the annual PM_{10} at national level. The input of the model was selected by genetic algorithm and variables are trained by ANN. In the result, ANN model showed the result effectively and accurately. The prediction obtained by ANN was three times high than compared to conventional models

(continued)

Table 2 (continued)

RF No	Algorithm or techniques	Pollutants	Main contribution
[26]	Back-propagation neural network MLP	$PM_{2.5}$	This study investigates the use of the LSTM recurrent neural network (RNN) as a framework for forecasting time series data of pollution and meteorological information in Beijing. The result conclude that by using LSTM framework, the prediction extends from a single time step out to 5–10 h into the future. This is promising in the quest for forecasting urban air quality and leveraging that insight to enact beneficial policy

Neural network model for air pollution: The subsequent research work focused to predict the air pollution using neural network models:

RF No	Algorithm or techniques	Pollutants	Main contribution
[26]	Artificial neural network (ANN)	Chlorophyll, dissolved oxygen, specific conductance, turbidity	To improve an air quality prediction techniques by using (AQ) air quality elements with the help of ANN, time-series breakdown
[27]	Artificial neural network-Markov Chain	Chlorophyll, dissolved oxygen, salinity, etc.	With the help of hybrid methods to predict the biochemical oxygen demand, termed to be a key element of air quality
[28]	Radial basis network function	residual chlorine, turbidity, pH, and temperature	To implement a unique technique by hybrid macro and detailed model to enhance the air quality constraints to target the diminishing time-consuming air quality models
[29]	Deep belief network	Dissolved oxygen, pH, and turbidity	To give an exact predictions for flexible data to estimate air quality
[30]	Decision tree model	(Ammonia nitrogen, nitrate nitrogen), concentration of hydrogen, Temp_C, biochemical oxygen demand, chemical oxygen demand)	To provide an organized data model technique with the help of decision tree to analyze air quality data

(continued)

Table 2 (continued)

RF No	Algorithm or techniques	Pollutants	Main contribution
[31]	Improved decision tree model	O, O_2, O_3	To demonstrate a developed decision tree learning method in order to make air quality forecast easily and predict accurately
[32]	Least square support vector machine model	Total organic carbon (TOC) criterion	To provide air quality forecasting established on spectrometry

This AQI value computations are describe as the second part of this paper. The study also reviewed the research-related air pollution analysis, predication, and computation of AQI. The factors are applied as per the national and international guidance for air quality index. In the process identified that CO is a vital pollutant that is ubiquitous for inner-city atmosphere. It is produce generally from sources ensuring incomplete ignition. The NO_2 is combustion processes present in rural and urban environments. The PM and ozone also involved in the analysis process. At the same time most of the research works are applied to the neural network model and its computation and predication. It stated that the research further focused on air pollution analysis and predication using neural network approaches. Hence, the best way to achieve the adequate air pollution analysis results is to use the neural network models in conjunction with a competent strategy of the environmental management.

References

1. Benvenuto, F., Marani, A.: Neural networks for environmental problems: data quality control and air pollution nowcasting. Glob. NEST Int. J. **2**(3), 281–292 (2000)
2. Lungu, E.: Development of a short-medium forecasting system for air pollution (in Romanian). Postdoctoral final research report, University Petroleum—Gas of Ploiesti, Department of Informatics (2007 Oct)
3. Oprea, M.: A case study of knowledge modelling in an air pollution control decision support system. In: AI Communications, IOS Press, vol. 18, No. 4 (2005)
4. Russo, A., Soares, A.O.: Hybrid Model for Urban Air Pollution Forecasting: A Stochastic Spatio-Temporal Approach. © International Association for Mathematical Geosciences (2013 July)
5. LeCun, Y., Bengio, Y., Hinton, G.: Nature. Deep Learn. **521**(7553), 436–44 (2015 May 28)
6. Hemlata, K., Usha Rani, K.: Advancements in multi-layer perceptron training to improve classification accuracy. Int. J. Recent Innov. Trends Comput. Commun. **5**(6), 353–357 (2017 June). ISSN: 2321-8169
7. Jamal, H.H., Pillay, M.S., Zailina, H., Shamsul, B.S., Sinha, K., Zaman Huri, Z., Khew, S.L., Mazrura, S., Ambu, S., Rahimah, A., Ruzita, M.S.: A study of health impact & risk assessment of urban air pollution in Klang Valley. UKM Pakarunding Sdn Bhd, Kuala Lumpur (2004)

 8. Kamal, M.M., Jailani, R., Shauri, R.L.A.: Prediction of ambient air quality based on neural network technique. In: 4th Student Conference on Research and Development, Selangor, 27–28 June 2006
 9. Junninen, H., Niska, H., Tuppurainen, K., Ruuskanen, J., Kolehmainen, M.: Methods for imputation of missing values in air quality data set. Atmos. Environ. **38** (2004)
10. Nasir, M.F.M., Juahir, H., Roslan, N., Mohd, I., Shafie, N.A., Ramli, N.: Artificial neural networks combined with sensitivity analysis as a prediction model for water quality index in Juru River, Malaysia. Int. J. Environ. Protect. **1**(3) (2011). http://dx.doi.org/10.5963/IJEP0103001
11. National Weather Service Corporate Image Web Team: NOAA's National Weather Service/Environmental Protection Agency—United States Air Quality Forecast Guidance. Retrieved 20 August 2015
12. Ott, W.R.: Environmental Indices: Theory and Practice. Ann Arbor Science Publishers Inc., Ann Arbor, Michigan, USA (1978)
13. Environment Protection Agency (EPA): USAir Pollution Index. Retrieved 20 Aug 2018. https://www3.epa.gov/airnow/aqi_brochure_02_14.pdf
14. Azid, A., Juahir, A., Latif, M., Zain, S., Osman, M.: Feed-forward artificial neural network model for air pollutant index prediction in the Southern Region of Peninsular Malaysia. J. Environ. Protect. **4**(12A), 1–10 (2013 Dec). https://doi.org/10.4236/jep.2013.412a1001
15. Marvin, H.: Green an air pollution index based on sulfur dioxide and smoke shade. J. Air Pollut. Control Assoc. **16**(12), 703–706 (1966). https://doi.org/10.1080/00022470.1966.10468537
16. Fensterstock, J.C., Goodman, K., Duggan, G.M., Baker, W.S.: The development and utilization of an air quality index. In: Proceedings of 62nd Annual Meeting of the APCA, New York, 1969; Paper 69–73 [15] (2004 Nov)
17. Siwek, K., Osowskia, S.: Improving the accuracy of prediction of PM_{10} pollution by the wavelet transformation and an ensemble of neural predictors. Eng. Appl. Artif. Intell. Eng. Appl. Artif. Intell. **25**(6) (2012 Sept)
18. Rahmana, N.H.A., Leea, M.H., Latifb, M.T., Suhartonoc, S.: Forecasting of air pollution index with artificial neural network. J. Teknol. (2013). eISSN 2180–3722 | ISSN 0127–9696
19. Rahman, P.A., Panchenko, A.A., Safarov, A.M.: Using neural networks for prediction of air pollution index in industrial city. In: IOP Conference Series: Earth and Environmental Science (2016). https://doi.org/10.1088/1755-1315/87/4/04
20. Bai, L., Wang, J., Ma, X., Lu, H.: Air pollution forecasts: an overview. Int. J. Environ. Res. Public Health (2018). https://doi.org/10.3390/ijerph15040780
21. Asghari Esfandani, M., Nematzadeh, H.: Predicting air pollution in Tehran: genetic algorithm and back propagation neural network. J. AI Data Min. 4(1), 49–54 (2016). https://doi.org/10.5829/idosi.jaidm.2016.04.01.06, (2015)
22. Narasimha Reddy, V., Mohanty, S.: Deep Air: Forecasting Air Pollution in Beijing, China (2017)
23. Barai, S.V., Dikshit, A.K., Sharma, S.: Neural Network Models for Air Quality Prediction: A Comparative Study (2007). https://doi.org/10.1007/978-3-540-70706-6_27
24. Catalano, M., Galatioto, F., Bell, M., Namdeo, A., Bergantino, A.S.: Improving the prediction of air pollution peak episodes generated by urban transport networks. Environ. Sci. Policy **60** (2016 June)
25. Antanasijević, D.Z., Pocajt, V.V., Povrenović, D.S., Ristić, M.Đ., Perić-Grujić, A.A.: PM_{10} emission forecasting using artificial neural networks and genetic algorithm input variable optimization. Sci. Total Environ. **443**, 511–519 (2013). ISSN 0048-9697
26. Feng, X., Li, Q., Zhu, Y., Hou, J., Jin, L., Wang, J.: Artificial neural networks forecasting of $PM_{2.5}$ pollution using air mass trajectory based geographic model and wavelet transformation. Atmos. Environ. **107**, 118–128 (2015). ISSN 1352-2310
27. Li, X., Song, J.: A new ANN-Markov chain methodology for air quality prediction. In: International Joint Conference on Neural Networks, pp. 12–17, July, 2015

28. Ma, L., Xin, K., Liu, S.: Using radial basis function neural networks to calibrate air quality model. World Acad. Sci. Eng. Technol. Int. J. Environ. Chem. Ecolog. Geolog. Geophys. Eng. **2**(2) (2008)
29. Aggarwal, S.H., Khare, K.: Predictive analysis of air quality parameters using deep learning. Int. J. Comput. Appl. **125**(9), 0975–8887 (2015). Access from Google Scholar, Sept. 2015
30. Jaloree, S., Rajput, A., Gour, S.: Decision tree approach to build a model for air quality. Bin. J. Data Min. Netw. **4**(1) (2014)
31. Liao, H., Sun, W.: Forecasting and evaluating air quality of Chao Lake based on an improved decision tree method. Procedia Environ. Sci. **2** (2010)
32. Yan-jun, L., Qian, M.: AP-LSSVM modeling for air quality prediction. In: Control Conference (CCC), 2012 31st Chinese. IEEE, New York (2012)

Implementation of Signal Processing Algorithms on Epileptic EEG Signals

Sasikumar Gurumoorthy, Naresh Babu Muppalaneni,
G. Chandra Sekhar and G. Sandhya Kumari

Abstract Epilepsy is a brain-related disorder of the central nervous system where the neurons of brain show the abnormal behavior at a certain instance of time. Electroencephalogram (EEG) signals play a significant role in the diagnosis of epileptic EEG signals. In the world, overall 50 million people are affected by epilepsy. It is very hard to determine the actions of brain EEG signal, because it contains artifacts or fluctuated information. EEG signal contains different artifacts like EOG, EKG, and ECG. ECG signal artifacts are produced by the function of heart. EOG signal artifacts are produced due to the movement of eyes and EMG signal artifacts are produced because of the muscles coordination. To solve these problems, this paper aims to remove the artifacts present in the EEG signal with the help of wavelet signal processing algorithms (WSA) in signal processing toolbox (Sptool). After removing the artifacts, the parameters have been calculated such as maximum peaks, mean, median, average frequency, variance, and standard deviation. Based on the number of maximum peaks of the EEG signal and with proposed parameters of EEG signal, it is possible to estimate the severity of the seizure.

Keywords Epilepsy · Sptoolbox (Sptool) · Wavelet signal processing algorithms (WSA)

S. Gurumoorthy · G. Chandra Sekhar · G. Sandhya Kumari
Sree Vidyanikethan Engineering College (Autonomous), Tirupati, India

N. B. Muppalaneni (✉)
Assistant Professor, Department of CSE, National Institute of Technology Silchar, Assam, India
e-mail: nareshmuppalaneni@gmail.com

© Springer Nature Singapore Pte Ltd. 2020   367
P. Venkata Krishna and M. S. Obaidat (eds.), *Emerging Research in Data
Engineering Systems and Computer Communications*, Advances in Intelligent
Systems and Computing 1054, https://doi.org/10.1007/978-981-15-0135-7_35

1   Introduction

1.1   Wavelet Signal Processing Algorithms

In signal processing, wavelets are used for signal de-noising, removing trends, and irregularities in signals. The continuous wavelet transform [1] is introduced as a signal processing tool for investigating time-varying frequency spectrum characteristics of non-stationary signals. The discrete wavelet transform is computed by passing a signal successively by way of a high pass along with a low pass air filter. The wavelet coefficient represents the similarity between frequency of a signal and a chosen wavelet function. Figure 1 represents the block diagram of system design.

2   Methodology

2.1   Frequency- and Time-Localized Reconstruction from the Continuous Wavelet Transform (CWT)

The constant wavelet change is a very unwanted change exactly where it utilizes internal items to determine the similarity between a signal as well as a considering purpose. In this paper, wavelet transformation techniques are used to identify epileptic spikes in the brain EEG. Figure 2 represents the flow chart of continuous wavelet transform. The computational resources required to compute the continuous wavelet transform and to store the coefficients are much larger than the discrete wavelet transform [2]. The constant wavelet completely transform computes the internal product or service of a signal $f(t)$, with translated as well as dilated types of a scrutinizing wavelet, $\psi(t)$.

The continuous wavelet transforms (CWT) compares the signal with shifted as well as compressed variation of a wavelet, the place that the compression functionality is called dilation. By evaluating the signal on the wavelet at different scales as well as positions, a function of two variables was obtained.

In case the wavelet is composed complex value, subsequently, the constant wavelet completely transforms (CWT) is a complex-valued feature of position and scale. In case the signal includes real value, subsequently the constant wavelet

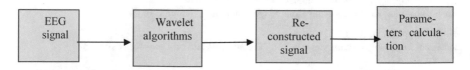

Fig. 1 Block diagram of system design

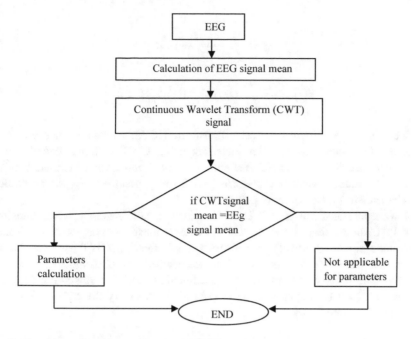

Fig. 2 Flow chart of continuous wavelet transform

change is a real-valued feature of position and scale [3]. For a scale parameter, $a > 0$, and position, b, the continuous wavelet transform is:

$$c(a,b) = \frac{1}{|a|^{\left(\frac{1}{2}\right)}} \int_{-\infty}^{\infty} x(t)\psi * \left(\frac{t-b}{a}\right) dt \tag{1}$$

If the signal is sampled with sampling time interval T_s then the continuous wavelet transform at the time step, $b = kT_s$ can be represented as:

$$C(a, kT_s) = T_s a^{(-1/2)} \sum_n s(nT_s)\psi * \left[\frac{(n-k)T_s}{a}\right] \tag{2}$$

2.2 Inverse Continuous 1-D Wavelet Transform (ICWT)

The inverse continuous wavelet transform used in the Wavelet Toolbox [4] utilized the Morse wavelet and L1 normalization. The inverse continuous wavelet transform (ICWT) can be represented in the double-integral form.

The wavelets with fourier transform that satisfies the admissibility condition and finite energy function $f(t)$, can be represented as:

$$c_\varphi = \int_{-\infty}^{\infty} \frac{\left|\widehat{\psi}(w)\right|^2}{|w|} dw < \infty \tag{3}$$

$$f(t) = \frac{1}{c_\psi} \iint \langle f(t), \psi_{a,b}(t) \rangle \psi_{a,b}(t) db \frac{da}{a^2} \tag{4}$$

The inverse continuous wavelet transform (ICWT) assumes that you have obtained the continuous wavelet transform using CWT with the default Morse (3,60) wavelet. So, the wavelet has symmetry of 3 and a time bandwidth of 60. Inverse continuous wavelet transform (ICWT) also assumes that the continuous wavelet transform uses default scales.

If wavelet transform is a 2-D matrix, inverse continuous wavelet transform (ICWT) assumes that the continuous wavelet transform was obtained from a real-valued signal. If wavelet transform is a 3-D matrix, ICWT assumes that the continuous wavelet transform was obtained from a complex-valued signal. The wavelet toolbox only supports real-valued functions and real-valued conditions.

The condition for the two-*wavelets* ψ_1 *and* ψ_2 to satisfy the admissibility condition can be represented as:

$$\int \frac{\left|\widehat{\psi_1}^*(w)\right| \left|\widehat{\psi_2}(w)\right|}{|w|} dw < \infty \tag{5}$$

where the operators less than(<), greater than(>) denotes the inner product, $\widehat{\psi_1}^*$ denotes the complex conjugate the dependency of ψ_1 and ψ_2 on scale and position has been suppressed for convenience. By summing the scaled continuous wavelet coefficients from select scales, we have obtained the approximation to the original signal. Figure 3 represents the flow chart of inverse continuous wavelet transform. The flowchart of inverse continuous wavelet transform explains about the EEG signal with different trending ranges [5]. Here, we have given the condition that, if the original signal mean equal to the reconstructed signal mean, then the signal is valid for the calculation of signal parameters.

2.3 Automatic 1-D De-noising Using Wavelets

De-noising is the process from which we have reconstructed the original signal from a noisy signal. Fourier filtering represents the features sharp, but it doesn't really suppress the noise. It is noted that soft thresholding provides smoother results when compared to hard thresholding [6]. Figure 4 represents the flowchart for automatic de-noising. Hard threshold is able to provide better edge preservation when compared to the soft thresholding. In this paper, three different wavelet de-noising algorithms [7] are applied to an original signal and the results of different techniques are displayed below:

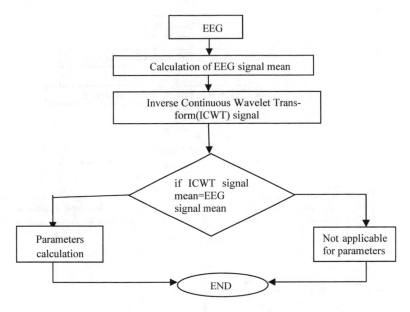

Fig. 3 Flow chart of inverse continuous wavelet transform

Donoho and Johnstone developed this method, where it offers quick and automated thresholding.

Shrinkage of the wavelet coefficients is estimated using the formula,

$$\lambda = \sigma\sqrt{2\log(n)} \tag{6}$$

where σ is the standard deviation of the interference of the noise amount n will be the test size.

A threshold level is given to every resolution degree of the wavelet completely transform. The threshold amount is selected by minimizing the Stein unbiased estimate of risk (SURE). This treatment is ideal for de-noising a broad range of features and in the threshold choice strategies, it can be required to calculate the regular deviation of the interference coming from the wavelet coefficients.

A common estimator is shown below:

$$\sigma = \frac{\text{MAD}}{0.6745} \tag{7}$$

where MAD is the median of the absolute values of the wavelet coefficients.

The flow chart of automatic de-noising explains that initially level 5 hard threshold wavelet decomposition and level 10 hard threshold wavelet decomposition has been applied to the epileptic EEG signal. The reconstructed signal of hard threshold algorithms consists of loss of original signal peak points, where half of the signal is completely linear.

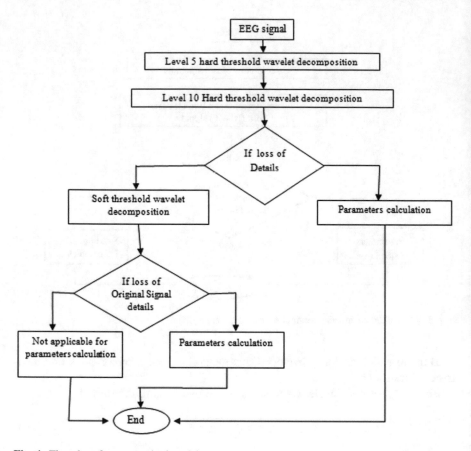

Fig. 4 Flowchart for automatic de-noising

For the calculation of signal parameters, the reconstructed signal peak points should be same like the original signal. So, soft threshold algorithms have been applied to the EEG signal, where the soft threshold builds the reconstructed signal without changing the peak points of the original signal. Finally, the soft threshold algorithm executed the output, which is valid for the calculation of the other required signal parameters.

3 Results

3.1 Frequency- and Time-Localized Reconstruction from the Continuous Wavelet Transform (CWT)

In this algorithm, the signal has been analyzed with band-pass filter of different frequency ranges [8]. Figure 5 explains the original EEG signal. Figure 6 explains

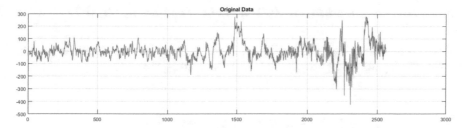

Fig. 5 Original EEG signal

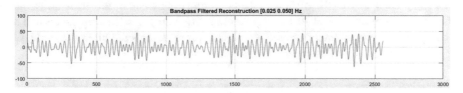

Fig. 6 Reconstructed signal with frequency range [0.025 0.050]

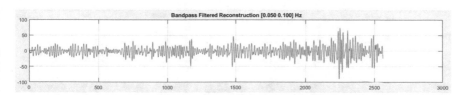

Fig. 7 Reconstructed signal with frequency range [0.050 0.100]

about the reconstructed signal with the band-pass filter of frequency range between [0.025 0.050]. Figure 7 explains about the reconstructed signal with the band-pass filter of frequency range between [0.050 0.100].

The mean of the original signal and the mean of the reconstructed signal with frequency range [0.025 0.050] is same. So, based on this it is clear that after the reconstruction of the signal with this frequency range, the peak points remains same and they are not modified with the filter. So, the parameters for the required EEG signal has been calculated with this reconstructed signal, whereas the mean of the signal with reconstructed frequency range of [0.050 0.100] is greater when compared to the signal with range of [0.025 0.050].

3.2   Inverse Continuous 1-D Wavelet Transform (ICWT)

Figure 8 represents the original signal. Figure 9 represents the signal for trending range 55. Figure 10 represents the signal for trending range 15.

3.3   Automatic 1-D De-noising Using Wavelets

The de-noised signal obtained by applying three different hard threshold selection rules [9], such as SURE, Minimax, Donoho and Johnstone's universal threshold. Figure 11 represents Level 10 wavelet decomposition output, which represents half of the signal completely linear and the remaining part represents the filtered signal [10].

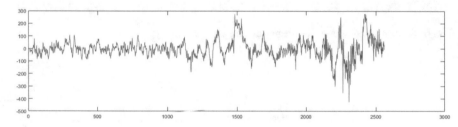

Fig. 8 Original signal

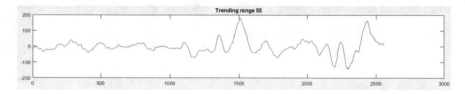

Fig. 9 Mean for trending range 55 is 1.3825

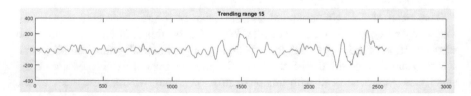

Fig. 10 Mean for trending range 15 is 1.3790

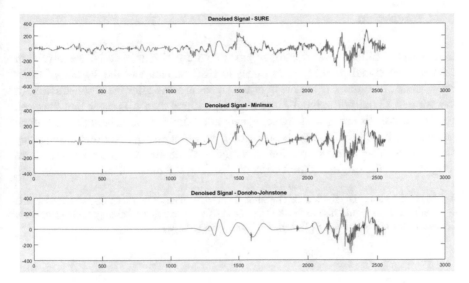

Fig. 11 Reconstructed signals with level 10

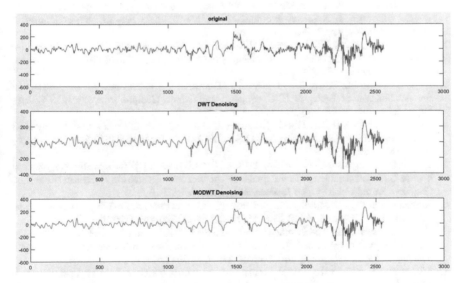

Fig. 12 Reconstructed signals with level 3 and threshold rule of maximal overlap discrete wavelet transform (MODWT)

Figure 12 represents the reconstructed signals with level 3 and soft threshold rule of maximal overlap discrete wavelet transform (DWT). The two different soft threshold selection rules like discrete wavelet transform (DWT), maximal overlap discrete wavelet transforms (MODWT) have been reconstructed the signal without disturbing the original signal coefficients, when compared to hard thresholding.

4 Conclusion

In this paper, different wavelet signal processing algorithms have been analyzed to calculate the signal parameters of epileptic EEG signals and the signals of other brain-related disorders. If the mean of the signal is same after the reconstruction of the original signal with signal processing algorithms then it is clear that the peak points are not disturbed and those signals are valid for the calculation of signal parameters. Based on the above-proposed signal processing algorithms, frequency and time-localized reconstruction technique of continuous wavelet transform reconstructed the original signal without changing the mean of the original signal.

Acknowledgements Dr. Sasikumar Gurumoorthy, Dr. M. Naresh Babu, G. Chandra Sekhar and G. Sandhyakumari would like to thank Department of Science and Technology (DST) Cognitive Science Research Initiative(CSRI). Ref. No.: SR/CSRI/370/2016.

References

1. Adelia, H., Zhoub, Z., Dadmehrc, N.: Analysis of EEG records in an Epileptic patient using wavelet transform. J. Neurosci. Methods **123**(1), 69–87 (2003 Feb 15). https://doi.org/10.1016/S0165-0270(02)00340-0
2. Fausta, O., Acharya, U.R., Adeli, H., Adeli, A.: Wavelet-based EEG processing for computer-aided seizure detection and Epilepsy diagnosis. Eur. J. Epilepsy (2015). https://doi.org/10.1016/j.seizure.2015.01.012
3. Kaya, Y., Uyar, M., Tekin, R., Yıldırım, S.: 1D-local binary pattern based feature extraction for classification of Epileptic EEG signals. Appl. Math. Comput. **243**, 209–219 (2014). https://doi.org/10.1016/j.amc.2014.05.128
4. Gurumurthy, S., Mahit, V., Ghosh, R.: Analysis and simulation of brain signal data by EEG signal processing technique using MATLAB. Int. J. Eng. Technol. (IJET) **5**(3) (2013 June–July). ISSN: 0975-4024
5. Das, A.B., Bhuiyan, M.I.H., Alam, S.M.S.: Classification of EEG signals using normal inverse Gaussian parameters in the dual-tree complex wavelet transform domain for seizure detection. Signal Image Video Process. **10**(2), 259–266 (2016 Feb)
6. Ambramovich, F., Sapatinas, T., Stilverman, B.: Wavelet thresholding via a Bayesian approach. J. R. Stat. Soc. Ser B (Stat Methodol) **60**(4), 725–749 (1998)
7. Cisar, P., Cisar, S.M.: The influence of thresholding method on 1D signal denoising using wavelet theory. Published in 2012 IEEE 10th Jubilee International Symposium on Intelligent Systems and Informatics. https://doi.org/10.1109/sisy.2012.6339492
8. EEG waves classifier using wavelet transform and fourier transform. World Acad. Sci. Eng. Technol. Int. J. Bioeng. Life Sci. **1**(3) (2007)
9. Walters-Williams, J., Li, Y.: A new approach to denoising EEG signals-merger of translation invariant wavelet and ICA. Int. J. Biom. Bioinf. **5**(2) (2011)
10. Yu, L.: EEG de-noising based on wavelet transformation. In: 3rd International Conference on Bioinformatics and Biomedical Engineering, 2009. ICBBE 2009, pp. 1–4, June 11–13, 2009

Big Data Analytics in Health Care

Tahmeena Fatima and Singaraju Jyothi

Abstract In today's world, data is growing exponentially and widespread accessibility of data led to analyze and visualize data effectively using analytical techniques in healthcare industry. Big data analytics play a vital role and provides long-term benefits in tremendously handling huge explosive data. In this paper, we present an overview of different big data platform tools and different technologies that support big data analytics in health care. It also describes different steps involved in big data analytics process and also presents ways to improve health care by considering various facts by using big data analytics. As big data analytics has the potential to provide useful insight in health care, this article uses a review methodology to categorize the uses of big data in health care. This study provides a baseline to assess the essential prospects of computational health informatics and the use of big data in health care in understanding different scopes of big data platforms.

Keywords Big data · Healthcare · Big data analytics · Review of big data platforms and tools

1 Introduction

In the current digital world, there is a rapid explosion of data in IT zone. Data is voluminous and extracted from the heap of data sets generated from different sources. The primary task is to extract useful data which has become a breakthrough in big data field. Big data can be analyzed from different perceptions for better decision making in diversified segments of various industries like health care

T. Fatima (✉)
Wadi Addawasir, Riyadh, Saudi Arabia
e-mail: tahmi.fatima18@gmail.com

S. Jyothi
Department of Computer Science, SPMVV, Tirupati, India
e-mail: jyothi.spmvv@gmail.com

© Springer Nature Singapore Pte Ltd. 2020
P. Venkata Krishna and M. S. Obaidat (eds.), *Emerging Research in Data Engineering Systems and Computer Communications*, Advances in Intelligent Systems and Computing 1054, https://doi.org/10.1007/978-981-15-0135-7_36

and education. Big data is an emerging technology that describes any huge amount of structured, semi-structured, and unstructured data that has the potential to be mined for information [1].

2 Big Data in Healthcare Analytics

Big data specifies the tools, methods, and processes which allow and enable any healthcare organization to handle, store and manage huge data sets. Big data in the healthcare industry embarks a potential in accumulation and integration of patient data in an effective manner. Demchenko et al. [2] explain big data by using five vs: Volume, velocity, variety, veracity, and value. Volume means the huge amounts of data in the medical field. Velocity indicates the speed at which new data is produced. Variety specifies the various levels of complexity in dealing with patient's data. Veracity indicates the validity of the generated data. The value specifies the quality of data that is produced.

2.1 Process of Big Data Analytics in Health Care

Big data analytics in healthcare industry plays a vital role to deal with the enormous data growth. There are various phases involved in the process of it. Below is the outline of the major phases of Big data analytics in health care.

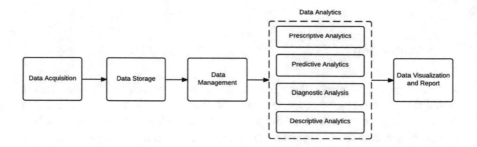

1. **Data Acquisition**: In the area of health care, big data can be available in various formats like structured, semi-structured, or unstructured. In this stage, the data is acquired from different sources. This collected data is provided as input to the system from various primary key sources like CPOE, clinical decision support systems (CDSS), EHR, EMR, and secondary sources like laboratories, government agencies, health pharmacies, and various health care insurance companies [3]. Input to system is provided from various external sources like different pharmaceutical companies, national health records. Machine-generated

data from wearable gadgets, social media data like Twitter feeds, Facebook status, web pages, blogs, articles, and many more also provide input [2].

1.1. *Electronic health records*: HERs maintain a complete medical history of the patients in digital form. Healthcare data sets in hospitals are made available with the digitization of medical histories from previous time span [4]. A multitude of various electronic medical records describing patient's information in healthcare industry form the foundation of personalized medicine for hospitals like clinical information, demographics, diagnosis, laboratory reports, X-ray, medical prescriptions from different physicians, and scans. Henceforth, EHR provides a valuable source of information in healthcare analytics.

1.2. *Biomedical image processing*: **Biomedical** images in the medical field are responsible and considered as the most important tool with respect to disease diagnosis and rise in the analysis of medical data. Various images, X-ray, and scans like ultrasound, magnetic resonance imaging (MRI), computed tomography (CT), photoacoustic imaging, positron emission tomography–computed tomography (PETCT), molecular mammography, and fluoroscopy are few examples of imaging that are well known in clinical operations [5]. Processing of images is quiet challenging as it includes superfluous and noisy data which needs to be removed that enables better decision making by clinical practitioners.

1.3. *Social media analysis*: Social media healthcare analysis can be done by collecting data from various networking sites and different social media like Facebook, Instagram, WhatsApp, Twitter, and LinkedIn This data collected from social media logs related to health is usually helpful in analyzing any epidemic disease [6]. The next phase is extracting information that affects healthcare industry by prediction about any infectious disease.

1.4. *Smart mobile phones*: Various apps installed in smart phones have become the most popular technological device and an important source of data in healthcare domain. Nowadays, mobile phones offer several features and different services which are a transformation from a basic communication tool into an intricate device. Various smart phones are currently equipped with different health-related apps like pedometers, fit bit, and sensors like satellite positioning services, accelerometer, monitoring pregnancy, and heartbeat rate and its frequency. There are many apps that monitor blood sugar level in diabetic patient. Enormous data is generated on a daily basis which contributes to the positron emission tomography–computed tomography which helps imaging services to be recognized in the medical field within clinical settings [5].

1.5. *Data sensors*: In the healthcare domain, there are various data sensors that are employed to monitor health-related issues and solutions. These devices provide the best way in monitoring a patient's health by measuring range of medical indicators like blood pressure, heartbeat, and body temperature. To monitor and ensure efficient health monitoring of various patients, various devices like microphones, CCTV cameras, and pressure sensors are installed.

2. **Data preparation**: Data preparation involves processing raw data after gathering data. Improper preparation routines lead to extra computational resources that may be in affordable in the big data context. Henceforth, it is required to prepare data appropriately by checking its accuracy, and entering it into a computer to develop your database and obtain accurate predictive models to improve data reliability. Data preparation further comprises of two steps.

   2.1 *Data Filtering*: The acquired data based on the priorities is filtered and corrupt is removed which makes no sense in the analysis. Data which is termed as "corrupt" may include missing or invalid data types. Irrelevant healthcare data is discarded based on certain defined criteria and analyzed, and the unfiltered data may be used in different types of analysis [7].

   2.2 *Cleaning of data*: Data cleaning is the process of removing corrupted data by filtering it based on the relevance and priorities from the gathered data. The data produced from sensors, social media, and medical prescriptions should be complete and in a structured manner [8]. Data cleaning involves different components like noise reduction, and normalization. Different algorithms like expectation maximization (EM) algorithm and multiple imputation algorithm are used to extract precise data that are deployed. Incomplete data should be handled with maximum precision to fetch good quality data in healthcare domain.

   2.3 *Noise treatment*: Data generated from various sources may encompass some superfluous data and noise. Thus, noisy data should be treated which involves two main methods. The first approach involves techniques to correct noisy values by polishing the data. But data polishing technique is only applicable when small amounts of noise exist. The second method of noise treatment involves noise filters in eliminating noise from noisy instances and these filters do not bring modifications on methods of data mining.

3. **Data storage**: Data storage is one of the challenging and vital tasks in big data framework in healthcare monitoring system. As there is a massive amount of data generated in healthcare industry, there is an urge for an efficient and large storage platform. Traditional data analysis methods are not capable of managing huge volumes of data. The voluminous data gathered is stored in the cloud rather than on physical storage disks because of the improvement in the cloud storage technology. Cloud computing is the most promising technology which provides elasticity to store enormous and complex data. Hashem et al. specified that cloud computing structure can assist as an actual platform to address the data storage essential to accomplish big data analysis [9]. Data gathered from diversified segments can be classified into structured data like EHR and EMR, semi-structured data like XML and JSON documents, and unstructured data like biomedical images are stored in appropriate databases. Various components like HDFS and NoSQL such as MongoDB are used currently as they are more scalable and enable voluminous data storage.

4. **Data management**: Data management in healthcare encompasses data mining, retrieval, cleaning, and organization data governance. The data management ensures removal of scrap superfluous data and validating healthcare data by removing it [10]. Wicks et al. [11] specify that retrieval of information is the process of examining large documents and retrieving medical texts and images. Enterprise health-related information should be governed for proper management of data like availability, usability, integrity, and security. Data management involves various legislation and government regulation to address health data privacy. Different data governance acts like HIPPA, HDI, GINA and Health Information Technology for Economic and Clinical Health (HITECH) are declared for the use of information knowledge and acceptance. Various issues like security and privacy of electronic health broadcast are addressed partly by civil and illegal application of HIPAA rules through different requirements of HITECH electronic healthcare systems that are implemented by all healthcare providers [12].

5. **Data Analytics**: The most challenging and the most interesting context which has the scope of data analytics is health care. Analytics in health care has the potential to reduce costs of treatment, predict epidemic outbreaks of certain disease, and elude inevitable diseases and progress in healthy life [13]. Data analytics is termed as a mechanical method of converting raw data collected from various resources into sensual data. Big data analytics in health care is categorized into descriptive, predictive, prescriptive, and diagnostic analytics [2].

Types of analytics	Operational definition
Descriptive	Detecting and examining information from the data set [14]
Predictive	Using historical data to predict the impending events [15]
Prescriptive	Support decision making by deploying various scenarios [15]

5.1. *Descriptive analytics*: Descriptive analysis deals with describing situations, reporting, and making decisions based on its past performance related to current and historical data. It is a data-driven approach. There are numerous techniques that are employed to perform this type of analytics which can classify, represent cumulative data, synthesize it supporting decision making, and analyze quality of healthcare data. Histograms, graphs, and charts are various descriptive statistics tools that are used among the techniques used in descriptive analytics. It is also termed as unsupervised learning.

5.2 *Predictive analytics*: Predictive analysis reflects the ability to predict future events by forecasting about what might happen in the future as it is probabilistic in nature. It is slightly advanced kind of analytics. For instance, what may occur? What are the forthcoming developments? It analyzes both past and current time data. It is also termed as supervised learning. It also helps in identifying trends and unknown patterns and determining probabilities of uncertain outcomes in multiple dimensions to group data in appropriate data sets. Predictive models are mostly built based on machine learning techniques. For example, one of its roles is to predict whether a patient can get complications or not.

5.3 **Prescriptive analytics**: The goal of prescriptive analysis is to offer appropriate actions leading to optimal decision making usually during healthcare problems. It inevitably produces big data before decision making while proposing appropriate actions based on different possible outcomes. It makes use of knowledge-related health and applied science in various medical domains like options related to prescription. An illustration of the role of prescriptive analysis may suggest rejecting a given treatment in the case of a harming side effect. This analysis assists the decision maker to execute further step. Prescriptive analytics is dominant compared to descriptive and predictive analytics. It supports evidence and personalized-based medicine.

5.4 **Descriptive analytics**: The major objective of diagnostic analysis is to explain the occurrence of certain events and the factors that triggered the events to occur. For instance, it uses historical data to predict the cause and the origin of a given problem and then analyze it. It uses expertise in understanding and discovering new medicines which may be unknown. For example, diagnostic analysis uses different methods like clustering and decision trees in the attempt to understand the reasons behind the regular readmission of some patients and helps in leading to the invention or discovery of former disorders.

The below-given figure presents the varied percentage of different types of analytics from various healthcare articles. Among various analytics descriptive analytics occupies a larger percentage based on its frequent use (48%), whereas predictive analysis has used 43% of articles. Prescriptive analytics is most uncommonly (9%) used analytics.

Percentage of Health Care Analytics

6. **Analysis and visualization of data**: Different tools are used to assess and evaluate useful data. Data analysis is the most labor-intensive step of the complete BDA process. The next step is visualizing data where the data is presented in a graphical arrangement like charts, graphs, tables, and various pictorial formats. This step of the BDA process helps for easy understanding and communication of data and enables effective decision making. Visualization enables to easily understand complicated data. It helps users understand pattern and correlation of diversified data. Various tools like R, Graphviz, NodeBox, Nephi, and Quadrigram are used to visualize the data.

3 Tools/Technologies of Big Data

Hadoop: Hadoop is a framework for dealing with big data for parallel processing in distributed environment. There are few sets of primitives available to do batch processing in Hadoop. Hadoop enables various applications to run on large clusters commodity hardware. Data motion to applications and reliability is provided with the help of Hadoop framework transparently [14].

Hadoop Distributed File System (HDFS): HDFS is a Hadoop-based cluster for storage. It enables the cluster to store huge data by apportioning data into small fragments and storing it in distributed nodes.

MapReduce: MapReduce is a processing technique in file system by distributing the tasks on it. This technique is used once the data is gathered from the acquisition process. This computational paradigm uses algorithm that has a mapper and reducer. Mapper takes a set of data and transforms it into another set of data, where individual elements are broken down into tuples (key/value pairs) [14]. It is a programming model for distributed computing based on the Java programming language.

PIG and PIG Latin: PIG is a high-level platform used with Hadoop to create MapReduce programs. The language used is termed as is Pig Latin. Structured or unstructured data can be accessed by Pig. Hadoop architecture enables execution of Pig with its own runtime environment [16].

Spark: Spark offers scalable data analytical platform in contrast to Hadoop which offers cluster storage mechanism. The in-memory computing available in Spark enables fast data retrieval by removing overhead of I/O. Spark is superior to Hadoop as it supports open source environment which improves the computing power. Various applications like machine learning algorithms and natural language processing (NLP) use Spark [17].

Hive: The Apache Hive uses SQL to read, write, and manage huge data sets stored in the Apache data warehouse by using an appropriate structure. Hive users are connected with the help of command line tool and JDBC driver [18].

Jaql: Jaql is a declarative functional language that works on large data sets to process parallel data by using query language. It converts high-level queries to low-level queries with MapReduce tasks to access information.

Zookeeper: Zookeeper provides integrated services by using large centralized infrastructure through distinct clusters. The integrated structure coordinates, manages, maintains, and configures huge number of hosts by encompassing synchronization and parallel processing in big data analytical health care.

HBase: HBase is a NoSQL-based approach which works on top of HDFS. In comparison to conventional databases that are row oriented, Hbase is a database system that works in column-oriented manner.

MongoDB: It is a NOSQL database which is capable of storing high volumes of data. It relies on Javascript Object Notation (JSON) for storing records. In addition, JSON scales better since join-based queries are not needed due to the fact that relevant data of a given record is contained in a single JSON document. Spark is easily integrated with MongoDB [19].

Cassandra: Cassandra is the current highly demanded topmost model that can deal with distributed big data across various utility servers. Hence, it is also termed as distributed database which uses distributed servers to provide reliable services without any failures. It is a robust system with NoSQL approach [20].

Oozie: Oozie is an open source project where workflow streamlining can be done by synchronization of multiple tasks.

A comparative study on tools and technologies of big data

Author	Tools/technologies	Advantages	Disadvantages
Prof. Jigna Ashish Patel Dr. Priyanka Sharma [17]	Hadoop Storm Spark	Significant insights of patient information and past data is presented with advanced tools and machine learning algorithms	Big data issues and challenges need to be addressed by these tools
M. Gowsalya, C. Villiyammai, K. Krushitha [21]	Hadoop MapReduce	Enables extraction of useful data from a huge database and enables better decision making	This tool can find patients with similar past records of health based on medication
S. Gomathi V. Narayani [22]	Big data	Exploration of diseases can be done in advance and appropriate treatment with length of stay (LOS)	Requires description of important attributes and methodologies in order to implement big data in health care
C. Imthyaz Sheriff Tawseef Naqishbandi [23]	IoT, big data analytics	Time and money are saved	Informatics and needs be resolved in health care
Thanga Prasad. S Sangavi. S, Deepa. A Sairabanu. F, Ra-gasudha. R [24]	Hadoop tool Components of Hadoop tool: Name node Data node HDFS client	Proposed system enables to calculate the stage of diabetes It majorly supports rural and urban areas	Prediction analysis can be done by using Hadoop tool
Mimoh Ojha Dr. Kirti Mathur [25]	Hadoop architecture	By various surveys in different fields of health care, time and money can be saved	Technologies like machine learning, A/B testing, visualization, and search-based application can be used for analytics
Gauri D. Kalyankar, Shivananda R. Poojara Nagaraj V. Dharwadkar [26]	ML algorithm in Hadoop MapReduce	Upcoming future occurrences are predicted by predictive analysis for non-communicable diseases like diabetes	Risk levels of diabetic patient can be predicted by employing pattern matching with various new patterns

(continued)

(continued)

Author	Tools/technologies	Advantages	Disadvantages
Ritu Chau-han Rajesh Jan-gade [5]	• Clinical prediction model • Time series data mining • Visual analytics for big data • Information retrieval	Effective and efficient patterns can be achieved with the big data challenges in future health industry	Privacy and security issues in big data Big data consistency Big data processing
Haritha Chennamsetty, Suresh Chalasani, Derek Riley [27]	Hive	Hive is used to generate reports form large data generated from EHR	Hive can get executed on different platforms like Hadoop
Hiba Asri, Hajar Mousannif, Hassan Al Moatassime, Thomas Noel [28]	Predictive analysis for various tools can be done with HER implementation and e-HPA tools	Big data solutions in health care can be discovered EHR implementation enables better data analysis	Companies need to invest more, require higher expertise and knowledge and different algorithms and techniques in data mining for big data analytical solutions

4 Conclusion and Future Work

Big data analytics in health care is an emerging and challenging field that led to the new insights in capturing data from voluminous data sets and support in better decision making. Though there are many challenges that need to be addressed, with the help of appropriate techniques and tools, various healthcare segments like clinical and emergency and various data repositories can have the potential to improve and transform. With the advent and tremendous growth of big data analytics, various issues and challenges in the healthcare industry can be addressed with more attention which can provide a major breakthrough in the medical domain.

References

1. Improved Approaches to Handle Bigdata through Hadoop KLEF University, India
2. Y. Demchenko, Z. Zhao, P. Grosso, A. Wibisono, C. de Laat: Addressing big data challenges for scientific data infrastructure. In: IEEE 4th International Conference on Cloud Computing Technology and Science (CloudCom 2012). IEEE Computing Society, based in California, USA, Taipei, Taiwan, pp. 614–617 (2012)
3. Al-Jarrah, O.Y., Yoo, P.D., Muhaidat, S., Karagiannidis, G.K., Taha, K.: Efficient machine learning for big data: a review. Data Res. 2, 87–93 (2015). https://doi.org/10.1016/j.bdr.2015.04.001

4. Chen, H., Fuller, S.S., Friedman, C., Hersh, W.: Medical Informatics: Knowledge Management and Data Mining in Biomedicine, 8. Springer Science & Business Media (2006)
5. Ritu, C., Jangade, R.: A robust model for big healthcare data analytics. Cloud System and Big Data Engineering (Confluence), 2016 6th International Conference. IEEE, New York (2016)
6. Sadilek, A., Kautz, H., Silenzio, V.: Modeling spread of disease from social interactions. In: Sixth AAAI International Conference on Weblogs and Social Media (ICWSM) (2012). http://www.cs.rochester.edu/~kautz/papers/Sadilek-KautzSilenzio_Modeling-Spread-of-Disease-from-SocialInteractions_ICWSM-12.pdf
7. Erl, T., Khattak, W., Buhler, P.: Big Data Fundamentals: Concepts, Drivers & Techniques. Prentice Hall. Part of the The Prentice Hall Service Technology Series from Thomas Erl Series (2016 Jan 5)
8. Agrawal, D., et. al.: Challenges and Opportunities with Big Data. Big Data White Paper-Computing Research Association (2012 Feb). Available http://cra.org/ccc/docs/init/bigdatawhitepaper.pdf
9. Hashem, I.A.T., Yaqoob, I., Badrul Anuar, N., Mokhtar, S., Gani, A., Ullah Khan, S.: The rise of "Big Data" on cloud computing: review and open research issues. Inf. Syst. 47, 98–115 (2014). https://doi.org/10.1016/j.is.2014.07.006
10. Archenaa, J., Mary Anita, E.A.: A survey of big data analytics in healthcare and government. Procedia Comput. Sci. 50, 408–413 (2015). Big Data, Cloud and Computing Challenges. Available: http://www.sciencedirect.com/science/article/pii/S1877050915005220
11. Wicks, P., Massagli, M., Frost, J., Brownstein, C., Okun, S., Vaughan, T., et al.: Sharing health data for better outcomes on PatientsLikeMe. J. Med. Internet Res. 12, e19 (2010). https://doi.org/10.2196/jmir.1549
12. U.S. Government, Department of Health and Human Services, Federal Register, Rules and Regulations, 74(2009)56123-56131. Available from: https://www.hhs.gov/sites/default/files/ocr/privacy/hipaa/administrative/enforcementrule/enfr.pdf
13. Sai Jyothi, B., Jyothi, S.: Doc-based modelling for medical big data. IADS SSRN: https://ssrn.com/abstract=3170156, Volume No. 01, Issue No. 03.[Scopus] (2018)
14. Russom, P.: Big data analytics; TDWI best practices report; Fourth Quarter; Report No.: 9.14.2011; TDWI: Renton, WV, USA (2011)
15. Mohammed, E.A., Far, B.H., Naugler, C.: Applications of the MapReduce programming framework to clinical big data analysis: current landscape and future trends. BioData Min. 7, 22 (2014)
16. The R project for statistical computing. http://www.r-project.org/
17. Patel, J.A., Sharma, P.: Big data for better health planning. In: IEEE International Conference on Advances in Engineering & Technology Research, August 2014
18. https://hive.apache.org/
19. Hows, D., Membrey, P., Plugge, E., Hawkins, T.: Introduction to mongodb. In: The Definitive Guide to MongoDB, 16, p. 1. Springer, Berkeley, CA (2015)
20. http://en.wikipedia.org/wiki/Apache_Cassandra
21. Gowsalya, M., Krushitha, K., Valliyammai, C.: Predicting the risk of readmission of diabetic patients using MapRe-duce. 2014 Sixth International Conference on Advanced Computing (ICoAC). IEEE, New York (2014)
22. Gomathi, S., Narayani, V.: Implementing big data analytics to predict systemic lupus erythematosus. In: 2015 International Conference on Innovations in In-formation, Embedded and Communication Systems (ICIIECS). IEEE, New York (2015)
23. Sheriff, C.I., Naqishbandi, T., Geetha, A.: Healthcare informatics and analytics framework. 2015 International Conference on Computer Communication and Informatics (ICCCI). IEEE, New York (2015)
24. Prasad, S.T., et al.: Diabetic data analysis in big data with predictive method. 2017 International Conference on Algorithms, Methodology, Models and Applications in Emerging Technologies (ICAMMAET). IEEE, New York (2017)

25. Ojha, M., Mathur, K.: Proposed application of big data analytics in healthcare at Maharaja Yeshwantrao Hospital. 2016 3rd MEC International Conference on Big Data and Smart City (ICBDSC). IEEE, New York (2016)
26. Kalyankar, G.D., Poojara, S.R., Dharwadkar, N.V.: Predictive analysis of diabetic patient data using machine learning and Hadoop. In: 2017 International Conference on I-SMAC (IoT in Social, Mobile, Analytics and Cloud) (I-SMAC). IEEE, New York (2017)
27. Chennamsetty, H., Chalasani, S., Riley, D.: Predictive analytics on Electronic Health Records (EHRs) using Hadoop and Hive. 2015 IEEE International Conference on Electrical, Computer and Communication Technologies (ICECCT). IEEE, New York (2015)
28. Asri, H., et al.: Big data in healthcare: challenges and opportunities. 2015 International Conference on Cloud Technologies and Applications (CloudTech). IEEE, New York (2015)

Analysing Human Activity Patterns by Chest-Mounted Wearable Devices

Jana Shafi, Amtul Waheed and P. Venkata Krishna

Abstract Internet of things (IoT) which is the invasion in the present era led to the emergence of various wearable devices which are known as smart wearable to monitor or measure different health attributes based on the activities. Today, wearable technology is constantly expanding in the market as it is highly in demand from the year 2010. Since then around 400 wearable have been developed, and 60% of them are activity trackers. Now, activity recognition is a promising research field, originated from ubiquitous, context-aware computing as well as multimedia. In recent times, recognizing daily activities is also part of the challenges for pervasive computing. Wearable devices are equipped with various sensors such as humidity sensor, gyroscope, accelerometer, and biosensors, for self-monitoring of routine physical actions. In this paper with the help of Python language, we analysed wearable devices data set which are positioned on chest in terms of wearable single chest-mounted uncalibrated accelerometer data sets collected from fifteen candidates separately performing seven different activities such as *Working at Computer, Standing Up, Walking and Going up/downstairs, Standing*, Walking, Going Up/Downstairs, Walking and Talking with Someone Talking while Standing*. These activities pattern analyses are helpful in measuring numerous health outcomes.

Keywords Smart chest wearable · IOMT · Wearable

J. Shafi (✉) · A. Waheed · P. Venkata Krishna
Sri Padmavati Mahila Visvavidyalayam, Tirupati, Andhra Pradesh, India
e-mail: janashafi09@gmail.com

A. Waheed
e-mail: w_amtul@yahoo.com

P. Venkata Krishna
e-mail: parimalavk@gmail.com

© Springer Nature Singapore Pte Ltd. 2020
P. Venkata Krishna and M. S. Obaidat (eds.), *Emerging Research in Data Engineering Systems and Computer Communications*, Advances in Intelligent Systems and Computing 1054, https://doi.org/10.1007/978-981-15-0135-7_37

389

1 Introduction

With the advancement of Internet of things, life becomes smarter with the help of many devices using wireless sensing technology employing good use of sensors including smart homes [1, 2]. Smart wearable is also the part of this technology which is focused on human's physical activity evaluation through the device which is worn on the body with the smartphone interface [3, 4]. Activity acknowledgement is an emergent research field derived from the bigger areas of ubiquitous and context-aware computing as well from multimedia [5]. Lately, daily activities recognition turns out to be one of the tasks for ubiquitous computing. Human activity recognition is one of the active research areas in computer vision for numerous environments like security investigation, human–computer interaction, and health care [6, 7]. It is also challenging in many applications of health care, smart environments, and native land security [8, 9]. Human activity tracking uses computer vision-based techniques with setup support requirement [10]. Also an effective method is to process the data from inertial measurement unit worn sensors on a user's body [11, 12] while tracking motion in a user's smartphone [13, 14].

Activity recognition is also a characteristic sorting problem. Here, the aim is to sense and identify daily human activities. An activity is regarded as a grouping of derived multiple sensors data. Chest position in human body is regarded as the utmost stable and correct signal gaining position for stability and validity. Chest signals are less exposed to artefact, and PTT from carotid arteries delivers the aortic arduousness. The server obtains the data for every measurement from the owner's smartphone which is connected wirelessly to the chest strap. The objective of activity recognition is to diagnose the actions and goals from successions of observations subject to the agents' actions and the environmental settings. In this paper, we are exploring, visualizing, and observing data of single chest-mounted accelerometer which is a smart wearable using two three-axis radio accelerometers stick to the human body and user's behaviour that rebuild and interpret activity patterns with the interface of smartphone wirelessly connected and used to measure and further classify the owner's daily activity patterns, as well we are going to analyse the chest wearable available in market with the prices with the help of data sets. The minimal low power usage and sensors equipped with IC (integrated circuit) components allow long-term continuous observations that give accurate movement developments (for days or weeks) with compact cost, longer battery life, compact sensor size, and weight. In the daily activities patterns such as sitting, walking, standing, and numerous alterations done under supervised environments by young and elderly, the ability to differentiate these patterns was examined using a single tri-axial accelerometer, mounted on the chest.

- Detecting numerous postural transitions.
- Detecting speed variations while walking.
- Classifying walking levels via stairs upwards or downwards with high sensitivity and specificity.
- The chest is a feasible site to wear accelerometer.

1.1   Smart Chest Wearable

Some of the available smart wearable which is specifically positioned at chest is displayed below.

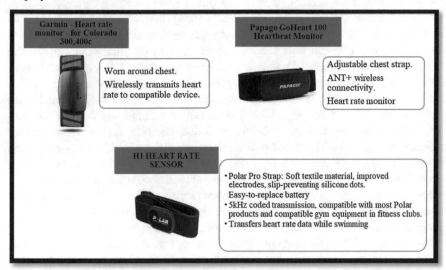

2   Related Work

Human physical daily activities patterns are recognized by collecting data sets from wearable device sensors with different methods and procedures which started recently. Initially, twenty-two wearable sensors (accelerometer, physiological sensors, GPS, and wrist-top computer) are engaged for seven activity movements including sitting, cycling, running, standing, etc. [15]. For data collection a personal computer, recorder placed at knapsack weighing five kg. Users feel inconvenient while wearing devices for long duration. For avoiding this, smartphones researchers replace the wearable which position in pants pockets [16] with three-axial accelerometer data is retrieved. The result was the WISDM data set. The data is sectioned into ten groups which is used in adaptive human activity pattern recognition research [17, 18]. The chance data set [19] is another evident work that intended to identify regular living activities using seventy-two sensors from ten diverse sorts of ambient (temperature, light, and power sensors for electrical services) and sensors wearable (accelerometer, smart movement jacket, microphone, etc.). Data collection is done with many candidates in smart lab for two hours session including various activities using PDAs for the GUI [20]. Anguita et al. also put out human activity recognition data set using accelerometers and gyroscopes in order to differentiate six different activity movements (standing, sitting, laying down, walking, and walking downstairs and upstairs.) They used support vector machine (SVM) classifier to evaluate the data set accuracy. The scholars in [21] worked on a combined data set of the smart watch and smartphone sensors to identify sign activities. They used a five phase (data assembly, pre-processing, breakdown, attributes extraction, and classification) procedure for their study. The chief objective was to investigate heterogeneous sensors effect in the above stated five phases. The research was carried out on thirteen different devices, and the consequences were calculated using four different classifiers (C4.5, Random Forest, KNN, SVM). Analysis indicated that deploying heterogeneous sensors let to ruin the performance system. Shoaib et al. [22] worked upon a data set of ten candidates aged between 25 and 35 years old. The activities motions are categorized into simple (i.e. standing, sitting, walking, etc.) and complex (i.e. drinking coffee, smoking, etc.) groups. The scholars used Samsung Galaxy S2 by positioning it on the wrist to simulate accelerometer and gyroscope data collected from a smart watch. Sztyler et al. [23] collected up-to-date data sets focused on position alertness in human activity recognition. The tests were carried out in specific subject (each member gathers her/his data) and cross-specific subject means (members with alike biometric features are clustered, and the learned model can be employed to the other group members). Thus, the last case is fit for users who are unable to gather initial working out data. Their data set is collected from 15 members for eight common activities motions. The other mechanisms focus on preparing a structure to gather the data free from the end device. These structures are usually comprised of two key units of gathering (responsibility of gathering data from several wearable) and linking (handling the connection to a storage unit) [24]. The authors in [25] displayed AWSense, a structure to gather the data from a Smart Apple Watch. It has two key portions of core and connects. The core part runs on Apple Watch and is in charge for gathering raw data from the sensors. The connect portion manages with data communication and message protocols amid the smartphone and the smart watch. Finally, the smartphones are exploited to position in a solo location which indicates the lack of spot-aware data sets.

3 Data Collection

The raw data which is derived with the frequency of 5 Hz (i.e. five measurements per second) for training from three-axial accelerometer mounted on the chest of the human comprises of timestamp, acceleration along X-, Y-, Z-axis. In this approach, every single instance is measured from accelerometer which retrieved the data in order to recognize following seven pattern activities:

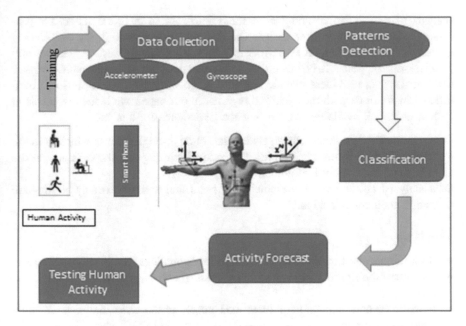

Fig. 1 Single chest accelerometer retrieving human daily activities data for training and classifying into patterns for further testing

(1) Working at Computer
(2) Standing Up, Walking and Going up/downstairs
(3) Standing
(4) Walking
(5) Going Up/Downstairs
(6) Walking and talking with someone
(7) Talking while standing

The retrieved data was separated into two developments stages. Each of them contained seven activities patterns. The first contributes for building the classification model as it was longer and had approx. 7000 instances. The second used for testing and analysing the results as it was shorter. The whole procedure of data collection, constructing the model and classification is presented in Fig. 1.

4 Methodology

Python language is used in our paper for analysing data sets which is a free open source, and it can be extended with the libraries functions. Pandas belong to the Python Data Analysis Library which consists of high-level data structures and manipulation tools (easy to load, index, classify, and group data) in order to make analysis of data informal and fast. Data is imported from Excel sheets for processing and analysing in time series. The data which is accessible by everyone is termed as open data which is permissible to make changes, restore without any permission violations. It is indeed an influential tool for multiple data analysis. Most of the data providers include agencies of governments, academics, and educational institutions with the specified regions. In our paper, we loaded data sets in Python notebook available at Data Science Experience for analysis.

Data sets loaded: We loaded data sets in data frame through browse which is in the form of CSV files following running the code which displays the data sets in the table form of the required columns.

Data analysis: The two variables which have to be analysed followed by plotting are defined in their relationships.

Matplotlib

- It introduces the fundamental Matplotlib functionalities, the basic types figure.
- It controls the figure attributes which consist of colour, thickness, patterns, and maps.
- It mark several figures with their axis range, phase ratio and mathematical coordinate system.
- It works with complex figures.

5 Data Analyses

5.1 Data Sets 1: Smart Wearable Commercially Available

This data set contains information on hundreds of wearable along with prices, company name and location, URLs for all wearable, as well as the location of the body on which the wearable is worn (582 rows). Chest wearable data sets (37 rows) are analysed with the help of Python data visualization with company name at X-axis and prices of the wearable at Y-axis in Fig. 2.

Observation: We can witness from the plotted graph

1. Maximum price is 199$ for chest-mounted wearable.
2. Minimum price is 19.99$.
3. The graph is detecting the highly consuming products in the range of 19–74$.
4. Polar, Garmin, and Papago are the popular companies and the choice for wearable.

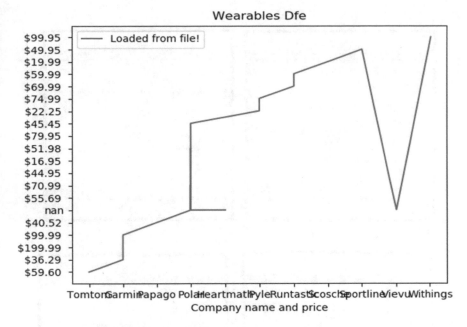

Fig. 2 Wearable data analysis with respect to price and manufacturer

5.2 Data Set 2: Human Activity Classification

The overall samples of uncalibrated single chest accelerometer data are retrieved from 15 candidates carrying out seven activities are counted as 1,926,896 samples which input to the classifier in diverse folds which led to the two million samples. Data set provides activities pattern.

Attribute Information

- Data is divided by participant.
- Each folder contains the following information (sequential number, x, y, z acceleration, tag).
- Tags are arranged by numbers as follows:

    1: Working at Computer
    2: Standing Up, Walking and Going up/downstairs
    3: Standing
    4: Walking
    5: Going Up/Downstairs
    6: Walking and Talking with Someone
    7: Talking while Standing

Analysing activities of 15 participants by Python data visualization plots is shown in graph plot in Fig. 3.

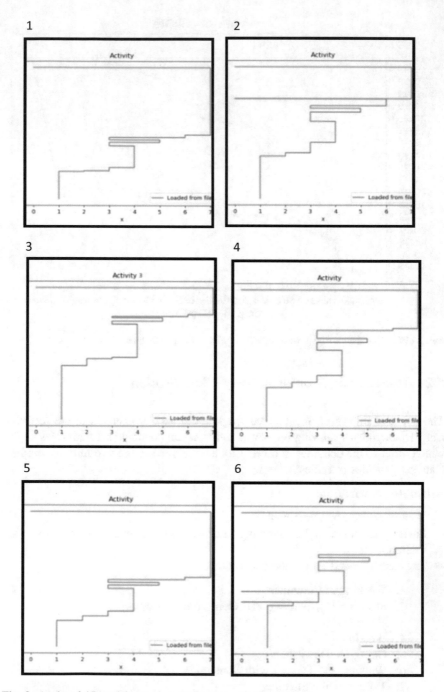

Fig. 3 Analysed 15 participants measured seven activities patterns

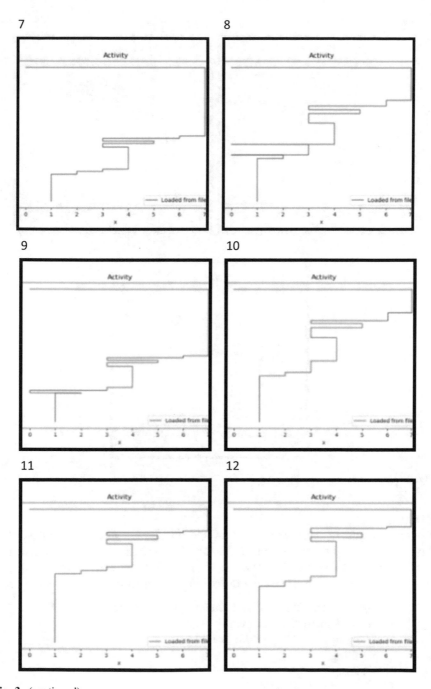

Fig. 3 (continued)

13   14

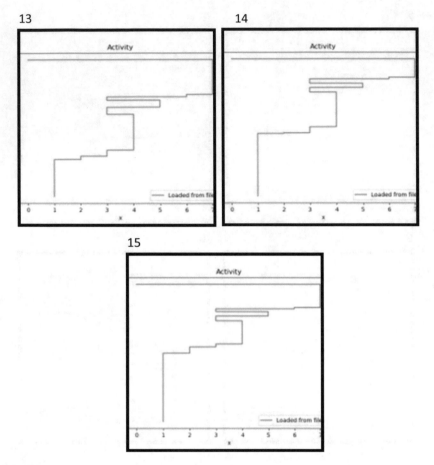

15

Fig. 3 (continued)

Evaluation

The data set class is a data frame as all of the variables are numeric. The median of the variable X is 1900. From the sorted result of the tags, it can be summarized that '**Talking while Standing (7) value**' is the most frequent activity of the human. Here is the sorted list in decreasing order (number represent activity pattern tag):

2: Standing Up, Walking and Going up\down staircases
5: Going Up\Down staircases
6: Walking and Talking with Someone
3: Standing
4: Walking
1: Working at Computer
7: Talking while Standing

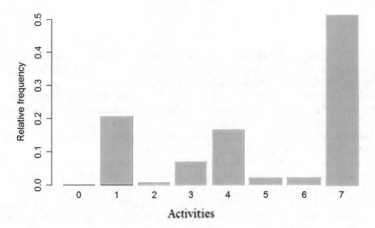

Fig. 4 Chart of the recorded activities

Observation

From the graph in Fig. 4, it can be clearly concluded that (Standing Up, Walking and Going up\downstairs) is the performed activity pattern followed by the maximum limit to the Talking while Standing.

Points which can be considered from the analysis are

- Lack of physical activity
- Effects on health parameters

6 Conclusion

Smart wearable and human activity recognition are prevailing areas of growth and evolution. In this paper, we took the opportunity to investigate into the data sets focused on smart chest wearable as the chest is the most stable position to provide the accurate signals which are able to record human regular physical activities and classify patterns accordingly. Single chest accelerometer wearable device is introduced and used as a model for data sets for analysing and observing vital sign wirelessly. We analysed the performance of different activity patterns recorded from the body as the instances data set and visualized the movements' samples which are tagged with values. The data sets of smart wearable specific for chest position also analysed with the price range and popular manufacturers.

Acknowledgements It must be clarified that in this chapter none of the human activity experiments is carried out, the data which is studied is taken from Google data sets and rest of the evaluation is done.

References

1. Ann, O.C., Theng, L.B.: Human activity recognition: a review. In: Proceedings of the 4th IEEE International Conference on Control System, Computing and Engineering, ICCSCE 2014, pp. 389–393, mys, November 2014. View at Scopus
2. Gubbi, J., Buyya, R., Marusic, S., Palaniswami, M.: Internet of Things (IoT): a vision, architectural elements, and future directions. Future Gener. Comput. Syst. 29(7), 1645–1660 (2013). View at Publisher • View at Google Scholar • View at Scopus
3. Billinghurst, M.: New ways to manage information. Comput. J. 32(1), 57–64 (1999). View at Publisher • View at Google Scholar • View at Scopus
4. Case, M.A., Burwick, H.A., Volpp, K.G., Patel, M.S.: Accuracy of smartphone applications and wearable devices for tracking physical activity data. J. Am. Med. Assoc. 313(6), 625–626 (2015). View at Publisher • View at Google Scholar • View at Scopus
5. Iosifidis, A., Tefas, A., Pitas, I.: Multi-view action recognition based on action volumes fuzzy distances and cluster discriminant analysis. Signal Process. 93(6), 1445–1457 (June 2013). Show Context CrossRef Google Scholar
6. Weinland, D., Ronfard, R., Boyer, E.: A survey of vision-based methods for action representation segmentation and recognition. Comput. Vis. Image Underst. 115(2), 224–241 (2011 Feb). Show Context CrossRef Google Scholar
7. Ali, S., Shah, M.: Human action recognition in videos using kinematic features and multiple instance learning. IEEE Trans. Pattern Anal. Mach. Intell. 32(2), 288–303 (2010 Feb)
8. Khan, A.M., Lee, Y.-K., Lee, S.Y., Kim, T.-S.: A triaxial accelerometer-based physical-activity recognition via augmented-signal features and a hierarchical recognizer. IEEE Trans. Inf. Technol. Biomed. 14, 5–1166 (2010)
9. Westerterp, K.R.: Assessment of physical activity: a critical appraisal. Eur. J. Appl. Physiol. 105, 6–823 (2009)
10. Poppe, R.: A survey on vision-based human action recognition. Image Vis. Comput. 28, 6–976 (2010)
11. Casale, P., Pujol, O., Radeva, P.: Human activity recognition from accelerometer data using a wearable device. In: Pattern Recognition and Image Analysis, 289. Springer (2011)
12. Krishnan, N. C., Colbry, D., Juillard, C., Panchanathan, S.: Real time human activity recognition using tri-axial accelerometers. In: Sensors, Signals and Information Processing Workshop, 2008
13. Kwapisz, J.R., Weiss, G.M., Moore, S.A.: Activity recognition using cell phone accelerometers. ACM SigKDD Explor. Newsl. 12, 2–74 (2011)
14. Kwapisz, Jennifer R and Weiss, Gary M and Moore, Samuel A, Cell phone-based biometric identification. In: 2010 Fourth IEEE International Conference on Biometrics: Theory Applications and Systems (BTAS) (2010)
15. Parkka, J., Ermes, M., Korpipaa, P., Mantyjarvi, J., Peltola, J., Korhonen, I.: Activity classification using realistic data from wearable sensors. IEEE Trans. Inf Technol. Biomed. 10(1), 119–128 (2006)
16. Kwapisz, J.R., Weiss, G.M., Moore, S.A.: Activity recognition using cell phone accelerometers. ACM SIGKDD Explor. Newsl. 12(2), 74–82 (2011)
17. Abdallah, Z.S., Gaber, M.M., Srinivasan, B., Krishnaswamy, S.: Adaptive mobile activity recognition system with evolving data streams. Neurocomputing 150, 304–317 (2015)
18. Abdallah, Z.S., Gaber, M.M., Srinivasan, B., Krishnaswamy, S.: Anynovel: detection of novel concepts in evolving data streams. Evolv. Syst. 7(2), 73–93 (2016)
19. Roggen, D., Troester, G., Lukowicz, P., Ferscha, A., Millán, J.d.R., Chavarriaga, R.: Opportunistic human activity and context recognition. Computer 46(2), 36–45 (2013)
20. Anguita, D., Ghio, A., Oneto, L., Parra, X., Reyes-Ortiz, J.L.: A public domain dataset for human activity recognition using smartphones. In: ESANN (2013)

21. Stisen, A., Blunck, H., Bhattacharya, S., Prentow, T.S., Kjærgaard, M.B., Dey, A., Sonne, T., Jensen, M.M.: Smart devices are different: assessing and mitigatingmobile sensing heterogeneities for activity recognition. In: Proceedings of the 13th ACM Conference on Embedded Networked Sensor Systems. ACM, pp. 127–140 (2015)
22. Shoaib, M., Bosch, S., Incel, O.D., Scholten, H., Havinga, P.J.: Complex human activity recognition using smartphone and wrist-worn motion sensors. Sensors **16**(4), 426 (2016)
23. Sztyler, T., Stuckenschmidt, H., Petrich, W.: Position-aware activity recognition with wearable devices. Pervasive Mob. Comput. **38**(Part 2), 281–295 (2017). special Issue IEEE International Conference on Pervasive Computing and Communications (PerCom) 2016
24. Lu, Y., Wei, Y., Liu, L., Zhong, J., Sun, L., Liu, Y.: Towards unsupervised physical activity recognition using smartphone accelerometers. Multimed. Tools Appl. **76**(8), 10701–10719 (2017)
25. Hänsel, K., Haddadi, H., Alomainy, A.: Awsense: a framework for collecting sensing data from the apple watch. In: Proceedings of the 15th Annual International Conference on Mobile Systems, Applications, and Services. ACM, pp. 188–188 (2017)

The Role of Big Data Analytics in Smart Grid Management

Bhawna Dhupia, M. Usha Rani and Abdalla Alameen

Abstract Data analytics is playing a vital role in the modern industrial era. Electricity is one of the industries that have adapted data analytics techniques to a great extent. The data collected in the smart grid through smart meters and other sensors installed is very huge. The processing of such a huge heterogeneous data is not possible without the use of big data analytics technique. Big data analytics and machine learning algorithms play a vital role in electricity transmission and distribution network for data collection, storage and analysis of the data, prediction for data forecasting, and maintenance of the system. These techniques can help to optimally deliver energy at a lower cost with high quality and can also improve the customer service as well as social welfare. This article will review the use of big data analysis techniques along with the machine learning for various applications that can be mapped for smart grid environment. This article will also discuss the various methods and algorithm to be used for the applications for smart grid. A comparative analysis will also be done to show the best methods and algorithm to be used for a particular application.

Keywords Smart grid · Machine learning · Applications in smart grid · Data analytics algorithms and methods

B. Dhupia (✉)
Department of Computer Science, SPMVV, Tirupati, India
e-mail: bhawnasgn@gmail.com

M. Usha Rani
Department of Computer Science, SPMVV, Tirupati, India
e-mail: musha_rohan@yahoo.com

A. Alameen
Prince Sattam Bin Abdulaziz University, Al-Kharj, Saudi Arabia
e-mail: a.alameen@psau.edu.sa

© Springer Nature Singapore Pte Ltd. 2020
P. Venkata Krishna and M. S. Obaidat (eds.), *Emerging Research in Data Engineering Systems and Computer Communications*, Advances in Intelligent Systems and Computing 1054, https://doi.org/10.1007/978-981-15-0135-7_38

1 Introduction

The electricity or power is considered to be the backbone of every country. It has been witnessed many recent developments in research and infrastructure to benefit the socioeconomic development at large in the field of power [1, 2]. With the technology progress in the area of energy, traditional power grids are transforming into an intelligent system called smart power grids. A smart grid is the next generation of electrical transmission and distribution. Smart grid is a combination of the features in existing power grids, the advanced information communication technology, and data analytics tools [3]. Altogether these characteristics contribute to the overall efficiency of the smart power system. The primary objective of the smart power system is to ameliorate the operation, dependability, security of the system along with reducing the loss of energy from the grid. Integration of renewable energy resource (RES) for the generation of energy has enhanced the capability of the smart grid to a higher level. Another big objective of a smart grid system is to concentrate on bringing down the price of energy in all aspects like generation, transmission, distribution, and use of energy. It focuses on the consumer by offering them better access to services and educates them on tariff-related information to make out their use of energy proactively [4]. Smart grid ensures an efficient system as compared to conventional energy system by providing a guaranteed proficient connection and utilization of all means of production resources at its maximum level. It also provides automatic and real-time management of two-way communication of electrical networks, monitors perfect quantity of energy consumption, increased level of dependability, henceforth leads to energy savings and lower costs [5].

Traditional electricity meters generate a very modest quantity of information. This information is mostly practiced to generate consumer utility bill, whereas the introduction of advanced metering technology (AMI) enabled smart grid to generate a variety of data that can be processed to get various information about the entire system. Hence, AMI is one of the significant ingredients of smart grid systems. This infrastructure is a combination of devices like smart meters, communication networks, and data management systems which enable two-way communication between utilities and customers. The grid can collect data on energy consumption, power quality, and load profiles of the customers in real time with the help of the AMI technology [4, 6]. There is a close relationship between AMI and smart meter in terms of communication network in smart grid. We call smart meters as an intelligent device because it measures the electricity usage almost in real time, electricity usage is controlled remotely, and disconnect–reconnect supply remotely when required. The consumer information collected through AMI is highly secret. As the privacy of the consumer, data is one of the major concerns; the admittance of this data is executed exclusively by the authorized individual. The data collected by the sensor devices at the confined collection point are of two types, i.e., energy consumption or usage data and event data. The reading from the smart meter is collected after every 15 min or 30 min. The unit for measuring the electricity consumption is recorded in kilowatt per hour (KWHs) [5]. The information collected at utility is used in many commercial enterprise applications such as customer profiling, energy usage pattern, consumer billing, forecasting, demand planning, and network monitoring. Along with the energy management, smart grid deals with other challenges, like high energy, velocity, high volume of storage, and advanced data analytics requirements [7]. In smart grid, huge amount of data is collected from various data resources. The type of data collected in smart grid is in heterogeneous form. To manage the data efficiently, variety of data analytics techniques such as optimization [8], forecasting [9, 10], classification, and clustering [11–13] are applied to the data of smart grid. These techniques help for energy optimization in real time, accurate electricity demand and response prediction, identify the energy consumption pattern, asset monitoring and customer side management of energy usage [14]. All the stakeholders of smart grid can get a great deal of support to take decision on crucial from the techniques applied through big data analytics. Main stakeholders of the smart grid are namely producer, operators, customers, and regulators.

Certainly, smart grid data needs to process with advanced methods, techniques, and algorithm to get refine results. Big data techniques offer a great deal of combination of machine learning, artificial neural network, and advanced statistical algorithm to manage this type of database. Smart grid processed data can help for the better understanding of customer behavior, conservation and demand of the energy, device estimation and monitoring, downtime prediction, etc. As stated above, big data analytics is the combination of different types of algorithms. So, the challenge is to identify which technique is best suited to some specific analysis method to transform the data into valuable information. Basically, the selection of procedure and methods depend upon the requirement of information by the users. In this context, this paper suggests the best-suited algorithm for particular task in smart grid management. But, before this, it discusses applications of smart grid that can be managed by big data analytics techniques.

2 Applications of Big Data Analytics in Smart Grid

2.1 Demand-Side Management System

Estimating the energy demand of the consumers in the energy system is one of the challenging tasks. It requires a highly advanced system designed with the combination of several advanced techniques. It has to be managed in such a way that it does not exceed the energy supply at any point of time. By proper planning and monitoring of the utility activities, the usage of the electricity by the customers can be controlled. For this, administrator has to analyze the usage pattern for each customer to understand the demand format. This pattern analysis provides the time and magnitude of utility usage by the customer. Main objective of doing demand-side management is to educate and encourage customers to decrease the usage of energy during peak time or to move the major energy consumption during off-peak hours [15]. This management is useful for the stakeholder, consumers, and providers also. The consumer can benefit by reducing their bill for the utility, and for provider, it will become easy to manage the quality of energy distribution and helps to flatten the demand curve also. Moreover, this management helps in optimizing energy and utilization of the asset installed to a great extent [16].

To make the grid reliable, seamless balance between supply and load is required throughout the energy distribution process [17]. Due to less control on demand-side, this task is really challenging. Consumption of energy is influenced by the weather conditions, price, and infrastructure. These data are interrelated, random, and heterogeneous in nature. It is therefore very complicated to predict the electricity consumption of the customers by analyzing their dynamic behavior. So, using big data technique, it is possible to analyze such kind of heterogeneous data for a perfect and accurate result [16].

2.2 Electricity Theft Detection

The non-technical loss of energy is usually initiated by electrical theft or errors in generation of bill due to wrong calculations. But, electrical theft is the main cause of non-technical energy loss. Theft of electricity can be defined as the usage of electricity from the companies without any contract and measurement of electricity consumed [18, 19]. Inquisitive actions with energy meters, tapping energy from distribution feeder, reduction of bills by influential approaches are few cause of non-technical energy loss [20]. Electricity theft is a major problem in both developed and developing countries. According to the government report, India bears a loss of approximately $4.5 billion due to energy theft [21] and for whole world it is around $89.3 billion every year [22].

Big data analytics can offer a great help to identify the theft detection problem efficiently. With the help of BDA techniques, usage pattern of each customer can be identified based on various criteria like number of persons, weather conditions, use of heavy electrical appliances for a particular interval of time. After collecting the usage of data pattern, it is fed into the system to track the consumption of electricity as a training data. After this, the current data can be compared with the usage pattern of the customer to analyze that the customer usage is normal or fraudulent [20, 23]. In the subsequent section of this paper, we will study about the algorithm implemented to capture the data of the customer and pattern analysis to identify the electricity theft.

2.3 Electric Devices State Estimation and Predictive Maintenance

Electricity is a vital component of our day-to-day life. Mismanagement in power transmission system can cause disastrous blackouts in power system. Therefore, many researchers have shown interest to make power grid smart and reliable [24]. The system reliability at the distribution level is the main motivation for distribution automation of smart grid. For reliability of the smart grid, it is important that the health of the equipment installed is checked and maintained regularly. The operational data from supervisor control and data acquisition (SCADA) or advanced metering infrastructure (AMI) are utilized to monitor the status of the device and to diagnose the faults in the devices [25]. The data collected for this purpose is huge as the reading from these devices are updated after every 15 min. Thanks to the big data analytics techniques, the huge data collected from the AMI is processed and analyzed so as to predict the device health. These predictions help the decision-making authority to plan the maintenance or shifting of load to some other machine in case of emergency proactively.

To stimulate the warning system, a consistent online detection of uncertain instances in power system is implemented through machine learning algorithms in an intelligent smart grid [24]. The data mining methods along with graphical alerts are used to show the potential failure prediction in the system. Reliability of the system is one of the most characteristics of smart grid system. That is the reason various companies are working toward increasing the reliability of the smart grids. Data analytics technology is utilized to investigate the working condition of renewable energy sources and inspect benchmark values to evaluate and detect the faults in the system instantly and remotely for early repair [16]. Early prediction of devices helps the company to repair them before any big loss due to poor working condition of the device.

2.4 Energy Production Optimization

Energy production optimization is strictly based upon the demand and response of electricity by the customers. To process the energy optimization, massive storage of data from smart grid is collected through various sensors installed in the system. The consumption of energy form the customer is continuously recorded and monitored. By analyzing this data, energy provider predicts that how much quantity of estimated energy is essential to generate to meet the demand of the customers. Calculation for demand–response is highly critical to optimize the production of the energy [26]. Leveling out the energy load with the help of peak clipping and valley filling has larger potential in energy savings. Smaller change in the pattern of energy consumption can help the energy provider to achieve the energy-based load, whereas decreasing the capacity for peak load. But these are activities cannot be achieved without the contribution of the consumer. Consumers must be involved in the complete process to reduce the peak load–average ratio by optimizing users' energy usage schedule and lowering the overall energy consumption by educating them on the time management for energy usage [16].

As discussed above, energy production optimization needs a huge calculation and analysis of the usage pattern of the customer. The data collected in smart grid needs a system which can process the data perfectly. For this reason, big data analytics techniques along with ANN and deep machine learning offer various algorithm and techniques for the data analysis process. The best way of predicting the accurate energy consumption is frequent pattern analysis based on big data time series on collected datasets. After categorizing the customers depending upon the pattern of usage, deep learning algorithm is applied to get a great result for energy production optimization [27].

2.5 Renewable Energy Forecasting

The environment-friendly and abundant renewable energy sources (RES) will be the prevailing energy source for the next generation of power grid. Regulatory bodies for energy in all the countries are count on the energy forecast to draft and instrument cost minimization and accurate and feasible energy policies. Demand and supply model of smart grid plays an important role to frame the requirement of RES forecasting [28]. If this forecast is not handled properly, then mismatch between the production and demand for energy costs high for provider and a nation too [29].

The renewable resources like wind, solar, biomass promote the greener environment by reducing the greenhouse emission [30]. The energy generated by RES is than integrated in smart grid to increase the energy capacity for distribution [4]. Demand and production of the energy in smart grid are affected by various factors including time of day, weather conditions, and fluctuations in climate, prices and renewable energy sources. Inadequate way of observing and mismanaged control on the flow of power leads to the increased possibility of failure due to overloading, load mismanagement, or congestion. Therefore, smart grid requires intelligent real-time techniques to detect the abnormal event in the system and to find the cause and location beforehand. These functions can be efficiently performed by applying big data analytics techniques along with artificial intelligence, simulation and modeling, and big data network management [31].

3   Methods of Data Analytics in Smart Grid

Smart energy meters are replacing conventional meters, hence providing all transaction data to the smart grid management tool [32]. The data collected serves for many decision on smart grid after processing. Data analysis is taken to plan energy distribution, plan strategies for current and future situations and to develop recommendations for energy policies. A great deal of data analytics algorithms and applications are being implemented focusing on data analysis for smart grid. The principal job of big data analytics in smart grid is to extract valuable information from historical data and to plan the operation and maintenance for future [24]. The vast amounts of data are gathered from various resources like smart meter, sensors, IOT-enabled devices are arranged with modern information management techniques. Integrating demand and response needs the capability to synchronize the communication of demand–response effect, and process data, such as analysis of predictive peak load's levels and the required capacity available for each demand–response [16]. To process accurate result for the data collected from smart grid, a great numbers of analytics algorithms and applications have been proposed by the researchers. These algorithms serve to forecast energy expenditure and demand, propose energy consumption plans, provide individual reports to consumers on patterns of energy usage, and help to design models for specific classes of customers [32]. Extract of the algorithm is extremely dependent on the type of data needed by the soul. Before choosing any particular technique, data analyst has to interpret the output desired.

The supervised learning and unsupervised learning are the two broad classifications in which data analysis methods for machine learning, data mining, and pattern identification are classified. Supervised learning algorithms are applied where the classification of data is predestinated. On the other hand, unsupervised learning algorithms are employed in situations where the data is not pre-categorized. Hence, to set up the data in unsupervised learning algorithm, clustering techniques are applied. Few supervised algorithms used in smart grid energy forecasting are artificial neural networks (ANNs), naïve Bayes, support vector machine (SVM), and deep learning algorithm [33, 18]. Unsupervised learning algorithms used for the classification and clustering of the data are K-means, hierarchical clustering, and Gaussian models [34]. Support vector machine (SVM) offers the great predictive capability for a limited amount of data under supervised machine learning algorithm. This is also employed for energy load prediction and benchmarking in energy management organization. Artificial neural network (ANN) can be used for data analysis to find out the energy demand prediction by using multivariable calculations [35, 36]. Gaussian process (GP) is used to predict the energy demand of the customers and depending upon the demand forecast for energy production is proposed [37]. The GP model is provided with the training data to estimate the pattern of energy consumption, and based on the evaluation result, energy requirement is predicted [38].

Table 1 Method/algorithms for different applications in smart grid

Applications	Methods/algorithms					
	Linear regression (LR)	Support vector machine (SVM)	Artificial neural network (ANN)	K-means	Naïve Bayes	Gaussian process
Demand-side management	√	√	√		√	√
Electricity theft detection		√		√	√	
Device state est. and management	√	√	√			
Energy prod. optimization	√	√	√		√	√
Renewable energy forecast		√	√		√	√

Table 1 describes a comparison of the most popular algorithm used for various applications for smart grid. This table will give a glance to choose the best-suited algorithm for several processes in smart grid management.

4 Conclusion

As energy is a critical component for the development of any country, many countries are working toward making the electricity as a priority field for research and development. This paper presented the role of big data analytical tools to advance analytical results in a smart grid. It also discusses the various applications that can be implemented by integrating various technologies with big data to make the management of smart grid efficient and intelligent. Integration of big data analytics with artificial intelligence and machine learning design is an efficient model for the smart grid management. The model designed is capable of managing the data collected in smart grid through various sensor and other data collection devices in the smart grid, hence resulting in smart decision-making system. By using combination of these advanced techniques, it has become possible to estimate the energy demand–response, device management and monitoring, energy forecasting, and energy production optimization efficiently. The paper concluded with a brief discussion about the best-suited algorithm for different applications in smart grid. For better understanding of the readers, most suitable methods or algorithm is presented in Table 1. Selection of most suited method to speed up the processing of data and give expected result in a short period of time.

References

1. Bigerna, S., Bollino, C.A., Micheli, S.: Socio-economic acceptability for smart grid development—a comprehensive review. J. Clean. Product. **131**, 399–409 (2016)
2. Dang-Ha, T.-H., Olsson, R., Wang, H.: The role of big data on smart grid transition. In: 2015 IEEE International Conference on Smart City/SocialCom/SustainCom (SmartCity), pp. 33–39. IEEE, New York (2015)
3. Jiang, H., Wang, K., Wang, Y., Gao, M., Zhang, Y.: Energy big data: a survey. IEEE Access **4**, 3844–3861 (2016)
4. Anjana, K.R., Shaji, R.S.: A review on the features and technologies for energy efficiency of smart grid. Int. J. Energy Res. **42**(3), 936–952 (2018)
5. Daki, H., El Hannani, A., Aqqal, A., Haidine, A., Dahbi, A.: Big data management in smart grid: concepts, requirements and implementation. J. Big Data **4**(1), 13 (2017)
6. Silipo, R., Winters, P.: Big data, smart energy, and predictive analytics. Time Ser. Predict. Smart Energy Data **1**, 37 (2013)
7. Smart Meters and Smart Meter Systems: A Metering Industry Perspective. A white paper from EEI-AEIC-UTC, March 2011, pp. 01–29
8. Zhou, K., Yang, S., Chen, Z., Ding, S.: Optimal load distribution model of microgrid in the smart grid environment. Renew. Sustain. Energy Rev. **35**, 304–310 (2014)
9. FanS, H.: Short-term load forecasting based on a semi-parametric additive model. IEEE Trans. Power Syst. **27**, 134–141 (2012)
10. Hong, W.-C.: Electric load forecasting by seasonal recurrent SVR (support vector regression) with chaotic artificial bee colony algorithm. Energy **36**, 5568–5578 (2011)
11. Ferreira, A., Cavalcante, C.A., Fontes, C.H., Marambio, J.E.: A new method for pattern recognition in load profiles to support decision-making in the management of the electric sector. Int. J. Electr. Power Energy Syst. **53**, 824–831 (2013)
12. Zhou, K., Fu, C., Yang, S.: Fuzziness parameter selection in fuzzy c-means: the perspective of cluster validation. Sci. China Inf. Sci. **57**, 1–8 (2014)
13. Zhou, K., Ding, S., Fu, C., Yang, S.: Comparison and weighted summation type of fuzzy cluster validity indices. Int. J. Comput. Commun. Control **9**, 370–378 (2014)
14. IBM: Managing big data for smart grids and smart meters (2014). http://www.bmbigdatahub.com/whitepaper/managing-big-data-smart-grids-and-smart-meters
15. Gelazanskas, L., Gamage, K.A.A.: Demand side management in smart grid: a review and proposals for future direction. Sustain. Cities Soc. **11**, 22–30 (2014)
16. Zhanyu, Ma., Xie, J., Li, H., Sun, Q., Si, Z., Zhang, J., Guo, J.: The role of data analysis in the development of intelligent energy networks. IEEE Netw. **31**(5), 88–95 (2017)
17. PdfKothari, D.P., Nagrath, I.J.: Modern Power Systems, 3rd edn. McGraw-Hill, New Delhi (2009)
18. Meng, F., Weng, K., Shallal, B., Chen, X., Mourshed, M.: Forecasting algorithms and optimization strategies for building energy management & demand response. Multidiscip. Digit. Publ. Inst. Proc. **2**(15), 1133 (2018)
19. Smith, T.B.: Electricity theft: a comparative analysis. Energy Policy **32**(18), 2067–2076 (2004)
20. Ahmad, T., Chen, H., Wang, J., Guo, Y.: Review of various modeling techniques for the detection of electricity theft in smart grid environment. Renew. Sustain. Energy Rev. (2017)
21. Overview of power distribution. Ministry of power, Govt. of India. http://www.powermin.nic.in⟩
22. PR Newswire: World Loses $89.3 Billion to Electricity Theft Annually, $58.7 Billion in Emerging Markets." [Online] (2014)
23. Jokar, P., Arianpoo, N., Leung, V.C.M.: Electricity theft detection in AMI using consumers consumption patterns. IEEE Trans. Smart Grid **7**, 216–226 (2016)
24. Zhang, Y., Huang, T., Bompard, E.F.: Big data analytics in smart grids: a review. Energy Inf. **1**(1), 8 (2018)

25. Wang, B., Fang, B., Wang, Y., Liu, H., Liu, Y.: Power system transient stability assessment based on big data and the core vector machine. IEEE Trans. Smart Grid **7**(5), 2561–2570 (2016)
26. Pérez-Chacón, R., Luna-Romera, J.M., Troncoso, A., Martínez-Álvarez, F., Riquelme, J.C.: Big data analytics for discovering electricity consumption patterns in smart cities. Energies **11** (3), 683 (2018)
27. Singh, S., Yassine, A.: Big data mining of energy time series for behavioral analytics and energy consumption forecasting. Energies **11**(2), 452 (2018)
28. Landa-Torres, Itziar, Iraide Unanue, Iñaki Angulo, Maria Rosaria Russo, Camillo Campolongo, Alession Maffei, Seshadhri Srinivasan, Luigi Glielmo, and Luigi Iannelli. "The application of the data mining in the integration of RES in the smart grid: consumption and generation forecast in the I3RES project. In: 2015 IEEE 5th International Conference on Power Engineering, Energy and Electrical Drives (POWERENG), pp. 244–249. IEEE, New York (2015)
29. Ma, J., Oppong, A., Acheampong, K.N., Abruquah, L.A.: Forecasting renewable energy consumption under zero assumptions. Sustainability **10**(3), 576 (2018)
30. Ak, R., Fink, O., Zio, E.: Two machine learning approaches for short-term wind speed time-series prediction. IEEE Trans. Neural Netw. Learn. Syst. **27**(8), 1734–1747 (2016)
31. Diamantoulakis, P.D., Kapinas, V.M., Karagiannidis, G.K.: Big data analytics for dynamic energy management in smart grids. Big Data Res. **2**(3), 94–101 (2015)
32. Liu, X., Golab, L., Golab, W.M., Ilyas, I.F.: Benchmarking smart meter data analytics. In: EDBT, pp. 385–396 (2015)
33. Liu, X., Golab, L., Golab, W., Ilyas, I.F., Jin, S.: Smart meter data analytics: systems, algorithms, and benchmarking. ACM Trans. Database Syst. (TODS) **42**(1), 2 (2017)
34. Haben, S., Singleton, C., Grindrod, P.: Analysis and clustering of residential customers energy behavioral demand using smart meter data. IEEE Trans. Smart Grid **7**(1), 136–144 (2016)
35. Rcn, Y., Suganthan, P.N., Srikanth, N.: Ensemble methods for wind and solar power forecasting—a state-of-the-art review. Renew. Sustain. Energy Rev. **50**, 82–91 (2015)
36. Gong, X.: Energy consumption control and optimization of large power grid operation based on artificial neural network algorithm. NeuroQuantology **16**(6) (2018)
37. Panda, M.: Intelligent data analysis for sustainable smart grids using hybrid classification by genetic algorithm based discretization. Intell. Decis. Technol. **11**(2), 137–151 (2017)
38. Alamaniotis, M., Chatzidakis, S., Tsoukalas, L.H.: Monthly load forecasting using kernel based Gaussian process regression, 60–8 (2014)

Fraudulence Detection and Recommendation of Trusted Websites

N. C. Senthilkumar, J. Gitanjali, A. Monika and R. Monisha

Abstract We live in the world of science where everything could be done possibly through online facilities. As the online services increase, the fraudulent sites also increase, however. Therefore, it enhances the difficulty of people to identify a fake or fraudulent or scam website. Fraudsters are highly intelligent in making or persuading fake products which seem exactly same as the original. Some scam websites use low prices to attract people. Most of the fraudulent sites use domain names that reference a popular brand or product name and design with mere change. Some websites will get your card details before you intend to buy a product. Fraudsters try to attract people's attention by advertising the fake products in many social websites such as Facebook, Instagram and Twitter, and when people wishes to view about the product and its details, they are forced to install an app of that site. So, here we have a solution to find those fraudulent products and recommend some trusted products for people to buy with good quality and at a fair price. We use the reviews or feedback or comments and the rating from the users who already purchased the product and shared their experience for identifying fraudulent products and giving a caution to the remaining people. This paper concentrates on analyzing the reviews and extracting the useful information to guide the customers using text mining. As a result of this, people will get a better idea of making a proper decision in buying the product through online.

Keywords Decision making · Fraudulent · Feedback · Rating · Recommendation · Reviews · Text mining

N. C. Senthilkumar (✉) · J. Gitanjali · A. Monika · R. Monisha
School of Information Technology, Vellore Institute of Technology, Vellore, India
e-mail: ncsenthilkumar@vit.ac.in

© Springer Nature Singapore Pte Ltd. 2020
P. Venkata Krishna and M. S. Obaidat (eds.), *Emerging Research in Data Engineering Systems and Computer Communications*, Advances in Intelligent Systems and Computing 1054, https://doi.org/10.1007/978-981-15-0135-7_39

1 Introduction

Everyone has different taste when it comes to shopping, to buy a single product, we will see thousands of products, consult hundreds of people, select top 10 brands and finally buy one product. This will take more time and in this modern world, no one has time to visit many shops, select the best brand and buy the product by analyzing all its features and quality. To facilitate the consumers, the idea of the supermarket came to the action. Instead of visiting many shops for buying the required items, people can buy all the products from one shop, yet at the same time the shoppers are not fulfilled. For the convenience of the people, using the technology, many online shopping sites have been evolved. Here, we can see any number of products and we choose to buy products that meet out our needs. People use the virtual basket to save their products for the later use. If they wish, they can buy it or even they can reject it permanently. The embarrassment of rejecting the selected product and then regretting missing the most liked products has been avoided. This resulted in an increase in more number of online websites. As the number of online websites increases, the chance of fraudulence also increases gradually. This virtual shopping has been successful because of the net banking and credit card payment options. This is the main stream used by the fraudsters to steal the consumer details and also empty the consumer accounts with their intelligence. Indeed, the security has been improved to protect our bank details from these deceivers. But still, the fraudulence is not controlled completely. Day-by-day, the number of online websites increases tremendously and with the most attractive advertisement, without any further investigation or analysis people tend to buy the products. The product's quality, confidentiality of the consumer details and the integrity of the website and also the original product are based on people's experience which is revealed out as reviews and comments. If there are thousands of comments or reviews for a particular product or website, it is too difficult to view all those comments to know about the product. The concepts like clustering, outlier detection, text mining and sentimental analysis are used to identify the characteristics of the product and to analyze about the experience of the consumer in using that product. This paper will help people to detect the fraudulent product and also to recommend the trusted product by analyzing the datasets obtained from the each product.

1.1 Motivation

The motivation is to enhance the predictive accuracy of fraudulent detection system by aligning all the information that are associated with some records such as purchase or credit application. The methodologies depicted here are applicable in the exact same way they are used in association with methods of text mining, except the unstructured data sources must be cleaned first using pre-processing to include them in the activities of data analytics. The reviews in the shopping sites can be viewed by the consumers, but they confuse the consumers. The main motive of this system is to provide the clear cut idea and suggestions about the product or websites based on the experience of the consumers.

1.2 Objective

The main objective of this system is to minimize fraudulent cases by providing a system for detecting the fraudulent products, identifying the pros and cons of the product using the customer review and recommending good products while buying products from online shopping sites and also to provide a user-friendly interface so that people can easily understand the system and work with the system.

1.3 Proposed System

The aim of the proposed system is to develop a website that has the capability to restrict customers from buying products from fraudulent websites. The existing system detects fraud after fraud has been occurred, but the proposed system detects the fraud before it occurs based on customer reviews about the product. Here, the customer reviews plays a vital role in detecting the fraudulent websites and recommending trusted websites.

1.4 Advantages of Proposed System

The proposed system will help in reducing the time spend for selecting the best product and viewing the reviews of the product for identifying about the product and also eliminate the risk of entering the bank details in the fraudulent websites.

2 Literature Survey

Kharrat et al. [1] proposed an approach that is useful for the recommendation of the product using the contextual analysis with the help of the Facebook comments. This approach has been implemented with the Java and LensKit framework. The proposed machine learning-based classifier is compared with the slope one and SVD algorithms to analyze the best solution. This resulted in 16% improvement when compared to the other analysis. The user profile is analyzed, and the comments are extracted for implementation using the collaborative filtering. The future work deals with the domain ontology to search several connected communities and enrich the knowledge about users. Using this acquired knowledge, this system will focus on detecting the fraudulent websites using the reviews and comments given by the users. These data will be analyzed using the clustering and classification concepts like k-means and naïve Bayes, and the data will be classified as positive, negative and neutral. With the analysis score, the trusted products or websites will be recommended. Forecasting is done using the supervised learning.

Boldt et al. [2] uses Nike's Facebook pages for collecting the datasets. AIDA framework is used for gathering the information about the likes, comments of Nike's Facebook page. With the collected datasets, case-by-case data analytics is achieved using the simple regression, multiple regression and correlation techniques. It is a unit-free measure of the extent to which two random variables move or vary together.

Ketcham et al. [3] uses the naïve Bayes classifier to analyze the positive and negative behavior of users in social media. Extraction of dataset from social media like Facebook, Twitter, etc. using API is clearly explained. Using the developer login, we can get access to data in the social media for research or analysis purpose. A large amount of data cannot be extracted, as it requires access permission from the social media authority.

Gotarane et al. [4] not only analyze the reviews but also the relationship between two people. The data is obtained from the social networking sites using the API, and the raw data is separated from the normal formal language. Using the separated data, the feature is generated and extracted to identify the presence of the linguistic variables like "happy," "sad," etc. Using lexicon-based classification, this data is evaluated and the result is summarized using the stronger relationship. The future work is based on analyzing the advertisements popping out on the social media websites.

Hanni et al. [5] concentrates on mining the opinion of the customer using the reviews based on Natural Language Processing. This framework is achieved through 3-step process. Step 1 involves in mining the reviews of the product where each feature and their strength, as well as weakness, is described by the customers. This raw data is analyzed using sentence breaker concept and categorized as positive, negative and neutral. Based on the categorization, the results are summarized and the product's quality is well determined. Here, feature extraction and probabilistic approach are followed. This study provides a more realistic and efficient result.

Nazeeh et al. [6] deal with an experiment that proposes a framework for fraud prevention by detecting the fraudulent websites automatically. This paper analyzes the available algorithms that can be used for extracting and classifying the data. Here, it is clearly proved that random forest tree algorithm is the best algorithm to identify the phishing websites. It covers three main areas: the automatic extraction of the website data and converting it to the structured format, classifying the data using the classification algorithms and finally summarizing the results. The future work focuses on identifying the more algorithms which is better than the RFT algorithm.

Raja et al. [7] examined the extraction of the opinion of each user about the product using the reviews and the comments. The multidimensional approach involves extracting the information from the emotions that is emoji. Nowadays, people use the emoji to express their experience with that product. This paper explains the method of converting the emotions and text reviews into the positive and negative results. The data extraction from the website has been achieved using the API framework and the Access Key. The raw data is converted into structured data using NLP concepts, and the analysis has been done. This helps in highlighting the positive feature of the product and recommendation of that product.

Gowri et al. [8] explains about the recommendation of a particular product based on the user preference and the product rating using cluster and collaborative filtering concepts. Here, the clustering technique enables the efficient searching mechanism without using the whole dataset. Using the nearest neighbor concept, collaborative filtering is facilitated for finding the users with the similar taste and preferences. The traditional method uses the collaborative filtering to identify the preferable recommendations. This combination of clustering and collaborative filtering has resulted in time-saving and also accurate recommendation according to the user preferences. Further work is about ensuring the avoidance of sparsity problem and evaluation of advanced clustering techniques.

Drew et al. [9] identified the replicated criminal websites using the combined clustering algorithm. Some fake websites uses the information related to the original websites. In this case, sometimes the research may end up in finding the wrong results. The scam websites are difficult to identify when it uses the design, logo and other information just like the original websites. Sometimes, the scam products look more promising and attractive than the original product. This paper finds a solution to solve the contradiction between the original and the scam products or websites using the combined clustering. Here, two types of datasets are collected from two different websites. These datasets were in the form of HTML, and it has been collected using wget. Phishing, counterfeit good shops and the duplicated content remain nearly identical. Many fraudsters use the website which has been closed due to some unavoidable reasons and renames it. These criminal websites lure the people and until any fraudulence is identified, these websites are kept active. The collected HTML data has been converted into the structured data by splitting the text into sentences using OpenNLP sentence breaker. And the HTML tags are used for calculating the distance, these distance matrices are combined and using the combined clustering (hierarchical and agglomerative) algorithm, the identification of criminal websites have been proved.

Kou et al. [10] explained the techniques used for detecting the different types of fraudulence clearly. As the technology evolves, the number of fraud also increases simultaneously. The outlier detection using the unsupervised learning is used to identify the credit card fraudulence, and neural networks are the major stream used widely to detect the fraudulence. The modern fraud and intrusion detection techniques not only detect the fraud but also provide to prevent the fraudulence and intrusion. This paper takes a survey on the various fraudulence and the techniques used to prevent and identify them. The main objective used in this paper is the number of the users and their behavioral analysis. The fraud or intrusion or account hacking is identified using the changes the user behavior and their frequency of usage of particular website. The future work deals with demonstrating these techniques on another kind of fraudulence.

Sandhu et al. [11] proposed an idea to prevent the fraudulence in finance paying portals. Based on the users and the amount, different algorithms are used to identify the fraudulent.

Vijayan et al. [12] explain about the clustering of the data using the parallel k-means and the adaptive k-means algorithm. In this paper, the number of iteration while clustering the data is effectively reduced.

3 Implementation of the Proposed System

This section concentrates on the detailed explanation of the implementation process. This involves scrapping the text, mining the scrapped text, analyzing the text and recommending the products using the comments and rating. This system uses the NLTK for processing the input data and to extract the required information.

3.1 System Architecture

The system architecture will explain the overall system using the diagrammatical representation (Fig. 1). Here, step-by-step processing can be easily recognized, and the flow of the events is visualized.

3.2 Module Description

This section explains each module in detail; it also includes the techniques and algorithms used to implement the proposed system.

Searching for the Product or Website. Sometimes, the user may have doubt in buying a product or buying in particular website. They will notify the admin about the website or product to detect about the website or product using the product ID. The admin will search for the domain website and identify the product. The fraud websites mostly will have the name of any old inactive websites. The background analysis will make sure whether that website was already active or not.

Fetching Data from Website. Once you get the product from the search result, enter the URL of that website to fetch data from the website. The data can be retrieved from the reviews or feedback which is below the product description of that website. Here, we use web crawling technique in data mining concept to fetch

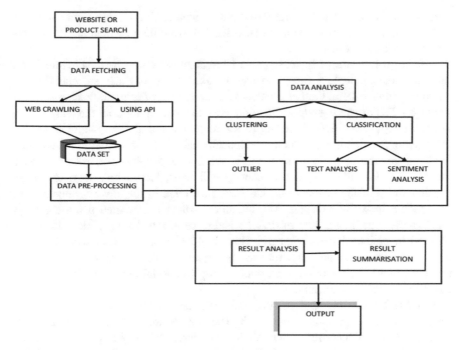

Fig. 1 System architecture

data effectively. Web crawling, which is also called as web spidering, allows the user to crawl and fetch data from websites. A crawler is a simple program that visits websites, reads the information in their pages, thereby creating entries for a search engine index. It uses the HTML tags to read the particular data. The admin will provide the headers of the particular tags, so that the crawler will read that required data. To fetch the data from the social websites like Facebook, Twitter, etc., we can make use of the API and gather the datasets. The fetched data will be stored in an excel sheet or in a CSV file. This is the input dataset used for the analysis. Each the dynamic input dataset is used based on the user request. Once the raw data is fetched, it can be used for the further processing and the analysis.

Data Pre-processing Technique. The data that has been fetched from the website will not be in a proper format or understandable format since real-world data is always inconsistent, incomplete and contain many errors. Data pre-processing is a proven method of resolving such issues. Data pre-processing is a data mining technique that transforms raw data into an understandable format. The null values are removed, and the noisy data is removed based on the neighbor values using the smoothing technology. The data consistency is checked so that the data can be used for further analysis. The consistent data is nothing, but the reviews or comments should not have any special symbols or other language words and the ratings should be in the correct format. Once this is solved, the data duplication will be identified. Since the characters are sensitive and each character had a specific ASCII value, the same word will be treated as the two different words. The smoothing technique is used to remove the unwanted and irrelevant data from the raw data and to convert into the consistent form.

Analyzing Data Using Clustering and Classification. Once the data is transferred into an understandable format, the data is taken for clustering and classification based on some criteria. Outlier concept is used to detect the fraudulent websites or products. The clustering algorithm will give the outlier data as the output. The cleaned data is classified according to the price and rating. The output is used to analyze the remaining data which is technically called as testing data. This will help in identifying the key points which will be useful to notify the users that product has the fake functionalities. The outlier concept will be used to identify the fake or scam functionalities of the product that has been illustrated in the advertisements about the product in social websites. The classified data is represented using the graph which represents the number of reviews and price or rating range. This cluster can be used to predict the result easily. This graphical representation will allows the user to understand the percentage of risk in buying the particular product. Instead of reading some 100 lines about a product, a simple pictorial representation can make the user to easily understand.

Result Prediction. The text mining and sentimental analysis also called opinion mining is used for predicting the result. The text mining will use the transformed data to break sentence into tokens or words. Here, Natural Language Processing like tokenization is used to classify the words as positive, negative and neutral. The tokenization is the predcfined package in the python which has particular inbuilt identification words for classifying the words in the sentences. The similar words are clustered together and identified using the numbering. The sentimental analysis is used to analyze the emotion or behavior of the user. The emotion is expressed in terms of the words. Again, python has an inbuilt package for the sentimental analysis using NLTK. Using this package, the words can be classified and the frequency of the each words and the each category like positive and negative can be identified. Naive Bayes classification is used to make the sentiment analysis more effective and the results more accurate. Once the data is clustered and classified, the result can be predicted by the graph. The result will provide the knowledge of positive, negative and neutral comments. Using this, we can predict if the product is trusted or not.

Recommendation or Opposition of Product. As a result of the review analysis, the most rated negative comments or reviews of the product will appear in front which is easy for the customer to view and take the decision whether to buy the product or not. The analysis results are represented using the graph, where the frequency of the positive and the negative reviews and their rating are clearly represented. If the analysis produces favorable positive results, then it is a trusted website and it recommends you to buy the product. If the result is not what we expected, then it is a scam or fraudulent product and so do not buy the product from that website. This also identifies the scam functionalities of the product and suggests the users to take the decision on it.

4 Experimental Results and Discussion

4.1 Scrapping Data from Website Result

Here, we scrap the reviews and ratings of particular product by specifying the product ID of the product from the website. The output of scrapped data is given in Fig. 2. Once the data is fetched, data cleaning is done to remove null values, or inconsistent values, or non-ASCII values. Then, the cleaned data has been transformed into usable format.

4.2 Text Mining Result

Text mining is done to breakdown each review sentence into tokens or separate words. Natural Language Processing technique is used to tokenize and classify the data. The outlier concept is used to classify the auspicious data. The output of text mining is given in Fig. 3.

4.3 Decision Graph Based on Text Mining Result

After text mining, using sentimental analysis, the positive and negative words are taken as keywords. Based on this analysis, a decision tree graph has been generated as given in Fig. 4.

REVIEWS	RATING
Lovely product.	5.0 out of 5 stars
Very good one	5.0 out of 5 stars
Used it just once so far but seems good n easy to use	5.0 out of 5 stars
Best hair straightener for beginners.	5.0 out of 5 stars
Old packagingit's old product	1.0 out of 5 stars
As of Now its working good	5.0 out of 5 stars
Good	3.0 out of 5 stars
Poor quality	1.0 out of 5 stars
Not a good one	1.0 out of 5 stars
Best product and works well	5.0 out of 5 stars
Good	3.0 out of 5 stars
Nice product worth buying	5.0 out of 5 stars
My experience is worst about this product.I didn't expect this	1.0 out of 5 stars
Good	5.0 out of 5 stars
Good product. No issues till date whatsoever.	5.0 out of 5 stars
Good Deal...	5.0 out of 5 stars

NEXT

Fig. 2 Sample scrapped data

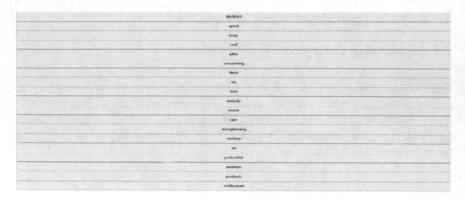

Fig. 3 Text mining data

Fig. 4 Decision graph for text mining

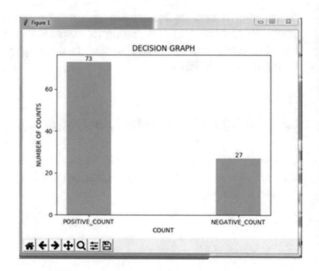

4.4   Rating Graph Based on Given Customer Ratings

The graph in Fig. 5 shows the number of 1 star, 2 stars, 3 stars, 4 stars and 5 stars ratings.

4.5   Decision Graph for Fraudulence Detection

The graph in Fig. 6 shows the percentage of fraudulence as well as non-fraudulence.

Fig. 5 Decision graph for rating

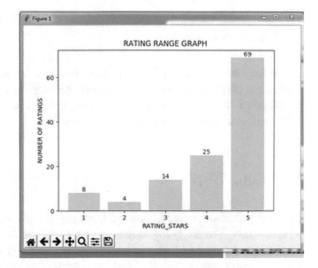

Fig. 6 Decision graph for overall result

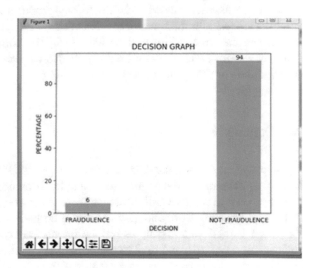

4.6 Fraudulence Detection Percentage

The fraudulence detection percentage will give the range of fraudulence detected in the analysis based on the reviews and ratings. This is used take decision on the product recommendation.

4.7 Final Result—Recommendation of Opposition of Product

If fraudulence percentage is more than non-fraudulence percentage, we oppose buying the product or else we recommend buying the product. The will reveal the overview of the comments or reviews posted by buyers for the product. Based on their feedback, the recommendation is done.

5 Future Work

Recommendation or opposition of buying products has been decided for some websites only. We are going to develop the project in such a way that it fetches data from all the product websites to conclude the result and we are going to create it as a user-friendly application that can be downloaded and installed in the mobile or computer to identify trusted websites to buy products in online. Now, individual products are used for the fraudulence detection. In the future, overall website will be tested, and the trust ability will be identified. The comparison of products under different category and different brands will also be made possible so that the users can find the suitable product and buy it.

6 Conclusion

In the proposed system, the data that has been gathered from the website using data scrapping technique and data pre-processing has been done on the customer reviews of the product to remove null or inconsistent or incorrect values. Then, text mining to break down each word and getting useful information from the customer reviews has been carried out. Then, using sentimental analysis, positive and negative reviews are separated, and final decision has been made whether to buy the product or not. The result is predicted in a graph which gives rating count and positive and negative reviews count which is easier for the user to understand. The fraudulence percentage has also been identified and a graph to indicate the percentage of fraudulence and non-fraudulence has also been generated for customer reference to make it easier for novice users to understand the working of the system in a better way.

References

1. Kharrat, F.B., Elkhleifi, A., Faiz, R.: Recommendation System Based Contextual Analysis of Facebook Comment. IEEE, New York (2016 Dec)
2. Boldt, L.C., et al.: Forecasting Nike's sales using Facebook data. In: 2016 IEEE International Conference on Big Data (Big Data). IEEE, New York (2016 Nov)
3. Ketcham, M., Ganokratanaa, T., Bansin, S.: The forensic algorithm on Facebook using natural language processing. 2016 12th International Conference on Signal-Image Technology & Internet-Based Systems (SITIS). IEEE, New York (2016 Nov)
4. Gotarane, V.R., Bojewar, S.: Relationship classification in online social network. International Conference on Emerging Technological Trends (ICETT). IEEE, New York (2016 Oct)
5. Hanni, A.R., Patil, M.M., Patil, P.M.: Summarization of customer reviews for a product on a website using natural language processing. In: 2016 International Conference on Advances in Computing, Communications and Informatics (ICACCI). IEEE, New York (2016 Sept)
6. Ghatasheh, N.: Fraud prevention framework for electronic business environments: automatic segregation of online phishing attempts. In: Cybersecurity and Cyberforensics Conference (CCC) (2016)
7. Raja, D.R.K., Pushpa, S., Naveen Kumar, B.S.: Multidimensional distributed opinion extraction for sentiment analysis—a novel approach. In: 2016 2nd International Conference on Advances in Electrical, Electronics, Information, Communication and Bio-Informatics (AEEICB). IEEE, New York (2016 Feb)
8. Gowri, R., Kumar, A., Arvind, M.J.: C2F: a clustering based collaborative filtering approach for recommending product to ecommerce user. In: 2015 International Conference on Computation of Power, Energy Information and Communication (ICCPEIC). IEEE, New York (2015 Sept)
9. Drew, J., Moore, T.: Automatic identification of replicated criminal websites using combined clustering. In: Security and Privacy Workshops (SPW), 2014 IEEE (2014 Nov)
10. Kou, Y., Lu, C.-T., Sirwongwattana, S., Huang, Y.-P.: Detecting the fraudulence using various big data concepts. In: 2014 IEEE International Conference on Big Data (Big Data). IEEE, New York (2014 Mar)
11. Sandhu, P.S., Senthilkumar, N.C.: Fraud prevention in paying portal. In: IOP Conference Series: Materials Science and Engineering, 263, 4, 042045 (2017)
12. Vijayan, E., Senthil Kumar, N.C., Agnihotry, S., Subuhani, M.: Effective clustering in big data for efficient knowledge discovery using parallel K-means and enhanced K-means algorithm. Int. J. Pharm. Technol. **8**, 18646–18652 (2016)

Detection and Localization of Multiple Objects Using VGGNet and Single Shot Detection

Kuppani Sathish, Somula Ramasubbareddy and K. Govinda

Abstract While profound convolutional neural systems (CNNs) have demonstrated an extraordinary accomplishment in single-mark picture characterization, take note of that true pictures for the most part contain numerous names, which could relate to various items, scenes, activities, and qualities in a picture. Conventional ways to deal with multi-name picture grouping learn free classifiers for every classification and utilize positioning or thresholding on the characterization results. These systems, albeit functioning admirably, neglect to expressly abuse the mark conditions in a picture. In this paper, we will utilize SmallerVGGNet, the Keras neural system design, which we will actualize and utilize for multi-mark classification. The VGG organize engineering was presented by Simonyan and Zisserman in their 2014 paper, Very Deep Convolutional Networks for Large Scale Image Recognition. This system is described by its straightforwardness, utilizing just 3×3 convolutional layers stacked over one another in expanding profundity. Diminishing volume gauge is managed by max pooling. Two totally related layers, each with 4096 center points are then trailed by a softmaxClassifier.

Keywords CNN · SSD · SGD · VGGNet

K. Sathish
CSE, Tirumala Engineering College, Narsaraopet, Guntur, India

S. Ramasubbareddy (✉)
Information Technology, VNRVJIET, Vignana Jyothi Nagar, Nizampet Rd, Pragathi Nagar, Hyderabad, Telangana 500090, India
e-mail: svramasubbareddy1219@gmail.com

K. Govinda
SCOPE School, VIT University, Vellore, Tamil Nadu, India

© Springer Nature Singapore Pte Ltd. 2020
P. Venkata Krishna and M. S. Obaidat (eds.), *Emerging Research in Data Engineering Systems and Computer Communications*, Advances in Intelligent Systems and Computing 1054, https://doi.org/10.1007/978-981-15-0135-7_40

1 Introduction

Picture question disclosure incorporates recognizing ricocheting boxes, embodying objects, and portraying each hopping boxes to see the concealed challenge grouping. Starting late, there has been mounting eagerness for the investigation system to distinguish distinctive dissents in an image using single shot detection strategies. These techniques effectively solidify region suggestion and request into a single development by the candidate box recommendation (or territory recommendation) module used by a couple of two-advance disclosure frameworks. Not simply this results in altogether speedier inquiry revelation, yet it furthermore upgrades exactness. The most exact and generally utilized algorithm, you only look once (YOLO), utilizes the single shot discovery procedures continuously to create exact and quicker yield outlines. SSDs are expected for inquiry and disclosure persistently. Faster R-CNN uses a zone recommendation framework as far as possible boxes and uses those cartons to gather objects. While it is seen as the start of-the-model inexactness, the whole method continues running at seven diagrams for each second. SSD quickens the technique by discarding the need of the territory suggestion orchestrates. To recover the drop estimation, SSD applies a few over-hauls including multi-scale features and default boxes. These upgrades empower SSD to facilitate the Faster R-CNN's precision using lower objectives pictures, which furthermore pushes the speed higher. According to going with a relationship, it achieves the progressing getting ready speed and even beats the accuracy of the Faster R-CNN. (Precision is evaluated as the mean ordinary precision mAP: the precision of the estimates.) SSD even though has the best results, its class comes with two central disadvantages: (a) SSD contemplates the detachment of each portion when the fat bob boxes are associated. Thusly, the semantic data of dynamic layers cannot be misused for better affirmation of test in the at an opportune time layers. At the region of the smallest request measure along the lines, SSD does not have a respectable execution at the region of the smallest request measure and that contains layers of introduction. (b) SSD building depends upon preloaded maps centers included in the standard IMAGEnet dataset [1, 2] without trying to get the rich guide parts of the significant party of untagged educational accumulations. An image will dynamically experience a couple of convolutional layers right when inputted in a neural framework. Convolutional layers truly gather the information contained by a couple of pixels together and in close proximity. In convolutional neural frameworks, each layer will expand its commitment on a component guide of reducing size. Therefore, the pixels information will be stuffed and compressed progressively by each layer. It is seen as a hypothesis in neural frameworks weight: As the framework consistently packs the information, it can find pixels structures that are similar in different pictures. In case a specific model is accessible in a couple of photographs of a comparable inquiry, it might be a marker that this image addresses vague challenge from well. The models found by the dynamic layers are continuously capricious in light of the fact that the open fields of areas on the part diagram are more prominent for each layer. For example, the important layers may find direct shading or shape plans while more significant layers may find complex ones like a wheel or a head defender. The neural framework will then use these models closeness information to make the request.

2 Background

When contrasted with SSD, some ongoing methodologies [3, 4] first take in a different bouncing box (or area) proposition organize, trailed by taking in a different arrangement over the proposition organize. In any case, such two-organize protest locators experience the ill effects of high memory use and poor induction time. In examination, SSD systems have been appeared to perform better and quicker. Moreover, the vast majority of the protest recognition strategies, including OverFeat, SPPnet [5], Fast R-CNN [4], Faster R-CNN, and YOLO, use just a solitary layer (ordinarily the best most layer) of a convolution system to recognize objects. This methodology does not abuse distinctive capabilities learned by various component maps at various scales [6], and accordingly is extremely constrained in distinguishing objects of various sizes and scales. In comparison, the state-of the-workmanship SSD systems use highlight maps from various layers so as to center around items that show up in specific sizes. Be that as it may, they work on each component outline without consolidating them in a significant way. Thus, these SSD systems do not especially perform well toward distinguishing proof of littler size items [7, 6]. A superior calculation that handles the issue of foreseeing precise jumping boxes while utilizing the convolutional sliding-window method is the YOLO calculation. YOLO stands for you just look once and was created in 2015 by Joseph Redmon, Santosh Divvala, Ross Girshick, and Ali Farhadi. It is prevalent on the grounds that it accomplishes high precision while running pro-gressively. This calculation is called so in light of the fact that it requires just a single forward proliferation to go through the system to make the forecasts. The calculation isolates the picture into lattices and runs the picture arrangement and restriction calculation (talked about under question limitation) on every one of the network cells. Using deconvolution setting and using highlight patterns at various levels in a system is what inspires our methodology. No earlier unsupervised methodologies figures out how to enhance SSD execution in any case, as appeared in Fig. 2 neither [6] nor any of the earlier versions. Our work abuses the distinction in highlights learned by various layers in both convolution and deconvolution squares. We are demonstrating that our methodology results in best in class exe-cution on benchmark datasets [1] by using SSD with unsupervised learning and conjunction of highlight maps from convolution and deconvolution squares. Our approach is partially inspired by [6] in terms of adding deconvolution context and utilizing feature maps at different layers in a network. However, as shown in Fig. 3 neither [3, 5, 6] nor any of the prior approaches explore unsupervised learning to

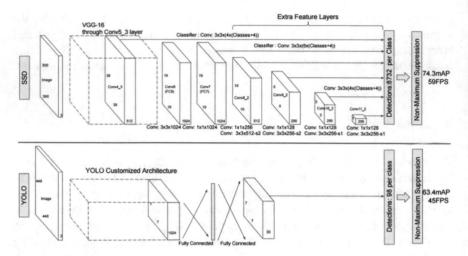

Fig. 1 Architecture of SSD

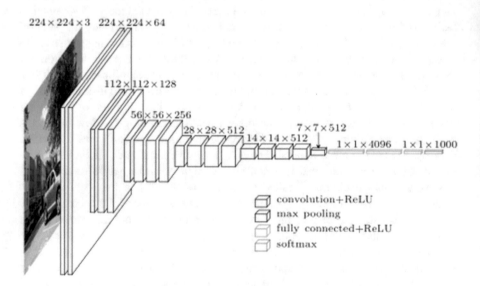

Fig. 2 Architecture of VGGNet

improve SSD performance. Moreover, none of the prior work exploits the difference in features learned by different layers in both convolution and deconvolution blocks. By refreshing SSD with unsupervised learning and confluence of feature maps from convolution and deconvolution blocks, we show that our approach results in state-of-the-art performance on benchmark datasets [1] (Fig. 1).

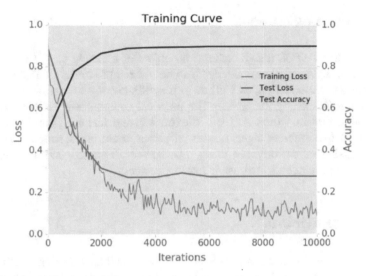

Fig. 3 Training curves between loss and accuracy

3 Proposed Method

3.1 Generalization

Object detection is sculpturesque as a grouping drawback. In spite of the fact that the characterization contemplates the expectation of the name of the object in an image, the recognition goes extra and conjointly finds the situation of those items. In order, it is expected that the object involves a major a piece of the picture. That the photographs, wherever absolutely different or completely different objects with various scales/measurements are detected in various positions, the recognition turns into extra important. In this way, it is a matter of discovering every one of the items objects in an image, anticipating their marks/classes, and task a bounding blob around these objects. In images, we tend to anticipate the odds of each class with the help of object recognition, and we tend to conjointly offer a bouncing box that contains the thing of that classification. Along these lines, the yield of the system should be:

1. Classification probability. (e.g., order)
2. Directions of the bouncing box. We tend to indicate them with one hundred ten (focus x arrange), Cy (focus y organize), h (question stature), and w (protest width)

Class chances ought to conjointly encapsulate an additional name that speaks to the foundation because of a few positions inside the picture do not coordinate any articles.

3.2 Image Classification

Learn how to tell if an image contains an object of a certain class (e.g., a dog, a mountain, or a person). The challenge is to be invariant to irrelevant factors such as viewpoint and illumination, as well as to the differences between objects (no two mountains look exactly the same). The practical covers using various deep convolutional neural networks (CNNs) to extract image features, learning an SVM classifier for five different object classes (airplanes, motorbikes, people, horses, and cars), assessing its performance using precision-recall curves, and training a new classifiers from data collected using Internet images.

3.3 Image Retrieval

Learn to recognize specific objects in images, such as the Notre-Dame cathedral or "Starry Night" by Van Gogh, by quickly matching a query to a large database. The challenge is to be invariant to changes in scale, camera viewpoint, illumination conditions, and partial occlusion. The practical covers matching images using sparse SIFT features, geometric verification, feature quantization and bag-of-visual-words, and evaluating a retrieval systems using mean average precision.

3.4 Object Detection

Learn to detect objects such as pedestrian, cars, traffic signs, in an image. The challenge is to not only recognize but also localize objects in images, as well as to enumerate their occurrences, regardless changes in location, scale, illumination, articulation, and many other factors. The practical covers using HOG features to describe image regions, building a sliding-window SVM object detector, operating at multiple scales, evaluating a detector using average precision, and improving it using hard negative mining.

3.5 General Object Detection Framework

Ordinarily, there are three stages in a question location structure.

1. Initial, a model or calculation is utilized to produce locales of intrigue or district proposition. These area recommendations are a vast arrangement of jumping boxes crossing the full picture (that is, a protest limitation part).

2. In the second step, visual highlights are separated for every one of the jumping boxes, they are assessed and it is resolved whether and which objects are available in the proposition dependent on visual highlights (i.e., a protest order part).
3. In the last post-handling step, covering encloses are joined to a solitary jumping box (that is, non most extreme concealment).

3.6 Single Shot Detection

SSD are expected for dissent ID consistently. Speedier R-CNN uses a region suggestion framework as far as possible boxes and uses those compartments to describe objects. While it is seen as the start of-the-craftsmanship in precision, the whole method continues running at seven traces for each second. Far underneath what a progressing dealing with requirements. SSD quickens the system by wiping out the need of the district suggestion orchestrates. To recover the drop in accuracy, SSD applies a few upgrades including multi-scale features and default boxes. These updates empower SSD to organize the Faster R-CNN's exactness using lower objectives pictures, which furthermore pushes the speed higher. As shown by the going with relationship, it achieves the progressing taking care of speed and even beats the precision of the Faster R-CNN.

3.7 VGGNet

This system is portrayed by its straightforwardness, utilizing just 3×3 convolutional layers stacked over one another in expanding profundity. Lessening volume measure is dealt with by max pooling. Two completely associated layers, each with 4096 hubs are then trailed by a softmaxclassifier. In 2014, 16 and 19 layer systems were viewed as profound (despite the fact that we presently have the ResNet design which can be effectively prepared at profundities of 50–200 for ImageNet and more than 1000 for CIFAR-10).

Simonyan and Zisserman discovered preparing VGG16 and VGG19 testing (particularly in regards to combination on the more profound systems), so as to make preparing less demanding, they originally prepared littler renditions of VGG with less weight layers (sections An and C) first.

The littler systems combined and were then utilized as instatements for the bigger, more profound systems. This procedure is called pre-preparing.

While seeming well and good, pre-preparing is an extremely tedious, dreary assignment, requiring a whole system to be prepared before it can fill in as an introduction for a more profound system.

4   Implementation

4.1   Unsupervised Pretraining

In SGD streamlining, one consistently begins, show weights unpredictably, and endeavors to go toward minimum cost by following the backward of incline of target work. For significant nets, this has not demonstrated a lot of advancement and it is acknowledged to be eventual outcome of to an extraordinary degree non-raised (and high-dimensional) nature of their objective work.

What Y. Bengio and others found was that, as opposed to starting weights carelessly and believing that SGD will take you to minimum reason for such an intense scene, you can pre-train each layer like an autoencoder. Here is the manner in which it works: you build an autoencoder with first layer as encoding layer and the transpose of that as decoder. Besides, you train it unsupervised, that is you train it to reproduce the data (suggest Autoencoders, they are uncommon for unsupervised part extraction assignments). Whenever set you up, settle weights of that layer to those you essentially found. By then, you move to next layers and repeat the comparable until you pre-train all layers of significant net (avaricious approach). Presently, you come back to the primary issue that you expected to light up with significant net (portrayal/backslide) and you streamline it with SGD yet starting from weights you basically got the hang of in the midst of pre-getting ready.

4.2   Combining Feature Maps

The underlying layers of a profound system need solid semantic data and react to just abnormal state highlights of a picture. Moreover, the change in gaining semantic data crosswise over back-to-back element maps is just minor, particularly in beginning layers of a system. In light of these perceptions, our second crucial change to SSD is to intertwine bland and semantic highlights to improve include maps. Not at all like earlier work, we join highlights from various layers of both convolution and deconvolution arrange.

5 Results

We test our methodology in two distinct spaces. The first space is a jumbled kitchen scene. The goal is to distinguish what's more, find all occasions of a specific nourishment item on a kitchen counter. The information of this area begins from unedited grayscale video film taken at 320×240 pixel goals with a handheld camera. The second area has the objective of recognizing phones in a profoundly jumbled work area scene. The information of this area was accumulated utilizing the inner camera of a moving SONY QRIO humanoid robot as it is delineated in Fig. 1b. The recordings recovered from the QRIO robot were at a 176×144 pixel goals and contained plainly obvious antiquities from a high JPEG pressure factor which made this an exceptionally difficult area.

5.1 Training Graphs and Error Graphs

We have to make a model with the best settings (the degree), yet we would prefer not to need to prop up through preparing and testing. There are no outcomes in our precedent from poor test execution, yet in a genuine application where we may play out a basic errand, for example, diagnosing malignancy, there would be not kidding drawbacks to conveying a defective model. We require a type of pre-test to use for model advancement and assess. This pre-test is known as an approval set.

A fundamental methodology is to utilize an approval set notwithstanding the preparation and testing set. This introduces a couple of issues, however, we could simply wind up overfitting to the approval set and we would have less preparing information. A more quick witted execution of the approval idea is k-crease cross-approval.

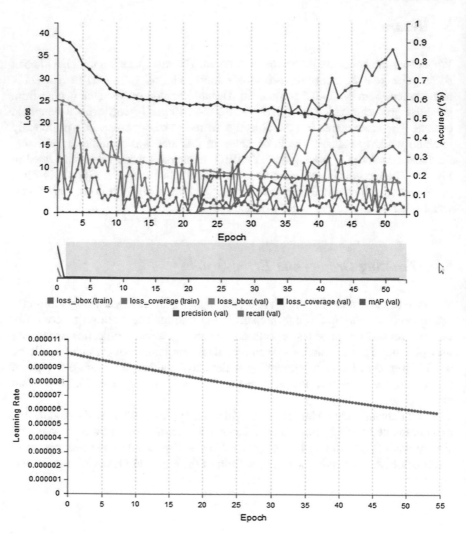

Figure 4 the x likelihood x of x location characterizes what number of right positive outcomes happen among every single positive example. The likelihood of false alert characterizes what number of off-base positive outcomes happens among every single negative example. ENVI characterizes administer pictures at n uniformly dispersed edges (where n is the number of focuses along the ROC bend) from a predetermined least and most extreme esteem. Every characterization is contrasted with the ground truth and turns into a solitary point on the ROC bend.

To achieve our multi-question following errand, we utilized OpenCV's cv2. MultiTracker_create work. This technique enables us to instantiate single protest

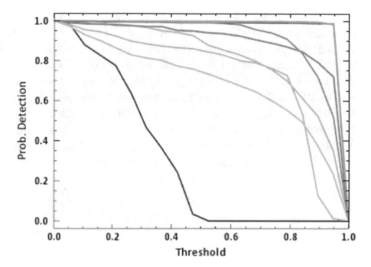

Fig. 4 Bends of the likelihood of identification (Pd, likewise called genuine positive rate) versus the edge. There is one bend for each standard (class) in the standard picture

trackers (simply as we did in a week ago's blog entry) and afterward add them to a class that refreshes the areas of items for us.

Remember, however, the cv2.MultiTracker class is an accommodation work — while it enables us to effectively include and refresh objects areas, it's not really going to enhance (or even keep up) the speed of our protest following pipeline.

To acquire quicker, more effective question following we'll have to use numerous procedures and spread the calculation load over various centers of our processor—I'll be demonstrating to you generally accepted methods to achieve this assignment in a future post in this arrangement.

6 Test Results

The mask region-based convolutional network (Mask R-CNN) utilizes the Faster R-CNN pipeline with three yield branches for every applicant question: a class mark, a bouncing box balance, and the protest cover. It utilizes Region proposal network (RPN) to create bouncing box recommendations and produces the three yields in the meantime for every region of interest (RoI). While getting ready on VOC07 + 12 trainval, we set up the entire framework with learning rate at 10^{-3} for 45 K gatherings, and a short time later with learning rate of 10^{-4} for 60 K bundles to execute unsupervised pretraining on the fundamental get ready dataset 4. In the midst of inquiry distinguishing proof setting we up, again alter the entire framework with learning rate of 2×10^{-3} for 40 K cycles, and 60 K emphasess with learning rate of 10^{-4}. Results over VOC07 test dataset are showed up in Tab. 1. To survey on VOC12 test dataset, as showed up in Tab. 2, we use VOC07 trainvaltest, VOC12 trainval for getting ready. We get ready CDSSD show for 65 K emphasess with 10^{-3} learning rate and 2×10^{-4} learning rate for 80 k cycles for unsupervised pretraining, and 10^{-3} and 10^{-4} learning rate for coordinated planning for 40 and 65 K cycles independently The underlying RoIPool layer utilized in the Faster R-CNN is supplanted by a RoIAlign layer. It evacuates the quantization of the directions of the first RoI and registers the correct estimations of the areas. The RoIAlign layer furnishes scale-equivariance and interpretation equivariance with the district recommendations (Figs. 5, 6).

7 Conclusion

Starting today, there are numerous forms of pre-prepared YOLO models accessible in various profound learning systems, including Tensorflow. The most recent YOLO paper is "YOLO9000: Better, Faster, Stronger." The model is prepared on 9000 classes. There are likewise various regional CNN (R-CNN) calculations dependent on particular territorial proposition, which I have not talked about. Detectron, programming framework created by Facebook AI additionally executes a variation of R-CNN, masked R-CNN.

Fig. 5 Output image with bounding boxes

Fig. 6 Output image with bounding boxes

References

1. Everingham, M., Van Gool, L., Williams, C.K.I., Winn, J., Zisserman, A.: The pascal visual object classes (voc) challenge. IJCV **88**(2), 303–338 (2010)
2. Fu, C.Y., Liu, W., Ranga, A., Tyagi, A., Berg, A.C.: Dssd: Deconvolutional single shot detector (2017). arXiv preprint arXiv:1701.06659
3. Cai, Z., Fan, Q., Feris, R.S., Vasconcelos, N.: A unified multi-scale deep convolutional neural network for fast object detection. In: ECCV (2016)
4. Girshick, R.: Fast r-cnn. In: ICCV (2015) 1. Bell, S., Zitnick, C.L., Bala, K., Girshick, R.: Inside-outside net: Detecting objects in context with skip pooling and recurrent neural networks. CVPR (2016)
5. Dai, J., Li, Y., He, K., Sun, J.: R-FCN: object detection via region-based fully convolutional networks. CoRR abs/1605.06409 (2016)
6. Fu, C., Liu, W., Ranga, A., Tyagi, A., Berg, A.C.: DSSD: Deconvolutional single shot detector. CoRR abs/1701.06659 (2017)
7. Bell, S., Zitnick, C.L., Bala, K., Girshick, R.: Inside-outside net: Detecting objects in context with skip pooling and recurrent neural networks. CVPR (2016)

Smart Security System Using IoT and Mobile Assistance

J. Indumathi, N. Asha and J. Gitanjali

Abstract The digital era has witnessed giant strides in the utilization of security systems, which perform certain pre-programmed tasks when a secured zone is breached. The prevailing traditional security systems have evolved embracing the cutting-edge technologies from time to time. The avalanche-like explosion of things and their connectivity through Internet have opened several new boulevards of exploration. The unification of security system based on the IoT platform has the potential of interrelating real time with the devices and makes them smart, with issues like privacy. To keep the issues in the bay, a new intrusion detection system is the need of the hour. Hence, a wireless intrusion detection smart security system is designed and developed to overcome the research gaps in the existing systems. The smart security system is built on the Internet of things. The system is a lightweight, low cost, extensible, flexible wireless IoT-based-smart security system which will permit the mobile devices and computers to remotely trail the happenings at the location, record the activities and save them in the prefixed cloud storage account. The system uses the hardware (things) to sense and fetch the data which is processed by the embedded software. On the occurrence of an abnormal activity (like breaching by an intruder), a valid signal is detected, and it sends signal to the board directing it to run the alert module. The alert module activates the GSM API to immediately alert the end-user by mobile. This system thus provides a wireless, incessant service to all the stakeholders by phone regarding the breaches occurring in the environment. This implemented smart security system using IoT with mobile assistance also integrates the various alarms, sends alert as pre-programmed, tracks the movements automatically, uses the state-of-the-art appropriate latest technologies to alert the concerned stakeholders and waits for actions to be taken.

J. Indumathi (✉) · N. Asha
Department of Information Science and Technology, Anna University, Chennai, India
e-mail: indumathi@annauniv.edu

N. Asha
e-mail: n.asha@vit.ac.in

J. Gitanjali
School of Information Technology, Vellore Institute of Technology, Vellore, India
e-mail: gitanjalij@vit.ac.in

© Springer Nature Singapore Pte Ltd. 2020
P. Venkata Krishna and M. S. Obaidat (eds.), *Emerging Research in Data Engineering Systems and Computer Communications*, Advances in Intelligent Systems and Computing 1054, https://doi.org/10.1007/978-981-15-0135-7_41

Keywords Smart security systems · Internet of things (IoT) · Global system for mobile communication (GSM) · Short message service (SMS)

1 Introduction

The ever-escalating astute needs have steered the custom of utilizing almost all the cutting-edge technologies in the real world for making human life simpler and comfortable. Among them, the Internet of everything, which is a cutting-edge technology, is finding its way into all the walks of life. In the recent literature, IoT is reflected from two [1] or three [2] chief perspectives (visions) such as "Internet "oriented, "Thing "oriented and "Semantic "oriented perspective.

The invasion of IoT into the real life has also raised criticism from the ethical perspective. Today's lifestyle necessitates both comfort and security from the IoT applications. The digital era has witnessed giant strides in the utilization of security systems. The prevailing traditional security systems have evolved embracing the latest technologies. The unification of security system based on the IoT platform has the potential of interrelating real time with the devices. There is a major risk for the security and safety of data, due to the prevailing conditions in the environment. A physical security alone is no longer adequate for the utilization of these services. It creates a lot of inconsistencies, generating adverse situations to the end-user when their privacy and security is compromised.

Lately, security cameras are being utilized in order to build safe and secure places in organizations. Individuals anticipate that information should be extricated and used to exhibit the plausibility and adequacy of innovation. Hence, these difficulties have to be encountered with the most sophisticated technology that could be an appropriate solution to these havocs. Several mechanisms and the relevant concepts of preserving security and privacy are existing in the literatures [3–18] (Gitanjali 2009a, 2009b; Prakash 2009; Satheesh Kumar 2008).

The rest of the paper is organized as follows.. Section 2 deals with the background, basics, IoT and smart environments. Section 3 deliberates the IoT research challenges and solutions. In Sect. 4, the issues in smart security system are explained. Section 5 pinpoints the research gaps identified. In Sect. 6, a comprehensive survey of the research work is discussed. Section 7 explains the work related to solving the security challenges of enterprise architectures. Section 8 formulates the problem statement and its solution. Section 9 describes the design requirements for smart IOT-based security system. In Sect. 10, the proposed smart security system using IoT with mobile assistance is explained. In Sect. 11, the proposed system architecture along with its implementation is discussed. Section 12 highlights the main conclusions and briefly describes the future work.

2 IoT and Smart Environments

The smart environment ensures that the life of Homo sapiens is more contented and more effective. The IoT-based smart environments permit the actual realization of smart objects. IoT objects are usually made up of embedded software and hardware. The software can be made of operating system, onboard application, etc., and the hardware comprises of electrical and mechanical components with embedded sensors, processors, connectivity antennas, etc. The sensors' objects detect and measure some events or changes in its environment, based on which they provide an output for future processing by performing various actions. These sensors that are embedded in the objects generally enable the value-added services based on IoT. Say, for instance, an IoT network is used to monitor and control the sensors remotely.

In smart environments, IoT permits objects to be detected or controlled remotely over the existing organizational infrastructure. It further coordinates the seamless, integration of the physical world into computer-based frameworks with forward productivity, exactness and financial advantage on expansion with decreased human intervention.

3 IoT Research Challenges and Solutions

IoT issues arise due to the assimilation of various technologies, heterogeneous environment, increased data storage and processing demands. Moreover, the sensors have partial processing abilities, and they forward the data for processing to another infrastructure (data storage) across the network, which is also a cause for the upsurge of issues. IoT constrained devices come with research challenges due to the requirements for device identity management, processing, memory, connectivity and energy capabilities which have led to the failure of several IoT projects [19].

Let us list a few of the challenges put forth to the research community by the IoT, viz. interoperability, mobility, massive scaling, management, energy efficiency, standardization, system architecture, interoperability and integration, availability, reliability, data storage and processing, scalability, management and self-configuration, performance and QoS, identification and unique identity, power and energy consumption, security and privacy and environmental issues. The next challenge is to enhance the sensors' energy efficiency and the design of low power sensors with extended battery life. Security and privacy issues are identified as key challenges in deployment of IoT solutions Babovic et al. [20] because there are numerous examples of threats, vulnerabilities and risks [21]. Several security models and threats taxonomy models for the IoT systems have been proposed [22, 23]. According to Hewlett Packard Enterprise research study [24], most of the device's privacy concerns raised due to: insufficient authentication and authorization, lack of transport encryption, insecure web interface, insecure software and firmware, etc.

One promising solution to the growing concern to the security problem is the use of alert system to the end-user about the abnormal activities. When IoT is expanded with sensors and actuators, the innovation gets to be an occurrence of the more common course of cyber-physical frameworks. Phone notifications are a good way to get alerted about an unusual activity. When we are developing an IoT solution, it is always good to send SMS/CALL to user's phone for certain activity as the smartphones are always within reach. It is not always possible for a user to monitor the data with a mobile app or a website. If they get a notification about a certain activity or a sudden variation in data, they will come to know this by time.

4 Issues in Smart Security System

A smart security system is used to customize, monitor and manage the security systems anywhere, at any time, from a remote gadget, for instance, a mobile phone. It gives the flexibility of controlling and accessing at a convenient time and place. It allows one to monitor, get alerts and notify in case of emergency from anyplace in the world using mobile application via cloud connectivity 24/7.

IoT security includes the protection of information that is stored, collected and transmitted from devices connected to the Internet [25]. In recent years, cyber threats have grown exponentially in both quantity and volume [26]. OWASP has identified the top ten such issues involved with IoT devices [24]

- Insecure web interfaces
- Insufficient authorization/authentication
- Insecure network services
- Lack of transport encryption
- Privacy concerns
- Insecure cloud interface
- Insecure mobile interface
- Insufficient security configurability
- Insecure software
- Poor physical security.

5 Research Gaps Identified

In the prevalent smart security systems with a fixed control plan, the alarm is triggered when a security is breached (i.e. intruder detected). This is useful if the owner is in the premises or is continuously watching the system; otherwise, it does not serve the purpose, and the system becomes ineffective. The existing system is more of a wired alert system using only the hardware sensors they do not provide any assistance to user about the situation status since lack of data provisions, system is developed based on the local alarm system that is present in the system which makes it easy for the intruder to locate the source and dismantle the system.

The existing framework utilizes individuals, things, innovation and security frameworks which turn up unreliable and cannot be trusted upon. Nevertheless, this know-how needs a person frequently to notice any problem in the frame taken from the camera. The CCTV cameras put at certain areas also contribute to the data flow of events at that point. The visuals from these CCTV cameras demand the continuous monitoring of the video/data by humans which at times becomes impossible. The current system cannot be trusted in places without proper lighting. When there is a power outage, the whole framework becomes a mere eyewash, and it doesn't fulfil the purpose of its design.

6 Related Work

Jivani [27] proposed solutions deprived of the security system in place averting from theft or fire or unsolicited situation. Yan and Shi [28] planned and prototyped a home lighting control system using Bluetooth grounded Android smartphone, and it did not have any smart security system in place [29]. The Bluetooth-based wireless home automation system using FPGA developed by Krishna [30] exchanges information wirelessly over short distance with convenience and control. David et al. [31] added more sensors and functionality. Manohar, Kumar [32] and Suryavanshi et al. [33] added email notification method. Anusha et al.'s [34] home automation and security are used to assist the disabled and old aged to manage home appliances and alert them in perilous situations maintained by the Android system methodology. Priyanka et al.'s [35] security system monitors an industry and residential space. The PIR sensing element is interfaced to the controller to announce the existence of intruders and instantaneously captures the image using camera linked to controller and headlongs through E-mail and intimates' others via a buzzer alert. Haque et al.'s [36] system controls the home appliances using Microsoft voice engine components for speech recognition purpose. Amrutha et al.'s [37] system uses the voice recognition method to control home appliances. The others could in the area of business [38].

7 Solving the Security Challenges of Enterprise Architectures

From the current state of the art, two reviews about IoT intrusion detection are taken for discussion: (i) Zarpelão et al. [39] propose an outline IoT-specific IDS and introduce a taxonomy to classify them. A detailed comparison between the different IDS for IoT with parameters like placement strategy, detection method and validation strategy is also discussed. However, Benkhelifa et al. [40] concentrate on progressions in intrusion detection practices in IoT. They review the recent state of the art with a special focus on IoT architecture and future directions for IoT.

IoT devices in the enterprise intercommunicate with the existing controllers, automation and manufacturing information networks and applications. The prevailing, existing security policies and approaches have to be adapted to embrace these new IoT security challenges. To address security concerns in an IoT setup, many solutions have been proposed along several dimensions like secure booting, access control, device authentication, firewalling and IPS, updates and patches.

Hence, appropriate solutions need to be designed and deployed, which are independent from the subjugated platform and able to pledge confidentiality, access control and privacy for users and "things", trustworthiness among devices and users, compliance with defined security and privacy policies.

8 Problem Statement

To design, develop and implement a lightweight, low cost, extensible, flexible wireless IoT-based smart security system with mobile assistance which has the following features, namely,

- It caters to more than one dimension like safety and security.
- It integrates the various alarms like burglar alarm, motion sensor alarms and sends alert as programmed.
- The tracking of movements is automated.
- Use the appropriate latest technologies to alert the concerned stakeholders.

9 Design Requirements for Smart IoT-Based Security System

An ideal smart IoT-based security system is expected to fulfil the following features, namely,

- *Authentication*—system to implement the commands sent by authenticated users with legal passwords,
- *Auto-configuration*—when an object comes under the visibility, it should be configured automatically, so that they will begin functioning as per the context/situation,
- *Centralized architecture*—satisfy objects discovery, objects management, event notifications/processing and real-time analytics,
- *Decentralized architecture*—satisfy objects-to-objects messaging, decentralized auditing and file sharing,

- *Develop open standards*—in order to use and extend the technologies available,
- *Energy efficient protocols*—to increase life time of objects, it is necessary to develop energy efficient protocols,
- *Heterogeneity*—to handle heterogeneous networks, sensor devices of diverse vendors and changing capabilities of devices,
- *Impartiality or transparency principle*—for privacy and data protection,
- *Large area coverage*,
- *Long battery life*,
- *Low cost per node/High node count*,
- *Low data throughput and mostly high latency*,
- *Programmability*—depending on the context/situation, the operations should change dynamically,
- *Robustness*—system should be fault tolerant and must have capable of providing services at any time,
- *Scalability* is the capacity of a system, network or process to tackle the increasing amount of work in the context of protocols and architectures—because large number of objects/things and number of objects that can be linked might vary at any time. For instance, system upgrade/downgrade by adding/eliminating hardware interface module must be informal and organized task,
- *Security*—provide the necessary access control and service access control,
- *Serviceability*—functionality/services of each subsystem/component must be independent of one another.

10 Proposed Smart Security System Using IoT with Mobile Assistance

The proposed *smart security system* (as shown in Fig. 1) is an embedded software system which is supported by the hardware to sense and fetch the data to execute the required functions. The system is a lightweight, low cost, extensible, flexible wireless IoT-based smart security system which will permit the mobile devices and computers to remotely trail the happenings at the location, record the activities and save them in the prefixed cloud storage account.

This system provides a wireless, incessant service to the stakeholders by phone regarding the breaches occurring in the environment. A wireless intrusion detection system is needed to detect the real-time human motion using Internet of things. This system embeds the security so as to interact with the surroundings and motions that are created by human disturbances. The principle of the system is to provide protocols remotely and access anything by the user. Say, for instance, the alerts and the status sent by the Wi-fi connected microcontroller managed system can be acknowledged by the user on his/her phone from any distance regardless of whether his/her mobile phone is connected to the Internet.

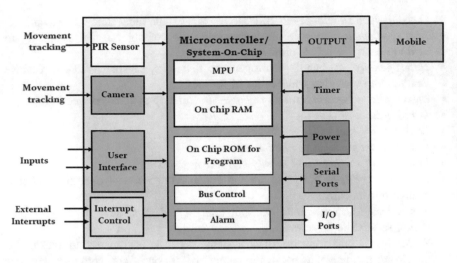

Fig. 1 System architecture—smart security system using IoT with mobile assistance

When we are developing an IoT solution, it is always good to send SMS/CALL to user's phone for certain activity as the smartphones are always within reach. It is not always possible for a user to monitor the data with a mobile app or a website. If they get a notification about a certain activity or a sudden variation in data, they will come to know this by time and can immediately check it with the mobile and can take the further steps.

The proposed system integrates the SMS/CALL module so that when an intrusion is detected or breached, it sends an SMS/CALL alert to the user for taking necessary action. Phone notifications are also a good way to get alerted about an unusual activity. The proposed system can be used in any scenario ranging from the simple to the complex systems, using the ultrasonic range sensors. This provides the data about the disturbances caused by the humans which is sent to the board to run the alert module and activate GSM API. The system runs on Internet of things using Arduino. A physical programmer circuit which could perform the code has been uploaded in it. The proposed system does not require the user to manually trigger an alarm; but still, it provides the user with the advantage of analysing the situation when an alarm is triggered by the security system from a remote place. This idea overcomes the common fault which causes unnecessary embarrassment by triggering security alarm due to the systems' inability to judge a special situation in which it should not have triggered the alarm.

11   System Architecture and Its Implementation

The main aim of this paper is to augment the traditional security system. The system comprises of a camera, voice sensor/microphone, motion/activity sensor and an LTE/Wi-fi module which are linked with the heart of the system, a microcontroller/system on chip.

The system architecture depicts the higher level of design of the software product, smart security system. The system recognizes unauthorized human entry (as shown in Fig. 1) into the room which is to be monitored. When there is an illicit entry, i.e. when a person enters, then there will be a warning sign and also an alarm will be triggered. If there is any system breach, then it should be able to provide alert to the admin by sending him a notification to his/her mobile via push notification. The system works on two sensors such as ultrasonic and a passive infrared sensor (PIR sensor) sensor which works on sensing and ranging.

End-to-end solution providers operating in vertical industries and delivering services using cloud analytics will be the most successful at monetizing a large portion of the value in IoT. While many IoT applications may attract modest revenue, some can attract more. The smart security system using IoT with mobile assistance will definitely be a boon to the operators who have the potential to open up a significant source of new revenue using IoT technologies (Fig. 2).

The proposed system is an embedded security system which will interact with the surroundings and any motions that are created by human are tracked. These motions will provide the data about the disturbances caused by the humans which is sent to the board, to run an alert module and activate the GSM API. The system runs on Internet of things using Arduino. The system utilizes the cloud services which retrieves data from the ultrasonic sensor to detect the range of the intruder via a private channel. The connectivity is used to launch an API that would use a push notification to personnel devices. Based on the charts, we could analyse the activities of the secure system. The proposed IoT implementation extracts the insights from data for analysis, where analysis is driven by cognitive technologies and the accompanying models that facilitate the use of cognitive technologies.

12   Conclusions and Future Work

The proposed smart security system is designed and developed to provide an alert service to the user about the intruder who sneaks in without his/her knowledge. The smart security system provides a continuous service to the user by utilizing the ultrasonic sensors. The proposed smart security system is built on the Internet of things. The system uses the hardware (things) to sense and fetch the data which is processed by the embedded software. On the occurrence of an abnormal activity (like breaching by an intruder), it is detected, and if it is above the threshold, it sends signal to the board directing it to run the alert module. The alert module

450   J. Indumathi et al.

Fig. 2 Flowchart depicting
the process of the proposed
smart security system

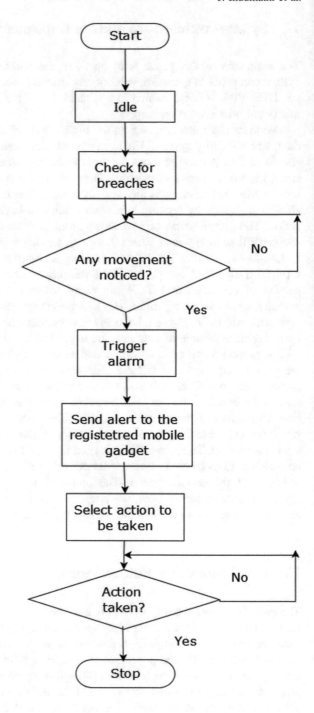

activates the GSM API to immediately alert the end-user by mobile. This system thus provides a wireless, incessant service to all the stakeholders by phone regarding the breaches occurring in the environment. This implemented smart security system using IoT with mobile assistance has the following features, namely, it accommodates more than one dimension like safety and security, and it integrates the various alarms like burglar alarm, motion sensor alarms and sends alert as pre-programmed, tracks the movements automatically and uses the state-of-the-art appropriate latest technologies to alert the concerned stakeholders.

A variety of enhancements can be made to this system to achieve greater accuracy in sensing and detection. For instance, change the way of automated notifications using the other technologies can make this system still more professional.

References

1. Gubbi, J., Buyya, R., Marusic, S., Palaniswami, M.: Internet of things (IoT): a vision, architectural elements, and future directions. Future Gener. Comput. Syst. 1645–1660 (2013)
2. Atzori, L., Iera, A., Morabito, G.: The internet of things: a survey. Comput. Netw. **54**(15), 2787–2805 (2010)
3. Gitanjali, J., Banu, S.N., Indumathi, J., Uma, G.V.: A panglossian solitary-skim sanitization for privacy preserving data archaeology. Int. J. Electr. Power Eng. **2**(3), 154–165 (2008)
4. Gitanjali, J., Banu, S.N., Geetha, M.A., Indumathi, J., Uma, G.V.: An agent based burgeoning framework for privacy preserving information harvesting systems. Int. J. Comput. Sci. Netw. Secur. **7**(11), 268–276 (2007)
5. Indumathi, J.: A generic scaffold housing the innovative modus operandi for selection of the superlative anonymisation technique for optimized privacy preserving data mining (Chap. 6). In: Karahoca, A. (ed.) Data Mining Applications in Engineering and Medicine, pp. 133–156, 335p. InTech (2012). ISBN: 9535107200 9789535107200
6. Indumathi, J.: Amelioration of anonymity modus operandi for privacy preserving data publishing (Chap. 7). In: Amine, A., Mohamed, O.A., Benatallah, B. (eds.) Network Security Technologies: Design and Applications, pp. 96–107, 330p. Tahar Moulay University/ Concordia University/University of New South Wales, Algeria/USA/Australia. Release Date: Nov 2013. Copyright © 2014 (2013a)
7. Indumathi, J.: An enhanced secure agent-oriented burgeoning integrated home tele health care framework for the silver generation. Int. J. Adv. Netw. Appl. **04**(4), 16–21 (2013b). Special Issue on "Computational Intelligence—A Research Perspective" held on "21st–22nd February, 2013
8. Indumathi J.: State-of-the-Art in reconstruction-based modus operandi for privacy preserving data dredging. Int. J. Adv. Netw. Appl. **4**(4), 9–15 (2013c). Special Issue on "Computational Intelligence—A Research Perspective" held on "21st–22nd February, 2013"
9. Indumathi, J., Uma, G.V.: Customized privacy preservation using unknowns to stymie unearthing of association rules. J. Comput. Sci. **3**(12), 874–881 (2007)
10. Indumathi, J., Uma, G.V.: Using privacy preserving techniques to accomplish a secure accord. Int. J. Comput. Sci. Netw. Secur. **7**(8), 258–266 (2007)
11. Indumathi, J., Uma, G.V.: A bespoke secure framework for an ontology-based data-extraction system. J. Softw. Eng. **2**(2), 1–13 (2008a)

12. Indumathi, J., Uma, G.V.: A new flustering approach for privacy preserving data fishing in tele-health care systems. Int. J. Healthc. Technol. Manag. **9**(5–6), 495–516(22) (2008b) Special Issue on: "Tele-Healthcare System Implementation, Challenges and Issues."

13. Indumathi, J., Uma, G.V.: A novel framework for optimized privacy preserving data mining using the innovative desultory technique. Int. J. Comput. Appl. Technol. **35**(2/3/4), 194–203 (2008c). Special Issue on: "Computer Applications in Knowledge-Based Systems" (2008, in press)

14. Indumathi, J., Uma, G.V.: An aggrandized framework for genetic privacy preserving pattern analysis using cryptography and contravening—conscious knowledge management systems. Int. J. Mol. Med. Adv. Sci. **4**(1), 33–40 (2008d)

15. Murugesan, K., Gitanjali, J., Indumathi, J., Manjula, D.: Sprouting modus operandi for selection of the best PPDM technique for health care domain. Int. J. Conf. Recent. Trends Comput. Sci. **1**(1), 627–629 (2009)

16. Murugesan, K., Indumathi, J., Manjula, D.: An optimized intellectual agent based secure decision system for health care. Int. J. Eng. Sci. Technol. **2**(5), 3662–3675 (2010)

17. Murugesan, K., Indumathi, J., Manjula, D.: A framework for an ontology-based data-gleaning and agent based intelligent decision support PPDM system employing generalization technique for health care. Int. J. Comput. Sci. Eng. **2**(5), 1588–1596 (2010)

18. Vasudevan, V., Sivaraman, N., SenthilKumar, S., Muthuraj, R., Indumathi, J., Uma, G.V.: A comparative study of SPKI/SDSI and K-SPKI/SDSI systems. Inf. Technol. J. **6**(8), 1208–1216 (2007)

19. Lin, N., Shi, W.: The research on internet of things application architecture based on web. In: Proceedings of Advanced Research and Technology in Industry Applications (WARTIA), pp. 184–187. IEEE Workshop (2014)

20. Babovic, Z.B., Protic, J., Milutinovic, V.: Web performance evaluation for internet of things applications. IEEE Access **4**, 6974–6992 (2016)

21. Qi, J., Vasilakos, A.V., Wan, J., Lu, J., Qiu, D.: Security of the internet of things: perspectives and challenges. Wirel. Netw. **20**(8), 2481–2501 (2014)

22. Babar, S., Mahalle, P., Stango, A., Prasad, N., Prasad, R.: Proposed security model and threat taxonomy for the Internet of things (IoT). In: Proceedings of Communications in Computer and Information Science, vol. **89**, pp. 420–429 (2010)

23. Kalra, S., Sood, S.K.: Secure authentication scheme for IoT and cloud servers. Pervasive Mob. Comput. **24**, 210–223 (2015)

24. OWASP: Internet of things top 10 projects. www.owasp.org/index.php/OWASP_Internet_of_Things_Top_Ten_Project (2015). Accessed Aug 2015

25. Weber, R.H.: Internet of things: privacy issues revisited. Comput. Law Secur. Rev. **31**(5), 618–627 (2015)

26. Hodgson, K.: The internet of [security] things. SDM magazine (2015). Available: http://www.sdmmag.com/articles/91564-the-internet-of-security-things

27. Jivani, M.N.: GSM based home automation system using appinventor for android mobile phone. Int. J. Adv. Res. Electr. Electron. Instrum. Eng. **3**(9) (2014)

28. Yan, M., Shi, H.: Smart living using bluetooth-based android smartphone. Int. J. Wirel. Mob. Netw. **5**(1), 65 (2013)

29. Govindraj, V., Sathiyanarayanan, M., Abubakar, B.: Customary homes to smart homes using internet of things (IoT) and mobile application. In: 2017 International Conference on Smart Technologies for Smart Nation (SmartTechCon), pp. 1059–1063. IEEE (2017)

30. Krishna, B.M., Nayak, V.N., Reddy, K., Rakesh, B., Kumar, P., Sandhya, N.: Bluetooth based wireless home automation system using FPGA. J. Theor. Appl. Inf. Technol. **77**(3), (2015)

31. David, N., Chima, A., Ugochukwu, A., Obinna, E.: Design of a home automation system using arduino. Int. J. Sci. Eng. Res. **6**(6), 795–801 (2015)

32. Manohar, S., Kumar, D.M.: E-mail interactive home automation system. Int. J. Comput. Sci. Mob. Comput. **4**(7), 78–87 (2015)

33. Suryavanshi, R., Khivensara, K., Hussain, G., Bansal, N., Kumar, V.: Home automation system using android and wifi. Int. J. Eng. Comput. Sci. **3**(10) (2014)
34. Anusha, S., Madhavi, M., Hemalatha, R.: Home automation using atmega328 microcontroller and android application. Int. Res. J. Eng. Technol. **2**, (2015)
35. Priyanka, P., Reddy, D.K.S.: Pir based security home automation system with exclusive video transmission. Int. J. Sci. Eng. Technol. Res. 2319–8885 (2015). ISSN
36. Haque, S., Kamruzzaman, S., Islam M. et al.: A system for smart home control of appliances based on timer and speech interaction. arXiv preprint arXiv:1009.4992 (2010)
37. Amrutha, S., Aravind, S., Ansu Mathew, S.S., Rajasree, R., Priyalakshmi, S.: Speech recognition based wireless automation of home loads-e home. Int. J. Eng. Sci. Innov. Technol. (IJESIT). **4**(1) (2015)
38. Saifuzzaman, M., Khan, A.H., Moon, N.N., Nur, F.N.: Smart security for an organization based on iot. Int. J. Comput. Appl. **165**(10), 33–38 (2017)
39. Zarpelão, B.B., Miani, R.S., Kawakani, C.T., de Alvarenga, S.C.: A survey of intrusion detection in Internet of Things. J. Netw. Comput. Appl. **84**, 25–37 (2017)
40. Benkhelifa, E., Welsh, T., Hamouda, W.: A critical review of practices and challenges in intrusion detection systems for IoT: towards universal and resilient systems. IEEE Commun. Surv. Tutor. **20**(4), 3496–3509 (2018)

Highway Toll Management and Traffic Prediction Using Data Mining

Pruthvi Kumar, Kotapalli Pranathi and J. Kamalakannan

Abstract Generally, we will confront many issues for intersection tollgates. Parcel of vehicles will be in a line to pay the toll for intersection the parkways. This is squandering a great deal of time and furthermore expanding in activity prompts burden amid well-being and different crises. And furthermore, there will be various approaches to achieve a goal yet picking a sit route with low movement one is another issue. To defeat this, an online toll administration framework and movement forecast investigation are created. This will have the capacity to pay the toll utilizing on the web installment gateway, clients can pay with approved certifications, and in view of passage of the vehicles into toll, movement status of specific thruway can be acquired.

Keywords H_DATABASE · AETC · RFID

1 Introduction

There are many expense gathering frameworks existing at tollgates, which are following ordinary philosophies like prepaid visa frameworks and RFID chips which are not easy to use and furthermore not efficient to the clients.

At first, both leave entryways are shut, and flag winds up red. At the point when vehicle comes surprisingly close to toll accumulation focus handset, it begins correspondence with ZigBee handset's vehicle. At the point when, vehicle is recognized methods for AETC began. AETC will check whether the vehicle is enrolled or unregistered in AETC or even the vehicle is kept and cautions to expert and toll

P. Kumar (✉) · K. Pranathi · J. Kamalakannan
VIT-Vellore, Vellore, Tamil Nadu, India
e-mail: pruthvikumar6055@gmail.com

K. Pranathi
e-mail: pranathikotapalli@gmail.com

J. Kamalakannan
e-mail: jkamalakannan@vit.ac.in

© Springer Nature Singapore Pte Ltd. 2020
P. Venkata Krishna and M. S. Obaidat (eds.), *Emerging Research in Data Engineering Systems and Computer Communications*, Advances in Intelligent Systems and Computing 1054, https://doi.org/10.1007/978-981-15-0135-7_42

gathered physically with punishment. Punishment is constrained with the goal that it will enlist vehicle in AETC and prepay sum routinely. On the off chance that the vehicle is enlisted in AETC explicitly stated the subtle elements of an auto, i.e., time and date of entry of vehicle. On the off chance that adequate sum is accessible, deduct the full aggregate sum pertinent for that particular vehicle from his record. The record is of prepaid kind. On the off chance that deficient sum accessible at that point confines an auto and gets toll in real money with relevant punishment to ensure that client will care for lacking asset in future and after that enable vehicle to encourage entryway. Send SMS on his versatile and furthermore alarm concerning the installment/derivation/adjust sum/prepayment caution if necessary. Turn flag green, open the comparing door, and enable vehicle to pass. Check whether the vehicle is passed or not or even hold up till vehicle experiences door. Close the entryway and turn flag red. As needs be the comparing path, leave door will stay open till vehicle passes, and leave entryway will stay shut till next vehicle is allowed to pass. Again seeking of an auto and identification process began.

ZigBee Trans-receiver cannot be installed to all the vehicles, one particular user is having many vehicles, and it is not economical. Money prepaid process is not entertained by all customers. This neglects users who are rarely using highways.

This framework can be viably executed on a parkway or road, where vehicle with a RFID tag will be permitted to go by deducting a sum from the label adjust. For the vehicles that do not have the tag, their recognizable proof will be sent alongside the depiction of the vehicle to the control focus distinguishing an illicit section, and in this way, move can be made. At that point, it should be possible that the specific vehicle not having the tag will be charged at their habitation or by means of mail. The previously mentioned misfortunes can put tremendous weight on Government and the nationals. Diminishing these misfortunes is the plentiful explanation behind which the requirement for ATCS is there. The loss of time puts in a great deal of dissatisfaction in everybody waiting for their swing to pay the duty. The majority of us need a fast transport with no obstacle. RFID chips cannot be introduced to every one of the frameworks since they are not efficient. Prepaid cash exchange will not be empowered by all clients.

The RFID tag is utilized to get to vehicle data with one of a kind id mounted on every vehicle contains a few data. The tag is perused by RFID peruser put at toll entryway. At the point when the vehicle goes to toll entryway, IR sensor will recognize which sort of vehicle goes to the toll door, and after that, according to the kind of vehicle, the particular sum will be deducted from client record, and exchange message will send on the clients enroll portable number with the assistance of GSM. At that point, toll entryway opens clockwise way with the assistance of engine drive. After the departure of the vehicle, within measure of the second toll, door turns anticlockwise way, and entryway will be closed. In the wake of intersection vehicle, the counter will be incremented by one, and result will be shown on LCD. RFID chips cannot be introduced to every one of the frameworks since they are not sparing. Assume if the vehicle is a recently gotten one, they cannot introduce all the equipment in a split second. Deducting cash straightforwardly from credits cards without our notice or OTP is not conceivable.

Here, when a vehicle comes into the toll court, the IR sensor gives the data to the microcontroller which is in the gadget, and the RFID peruser peruses the data from the tag in the vehicle. Microcontroller forms it, and the exchange happens at the toll court. The points of interest will be shown on the LCD, and also, it is sent to the client's portable through the GSM. At the point when individual enters the toll square with an invalid card, with a card having no adjust or with no card, the microcontroller underwear the engine to close the door at the court. By along these lines, we can control the wrong going in the toll court. RFID chips cannot be introduced to every one of the frameworks since they are not practical. Prepaid cash exchange will not be supported by all clients.

To conquer the overwhelming line process for intersection tollgates, in this paper, an online installment framework which will have the capacity to make installments for toll doors for voyaging throughways are created. And furthermore, an activity expectation examination will be made to know the movement status in the specific expressway. In this framework, client will enter to the parkway toll, web-based interface will pick the thruway he needs to travel and subsequent to entering every one of the particulars and subtle elements he will go for installment, and installment will be finished by utilizing the bank servers. And furthermore, in a similar gateway, there will be an activity expectation status for all the parkways utilizing this client can pick which interstate he needs to travel. Furthermore, after installment is done online, receipt will be produced. And, keeping in mind that client is making a trip to toll entryway utilizing the vehicle number his payment status will be distinguished, and he will be permitted to go through the toll door.

Starting at now, all the toll administration specialists are utilizing the customary model for toll installments where every one of the travelers alongside vehicles are utilized to take after a line, hold up until the point when their turn is sought installments, this is taking a ton of time and furthermore a ton of man work is likewise required, and in a few places, some prepaid cards are utilized to install-ments yet which are utilized futile for traveler who is not utilizing it consistently. Subsequently, there are gigantic movement issues and furthermore time squander in this conventional model.

2 Proposed Methodology

The proposed framework will be such a path, to the point that client will have the capacity to do installments by means of online gateway, client can enter to the entry, and he can scan for the interstate he needs to enter and he can see the activity status of the parkway and subsequent to giving every one of the subtle elements he can have the capacity to cross the toll door. This will be useful to diminish the activity and furthermore time squander at the toll doors, while the client is entering to the toll entryway utilizing his vehicle number, his installment status will be checked, and he will be permitted to go through the toll entryway math formulae.

Use equation editor to type all the formulae.

3 Architecture

All the installment points of interest will be put away in H database from which activity can be anticipated. At the point when a vehicle is entering the toll entryway, its enrollment number will be examined with help of a camera, and utilizing content extraction calculation, the installment subtle elements will be cross-checked, and vehicle is permitted into the roadway (Fig. 1).

Flow diagram (see Fig. 2).

4 Working Model

Client needs to enroll to the entry with the end goal that his record will be made, he can include his vehicles subtle elements, in the event that he has more than one vehicle he can include them additionally, points of interest incorporate, driving permit id, aadhar id, vehicle enlistment number, and enlistment year. At the point when client needs to go through an expressway, he can log in the gateway, he can scan for the roadway, he can foresee the movement status, and he can choose the vehicle from spared rundown or he can include another vehicle, at that point installment page will be diverted at that point, installment will be done lastly an e receipt will be created.

While the client is entering the toll entryway utilizing webcams, vehicle enlistment number will be removed, it will be checked with the installment

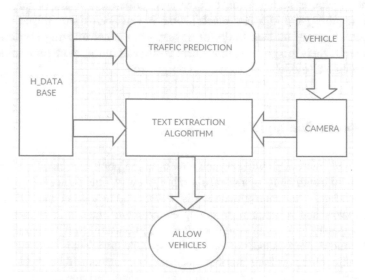

Fig. 1 Block diagram

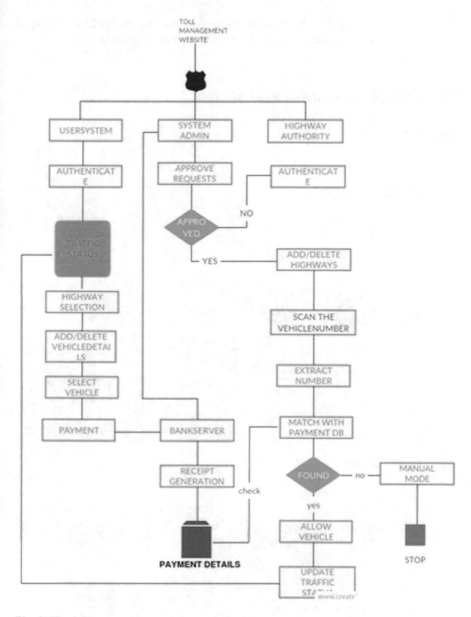

Fig. 2 Flow chart

database, if the number is coordinated, then the vehicle will be permitted to pass the toll door else the vehicle will be permitted to do a manual installment.

In the meantime the vehicle is entering to the parkway, the tally of kind of vehicle will be expanded utilizing the vehicle tally, and once the vehicle is out from the interstate, the check will be diminished. From this, we can anticipate the status of movement on the expressway utilizing information mining systems.

4.1 Registration and Payment Module

See Fig. 3.

4.2 Checking module

See Fig. 4.

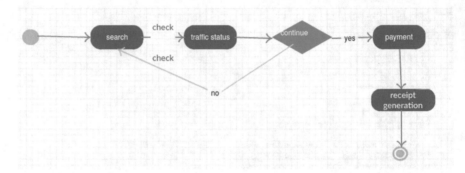

Fig. 3 Registration flow

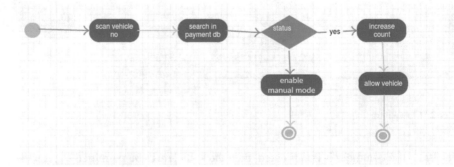

Fig. 4 Checking module

5 Implementation

Client needs to join to the utilizing legitimate accreditations required. The framework ought to contain three sections clients, oversee, expressway experts, and parkway specialists, can be gotten to the framework with the direct authorizations. In the event that another expressway specialist needs to enter the framework, they ought to get consents from the legislature and the regulate, and they can make another interstate toll by entering every one of the insights about the thruway, for example, toll rates, length of the roadway, and course delineate. Administer will screen the points of interest.

Client can enroll to the system with his legitimate certifications, for example, mail and telephone number, once it is done, he can utilize the framework. Client can look through the roadway utilizing his form and goal address, he can choose the interstate which they need to go by foreseeing activity which is given for each parkway once he chooses the thruway he should choose the vehicle from the spared information or he can include another vehicle by giving every one of the points of interest required, at that point the framework is diverted to installment mode, and installment should be possible there, a receipt will be produced and include will be sent to telephone and mail id.

While the vehicle is entering to the framework utilizing a webcam, each vehicle enrollment number will be distinguished utilizing picture handling procedures, the number will be checked with the installment database, and if the match is discovered, then the vehicle will be permitted to go through the framework else the vehicles will be coordinated to do manual installment.

Once the vehicle is entered to the interstate, in light of the sort of vehicle, tally will be expanded, and movement examination will be made.

6 Algorithms

6.1 Traffic Prediction Algorithm

Based on the entries of the vehicles into highway, we can predict the traffic using data mining techniques, once the vehicle is entered to the highway, it will be counted according to its type, and usually in national highways, there will be different lanes used by different types of vehicles so it will be easy to predict traffic if we keep trace of types of vehicles and their count. Traffic status can be obtained by considering length of the highway to the minimum speed limit of the vehicles and type of vehicle. Suppose consider types of vehicles are two-wheelers, cars and trucks, and heavy vehicles. Minimum speed limit for two-wheelers be 40, cars and trucks be 80, and heavy vehicles be 30. Consider the length of the highway be 100 km, a car will be a length of 10 ft on an average and needs 10 foot space to create a minimum speed of 80 km/h such that a highway with 100 km length cannot accommodate more than 1000 cars at a particular instant of time, and if the entries are reaching more that the limit, then the traffic situation gets into worse.

- Prediction (p) = number of vehicles by type (n)/length of the highway (l).
- Let average speed that has to be maintained be sp and traffic status be ts.
- If $p < 1/2(\text{sp})$, then ts = low.
- If $p > \text{sp}$, then ts = high.
- If $p > 1/2(\text{sp})$ and $p < \text{sp}$, then ts = moderate.

Using this algorithm, we can predict the traffic conditions and can be displayed to the users.

6.2 Image to Text Extraction Algorithm

Text can be extracted from the image using Java code, and it will take image as an input and provide the text in that image as an output.

- At first, set SWT = ∞.
- Discover edge by watchful edge finder.
- Take after the beam $r = p + n \cdot dp$, $n > 0$ until the point that another edge is found.
- On the off chance that $dq = -dp \pm \Pi/6$, then SWT = $|p - q|$ and dp++ else dispose of the beam.
- In the event that SWT proportion ≤ 3 at that point gathers neighboring pixels.
- On the off chance that two letters are having comparable stroke width, they can be gathered.
- The yield is an arrangement of rectangles assigning bouncing boxes for distinguished words.
- Hunt the content on web or in database.
- Match the word and recover the related data.
- Show recovered data on screen.

7 Results

A graph is plotted between the number of vehicles at the toll gate and time taken for payment, and it shows the differences between the vehicles entering and time taken to cross the toll gates in existing system and proposed system.

On the x-axis—number of vehicles and y-axis—n time taken to cross toll gate.

7.1 Graphs

Time Taken in Conventional System

See Fig. 5.

Time Taken Using the New System

See Fig. 6.

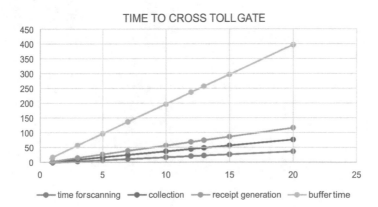

Fig. 5 Result graph 1

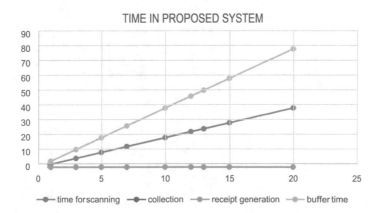

Fig. 6 Result graph 2

8 Conclusion

In this paper, a highway toll payment system which helps users to know the traffic status before entering the highway is implemented. It also helps to stop all the traditional methods we are having till the date to cross the highways by implementing a user-friendly system which helps user to pay the toll before he enters the toll gate and helps users in decreasing the waiting time at toll gates. Users can also have a facility to check the traffic status in a particular highway and can select according to it. This traffic status can be obtained by using mining algorithm.

Machine Learning Usage in Facebook, Twitter and Google Along with the Other Tools

K. Umapavankumar, S. V. N. Srinivasu, S. Siva Nageswara Rao
and S. N. Thirumala Rao

Abstract The trend in the current industry and academics is machine learning (ML), and it is not hype but the reality. The market requirements are not limited by the earlier computing and storage models if those can be integrated with ML algorithms. The requirements of current industry, research and other domains like banking, finance, retail and medical have to depend on the collection of huge amounts of the data and analyse the data to better serve the stakeholders. The organizations have to focus on better storage models and advance logics of ML to meet the current needs of the people and usage of the limited amount of time to handle the huge amounts of the data. The ML research is moving in a high potential with the involvement of the social media like Facebook, Twitter and Google. These are not only using the ML perspective in their applications but also contributing to the development of new algorithms and API in the context of big data, ML and deep learning. The paper objective is to walk through the ML, big data along with deep learning fundamentals with the specification of various algorithms in the reference of above-mentioned social media with their contributions in the development of ML landscape along with the API (TensorFlow), services like priority inbox and the algorithms like RankBrain. The discussion will help the researchers and academicians so as to get the overview and detailed significance of ML research and the importance of ML in various dimensions. All these algorithms, tools and services are indirectly dependent on the artificial intelligence (AI) to frame the rules and to fine-tune the rules as per the requirements.

K. Umapavankumar (✉) · S. V. N. Srinivasu (✉) · S. Siva Nageswara Rao (✉) ·
S. N. Thirumala Rao (✉)
Department of CSE, Narasaraopeta Engineering College, Narasaraopet,
Andra Pradesh, India
e-mail: dr.kethavarapu@gmail.com

S. V. N. Srinivasu
e-mail: drsvnsrinivasu@gmail.com

S. Siva Nageswara Rao
e-mail: profssnr@gmail.com

S. N. Thirumala Rao
e-mail: nagatirumalarao@gmail.com

© Springer Nature Singapore Pte Ltd. 2020
P. Venkata Krishna and M. S. Obaidat (eds.), *Emerging Research in Data
Engineering Systems and Computer Communications*, Advances in Intelligent
Systems and Computing 1054, https://doi.org/10.1007/978-981-15-0135-7_43

Keywords Artificial intelligence · Big data · Deep learning · Machine learning · RankBrain · TensorFlow

1 Introduction

The term AI is an umbrella kind of activities like ML and DL, and the usage of these algorithms and services on top of big data gives rise to analytical aspects such as better recommendations in case of online shopping, usage of good financial plans based on the age and income of the customers of an insurance company. The term big data is used as a reference of huge amounts of the data and provides a solution as Hadoop to store this data and apply some kind of the algorithms in the form of MapReduce, Hive or else Pig Latin to process the data in parallel and distributed manner which in turn improves the productivity by saving time and leverage the capacity of the data processing.

The AI perspective is to add common sense reasoning to the machine, and any experiment to adopt smartness to the machine comes to the application of AI. The best test is Turing test which is a test made with some questionnaire given to both computer and a normal person testing the reasoning and analytical thinking between. The form of data processing in AI is to frame the rules related to a context of huge amounts of the data and to expect the desired the answer. The most importance point in this AI cstablishment is the rules could not learn or adapt to new data. The addition of new rules is possible to handle new data, but there is no learning in the AI perspective.

Machine Learning is a new way of AI implementation with the capacity of learning by the machine without being programmed externally. The task running usually gets some experience, and every experience improves the performance of the result given by the task running. The algorithms learn from the data rather than the task-driven mode of the operation unlike in case of other algorithms. ML is a combination of various classifiers along with set of algorithms that can learn them to evolve and adapt when exposed to new data as input. The algorithms such as classification, clustering and recommender systems are very much helpful to come up with meaningful insights of the data.

Deep learning is a particular branch or technique of ML uses artificial neural networks (ANN) with many layers to learn the optimal usage of parameters in the construction of accurate models. In the task execution, with the usage of DL, the automatic detection of various features to efficiently achieve better results in the task execution. The applications like image captioning, video captioning, object detection and object tracking are some of the major areas in DL. The concepts like convolutional networks (CNN), recurrent neural networks (RNN) and attention networks are most commonly used kind of the models used in the DL usage.

The architecture of the paper is as follows. In Sect. 2, the ML architecture along with some set of the algorithms with the available packages in Python and R were discussed. Some of the use cases in Facebook, Twitter and Google were also mentioned. In Sect. 3, the deep learning importance along with some most important algorithms was mentioned. The DL packages and algorithms used in R and Python with sample code were described. In Sect. 4, the description about the ML use cases in the various scenarios along with future research directions in the fields of ML and DL. The section also covers some of the AI, DL and ML platforms which are popular and allowing the researcher to develop the applications and solutions for various current issues in the industry. In Sect. 4, the conclusion of the entire discussion was mentioned.

2 Machine Learning Usage in Social Media (FB, Twitter and Google) with Other Algorithms

The ML usage in real time is limit less and obvious, especially the usage of ML in social media, and search engine is quite interesting. The companies like Facebook and Google are contributing much in the development of ML research to serve the users and improve the business. The current section deals with the usage of ML in Facebook and how Facebook is contributing and developing various algorithms along with certain applications. The section also covers the usage of ML in Twitter and Google and the contribution of Google in the design of advanced ML algorithms.

Figure 1 shows the AI ecosystem usage in Facebook involves the embedding of core ML software such as Caffe2, PyTorch and ONNX. The second layer contains workflow management and deployment FB learner, and third layer is the combination of server management, storage handling and network strategy.

Caffe 2 is a platform of DL that allows users to develop and deploy various models and algorithms in compact manner even the developer can leverage the development up to (GPU) graphics processing unit is stream or vector processor.

TensorFlow supports the numerical computations in the large-scale ML platform, and both DL and ML models and algorithms supported by TensorFlow. It follows the convention of define-by-run framework where the conditions can be

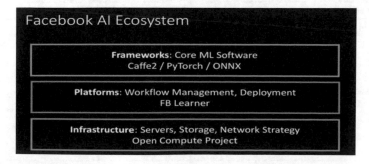

Fig. 1 AI in Facebook

defined and iterated in the graph structure. The best application of TensorFlow is automatic image captioning.

PyTorch supports natural way of coding and follows the convention of define-by-run framework, and the graph structure is defined on the fly during computation. The most common use of PyTorch is found in varying loads of the dynamic graphs computing.

Facebook is the most frequently used social media nowadays, lot of data getting populated daily by the users of FB, and FB has to store process and analyse the data along with friend suggestions, trending and Facebook recommendations of videos and events based on the user profiles and interest areas. Table 1 illustrates the kind of activity along with the ML/DL algorithm used by FB.

Twitter is the second highest used social media after Facebook. Even Twitter uses the DL algorithms in the efficient summarization of the news tagging and news feed to the users, and without much data and by considering the short span of the time, the entire happenings which were taken place in the absence of user and getting the most interesting news by prediction based on the user profile can be visualized to the users. In Twitter news feeds, one interesting feature is to summarize the news if any user is logging into Twitter after some time gap between (Fig. 2).

The Open Neural Network Exchange is a collaborative project between Microsoft and Facebook is a format for open container which helps to exchange the NN models between different frameworks, supports framework interoperability and hardware optimizations with the power of AI exchange formats and can select the best-integrated models. The other tools like Caffe2, PyTorch, Cognitive Toolkit (CNTK), TensorFlow, Apache CoreML and SciKit-Learn are also supported. The tools supported by ONNX involve verifying correctness and comparing various models to estimate performance and visualization.

Sigma is a weapon developed by Facebook to handle spam, malware and other kind of abuse. The main purpose of Sigma is to identify malicious actions on Facebook like phishing attacks, posting links to malware, etc.

Table 1 Usage of ML/DP in Facebook

Kind of activity	ML/DL algorithm used
Newsfeed, ads, search and sigma	Multilayer perceptron
Sigma	General Classification and Gradient Boosted Decision Tree
Lumos	Convolutional Neural Networks (CNN)
Text understanding, translation and speech recognition	Recurrent Neural Networks (RNN)
Recommendation to users	Collaborative Filtering
Textual analysis	Tool Called DeepText
Facial recognition	DeepFace
Targeted advertising	Deep Neural Networks (DNN)

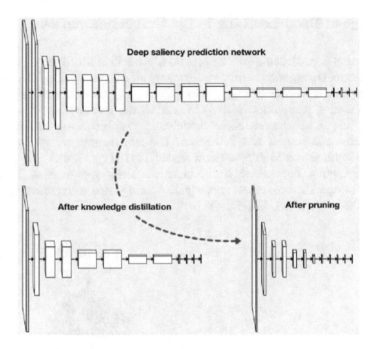

Fig. 2 Twitter news feed (summary of interesting news)

Table 2 Most important libraries in Python in the context of ML

Library	Usage
TensorFlow	Google voice search, parallelism is major advantage
Numpy	Mathematical and scientific library can be used to express images, sound waves
Keras	Easier way to express NN, visualizations and compilation of models
Pandas	Used with labelled and relational data can be used to manipulate the data along with aggregation and visualization
Seaborn	Visualizations
SciKit-Learn	Complex math operations provides consistent interface
NLTK	NLP-related tools like text tagging, classification, sentiment analytics and automatic summarization
Gensim	Vector space modelling, topic modelling, used for raw and unstructured digital texts

Lumos is platform of computer vision developed by Facebook to unlock all the images in the searching of contents by the users. It is basically AI-based platform to provide the ability of searching various images (Tables 1, 2).

3 Usage of Deep Learning in the Current Scenarios

Deep learning is particular aspect of ML, does better than ML if the input data are huge in nature. DL algorithms involve colorizing the images for the given black and white images, automatically tagging the photos and gender classification in the images. Emotion recognition with CNN, extract the text consists of smart construction, Google translations, object detection in complex scenes the development of DL starts with neuron and Perceptron. The term neuron coined by Heinrich Wilhelm Gottfried von Waldeyer-Hartz around 1891 (Figs. 3, 4).

The perceptron be able to learn, make decisions and translate languages. Handling of speech recognition and winning more visual recognition challenges leverage the usage of DL in (Fig. 5).

Fig. 3 Model of simple neuron

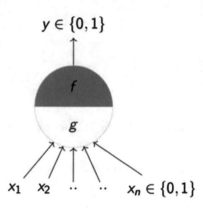

Fig. 4 Model of perceptron

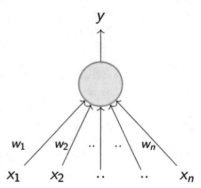

Fig. 5 Visual recognition challenge

Network	Error	Layers
AlexNet	16.0%	8
ZFNet	11.2%	8
VGGNet	7.3%	19
GoogLeNet	6.7%	22
MS ResNet	3.6%	152!!

The most commonly used DL networks are CNN used to recognize handwritten digits with back propagation, and RNN is used to refer dynamic temporal behaviour for a time sequence. RNN uses internal state to process sequences of inputs.

4 Conclusion

The usage of ML and DL in Facebook and other social media was described along with models and algorithms. The libraries and platforms developed by Google, Facebook and Microsoft were also covered. The applications like face recognition, image captioning and identification of important postures of videos are some of the current developments of deep learning. The next paper is going to cover the discussion of various ML algorithms like classification, clustering and recommender systems along with implementation details in Python-based libraries.

References

1. Hazelwood, K.: Applied machine learning at Facebook: a datacenter infrastructure perspective (2017)
2. Reagen, B., Adolf, R., Whatmough, P.N., Wei, G., Brooks, D.M.: Deep Learning for Computer Architects. Series Synthesis Lectures on Computer Architecture, Morgan & Claypool Publishers (2017)
3. Pino, J.M., Sidorov, A., Ayan, N.F.: Transitioning Entirely to Neural Machine Translation, (Aug 2017). https://fb.me/pino
4. www.wikipedia.org
5. https://algorithmia.com/tags/deep-learning
6. https://medium.com
7. www.analyticsvidya.com
8. www.udemy.com
9. www.kaggle.com
10. www.github.com
11. www.quora.com
12. www.forbes.com
13. https://machinelearningmastery.com/best-machine-learning-resources-for-getting-started/
14. www.courseera.org
15. https://www.kdnuggets.com

Automatically Labeling Software Components with Concept Mining

A. S. Baby Rani and A. R. Nadira Banu Kamal

Abstract In the software development life cycle, software maintenance phase needs much time and effort of programmer analysts in maintaining the software. With the proliferation of software applications in the agile development environment, it is highly challenging and vital to cope up with software evolution and maintenance. Software maintenance could be carried out using static and dynamic analysis of source code. Recently, text mining techniques are widely used in static analysis of source code. In this work, a tool is designed to automatically label the components with the classes referred by them using text mining and formal concept analysis. This tool can be deployed in the software engineering tasks like architecture recovery and change impact analysis.

Keywords Text mining · Clustering · Formal concept analysis · Software maintenance

1 Introduction

In agile development environment with a lot of cloud services, mobile and web applications and open-source software, a large number of applications are being deployed every day. Programmers often spend much time in browsing through the source code to understand the software system. It is indeed a formidable task for them to understand a software system quickly in such a rapid development environment. Due to its inherent complexity, maintenance of large software systems

A. S. Baby Rani (✉)
Computer Science, Sri Meenakshi Govt. Arts College for Women(Autonomous),
Madurai, Tamil Nadu, India
e-mail: asrani18@yahoo.com

A. R. Nadira Banu Kamal
Mohammed Sathak Hamid College of Arts and Science for Women,
Ramanathapuram, Tamil Nadu, India
e-mail: nadirakamal@gmail.com

© Springer Nature Singapore Pte Ltd. 2020
P. Venkata Krishna and M. S. Obaidat (eds.), *Emerging Research in Data
Engineering Systems and Computer Communications*, Advances in Intelligent
Systems and Computing 1054, https://doi.org/10.1007/978-981-15-0135-7_44

473

presents many challenges. Firstly, applying static or dynamic analysis in a large software system is cumbersome as the graph models—call graph or control-flow graph that represents the software grow vast, and it will exhibit complex relations among the entities that are difficult to understand. Cross-cutting concerns or aspects that are implemented across classes may be overlooked in these formal approaches. Secondly, there is a semantic gap between the specification that describes the features of a software system and the functional elements of the source code that implement the features. Thirdly, in a rapid application development environment, proper reference documentation of software systems will not be available, or it may be incomplete. Precisely, software development adopts collaborative multi-lingual platform that imposes limitation in applying formal methods or adopting a single technique in documenting the software. As stated by Kazman [1], "Open Source Communities and agile software development communities tend to emphasize the costs and downplay the benefits and hence tend to lead developers away from documenting software architecture." The lack of architectural documentation may inhibit developers in carrying out the maintenance task and hence a reference documentation is vital to support developers so as to sustain the existing large software systems.

There are several approaches researched by Software Engineering (SE) community in supporting developers in software maintenance activity [2]. Code summarization is a technique that precisely describes the software components which can help the developers to understand the software system in a better manner. Earlier, profilers were used to document the software using static analysis, which encompasses the process of collecting the implementation details and instrumenting the source code. For example, Javadoc utility can generate reference documentation from the comments and annotations of Java application. Program slicing is a technique to extract the features in the programmers' context from the graph models of a software system. Auto folding [3] is another technique that is used to suppress the insignificant code segments so that developers can browse through only the important source code regions that implement the application logic. Code summarization otherwise referred to as program slicing [4] that hides irrelevant lines of code. Code fragments are summarized by a list of keywords, and McBurney [5] takes this idea further and present the keywords in a navigable tree structure.

Labeling software components [6] with the entities and information embedded in the underlying source code can support developers in program comprehension, code review, code refactoring, etc. Named software entities are identified in social forum like Stack Overflow and Quora to label the responses [7]. Software domain ontologies are also used in understanding software systems. Recently, topic modeling, an unsupervised machine learning technique, is widely used [8] in program comprehension [9–14] and maintenance tasks. It explores topics or the most relevant keywords to represent a source code on contextual basis rather semantically. But the drawback is that there is no agreed-upon technique to label the documents with topics so that the topics can be understood in the context. Further, the existing methods do not provide any relation among the topics.

The task of natural language summarization has been studied extensively [5], mostly focusing on extractive summarization, the problem of extracting the most relevant text segments from source code documents. Despite much research, there is still a large need for better tools that aid program comprehension activity, thereby reducing the cost of software development. Hence, in this work, a software tool implemented in Java is developed to label the software components automatically with type parameters or identifiers using concept mining or formal concept analysis (FCA). The concept mining is applied on the data modeled with text mining technique called Latent Dirichlet Allocation (LDA) modeling, also termed as topic modeling to label the documents with features. The LDA model is validated before applying labeling to ensure that the meta-model used is a good one. It is validated with the cluster validation measure, and its cohesiveness is compared against the cohesive property of the components which are analyzed with static analysis using call graph.

The following sections are organized as follows. Section 2 provides the necessary details about LDA, FCA and validation measures. Section 3 describes the methodology, and Sect. 4 explains about the case study undertaken. Section 5 discusses the results obtained and concludes with Sect. 6.

2 Background

2.1 Latent Dirichlet Allocation Model (LDA)

Latent Dirichlet Allocation modeling [8], also termed as topic modeling, is a generative probabilistic model to extract terms out of a corpus in an unsupervised machine learning approach based on natural language processing technique. The basic idea is that documents are represented as random mixtures over latent topics where each topic is characterized by a probability distribution over terms or words. The semantics of the words are derived from the contexts that address synonymy and polysemy issues too. LDA method can be referred in [15] for better understanding.

LDA assumes the following generative process for each document $\mathbf{w}$ in a corpus D:

1. Choose $N \sim$ Poisson (ξ).
2. Choose $\theta \sim$ Dir (α).
3. For each of the N word w_n,
(a) Choose a topic $z_n \sim$ Multinomial (θ)
(b) Choose a word w_n from $p(w_n| z_n, \beta)$, a multinomial probability conditioned on the topic z_n.

$\mathbf{w}$ represents a document (a vector of words) where $w = (w_1, w_2, \ldots w_N)$

α is the parameter of the Dirichlet distribution, technically $\alpha = (\alpha_1, \alpha_2, \ldots, \alpha_k)$, but unless otherwise noted, all elements of α will be the same.

β is a $k \, x \, V$ topic by term probability matrix for each topic (row) and each term (column), where $\beta_{ij} = p(w^i = 1|z^j = 1)$

z represents a vector of topics, where if the ith element of z is 1, then w draws from the ith topic.

Probabilistic Latent Semantic Indexing (pLSI) [16], Relational Topic Model and Variational Inference Model are the other topic modeling applied for information retrieval and text mining [17]. In applying topic modeling, it is important to tune the model parameters otherwise the resulting model will not be good and hence validating the model that is derived is crucial. In order to validate the model, first, the documents are grouped together into clusters based on dominant topics or using clustering algorithms—k-means, k-Medoid, agglomerative or divisive hierarchical clustering and the cluster validation measure are used to validate the topic model. Many cluster validation measures [18] are proposed in the literature to validate the clusters. In this work, silhouette coefficient, a measure that considers both the inter-cluster distance and intra-cluster distance, is used to validate the clusters of LDA model clustered with K-means clustering algorithm.

2.2 Formal Concept Analysis (FCA)

FCA [19] is a technique for data analysis and knowledge representation based on lattice theory. Concept lattices are core structures of FCA for extracting a set of concepts from a dataset, called a formal context, composed of objects described by attributes. A formal context is defined as a triple $K = (O, A, R)$ where O is a set of objects, A is a set of attributes, and R is a binary relation between objects and attributes, indicating which attributes are possessed by each object.

More formally, a concept is a pair of sets (X, Y) such that:

$$X = \{o \in O | \forall a \in Y : (o, a) \in R\}$$
$$Y = \{a \in A | \forall o \in X : (o, a) \in R\},$$

where X is considered to be the extent of the concept, and Y is intent of the concept. This set of concepts is called a complete partial order where some concepts are super- or sub-concepts with respect to others. The set of all concepts and the hierarchical relation among the concepts constitute a concept lattice, and there are several algorithms to compute concepts and concept lattices from a given formal context. The details on these algorithms, as well as more complete description on FCA, can be referred in [20]. FCA is applied in feature location [21] and reverse engineering [12].

2.3 Validation Measures

Silhouette Coefficient

Silhouette coefficient—an internal validation measure—is based on *cohesion* (similarity) measure, which determines how closely related the documents in a cluster are, and *separation* (dissimilarity) measure, which determines how distinct (or well-separated) a cluster is from other clusters. It combines both cohesion and separation in only one scalar value. The silhouette coefficient is computed for each document using the concept of centroids of clusters. Formally, let C be a cluster; its centroid is equal to the mean vector of all documents belonging to C:

$$\text{Centroid}(C) = \sum_{d_i \in C} d_i / |C|.$$

Starting from the definition of centroids, the computation of the silhouette coefficient consists of the following three steps:

(1) For document d_i, calculate the maximum distance from d_i to the other documents in its cluster. We call this value $a\,(d_i)$.
(2) For document d_i, calculate the minimum distance from d_i to the centroids of the clusters not containing d_i. We call this value $b\,(d_i)$.
(3) For document d_i, the silhouette coefficient $s(d_i)$ is:

$$s(d_i) = \frac{b(d_i) - a(d_i)}{\max(a(d_i), b(d_i))}$$

The value of the silhouette coefficient ranges between −1 and 1. A negative value is undesirable because it relates to the case where $a\,(d_i) > b\,(d_i)$, i.e., a document in one cluster is closer to a document belonging to another cluster and hence it implies poor clustering.

For measuring the distance between documents, the squared Euclidean distance is used. The overall measure of the quality of clustering $C = \{C_1, C_2, \dots C_k\}$ can be obtained by computing the mean silhouette coefficient of all documents as follows.

$$S(C) = \frac{1}{n} \sum_{i=1}^{n} s(d_i)$$

This measure is used to evaluate the quality of LDA configuration.

Method Community Cohesion (MCC)

Software system quality can be assessed with cohesion measure or lack of cohesion measures—LCOM, LCOM2, LCOM5. Yu Qu, et al. proved in their work [22] that Method Community Cohesion Measure results in better consistency and hence MCC is used in this work which is explained below.

Given a class Cl with m methods located in a connected component of a call graph that has N connected components, m methods distributed in N communities or clusters. For the ith cluster, there are ni methods belonging to Cl $(1 \leq i \leq N)$. Let $n\mathrm{max} = \max\{ni\}$.

$$\mathrm{MCC}(Cl) = \begin{cases} 1, & \text{if } m = 1 \\ 0, & \text{if } n_{\max} = 1 \text{ and } m > 1 \\ \frac{n_{\max}}{m} & \text{Otherwise} \end{cases}$$

The definition of MCC describes that the largest portion of its methods that reside in the same community, which represents more cohesive relation than the rest of the methods in Cl. The second line in Eq. (2) [22] is proposed to make sure that the lower bound of MCC is consistent for different classes and is not influenced by the number of methods in a class. For instance, suppose class C_1 has three methods that are distributed in three clusters, and class C_2 has four methods that are distributed in four clusters, then the cohesiveness of these two classes should all reach the lower value 0 rather 1/3 for C_1 and ¼ for C_2. And MCC measure is used to compare textual analysis against static analysis to get an insight about how good the clusters formed with text mining technique.

3 Proposed Model

In this work, text mining technique is used for source code analysis, and concept mining or formal concept analysis is applied to label the source code components. The unsupervised machine learning technique, LDA modeling applied to get the meta representation of the text corpus, and the model derived is evaluated with the cluster validation measure—silhouette coefficient. Method call graph analysis is also performed to find the cohesiveness of the components, and this measure is used as the benchmark to test the cluster quality where the software components are grouped on the basis of lexical similarity. The schematic diagram of the proposed model is shown in Fig. 1

3.1 Source Files

First, the Java software system under experimental study is downloaded from the open-source software repository.[1] Source code files are pooled out from the '.jar' and '.java' files located in various subdirectories under the project root directory.

---

[1]https://www.github.com.

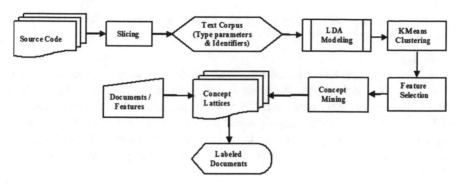

Fig. 1 Proposed model

3.2 Slicing

The software system can be sliced at package level, class level, method and block level for source code analysis. In this work, software system is sliced at method level granularity, and a source code document corresponds to a method defined in a class and henceforth document and component will be referred interchangeably.

3.3 Preprocessing

Generally, preprocessing in text mining process involves text corpus creation, text normalization with case conversion, noise reduction with stop word removal and lemmatization with stemming. As this work focuses on labeling the documents, only the terms or entities of interest are extracted and no further processing is carried out which may affect the readability.

Text Corpus
Text corpus can be prepared by extracting the features or the terms of source code documents. The features of a source code document may correspond to class and method names, type parameters, identifiers, comments and method calls or maybe a combination of two or more of these entities. In many research work [9, 23], identifiers and comments are considered as features with the intuition that the semantic information will be hidden in those elements. But the outcome depends on how much information is actually available in such sources. The programmers' expertise will have direct implication on choosing appropriate names for identifiers and putting proper annotations in the comments. As there is a degree of uncertainty in adopting identifiers and comments as features, in this work, the concrete element -type parameters (datatypes/classes) and identifiers are treated as features. Common datatypes and library classes are ignored while generating the text corpus. In the

context of this application or SE task undertaken, stop word removal and stemming are not performed and the entities are considered as it exists in the document.

Indexing
Following text corpus creation, the vocabulary of the software system is indexed, and each document is represented as the weighted term frequency vector. The documents are generally represented in a high-dimensional feature space, and it is difficult to apply any text mining algorithm like classification or clustering in such a high-dimensional space. In order to reduce the dimensionality and find the semantics contextually rather linguistically, LDA modeling is applied, which is described below.

3.4 LDA Modeling

LDA modeling (see Sect. 2.1) is used to transform the source code documents into its intermediate form which facilitates further analysis. Fast Gibbs sampling algorithm [24] is applied using GibbsLDA ++ on the term-by-document matrix which reduces the term-by-document space of the text corpus into topic-by-document space. It also provides the vector of topic assignment to each term and the topic-by-term matrix that constitutes the probability distribution of terms over a particular topic. In LDA modeling, each document is characterized as the probability distribution of topics, which in turn is represented as the probability distribution of terms or words. In order to validate the derived model, the documents are clustered and the cluster validation measure—silhouette coefficient is computed.

3.5 K-Means Clustering

In LDA modeling, clusters can be formulated with dominant topics, but to consider the relative significance of all the topics in a document, K-means clustering is applied. The related documents are clustered with K-means clustering using the topic-by-document representation of the source code documents with a fixed cluster size of 200.

3.6 Concept Mining

Each cluster is anlaysed with concept mining or FCA using formal contexts (documents are objects and type parameters & identifiers are attributes) pertaining to the document set and the associated features of the particular cluster. In this work, concept lattices are generated statically in offline, but it could also be generated dynamically, provided, few features (10–20) are used for defining the formal context. In a large complex software system, there could be around 1000 features to the maximum in a cluster. Due to the exponential time complexity of the concept lattice construction algorithm, feature selection is made to filter orthogonal features using co-variance measure [17]. Concept lattices are constructed with the limited number of features (10–50) that exhibits higher variance among the features in the cluster.

Using concept lattice construction algorithm [19], document sets are labeled by its shared feature set. The hierarchical partial order relation among the concepts are represented with the Directed Acyclic Graph (DAG), with concepts of general or common features at the top and concepts with unique or specific features at the bottom. Each node in the concept lattice holds a set of documents and common features shared by them.

A graph traversal algorithm—breadth-first search or depth-first search can be applied to identify the unique features of a document and other shared features. Similarly, given an attribute, it is easy to find whether it is a common or unique feature from the concept hierarchy of the concept lattice. The major limitation in applying FCA is its exponential time complexity. The experiments are conducted with the varying number of objects and attributes and the time taken for constructing concept lattices is presented.

4 Experimental Study

Apache POI open-source software project is used for the experimental study. Apache POI project provides Java API for manipulating various file formats based upon the office open XML (OOXML) and Microsoft's Apache OLE2 compound document format. With POI, MS Word, Excel and Power Point files can be read or written. Apache POI Version 3.17 of size 167 KLOC—source files are downloaded from the Apache Web site.[2] Text mining is applied on source code files and to validate the resulting model, static analysis with call graph is also employed in this study. Preprocessed dataset of this project, class-hierarchy graph and method call graph in Excel file format, which is available at the link[3] is used for call graph analysis.

As explained in Sects. 3.2 and 3.3, source code files are sliced and preprocessed. The characteristics of the software system following preprocessing are tabulated in Table 1.

The term-by-document matrix dimension is $12{,}649 \times 30{,}385$. This large feature space is reduced to $400 \times 30{,}385$ with LDA model input parameters $[K, \alpha, \beta]$ assigned as $[400, 0.5, 0.5]$. K-means clustering is applied to cluster the documents

---

[2]https://poi.apache.org.
[3]https://zendo.org/record/1195817#.XHu2nAzbIU.

Table 1 System characteristics

System	Size KLOC	# Methods	# Datatypes	# Identifiers	# Terms (Datatypes & Identifiers)
Apace POI	167	30,385	1393	11,256	12,649

with cluster size 200 on topic-by-document matrix produced by LDA model, and the document distribution in various clusters are depicted as Histogram chart in Fig. 2. Clusters obtained are validated with the internal cluster validation measure, silhouette coefficient.

To compare the text mining approach against the formal method—static analysis, call graph analysis is performed and the cohesion measures are compared. Call graph analysis is carried out by constructing the call graph from the Excel file in which each entry holds the caller and callee method names and the type of method calls. The process of grouping the entities and comparing its cohesiveness with text mining and static analysis is shown in the block diagram—Fig. 3

In a call graph, each node represents a method, and nodes are connected by edges if a method calls another method. For instance, methods—m_1 & m_2 are connected by directed edges, if m_1 calls m_2 and vice versa. When methods are cohesive, it could be reachable from one node to the other node through graph traversal. Based on this heuristics, connected components of the graph are discovered and also the biconnected components are listed out. Software system characteristics in the context of call graph analysis are shown in Table 2. By treating a connected component as a cluster, Method Community Cohesion (MCC) of each class is evaluated. Similarly,

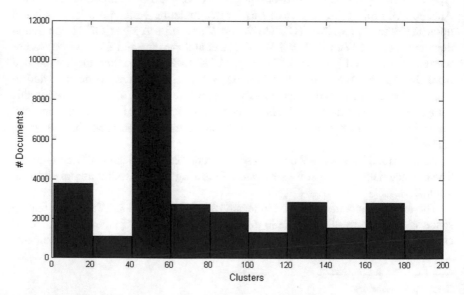

Fig. 2 Histogram chart of document distribution in various clusters

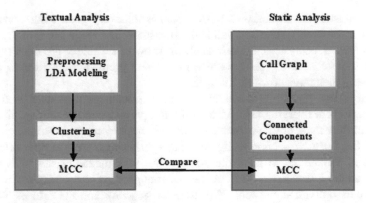

Fig. 3 Comparing method cohesion

Table 2 System characteristics—call graph analysis

No.	Data source	Element type	Total no. of nodes	Clusters/Connected components
1.	Method call graph	Class instances	8976	527
		Method instances	87,323	
2.	Class call graph	Classes and interfaces	5989	1

in text mining, MCC is computed by treating each cluster as cohesive set of components. The cohesive measure of all the classes having more than one method is shown as boxplot. The mean MCC value of text mining approach is comparable to that of static analysis. This MCC value as well as the mean and standard deviation of silhouette coefficient substantiates the application of text mining in SE tasks. Now, using concept mining explained in Sect. 2.2, a tool is developed to automatically label the software components with concept mining applied on individual clusters that are obtained from K-means clustering. Given a class or component name, the set of features that are labeled will be displayed.

5 Results and Discussion

LDA modeling is applied on the software system under study, and it results with the validation measure—mean silhouette coefficient as **0.1216** and Standard Deviation **0.1607**. The silhouette coefficient can be in the range between -1 and $+1$, and a positive value signifies that the inter-cluster distance is higher than the intra-cluster distance and hence the documents within the cluster are cohesive. Also, the standard deviation also indicates that there is no much variation of silhouette measure in different clusters. The quality of LDA model is relatively good while considering size of the software system and validation measures. The mean MCC of the classes of both textual analysis and static analysis shown in Table 3 provides (approximately) same cohesive measure which indicates a promising direction to employ text mining techniques for the SE tasks.

On applying clustering, the total number of documents that exist in the individual clusters are varying from 10 to 1000. In applying concept mining, size of the cluster and the number of attributes will impose limitations in constructing concept lattices. So, feature selection using covariance measure is used to filter orthogonal features of limited set to label the documents. As *K*-means clustering is a partitioning method, document sets are non-overlapping, whereas features might be repeated in individual clusters. The time to generate concept lattices with the complete partial order relations among the concepts with varying number of objects and attributes are shown in Fig. 4

The empirical study suggests that document set of size 100 with feature set ranging from 30 to 50 can generate concept lattices faster. But the time factor is less significant while considering the benefits of concept mining like feature analysis, extracting general and specific features, finding association or implications between the documents and features. In addition, concept mining is carried out statically offline, and the concept lattices are available persistently. A sample of source code document with the labeled features as highlighted text is shown in Fig. 5. Hence, while searching through the concept lattice with these features, it will provide the document and other documents in the hierarchy that share the features.

Table 3 Method cohesion measure

Analysis type	MCC
Textual	0.25
Call graph	0.26

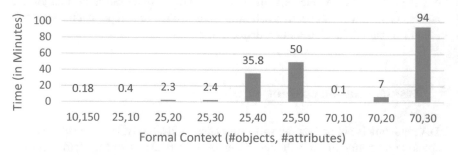

Fig. 4 Time to construct concept lattices

Fig. 5 Sample *Source* Code Document

```
// TestXSLFAutoShape

public void testTextParagraph() throws IOException {

XMLSlideShow ppt = new XMLSlideShow();

XSLFSlide slide = ppt.createSlide();

assertTrue(slide.getShapes().isEmpty());


XSLFAutoShape shape = slide.createAutoShape();

assertEquals(0, shape.getTextParagraphs().size());

XSLFTextParagraph p = shape.addNewTextParagraph();

assertEquals(1, shape.getTextParagraphs().size());
```

6   Conclusion

Software support tools are necessary to help the programmer analysts in under-standing large software system and to carry out various SE tasks like Feature Location, Bug Localization, Change Impact Analysis, etc. In this work, a tool is developed to automatically label the source code documents with the extracted features using concept mining. The model implemented is validated with the sta-tistical measures. The features or documents can be explored with a simple graph traversal algorithm. The relevance of features selected will have an impact on its applicability. In this work, in addition to identifiers, the concrete elements, i.e., type parameters are used as features that are required for fixing bugs, carrying out software changes, etc. Hence this tool will be useful to the programmers in software maintenance tasks.

References

1. Kazman, R., Goldenson, D., Monarch, I., Nichols, W., Valetto, G.: Evaluating the effects of architectural documentation: a case study of a large scale open source project. IEEE Trans. Softw. Eng. **42**(3), 220–260 (2016)
2. Dit, B., Revelle, M., Gethers, M., Poshyvanyk, D.: Feature location in source code: a taxonomy and survey. Wiley J. Softw. Evol. Process **25**(1), 53–95 (2013)
3. Fowkes, J., Chanthirasegaran, P., Ranca, R., Allamanis, M., Lapata, M., Sutton, C.: Autofolding for source code summarization. In: IEEE/ACM 38th International Conference on Software Engineering Companion (ICSE-C) (2016)
4. Kuhn, A.: Automatic labeling of software components and their evolution using log-likelihood ratio of word frequencies in source code. In: 6th IEEE International Working Conference on Mining Software Repositories (2009)

5. McBurney, P., Liu, C., McMillan, C., Weninger, T.: Improving topic model source code summarization. In: Proceedings of the 22nd International Conference on Program Comprehension, pp. 291–294 (2014)
6. Ma, X., Liu, C., Ye, X., Shen, H., Bunescu, R.: From word embeddings to document similarities for improved information retrieval in software engineering. In: IEEE/ACM 38th International Conference on Software Engineering (ICSE) (2016)
7. Ye, D., Xing, Z., Foo, C.Y., Ang, Z.Q., Li, J., Kapre, N.: Software-specific named entity recognition in software engineering social content. In: IEEE 23rd International Conference on Software Analysis, Evolution, and Reengineering (SANER) (2016)
8. Blei, D.M., Ng, A.Y., Jordan, M.I.: Latent dirichlet allocation. J. Mach. Learn. Res. **3**, 993–1022 (2003)
9. Panichella, A., Dit, B., Oliveto, R., Di Penta, M., Poshynanyk, D., De Lucia, A.: How to Effectively Use Topic Models for Software Engineering Tasks? An Approach Based on Genetic Algorithms, pp. 522–531, IEEE ICSE (2013)
10. Dasgupta, T., Grechanik, M., Moritz, E., Dit, B., Poshyvanyk, D.: Enhancing software traceability by automatically expanding corpora with relevant documentation. In: IEEE International Conference on Software Maintenance, ICSM (2013)
11. McMillan, C., Grechanik, M., Poshyvanyk, D., Fu, C.: Exemplar: a source code search engine for finding highly relevant applications. IEEE Trans. Softw. Eng. **38**(5) (2011)
12. Al-Msie'deen, R., Huchard, M., Seriai, A.-D., Urtado, C., Vauttier, S.: Automatic documentation of [Mined] feature implementations from source code elements and use-case diagrams with the REVPLINE approach. Int. J. Softw. Eng. Knowl. Eng. **24**(10), 1413–1438 (2014)
13. Poshyvanyk, D., Guéhéneuc, Y.G., Marcus, A., Antoniol, G., Rajlich, V.: Combining probabilistic ranking and latent semantic indexing for feature location. In: Proceedings of 14th IEEE International Conference on Program Comprehension, pp. 137–148 (2006)
14. Poshyvanyk, D., Gueheneuc, Y.G., Marcus, A., Antoniol, G., Rajlich, V.: Feature location using probabilistic ranking of methods based on execution scenarios and information retrieval. IEEE Trans. Softw. Eng. **33**(6), 420–432 (2007)
15. Reed, C.: Latent dirichlet allocation: towards a deeper understanding (2012)
16. Deerwester, S., Dumais, S.T., Furnas, G.W., Landauer, T.K., Harshman, R.: Indexing by Latent Semantic Analysis. J. Am. Soc. Inf. Sci. **41**(6), 391–407 (1990)
17. Berry, M.W.: Survey of Text Mining, pp. 81–83. Springer (2014)
18. Liu, Y., Li, Z., Xiong, H., Gao, X., Wu, J.: Understanding of internal clustering validation measures. In: IEEE International Conference on Data Mining (2010)
19. Mens, K., Tourwe, T.: Delving source code with formal concept analysis. Elsevier J. Comput. Lang. Syst. Struct. **31**(3–4), 183–197 (2005)
20. Gregor, S.: Concept lattices in software analysis. In: Lecture Notes in Computer Science book series Formal Concept Analysis, LNCS, vol. 3626, pp. 272–287. Springer (2005)
21. Eisenbarth, T., Koschke, R., Simon, D.: Locating features in source code. IEEE Trans. Softw. Eng. Arch. **29**(3), 210–224 (2003)
22. Qu, Y., Guan, X., Zheng, Q., Liu, T., Wang, L., Hou, Y., Yang, Z.: Exploring community structure of software Call Graph and its applications in class cohesion measurement. J. Syst. Softw. **108**, 193–210 (2015)
23. Lukins, S.K., Kraft, N.A., Etzkorn, L.H.: Bug localization using latent Dirichlet allocation. Inf. Softw. Technol. **52**(9), 972–990 (2010)
24. GibbsLDA ++. http://gibbslda.sourceforge.net/. In: International Conference on Research and Development in Information Retrieval, pp. 433–434. Toronto, Ontario, Canada (2003)

The Avant-Garde Ways to Prevent the WhatsApp Fake News

J. Indumathi and J. Gitanjali

Abstract The birth of the information era has perceived many machineries which have throttled the Communication Engineering applications. Amid the zillion instant messaging applications, WhatsApp has its share of highlights and challenges. In India, chat messaging app—WhatsApp—is seen trapped in multiple court cases relating to the spread of misinformation, encrypted messages and fake news. Unless the liabilities are restricted, it will no longer be an asset but rather will become a curse on the user community. Over seventy billion messages are spread on WhatsApp daily, and the false rumours are spread at lightning speed. They have encompassed conspiracy theories, anti-vaccination misinformation and panicked rumours and have led to fatal lynchings worldwide. Many have exploited WhatsApp for dispersal of falsehood and abhor discourse. The inspiration behind this paper is to decrease the spread of fake news in WhatsApp. This paper will recognize the fake news, depending on the data given from the client end. Client can propose just those messages as fake, that are approved for the proposition. Service provider will accumulate all these data and the need for each fake news will be given based on the count of proposal and furthermore dependent on the priority level of the clients. For every priority level, a star rating will be given. Every time when the count reaches the red label level (most elevated amount), that specific message will be blocked, and it can't be sent to anybody. This technique is incontestable and can guarantee a fake free WhatsApp.

Keywords WhatsApp · Service provider · Red label level · Star rating · Misinformation · Content dissemination · Textual information

J. Indumathi (✉)
Department of Information Science and Technology, Anna University, Chennai, India
e-mail: indumathi@annauniv.edu

J. Gitanjali
School of Information Technology, Vellore Institute of Technology, Vellore, India
e-mail: gitanjalij@vit.ac.in

© Springer Nature Singapore Pte Ltd. 2020   487
P. Venkata Krishna and M. S. Obaidat (eds.), *Emerging Research in Data Engineering Systems and Computer Communications*, Advances in Intelligent Systems and Computing 1054, https://doi.org/10.1007/978-981-15-0135-7_45

1   Introduction

With the dawn of the numerous social networking platforms and apps, there are umpteen ways to be socially active and exchange information on the Internet. The social media sites can be categorized as anonymous social networks, blogging and publishing networks, bookmarking and content curation networks, consumer review networks, discussion forums, interest-based networks, media sharing networks, sharing economy networks, social networks and social shopping networks. Amongst them, the instant messaging applications are the most popular with a larger user base.

The rest of the paper is organized as follows. Section 2 deals with the basics of instant messaging applications and WhatsApp. Section 3 deliberates the Fake news and WhatsApp. Section 4 provides a comprehensive survey of the research work. Section 5 pinpoints the approaches utilized to fight back the fake news. Section 6 deals with how WhatsApp fights back the fake news. Section 7 explains the proposed system architecture along with its implementation, to curtail the fake news in WhatsApp. Section 9 highlights the main conclusions and briefly describes the future work.

2   Instant Messaging Application

Instant messaging applications have replaced the traditional short messaging service (SMS) and multimedia messaging service (MMS) due to the simplicity to learn and use.

The popularity of instant messaging applications is also credited to-no limit on the length or the number of messages, they routinely import contacts on mobile phone or another service they support, and they offer user's to create richer profiles. Users can send text, audio messages, different types of attachments such as photos, videos and contact information to their contacts in real time. The instant messaging applications offer two types of communication, namely the peer-to-peer and group chat. Instant messaging applications have replaced the traditional short messaging service (SMS) and multimedia messaging service (MMS) due to the simplicity to learn and use.

WhatsApp use finds its usage in various applications [1–8]. The key reason behind this popularity is that instant messaging applications use Internet instead of short message service technical realization (GSM); they are free to use, and they only require Internet connection which is the most common way of communication today.

According to Sanchez [9], most of these highly popular instant messaging applications including WhatsApp have some known vulnerabilities. The WhatsApp with its huge user base in the mobile has expanded itself on to the web and PCs. Given the evolving significance of this IM network, it is not startling that there is a rising interest in investigating it, including user studies about people's WhatsApp use and possible applications [1–8, 10, 11]. As every flow has its ebb, WhatsApp has its ebb, namely the fast spreading fake news. Spreading of fake news in WhatsApp is one of the most concerning issues in India. WhatsApp is under flame over reports of swarm brutality and lynching episodes in India, because of the spread of fake news. The battle against fake news is not simple because of the quantity of individuals who utilize WhatsApp all around or in other words billion.

3 Fake News

The term fake news has existed all over the human history in the form of gossip, rumour and misinformation [12]. The gossip, rumour and misinformation have been used to impact and manipulate other people, community and society for their profit [13]. The medium used in the pre-Internet and pre-social media (print media, visual media like radio or televisions) had multiple restrictions and limitations, and hence, the impact of fake news was not alarming or extensive but with the development in technology, ease of Internet access and social media; content creation, curation, distribution (audience reach) and consumption have made it a lot easier and cheaper to create and spread fake news (Róisín Kiernan). With technological advances, fake news creators are using diverse types of content formatting and structure to seize the reader's attention [14, 15].

4 Fake News and Whatsapp

Social media hastens fake news broadcasting as it breaks the physical distance barrier amid individuals, offers rich platforms to share, forward, vote and review to inspire users to participate and discuss online news. In India, 250 million clients are on WhatsApp. WhatsApp has been deploying measures to alleviate this problem. The solutions adopted by WhatsApp comprise dipping the limit for forwarding a message to at most five users at once. One of the studies done by the researchers at Federal University of Minas Gerais in Brazil and Massachusetts Institute of Technology (MIT) noted that using an epidemiological model and real data, it has assessed the impact of limiting virality features in this kind of network. The current efforts deployed by WhatsApp can offer noteworthy delays on the information spread, but they are unsuccessful in blocking the propagation of misinformation campaigns through public groups when the content has a high viral nature. With the end goal to maintain a strategic distance from the spread of fake news, WhatsApp started WhatsApp sent mark. Utilizing this, you can forward the message to the most extreme of five individuals or groups. This works for photographs, recordings and in addition instant messages. This strategy is not that productive to keep the spread of fake news where numerous individuals do not comprehend its motivation.

5 Related Work

Rosenfeld et al. [16] have proposed a broad investigation of the use of the WhatsApp informal organization and web informing application that is rapidly supplanting SMS informing. They have additionally performed broad factual and numerical investigation of the information and have discovered huge contrasts in WhatsApp use crosswise over individuals of different gender and diverse ages.

Garimella et al. [17] have proposed the work on gathering and examination of WhatsApp bunch information. Their essential objective was to investigate the practically of gathering and utilizing WhatsApp information for sociology look into. They perform a factual investigation to enable specialists to comprehend what open WhatsApp establishment information might be accumulated and how these records can be utilized.

Bhatt et al. [18] have proved the hypothesis that WhatsApp is a major cause for social seclusion and reassures only virtual relationship as a substitute of real relationship. It not only creates interventions in daily routine and privacy of the youth, and it has the negative impact on the study and encourages the grammatical mistakes, error in sentence constriction, lectures bunking. Kumar et al. [19] have explored the utilization and impacts of WhatsApp. Seufert et al. [20] WhatsApp is an extremely well-known versatile informing application, which overwhelms todays portable correspondence. Mueller et al. [21] have gathered data about how security and protection highlights have developed in new applications when contrasted with other more seasoned applications. The outcomes have demonstrated that there is a dominating advancement in new applications contrasted with the more seasoned ones.

6 Fighting Back Fake News

Conventional methods for spotting fake news are not appropriate and active for online social networks [22]. Therefore, several approaches are accessible in the literature using modern tools and techniques to tackle the problem of fake news on social media platforms. Jin et al. [23] built a trustworthiness network with contradictory relations for detection of fake news spread via tweets. Tacchini et al. [24] proposed an approach for fake news detection using machine learning techniques and boolean crowdsourcing algorithms. Shu et al. [25] proposed a network analysis approach for detection and mitigation of fake news. Also, several other mechanisms and the relevant concepts of preserving security and privacy are existing in the literature, whose techniques can be customized for curbing fake news [26–41] (Gitanjali 2009a, b; Satheesh Kumar 2008). The inventor of the World Wide Web, Sir Tim Berners-Lee, set out a five-year strategy that the solutions will not be simple. Certain initiatives are taken to stop the spread of fake news. These include:

a. **Human intervention to validate information veracity**: with reference to the International Fact-Checking Network (IFCN) that permits American and German Facebook users to flag deliberately false articles. Fact checkers from media organizations such as the Washington Post and Snopes.com are flagging the fake news. A fact-checking unit called "Les Decodeurs" (The Decoders) is developed as a web extension called Decodex to verify the fake news. The same methodology is being applied by the French newspaper Le Monde, and its fact-checking unit called "Les Decodeurs" (The Decoders), who have developed a web extension called Decodex.

b. **Using algorithms to fight algorithms**: algorithms are part of the fake news, and it has to identify fake content and validate the information sources. The requirement is to develop a robust algorithm to perform a reliable verification of which information is false or not. The methods are: (i) algorithms that are created on the content; (ii) algorithms that are created on the diffusion dynamics of the message and (iii) hybrid algorithms that are based on a weighted sum or a group of features feeding a learning algorithm.

7 Whatsapp—Fighting Back Fake News

WhatsApp totally rehauled their authentication and messaging protocols subsequent to the research of Schrittwieser et al. [42]. The verification code (6 digits) is no longer sent to the device permitting for easy parody and capture, but relatively, the entered code is sent to the server and checked for validity there.

To limit the spread of fake news/misinformation, WhatsApp has used several techniques, by limiting the number of times a unique message can be advanced by the same user. One step is to identify the characteristics of messages containing misinformation that differentiate them from regular content. Earlier studies about WhatsApp focused on noting the general patterns of how users interact with the application [43, 44] besides its use on specific tasks (e.g. educational tasks and medical information exchange) [1, 45]. In a recent work [46], we have studied the dissemination of images in political public groups in WhatsApp, stressing some differences in images encompassing earlier known misinformation from the rest.

8 Solution Proposed for Curbing Fake News in Whatsapp

'WhatsApp' started WhatsApp sent mark with the aim of curtailing the spread of fake news. Applying this, a user can forward the message (like photographs, recordings and in addition instant messages) at the most to five individuals or groups. This approach is not that fecund to keep the spreading of fake news at bay, because numerous individuals do not realize its motivation.

A better technique (Fig. 1) is used in this paper to maintain the strategic distance from the spread of fake news. Service provider will channel the message depending on certain criteria. Service provider will check if the number of characters surpasses the point of confinement; if exceeded, then the keywords present in that message are checked. If the keyword identified matches with the keywords present in the list of basic spam keywords, next check the number of individuals to whom the specific message is shared. If the quantity surpasses the threshold, then those messages are qualified for the proposition of the clients. Each time the client thinks that a specific news is fake news, then the client can suggest that specific message as fake news.

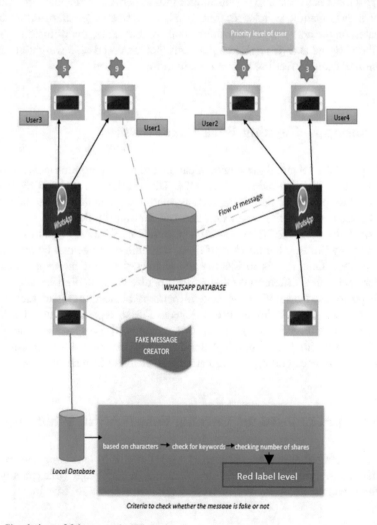

Fig. 1 Circulation of fake news in Whatsapp

Service provider will accumulate all the data and the number of individuals who have really proposed that the news is fake. Checking a message whether it is fake news or not will be done in the local database dependent on the three criteria made reference to. The necessity for each fake news will be identified based on the quantity of individuals who have really recommended that it is a fake news; and furthermore, it is also dependent on the priority level of the client.

There are ten priority levels for the clients. For each level, there will be an increment of 0.25%. At whatever point a client with higher priority level is contradicting a specific message as fake, the count of that specific message will increment when contrasted with the client with less priority level that is restricting. The priority level for every client is given dependent on the number of right fake news they have found. Each time they continue announcing the fake news, their priority level will continue increasing. Similarly, in this paper, a level called **red label level** is incorporated to show the danger of that specific message. Each time the count achieves the red label level, that specific message will be blocked so that further dissemination is curbed.

To recognize whether a specific message is fake or not, we will initially check the number of characters present in that specific message. If the number of characters is over 125, then we will check whether the message has regular spam keywords like share free, unlimited, cash back, etc. and so forth, and we will likewise check to what number of individuals that specific message is shared.

```
public void send_Message()
{
Message initialMessage=create();
IntialMessage.setFrom(currentUser.getName());
currentUser.sendMessage(initialMessage);
if(initialMessage.getData().length()>125)
{
if(verifyValidMessage(initialMessage))
{
initialMessage.setkeyword();
}
}
```

Code to check the number of characters

Utilizing this calculation, we will check whether the quantity of characters in a message is more than 125, on the off chance that in the event that is in excess of 125 then it will be proceeded onward to the following level of watching that is the keyword checking.

```
private boolean searchForkeyWord(String message, String key)
{
String b = "\\b";
String word = b+key+b;
Pattern keyWordPattern=Pattern.compile(word);
```

```
Matcher messageMatcher=keyWordPattern.matcher(message);
return messageMatcher.find();
}
public boolean verifyValidMessage(Message a)
{
String data = a.getData(().toUpperCase();
for(int keyWordCount=0;keyWordCount<keyWords.size();keywordCount++)
{
if(searchForkeyWord(data,keyWords.get(keyWordCount)))
{
return true;
}
}
return false;
}
```

Code for checking the spam keywords

This code is to check whether the given message has any spam keywords that matches with the keywords present in the spam keywords list. At whatever point the keyword matches, then that specific message will be proceeded onward to the following level of checking.

```
public void addkeyWords()
{
String[] keys={
"SHARE",
"DON'T DELETE",
"SEND",
"URGENT",
"FREE",
"GET IT NOW",
"CLICK BELOW",
"CLICK HERE",
"FREE CASH",
"EARN MONEY",
"CASH BACK",
"YOU ARE A WINNER",
"GREAT OFFER",
"MONEY BACK",
"SIGN UP FOR FREE",
"OFFER EXPIRES",
"UNLIMITED",
```

```
"PLEASE READ",
"OFFER CUPPON",
"FORWARD TO MANY",
};
keyWords=new ArrayList<>(Arrays.asList(keys));
}
}
```

Code for adding a list of common spam keywords

This code is to include the list of spam keywords with the goal that we can distinguish the fake message dependent on all these common keywords that are widely utilized in a fake message. Utilizing all these codes, we will discover the messages that can mostly be fake.

9   Conclusions and Future Work

Every invention has praiseworthy aspects and shortcomings. Amongst the most popular instant messaging applications, WhatsApp also has its share of pro et contra. Unless the liabilities, like spreading of fake news are curbed, it will no longer be an asset but rather will become a curse. Over seventy billion messages are spread on WhatsApp daily, and in current years, false rumours have spread at a frightening speed. These have encompassed conspiracy theories, anti-vaccination misinformation and panicked rumours and have led to fatal lynchings worldwide. Many have lost their livelihood, goodwill, etc. Political gatherings utilize WhatsApp for dispersal of falsehood and detest discourse. There is a five-forward limit introduced by WhatsApp which has slowed the spread of content by one order of magnitude; giving fact checkers far more time to verify the truth of a piece of content. But this delay is subject to the virality of the content—how probably users share an image after seeing it. For highly viral content, the restrictions were not operative in stopping it from quickly reaching a large portion of the network. This technique is an attempt to confine the quantity of advances which is not that captivating to control the spread of fake news. In this paper, a methodology has been proposed to control the spread of fake news in WhatsApp. With the end goal to conquer this issue, it is endeavoured to hinder the most extreme fake message by getting the input form the clients itself. By giving the clients star rating and priority levels, we are ensuring that each client will put an additional push to report the fake message. Our principle objective is to work with others in the public arena to enable protection to each individual using WhatsApp. Beyond dispute, this technique can guarantee a superior security for WhatsApp.

References

1. Bouhnik, D., Deshen, M.: Whatsapp goes to school: mobile instant messaging between teachers and students. J. Inf. Technol. Educ. Res. **13**, 217–231 (2014)
2. Montag, C., Błaszkiewicz, K., Sariyska, R., Lachmann, B., Andone, I., Trendafilov, B., Eibes, M., Markowetz, A.: Smartphone usage in the 21st century: who is active on whatsapp? BMC Res. Notes **8**(1), 1–6 (2015)
3. Church, K., De Oliveira, R.: What's up with whatsapp?: Comparing mobile instant messaging behaviors with traditional sms. In: Proceedings of the 15th International Conference on Human-computer Interaction with Mobile Devices and Services, MobileHCI'13, pp. 352–361 (2013)
4. Fiadino, P., Schiavone, M., Casas, P.: Vivisecting whatsapp through large-scale measurements in mobile networks. In: SIGCOMM 2014 (2014)
5. Johnston, M., King, D., Arora, S., Behar, N., Athanasiou, T., Sevdalis, N., Darzi, A.: Smartphones let surgeons know WhatsApp: an analysis of communication in emergency surgical teams. Am. J. Surg. **209**, 45–51 (2015)
6. O'Hara, K.P., Massimi, M., Harper, R., Rubens, S., Morris, J.: Everyday dwelling with WhatsApp. In: Proceedings of the 17th ACM Conference on Computer Supported Cooperative Work and Social Computing, CSCW'14, pp. 1131–1143 (2014)
7. Pielot, M., De Oliveira, R., Kwak, H., Oliver, N.: Didn't you see my message?: Predicting attentiveness to mobile instant messages. In: Proceedings of the SIGCHI Conference on Human Factors in Computing Systems, CHI'14, pp. 3319–3328 (2014)
8. Mudliar, P., Rangaswamy, N.: Offline strangers, online friends: bridging classroom gender segregation with whatsapp. In: Proceedings of the 33rd Annual ACM Conference on Human Factors in Computing Systems, CHI'15, pp. 3799–3808 (2015)
9. Sanchez, J.: Malicious threats, vulnerabilities and defenses in WhatsApp and mobile instant messaging platforms (2014)
10. Jain, J., Luaran, J.E., Binti Abd Rahman, N.: Learning Beyond the Walls: The Role of Whatsapp Groups, pp. 447–457 (2016)
11. Gulacti, U., Lok, U., Hatipoglu, S., Polat, H.: An analysis of whatsapp usage for communication between consulting and emergency physicians. J. Med. Syst. **40**(6), 1–7 (2016)
12. Burkhardt, J.M.: Combating Fake News in the Digital Age, vol. 53, 8th edn. ALA TechSource (2017)
13. Ahmed, H., Traore, I., Saad, S.: Detecting opinion spams and fake news using text classification. Secur. Priv. **1**(1), (2017). https://doi.org/10.1002/spy2.9
14. Lazer, D., Matthew, B., Benkler, Y., Berinsky, A., Greenhill, K., Menczer, F., Metzger, M., Nyhan, B., Pennycook, G., Rothschild, D., Schudson, M., Sloman, S., Sunstein, C., Thorson, E., Watts, D., Zittrain, J.: The science of fake news. Dissemination in WhatsApp: gathering, analyzing and countermeasures. In: Proceedings of the Web Conference (2018)
15. Explained.: What is Fake News? Social Media and Filter Bubbles. 02 July 2018. Retrieved from https://www.webwise.ie/teachers/what-is-fake-news/
16. Rosenfield, A., Sina, S., Sarne, D., Avidov, O., Kraus, S.: WhatsApp usage patterns and prediction models. In: ICWSM/IUSSP Workshop on Social Media and Demographic Research (2016)
17. Garimella, K., Tyson, G.: WhatsApp, doc? A first look at whatsapp public group data (2018)
18. Bhatt, A., Arshad, M.: Impact of WhatsApp on Youth: A Sociological Study (2016)
19. Sharma, S., Kumar, N.: Survey Analysis on the Usage and Impact of WhatsApp Messenger (2017)
20. Seufert, M., Hobfeld, T., Schwind, A., Burger, V., Tran-Gia, P.: Analysis of Group-based Communication in WhatsApp (2016)
21. Mueller, R., Schrittwieser, S., Fruehwirt, P., Kieseberg, P., Weippl, E.: What's new with WhatsApp & Co.? Revisiting the security of smartphone messaging applications (2014)

22. Shu, K., Sliva, A., Wang, S., Tang, J., Liu, H.: Fake news detection on social media: a data mining perspective. ACM SIGKDD Explor. Newsl. **19**(1), 22–36 (2017)
23. Jin, Z., Cao, J., Zhang, Y., Luo, J.: News verification by exploiting conflicting social viewpoints in microblogs. In: AAAI, pp. 2972–2978 (2016)
24. Tacchini, E., Ballarin, G., Della Vedova, M.L., Moret, S., de Alfaro, L.: Some like it hoax: automated fake news detection in social networks. Preprint at arXiv:1704.07506 (2017)
25. Shu, K., Bernard, H.R., Liu, H.: Studying fake news via network analysis: detection and mitigation. In: Emerging Research Challenges and Opportunities in Computational Social Network Analysis and Mining, pp. 43–65. Springer (2019)
26. Gitanjali, J., Banu, S.N., Indumathi, J., Uma, G.V.: A panglossian solitary-skim sanitization for privacy preserving data archaeology. Int. J. Electr. Power Eng. **2**(3), 154–165 (2008)
27. Gitanjali, J., Banu, S.N., Geetha, M.A., Indumathi, J., Uma, G.V.: An agent based burgeoning framework for privacy preserving information harvesting systems. Int. J. Comput. Sci. Netw. Secur. **7**(11), 268–276 (2007)
28. Indumathi, J.: A generic scaffold housing the innovative modus operandi for selection of the superlative anonymisation technique for optimized privacy preserving data mining (Chap. 6). In: Karahoca, A. (ed.) Data Mining Applications in Engineering and Medicine, vol. 335, pp. 133–156. In Tech (2012). ISBN: 9535107200 9789535107200
29. Indumathi, J.: Amelioration of anonymity modus operandi for privacy preserving data publishing (Chap. 7). In: Amine, A., Mohamed, O.A., Benatallah, B. (eds.) Network Security Technologies: Design and Applications, vol. 330, pp. 96–107. Tahar Moulay University, Algeria, Concordia University, USA, University of New South Wales, Australia (2013a). Release Date: November, 2013
30. Indumathi, J.: An enhanced secure agent-oriented burgeoning integrated home tele health care framework for the silver generation. Int. J. Adv. Netw. Appl. **04**(04), 16–21 (2013b). Special Issue on "Computational Intelligence—A Research Perspective" held on 21st–22nd Feb 2013
31. Indumathi, J.: State-of-the-Art in reconstruction-based modus operandi for privacy preserving data dredging. Int. J. Adv. Netw. Appl. **04**(04), 9–15 (2013c). Special Issue on "Computational Intelligence—A Research Perspective" held on 21st–22nd Feb 2013
32. Indumathi, J., Uma, G.V.: Customized privacy preservation using unknowns to stymie unearthing of association rules. J. Comput. Sci. **3**(12), 874–881 (2007)
33. Indumathi, J., Uma, G.V.: Using privacy preserving techniques to accomplish a secure accord. Int. J. Comput. Sci. Netw. Secur. **7**(8), 258–266 (2007)
34. Indumathi, J., Uma, G.V.: A bespoked secure framework for an ontology-based data-extraction system. J. Softw. Eng. **2**(2), 1–13 (2008)
35. Indumathi, J., Uma, G.V.: A new flustering approach for privacy preserving data fishing in tele-health care systems. Int. J. Healthc. Technol. Manag. **9**(5–6), 495–516 (2008b). Special Issue on: "Tele-Healthcare System Implementation, Challenges and Issues" (22)
36. Indumathi, J., Uma, G.V.: A novel framework for optimized privacy preserving data mining using the innovative desultory technique. Int. J. Comput. Appl. Technol. **35**(2/3/4), 194–203 (2008c). Special Issue on: Computer Applications in Knowledge-Based Systems (2008 in press)
37. Indumathi, J., Uma, G.V.: An aggrandized framework for genetic privacy preserving pattern analysis using cryptography and contravening—conscious knowledge management systems. Int. J. Mol. Med. Adv. Sci. **4**(1), 33–40 (2008)
38. Murugesan, K., Gitanjali, J., Indumathi, J., Manjula, D.: Sprouting modus operandi for selection of the best PPDM technique for health care domain. Int. J. Conf. Recent. Trends Comput. Sci. **1**(1), 627–629 (2009)
39. Murugesan, K., Indumathi, J., Manjula, D.: An optimized intellectual agent based secure decision system for health care. Int. J. Eng. Sci. Technol. **2**(5), 3662–3675 (2010)
40. Murugesan, K., Indumathi, J., Manjula, D.: A framework for an ontology-based data-gleaning and agent based intelligent decision support PPDM system employing generalization technique for health care. Int. J. Comput. Sci. Eng. **2**(5), 1588–1596 (2010)

41. Vasudevan, V., Sivaraman, N., SenthilKumar, S., Muthuraj, R., Indumathi, J., Uma, G.V.: A comparative study of SPKI/SDSI and K-SPKI/SDSI systems. Inf. Technol. J. **6**(8), 1208–1216 (2007)

42. Schrittwieser, S., Frühwirt, P., Kieseberg, P., Leithner, M., Mulazzani, M., Huber, M., Weippl, E.R.: Guess who's texting you? Evaluating the security of smartphone messaging applications. In: NDSS (2012)

43. Garimella, K., Tyson, G.: Whatapp Doc? A first look at whatsapp public group data. In: Twelfth International AAAI Conference on Web and Social Media (2018)

44. Caetano, J.A., de Oliveira, J.F., Lima, H.S., Marques-Neto, H.T., Magno, G., Meira Jr, W., Almeida, V.A.: Analyzing and characterizing political discussions in WhatsApp public groups. Preprint at arXiv:1804.00397 (2018)

45. Wani, S.A., Rabah, S.M., AlFadil, S., Dewanjee, N., Najmi, Y.: Efficacy of communication amongst staff members at plastic and reconstructive surgery section using smartphone and mobile WhatsApp. Indian J. Plast. Surg. Off. Publ. Assoc. Plast. Surg. India **46**(3), 502 (2013)

46. Resende, G., Melo, P., Sousa, H., Messias, Marisa, J., Vasconcelos, M., Almeida, J., Benevenuto, F.: (Mis)Information (2019)

47. Najmi, Y.: Efficacy of communication amongst staff members at plastic and of India. **46**(3), 502 (2013)

Statistical Approaches to Detect Anomalies

G. Sandhya Madhuri and M. Usha Rani

Abstract The term anomaly is derived from a Greek word *anomolia* meaning uneven or irregular. Anomalies are often referred to as outliers in statistical terminology. For a given set of data if we plot a graph and observe, all the data points that are relative to each other will be plotted densely, whereas some data points which are irrelevant to the data set will be lied away from the rest of the points. We call those points as outliers or anomalies. Anomaly detection is also called as *deviation detection,* because outlying objects have attribute values that are significantly different from expected or typical attribute values. The anomaly detection is also called as *exception mining* because anomalies are exceptional in some sense. Anomalous data object is unusual, irregular or in some way, inconsistent with other data objects. In this case, unusual data object or irregular patterns need not be termed as not occurring frequently. If we take a large data set or a continuous data stream, then an unusual data object, that occurs 'one in a thousand' times, can occur millions of times in billions of events considered. To find out the anomalies in data sets, we have many approaches like *statistical, proximity–based, density–based and cluster–based.* Statistical approaches are model-based approaches where a model is created for the data and objects are calculated with respect to how they are relative with all other objects. In this paper, we will be discussing various statistical approaches to detect anomalies. Most statistical approaches to outlier detection are based on developing a probability distribution model and considering how probable objects are under that model.

Keywords Deviation · Exceptions · Data stream · Anomalies · Statistical approaches

G. S. Madhuri (✉) · M. Usha Rani (✉)
Department of Computer Science, SPMVV, Tirupati, India
e-mail: sandhyamadhuri@gmail.com

M. Usha Rani
e-mail: musha_rohan@yahoo.com

P. Venkata Krishna and M. S. Obaidat (eds.), *Emerging Research in Data Engineering Systems and Computer Communications*, Advances in Intelligent Systems and Computing 1054, https://doi.org/10.1007/978-981-15-0135-7_46

1 Introduction

In anomaly detection, the basic idea is to find objects that are different from most other objects. Regularly, anomalous objects are called as outliers, because when data is plotted, these points occur far away from other data points [1].

Typically, abnormal information will be connected to some reasonable drawback or rare event like, e.g. bank fraud, medical issues, structural defects, defective instrumentation, etc. This association makes it attention-grabbing to be able to discover that information points will be thought of as anomalies [2].

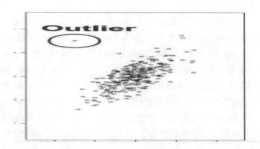

1.1 Statistical Approaches to Find Anomalies

Let us observe the statistical approaches in detail.

Statistical approaches are having a firm foundation in outlier identification if sufficient knowledge is available about data and type of tests to be applied. Then these approaches give effective results.

Statistical approaches are model-based approaches that mean a model or a prototype is created for the data, and objects are evaluated with respect to how good they fit in the model.

Most statistical approaches to find anomalies are based on building a probability distribution model and considering how probable objects are under that model.

Probabilistic definition of an Outlier: An outlier is an object that has a low probability with respect to probability distribution model of data.

A probability distribution model is created from the data estimating the parameters of a user–specified distribution. If the data is supposed to have a Gaussian distribution, then the mean and standard deviation of the basic distribution can be assessed by calculating the mean and standard deviation of the data. The probability of each entity under the distribution can then be assessed [3, 4].

A wide variety of statistical tests based on the probabilistic definition of the data have been developed to detect outliers, or discordant observations, as they are often called in terms of statistical literature.

Many of these discordancy tests are highly specialised and assume a level of statistical knowledge beyond the scope of this paper. Thus, we illustrate the basic ideas with a few examples.

2   Detecting Outliers in a Univariate Normal Distribution

The Gaussian (normal) distribution is one of the most commonly used distributions in statistics. Let us use it to define a simple approach to statistical outlier detection. This distribution has two parameters, μ and σ, which are mean and standard deviation, respectively, and is represented using the notation $N(\mu, \sigma)$. The figure below shows the density function of $N(0, 1)$ (Fig. 1).

There is a meagre chance that an object or a value from a $N(0, 1)$ distribution will occur in tails of the distribution. For instance, there is only a probability of 0.0027 that an object lies beyond the central area between $\pm$ 3 standard deviations. More generally, if c is a constant and x is the attribute value of an object, then the probability that $|x| \geq c$ decreases rapidly as c increases.

Let $\alpha = \text{prob}(|x| \geq c)$. Below table shows some sample values for c and the corresponding values for α when the distribution is $N(0, 1)$. Note that a value that is more than four standard deviations from the mean is a one–in–ten—thousand occurrences.

Let us observe some sample pairs

$$(c, \alpha), \alpha = \text{prob}(|x| \geq c)$$

Fig. 1 Probability of a Gaussian distribution whose a mean is 0 and standard deviation 1

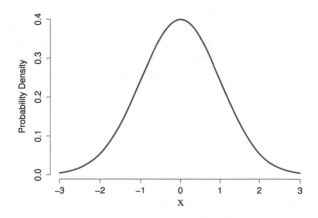

for a Gaussian distribution with mean 0 and standard deviation 1

c	α for $N(0, 1)$
1.00	0.3173
1.50	0.1336
2.00	0.0455
2.50	0.0124
3.00	0.0027
3.50	0.0005
4.00	0.0001

To detect whether an object (value) is an outlier or not, if we consider the fact that the value's distance(c) from the centre of the distribution $N(0, 1)$ is directly related to the value's probability, then we can take it as the criteria to find probability.

Let us examine the definition below to find out anomalies or outliers.

Outlier for a Single $N(0, 1)$ Gaussian Attribute: An object with attribute value x from a Gaussian distribution with mean of 0 and standard deviation 1 is an outlier if $|x| \geq c$, where c is a constant chosen so that $\mathrm{prob}(|x|) \geq c = \alpha$.

To use the above definition, it is essential to state a value for α. From the point of view that unusual objects indicate a value from another different distribution, α indicates the probability that we erroneously categorise an object from the given distribution as an outlier. From the viewpoint that an outlier is a rare value of a $N(0, 1)$ distribution and α represents the degree of rareness.

Now, if a distribution of an attribute of interest for normal objects has a Gaussian distribution with mean μ and standard deviation σ, i.e. a $N(\mu, \sigma)$ distribution, then to use definition mentioned above, we have to change the attribute x to a new attribute t, which has a $N(0, 1)$ distribution.

In precise, the approach is to set $t = (x-\mu)/\sigma$. (t is typically called as t-score). Here, μ and σ are unknown and are assessed using the sample mean x and standard deviation S_x. In practice, this works well when the number of observations is large. However, we note that the distribution of z is not actually $N(0, 1)$. A more sophisticated statistical procedure is Grubb's test.

3 Grubb's Test to Detect Outliers in Statistical Approaches

The Grubb's test is statistically refined procedure for identifying outliers. It is iterative and contemplates the t-score that does not have a normal distribution. This algorithm calculates t-score of all the objects (values) based on the sample mean and standard deviation of the existing set of values. The objects (values) which have high magnitude *t*-score will be termed as outliers (discarded). This process is repeated till there are no values to be eliminated or discarded [5].

```
Grubb's Algorithm:
1: Input the values and α
{m is the number of values, α is a parameter, and t_c is the
value chosen so that,
α = prob(x ≥ t_c) for a t distribution with m-2 degrees of
freedom.}
2: repeat
3: Compute the sample mean (x̄) and standard deviation (s_x).
4: Compute a value g_c so that prob(|z| ≥ g_c = α
```

(In terms of t_c and m, $g_c = \frac{m-1}{\sqrt{m}}\sqrt{\frac{t_c^2}{m-2+t_c^2}}$)

```
5: Compute the t-score of each value, i.e., t = (x - x̄)/s_x
6: Let g = max|z|.i.e., find the t-score of largest magnitude
and call it g.
7: If g > g_c then
8: Eliminate the value corresponding to g.
9: m ← m-1
10: end if
11: until No objects are eliminated
```

Temperature	Gender	Heart rate
98.4	Male	84
98.4	Male	82
98.2	Female	65
97.8	Female	71
98	Male	78
97.9	Male	72
99	Female	79
98.5	Male	68
98.8	Female	64
99	Male	75
98.9	Male	80
98.7	Female	72
98	Male	71
97.1	Male	75
98	Female	73
100.8	Female	77
99.5	Male	75
98.8	Male	78
98	Male	67
97.4	Male	78

Fig. 2 The points that are
distantly away from the centre
points in the normal
distribution are outliers

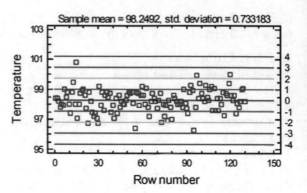

Let us try and apply Grubb's test on sample data in the below
Table 1.1 collected from https://github.com/droglenc/NCData/blob/master/
BodyTemp.csv

Let us see the results using Grubb's test (Fig. 2).

4   Outliers in Multivariate Normal Distribution

For multivariate Gaussian observations, lets us see such an approach similar to that
given for a univariate Gaussian distribution.

The points will be considered as outliers if they have low probability with
respect to the estimated distribution of the data. This can be established by a simple
test; for example, distance of a point from the centre of the distribution.

Because of the correlation between the different variables, a multivariate normal
distribution is not symmetrical with respect to its centre [5].

Below figure shows the probability density of a two-dimensional multivariate
Gaussian distribution with mean (0, 0) and a covariance matrix of

$$\sum = \begin{bmatrix} 1.00 & 0.75 \\ 0.75 & 3.00 \end{bmatrix}$$

If we have to use a simple threshold to find out whether an object is an outlier,
then we will need a distance measure that looks like data distribution into account.
The Mahalanobis distance is such a distance. This is the distance between a point
x and the mean of the data $\bar{x}$ is shown in the below equation.

$$\text{Mahalanobis}(x, \bar{x}) = (x - \bar{x})S^{-1}(x - \bar{x})^T$$

where S means the covariance matrix of the data.

It is simple to show that the Mahalanobis distance of a point to the mean of the
underlying distribution is directly related to the probability of the point.

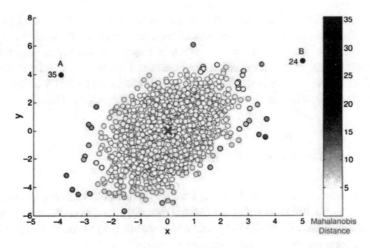

Fig. 3 Mahalanobis distance from the centre of a two-dimensional set of 2002 points

In particular

$$\text{Mahalanobis distance} = \log$$
$$(\text{probability of a density of the point} + \text{a constant})$$

As an example, check the figure below which shows the Mahalanobis distance from the mean of the distribution (Fig. 3).

Here, the points $A(-4, 4)$ and $B(5, 5)$ are outliers. The Mahalanobis distance of points A and B is given as 35 and 24, respectively [5].

As both A and B unlike all other 2000 points have large Mahalanobis distance, we consider them as outliers. When the Euclidean distance of A and B from centre is calculated, A is closer when compared to B but Mahalanobis takes the shape of the distribution into consideration so their distances are far from other 2000 points.

Let us further observe other kinds of approaches for detecting anomalies.

5 Hypothesis Testing

A simple statistical technique for anomaly detection can be based on finding whether the test sample(s) come from the same distribution as that of the training data or not.

For example, let us see how Ruotolo and Surace used this approach using a t-test to find damaged beams. It starts by taking n_1 measurements of the first n_f natural vibration frequencies of the structure at the beginning stage, i.e. as soon as the structure is built. After that, they intermittently take n_2 measurements of the same natural frequencies and use a t-test to compare n_1 and n_2. If the test shows note-worthy difference between the two sets of measurements, then damage is present.

The approach was tested on a case study including a continuous beam. The experimental data of the undamaged and damaged beam was obtained running a finite element program where the cracks are simulated. The test was executed at 0.05 and 0.01 significance levels and showed encouraging results.

6 Parzen Density Estimation

Parzen windows method can be used for non-parametric data density estimation. Yeung and Chow follow a well-established anomaly detection approach, centred on assessing the density of the training data and discarding patterns. The authors put on their technique on an intrusion detection problem.

The authors have selected Gaussian kernel functions for two reasons. Initially, the Gaussian function is smooth and hence the density estimation also varies easily and another, if a centrifugally symmetrical Gaussian is assumed, the function can be totally identified by a variance parameter only. The uniqueness threshold is set using a distinct training set called 'threshold determination set' and it is used on the unconditional probability $p(x)$ of a test pattern x based on the modelled distribution.

7 *kNN* Based Approaches

The *k-Nearest Neighbour* algorithm is another technique for calculating the density function of data. This technique eradicates some of the difficulties of Parzen win-dow. It does not involve a smoothing parameter. As an alternative, the width parameter is set as an outcome of the position of the data point in relation to other data points by considering the k-nearest patterns in the training data to the test pattern [6].

The problem with this technique is that for large-sized data sets, a large number of computations have to be performed. For anomaly detection, the distribution of normal vectors is described by a small number of spherical clusters placed by the k-nearest neighbour technique. Anomaly is evaluated by measuring the normalised distance of a test sample from the cluster centres. A number of studies have used such an approach to anomaly detection as detailed below. Hellman used the nearest neighbour (NN) classifier for discarding patterns with higher risk of being mis-classified same as Chow used threshold. Chow though supposed that for a given data, a priori probabilities and conditional probability density are known. In most real-world applications, conversely these statistics are indefinite and have to be concluded from a set of characterised examples.

The advantage of this approach is that it makes no suppositions regarding these original statistics. Moreover, Cover and Hart have shown that as the number of training samples inclines to infinity, the nearest neighbour risk is no way better than twice the Bayes risk, irrespective of the underlying statistics [7].

Hellman elucidated that based on single NN, not enough information is there to reject the samples. More information is essential and it is given by considering two NNs. If both come from the same class, then the pattern is categorised otherwise it is excluded. Using the new rule, the two NNs, more samples are rejected.

Therefore, the new rule has always lower cost when compared to the single NN. Hence, this rule can be extended to examine k NNs and categorise only when all the neighbours agree. This approach is to examine the k NNs, and if all neighbours agree, classify the test pattern else discard it.

8 A Mixture Model Approach for Anomaly Detection

In this model, basically the data is modelled as mixture of two distributions; one normal data and one for outliers.

Our goal is to estimate the parameters of the distributions so that the probability of the data can be maximised. At the outset, all the objects are put in a set of normal objects and the set that has anomalous objects is kept empty.

Now, all the objects are transferred using an iterative process from normal set to the anomalous set as long as the likelihood of the data is increasing.

Let us consider that D is the data set that contains mixed objects (both normal and anomalous) from a mixture of two probability distributions. M is the distribution that has normal objects majorly and A is the distribution of anomalous objects. The overall probability can be represented as:

$$D(x) = (1 - \lambda)M(x) + \lambda A(x)$$

where x is an object and λ is the number between 0 and 1 that gives the expected fraction of outliers.

Algorithm: Likelihood—based outlier detection

1. Initialization: At time t = 0, let M_t contain all the objects, While A_t is empty.
   Let $LL_t(D) = LL(M_t) + LL(A_t)$ be the log likelihood of all the data.
2. for each object x that belongs to M_t do
3. Move x from M_t to A_t to produce new datasets A_{t+1} and M_{t+1}
4. Compute the new log likelihood of D,
   $LL_{t+1}(D) = LL(M_{t+1}) + LL(A_{t+1})$
5. Compute the difference, $\Delta = LL_t(D) - LL_{t+1}(D)$
6. if $\Delta > c$, where c is some threshold then

7 $\times$ is classified as an anomaly, i.e., A_{t+1}, M_{t+1} are left
   unchanged and become the current and normal and anomaly
   sets.
8 end if
9 end for

9 Issues with Statistical Approaches of Anomaly Detection

Identifying the suitable distribution: Almost all types of data can be described by commonly used distributions like Gaussian, Poisson or binomial distributions. There are certain data sets that are with non-standard distributions. In this scenario, if a wrong model is chosen, then an object can be erroneously classified as outlier. For instance, the data may be modelled as coming from Gaussian or Poisson distributions but is actually having high probability than the later; then values will obviously be far from mean. Statistical distributions with such behaviour are called heavy-tailed distributions.

Number of Attributes: Almost all statistical outlier detection techniques apply to a single attribute; only some techniques are for multivariate data [5].

10 Conclusion

In this paper, we have discussed statistical approaches for anomaly detection. There is further scope for defining techniques for multivariate data. The techniques suitable for mixture models can be explored further.

References

1. Chandola, V., Banerjee, A., Kumar, V.: Anomaly detection: a survey. ACM Comput. Surv. **41**(3) (2009)
2. Goldstein, M., Uchida, S.: A comparative evaluation of unsupervised anomaly detection algorithms for multivariate data. PLoS ONE **11**(4), 31 (2016)
3. Markou, M., Singh, S.: Novelty detection: a review—part 1: statistical approaches. Sig. Process. **83**(12), 2481–2497 (2003)
4. Hawkins, J., Ahmad, S., Lavin, A.: Biological and Machine Intelligence. Available at: http://bit.ly/2qJzvjO (2016)
5. P-N Tan.: Introduction to *"Data Mining"*, Michigan State University, Michael Steinbach, *University of Minnesota*, Vipin Kumar, University of Minnesota and Army High Performance Computing Research Center
6. https://journals.plos.org/plosone/article?id=10.1371/journal.pone.0152173

7. Song, X., Wu, M., Jermaine, C.: Conditional anomaly detection. IEEE Trans. Knowl. Data Eng. **19**(5), 631–645 (2007)
8. The data source for the Table 1.1 is https://github.com/droglenc/NCData/blob/master/BodyTemp.csv

An Extensive Survey on Some Deep-Learning Applications

Jabeen Sultana, M. Usha Rani and M. A. H. Farquad

Abstract Deep learning prospered as a distinct era of research and fragment of a wider family of machine learning, based on a set of algorithms that strengthen to model high-level abstractions in data. It tries to imitate the human intellect and learns from complicated input data and resolve different types of difficult and complex tasks. Because of Deep Learning, it was successful to deal with different input data types such as text, sound, and images in various fields. Improvement in deep-learning research has already influenced the search for speech recognition, automatic navigation systems, parallel computations, image processing, ImageNet, natural language processing, representation learning, Google translate, etc. Here, we present a review of DL and its applications including the recent development in natural language processing (NLP).

Keywords Deep learning (DL) · Natural language processing (NLP) · Image processing (IP) · Speech recognition · Parallel computations

1 Introduction

The performance of machine-learning techniques is highly inclined on the basis of data representation selection, largely a product of machine-learning methods to which they are applied [1]. Preprocessing pipelines and data transformations is the by-product of the effort in strategic spreading out of machine-learning algorithms. They contribute in data representation to provide for active machine learning, and focus the drawbacks of current learning algorithms [2]. This drawback can be counterbalanced by feature engineering, which helps to overcome to extract and organize the discriminative information from the data by compelling precise benefit

J. Sultana · M. Usha Rani (✉)
Department of Computer Science, SPMVV, Tirupati, India
e-mail: musha_rohan@yahoo.com

M. A. H. Farquad
InI Labs Inc., Waterloo, Ontario, Canada

© Springer Nature Singapore Pte Ltd. 2020
P. Venkata Krishna and M. S. Obaidat (eds.), *Emerging Research in Data Engineering Systems and Computer Communications*, Advances in Intelligent Systems and Computing 1054, https://doi.org/10.1007/978-981-15-0135-7_47

511

of the social inventiveness and prior knowledge [3]. In order to make progress toward artificial intelligence, it is highly recommendable to reduce the dependency of feature engineering which would also aid in constructing/making novel applications, thereby simplifying applications of machine learning and contributing to broaden the scope. DL provides a perception to analyze multiple-layered input models and neural nets comprising several stages of non-linear operations. Investigating one of the deep architectures parameter space was a difficult task earlier to 2006. To counter this problem, recently proposed DL algorithms have achieved success because of tremendous advancement in the technology [4]. DL is an unsupervised pre-training method where features are learnt from top and are arranged in hierarchical manner [5]. Basically, feature learning implies to reconstruct the original data by learning new transformations over the features learned in the past at each level [4, 6, 7]. Deep generative model or neural-net classifier such as a Deep Boltzmann Machine ensures the groups of layers trained with weights are heaped to predict a deep monitored network [8].

2   Importance of Deep Learning Over Other Techniques

As Carlos said, ANNs [9] are strong in a way that they are global approximations, however that not the only plausible spot as to why DL is a strong tool. However, precisely keeping in mind, deep learning is not first heaping a bunch of neurotic layers & training thereby, it also includes the amazing starters that come into sight while considering compositionality (that the world around is comprised of complex characteristics which are in turn constructed out of smaller simpler features) which are beneficial while doing automaton learning in general. Two important points to be noted here are

- Distributed Representations: Classical methods like KNN, SVMs, decision tress, etc., can divide the space of features proportional to their parameters. Keep this in mind, when you go to higher dimensions, due to curse of dimensionality, you need exponentially many examples to achieve the same. However, in deep-learning models, you can attain this approximately in a linear number of instances.
- The power of depth: Again, let us take a similar example. Representing an arbitrary Boolean function in a single-hidden layer will take exponential number of nodes, whereas if you increase the depth, you will need a linear number of nodes, subsequently you can just condense the most repeated computations in the lower layers. Henceforth reduced number of parameters as well. However, these two are just most basic priors that make deep learning special and much dominant compared to other methods, but there are innumerable others that remain unexplored.

3 Deep-Learning Applications

A. DL in natural language processing

Apart from recognizing the speech, DL has a lot of implications to several applications of natural language processing. Word embedding is one of the most important applications. The concept of representing the figurative data through distributed representation was initiated by Hinton [10]. The first development for statistical linguistic modeling based on contexts by Bengio et al. [11]. In distributed presentation learning, word embedding is implemented to map set of words to vectors using deep convolutional network by Collobert et al. [12, 13]. He later on researched which resulted in the development of SENNA system, which shares the representations over varied NLP tasks [14]. The results make evident that the DL approach is far better than traditional approaches.

One of the important contributions of Colbert's work is away from tasks specific "man-made" feature engineering related to learn multiplicity and combined characteristics in variably from deep learning. The process explained in variably learns inner representations [12, 13] through huge amounts of usually unlabeled training data [5, 15]. It describes competence features important to the numerous distinguished NLP tasks, also comprises of part of speech, tagging, chunking labeled entity given much defined previous knowledge. These tasks are united into unique structure and combindly trained. With exception to the language model, they rest of the tasks are monitored tasks with labeled training data.

Concerning the approaches, DL approach is distinct from the conventional NLP approach. Therefore, the choice of features is largely grounded on trial and error like empirical process and the selection of features relies on that task, signaling extra research for each NLP tasks. NLP tasks such as POS have achieved success through this. However, complicated tasks like SRL are in need of a large number of possibly complex (ex- extracted or filled out of a parts tree) that results in making systems slow and inflexible to a large-scale applications. DL methods have been applied for image caption generation [15] and handwriting generation [16].

B. DL in Speech Recognition

According to the surveys by Yann, Bengio and Hinton [17], it has been stated that deep-learning systems have inevitably resulted in high accuracy improving recognition of speech, and numerous profound deep architectures and learning approaches have been advanced with matchless gains [18, 19]. Inputs are extensively layered in characteristics and are non-linear classification problems [20]. In 2011, neural network was designed and trained with deep-learning algorithms in the project named as Google Brain. Just after watching YouTube videos, it identified complex level notions resembling cats, without being told what a "cat" is surprisingly. In order to enhance the recognizing capability of faces and objects in the photographs and uploaded videos [21], Facebook started generating solutions using deep-learning expertise, respectively [22–24].

C. DL in Automatic Navigation Systems

A novel deep-learning-based obstacle detection framework was suggested to advance the appearance and appropriate cues. One of the vital capabilities for self-driving cars is to detect the minor highway dangers, such as misplaced cargo, exciting, and hardly ever appearances problem with a visualization method that influences presence background, as well as geometric cues [25]. To foresee a semantic labeling pixel-wise at variant level, a fully convolutional network was proposed. A modern detection approach was drawn in order to predict obstacles from stereo input images. In order to define hurdles in a smooth, dense, and strong manner, the intermediate pixel demonstration is utilized. Very exciting scenes were obtained by evaluating the novel obstacle discovery method on lost and found dataset.

For autonomous driving, a new direct perception deep-learning algorithm was introduced in this paper [26]. Contrasting previous efforts that focused on the feature extraction capabilities of a given CNN, the authors considered autonomous driving performance for different deep-learning architectures. For example, here, the algorithm works without making assumptions about the speed of the car, instead the algorithm uses a more realistic model by assuming the availability of additional sensors that provide distances to the cars around the autonomous vehicle and comparison was drawn to measure the performance of the top three CNNs in road feature extraction; the results [26] showed that GoogleNet is the most accurate CNN for that task. In addition to the feature extraction performance, it was suggested to use additional parameters to evaluate the performance of autonomous driving. Finally, the suggested parameters to compare our algorithm against the previous direct perception algorithm found improvement in the performance of the model relative to the previous effort and struggles to navigate the whole track. The reason for this improvement is due to removing the overlapped and redundant affordance parameters that are used in [3] in addition to selecting a CNN model that is proficient in digging out road features more accurately than the model they used.

D. DL in Parallel Computations

When dealing with huge real life time-consuming datasets like audio, visual natural language processing. To solve time-consuming problems faced by deep neural networks, parallel algorithms have been proposed in the recent years. Implemention of the algorithm becomes indispensable. The demand to compute high-end problems with maximum speed lead to the advent of intensive algorithms in combination with massive datasheets for applications [27]. There is an intensive training process in deep neutral networks. The process, as a result, becomes more and more time consuming when the data size increases. Therefore, ascends the necessity for exploring competent ways in parallel execution.

GPU and CPU are used in GPU-accelerated computing. Using compute unified device application (CUDA) , which is a programming model. The major computational part of the program runs on GPU and the minor sequential program executes on CPU. Parallel virtual machine, message-passing interface, and GPU's for large-scale neural networks training [28–30] are some of the parallel programming software tools utilized here. DistBelief permits parallelism using multi-threading concept and message-passing concept is applied across machines inside a machine [31]. Deep neural networks execution part on GPU's has been reported in [32].

E. DL in Image Processing

In this paper, taking into account the difficulty of identifying human actions out of the still images, a new approach was suggested which can treat the posture of the individual in the photograph (image) as hidden fluctuations (variables) that aids identification [33]. It is unique from other works that learns each system for posture estimation and identification of the action, later combining them in an ad hoc style, the system is thereby trained in an combined style that together takes into account postures and actions. The learning objective is framed to directly correct the poster information for identifying the action. The experimental result shows that by predicting the hidden postures, we can improvise the resultant action identification results.

The scope of video interpretation addresses recognition of human actions. To recognize action in the still images, bag-of-features methods and grouping with the latent SVM approach (part-based)can be applied [6]. Authors analyzed action recognition in still images, and improved action recognition performance can be achieved by investigating the role of background scene context on the datasets [8, 21]. Subsequent proposal stated here is to examine and detect real persons [4, 6] into the classifier.

F. DL in ImageNet

A wide range of complex tasks achieved state-of-art level of performances by deep neural networks (DNNs). Recently, the research is taking place to analyze black box features of DNN and to comprehend the learning mannerisms, tendency, and loop holes of DNNs [34]. The authors have gone deep inside the limitation of DNNs in image classification tasks and scrutinized it with the method inspire by cognitive psychology also, authors hypothesize that DNN's do not sufficiently learn to combine related classes of objects to crosscheck how DNN's comprehend the relatedness between objects of classes they conducted experiments. Also, [34] author observed that DNN's exhibit limited performance in establishing relatedness among object classes. Over and above, the DNN's showcase a better performance in finding relatedness based on sameness; however, they perform poorer in finding relatedness based on association. Through these experiments, a new analysis of learning behavior of DNN's is granted and the drawback which needs to be overcome is suggested accordingly.

G. DL in Representation Learning

One of the key benefits of deep learning is its representation learning. There are of course many incredible and stunning ideas using deep learning in the field of representation learning. In a sudden dramatic change in 2006, the initiation of deep learning and feature learning by Geoff Hinton lead to prompt follow up in the same year [6, 7] and were soon widely reviewed and discussed in [4, 34].

Deep Boltzmann machine, a deep generative can be formed by linking the set of layers to initialize a deep supervised predictor, such as a neural network classifier [8]. Layer wise heaping of characteristics improved representation. Deep architectures can be represented by using feature extraction and reduce classification error [35, 36]. This layer-wise procedure may also be implemented with the help of greedy layer-wise supervised pre-training [7].

Deep-learning methods have played a massive role in natural language processing. Most of the deep learning tasks in NLP have been oriented toward methods which using word vector representations [37]. Constant vector representations of words algorithms such as Word2Vec and Glove are deep learning techniques, which can convert words into meaningful vectors. The vector representations of words are very useful in text classification, clustering, and information retrieval. Word embedding's techniques have some advantages compare to bag-of-words representation. For instance, words close in meaning are near together in the word embedding space. Also, word embeddings have lower dimensionality than the bag-of-words [38]. The accuracy of the Word2vec and Glove depends on text corpus size. Meaning, the accuracy increases with the growth of text corpus. A method was proposed to learn continuous word representations for sentiment analysis on Twitter, which is a large social networks dataset [39]. Authors used Word2Vec method to learn the word embedding's on 50 M tweets and applied generated pre-trained vectors as inputs of a deep-learning model [40]. Recently, Lauren et al. [41] have proposed a discriminant document embedding's method which has used skip-gram for generating word embedding's of clinical texts. Fu et al. [42] applied Word2Vec for word embedding's of English Wikipedia dataset and Chinese Wikipedia dataset. The word embedding's used as inputs of recursive auto encoder for sentiment analysis approach. Ren et al. [43] suggested a new word representation method for Twitter sentiment classification. They used Word2Vec to generate word embedding's of some datasets in their method. Qin et al. [14] trained Word2Vec algorithm by English Wikipedia corpus, which has 408 million words. They used those pre-trained vectors as inputs of convolutional neural networks for data-driven tasks. Nevertheless, as already mentioned, these word embedding algorithms need a huge corpus of texts for training [44] and most of them ignore the sentiment information of text [39]. Because of the limitations and restrictions in some corpuses, researchers prefer to use pre-trained word embedding's vectors as inputs of machine-learning models. Kim [45] has used pre-trained Word2Vec vectors as inputs to convolutional neural networks and has increased the accuracy of sentiment classification.

4 Conclusion

Deep-learning paradigm has been deliberated above and critical review suggests the recent developments some deep-learning applications. DL has attained wide-ranging interests in diverse research areas and is an innovative field of machine learning, and since it can give world record brings about various characterization and relapse issues and datasets. Numerous partnerships including Google, Microsoft, and Nokia revise it effectively. Seeing deep adapting admirably requires scientific development and great information of probabilistic demonstrating. Learning calculations are problematical, and great instatement is critical. DL is advancing with innovative structures and learning techniques presented constantly. These include NLP, automatic navigation systems. In addition to this, most of the investigations of deep learning are practical, dense hypothetical basics of DL requirements to be recognized. Hybridized deep neural network model is also a potential research area, which needs to be identified.

References

1. Arel, I., Rose, D.C., Karnowski, T.P.: Deep machine learning—A new frontier in artificial intelligence research [research frontier]. Comput. Intell. Mag.-IEEE **5**(4), 13–18 (2010)
2. Schmidhuber, J.: Deep learning in neural networks: an overview. Neural Netw. Elsevier **61**, 85–117 (2015)
3. Coates, A., Lee, H., Ng, A.Y.: An analysis of single layer networks in unsupervised feature learning. AISTATS **15**, 215–223 (2011)
4. Bengiom Y.: Learning deep architectures for AI. Found. Trends Mach. Learn. **2**(1), 1–127 (2009)
5. Bengio, Y., Courville, A.: Representation learning: A review and new perspectives. IEEE Trans. Pattern Anal. Mach. Intell. **35**(8), 1798–1828 (2013)
6. Hinton, G.E., Osindero, S.: A fast learning algorithm for deep belief nets. Neural Comput. **18**(7), 1527–1554 2006
7. Bengio, Y., Lamblin, P.: Greedy layer wise training of deep networks. Adv. Neural. Inf. Process. Syst. **19**, 153 (2007)
8. R. Salakhutdinov, G. E. Hinton, "Deep Boltzmann machines, "*International Conference on Artificial Intelligence and Statistics, 2009*
9. Carlos: http://videolectures.net/deeplearning2015_montreal/
10. Hinton, G.E.: Learning distributed representations of concepts. In:: Proceedings of the Eighth Annual Conference of the Cognitive Science Society, Amherst, MA (1986)
11. Bengio, Y., Ducharme, R.: A neural probabilistic language model. J. Mach. Learn. Res. **3**, 1137–1155 (2003)
12. Collobert, R., Weston, J.: A unified architecture for natural language processing: deep neural networks with multitask learning. In: Proceedings of the 25th International Conference on Machine Learning, pp. 160–167. ACM (2008)
13. Collobert, R., Weston, J.: Natural language processing (almost) from scratch. J. Mach. Learn. Res. **12**, 24932537 (2011)
14. Qin, P., Xu, W., Guo, J.: An empirical convolutional neural network approach for semantic relation classification. Neuro Comput. **190**, 1–9 (2016)
15. Mohamed, A.: Deep belief networks for phone recognition. In: Nips Workshop on Deep Learning for Speech Recognition And Related Applications, vol. 1, pp. 635–645 (2009)
16. Deng, L., Yu, D.: Deep learning: methods and applications. In: Foundations and Trends® in Signal Processing, vol. 7, pp. 197–387 (2014)
17. Yann, L., Bengio, Y., Hinton, G.: Deep learning. Nature **521**(7553), 436–444 (2015)
18. Edgbaston: Formative Assessment: A Key to Deep Learning, Birmingham, UK (2007)

19. Sree, P.K., Babu, I.R., Devi, N.U.: A fast multiple attractor cellular automata with modified clonal classifier. promoter region prediction. J. Bioinf. Intell. Control **3**, 1–6 (2014). https://doi.org/10.1166/jbic.2014.1077

20. Karhunen, J., Tapani, R., Cho, K.H.": Unsupervised deep learning: a short review. In: Advances in Independent Component Analysis and Learning Machines, vol. 125 (2015)

21. Alison Rushton, A.: University of Birmingham

22. Sree, P.K., Babu, I.R, Devi, N.U.: Investigating an artificial immune system to strengthen the protein structure prediction and protein coding region identification using cellular automata classifier. Int. J. Bioinf. Res. Appl., **5**(6), 647–662 (2009) (Inderscience Journals, UK)

23. Sree, P.K., Babu, I.R., Devi, N.U.: Identification of promoter region in genomic DNA using cellular automata based text clustering. Int. Arab. J. Inf. Technol. (IAJIT) **7**(1), 75–78 (2010). ISSN: 1683-3198H Index (Citation Index): 05 (SC Imago, www.scimagojr.com)(Eleven Years Old Journal)(SCI Indexed Journal)

24. Sree, P.K., Babu, I.R., Devi, N.U.: Fast multiple attractor cellular automata with modified clonal classifier for coding region prediction in human genome. J. Bioinform. Intell. Control **3**, 16 (2014). https://doi.org/10.1166/jbic.2014.107 (American Scientific Publications, USA)

25. Stefan, G.: Detecting unexpected obstacles for self-driving cars: fusing deep learning and geometric modeling

26. Mohammed, A., Barjasteh, I., Al-Qassab, H., Radha, H.: Fellow, IEEE "Deep Learning Algorithm for Autonomous Driving using GoogleNet"

27. Soniya: A review on advances in deep learning. In: IEEE Workshop on Computational Intelligence: Theories, Applications and Future Directions (WCI) (2015)

28. Geist, A., Beguelin, A., Dongarra, J., Jiang, W., Manchek, R., Sunderam, V.: PVM: Parallel Virtual Machine—A Users' Guide and Tutorial for Networked Parallel Computing. MIT Press (1994)

29. Quinn, M.J.: Parallel Programming in C with MPI and Open MP. Tata McGraw-Hill Higher Education (2003)

30. Krizhevsky, A.: One weird trick for parallelizing convolutional neural networks. Comput. Res. Repository (CoRR), **abs/1404.5997** (2014)

31. Dean, J., Corrado, G., Monga, R., Chen, K., Devin, M., Le, Q.V., Mao, M.Z., Ranzato, M., Senior, A.W., Tucker, P.A., Yang, K., Ng, A.Y.: Large scale distributed deep networks. In: Advances in Neural Information Processing Systems 25: 26th Annual Conference on Neural Information Processing Systems 2012. Proceeding of a meeting held December 3–6, 2012, pp. 1232–1240, Lake Tahoe, Nevada, United States (2012)

32. Krizhevsky, A., Sutskever, I., Hinton, G.E.: Image Net classification with deep convolutional neural networks. In: Advances in Neural Information Processing Systems 25: 26th Annual Conference on Neural Information Processing Systems 2012. Proceedings of a meeting held on December 3

33. Yang, W.: Recognizing human actions from still images with latent poses. In: IEEE Conference on Computer Vision and Pattern Recognition (CVPR) (2010)

34. Lee, H.: Can Deep Neural Networks Match the Related Objects?: A Survey on Image Net-trained Classification Models (2017)

35. Larochelle, H., Bengio, Y., Louradour, J., Lamblin, P.: Exploring strategies for training deep neural networks. J. Mach. Learn. Res. **10**, 1–40 (2009)

36. Erhan, D., Bengio, Y., Courville, A., Manzagol, P.A., Vincent, P., Bengio, S.: Why does unsupervised pre-training help deep learning. *J. Mach. Learn. Res.* **11**, 625– 660 (2011)

37. Araque, O., Corcuera-Platas, I., Sánchez-Rada, J., Iglesias, A.: Enhancing deep learning sentiment analysis with ensemble techniques in social applications. Expert Syst. Appl. **77**, 236246 (2017)

38. Mikolov, T., Chen, K., Corrado, G., Dean, J.: Efficient estimation of word representations in vector space. ICLR Workshop (2013)

39. Tang, D., Wei, F., Yang, N., Zhou, M., Liu, T., Qin, B.: Learning sentiment-specific word embedding for twitter sentiment classification. In: Proceedings of the 52nd Annual Meeting of the Association for Computational Linguistics, vol. 1, pp. 1555–1565 (2014)

40. Severyn, A., Moschitti, A.: Twitter sentiment analysis with deep convolutional neural networks. In: Proceedings of the 38th International ACM SIGIR Conference on Research and Development in Information Retrieval, pp. 959–962 (2015)
41. Lauren, P., Qu, G., Zhang, F., Lendasse, A.: Discriminant document embedding's with an extreme learning machine for classifying clinical narratives. Neurocomputing 1–10 (2017)
42. Fu, X., Liu, W., Xu, Y., Cui, L.: Combine How Net lexicon to train phrase recursive auto encoder for sentence-level sentiment analysis. Neuro comput. **241**, 18–27 (2017)
43. Ren, Y., Wang, R., Ji, D.: A topic-enhanced word embedding for Twitter sentiment classification. Inf. Sci. **369**, 188–198 (2016)
44. Giatsoglou, M., Vozalis, M., Diamantaras, K., Vakali, A., Sarigiannidis, G., Chatzisavvas, K.: Sentiment analysis leveraging emotions and word embedding's. Expert Syst. Appl. **69**, 214–224 (2017)
45. Kim, Y.: Convolutional neural networks for sentence classification. In: Proceedings of the 2014 Conference on Empirical Methods in Natural Language Processing (EMNLP). pp. 1746–1751 (2014)
46. Du, T., Shanker, V.: Deep Learning for Natural Language Processing. *Eecis. Udel. Edu,* pp. 1–7 (2009)

A Cluster-Based Improved Expectation Maximization Framework for Identification of Somatic Gene Clusters

Anuradha Chokka and K. Sandhya Rani

Abstract The early identification of cancer disease is the key objective of this paper. Machine learning algorithm's contribution is significant for early recognition of somatic mutations in cancer patients. So, the study of classification and clustering play a vital role in predicting the somatic mutations patterns. As the size of gene variants and somatic mutation patterns in the tumor increases, it is essential and effective to predict the disease patterns using the machine learning models. In this proposed work, a novel framework is designed and implemented on the somatic cancer datasets. In this work, somatic mutation patterns are clustered using the related features of gene sequences by using the proposed improved expectation maximization (IEM) model. On each cluster, AdaBoost classifier is applied to classify somatic mutations patterns. Experimental results proved that the proposed clustering algorithm IEM is better than the traditional approaches in terms of cluster quality rate. The overall classification accuracy for all the clusters is also satisfactory.

Keywords Somatic mutations patterns · Improved expectation maximization · Classification and clustering

1 Introduction

Somatic mutations presence constitutes a major part in tumor cells. Somatic mutations association provides vision into the mechanism of mutation development, which consequences a cancer therapy. Cancer is caused by the combination of conservational, genetic and lifestyle aspects. The main work concentrated on the

A. Chokka · K. Sandhya Rani (✉)
Department of Computer Science, Sri Padmavati Mahila Visvavidyalayam,
Tirupati, AP, India
e-mail: sandhyaranikasireddy@yahoo.co.in

A. Chokka
e-mail: akshayagokul2009@gmail.com

© Springer Nature Singapore Pte Ltd. 2020
P. Venkata Krishna and M. S. Obaidat (eds.), *Emerging Research in Data Engineering Systems and Computer Communications*, Advances in Intelligent Systems and Computing 1054, https://doi.org/10.1007/978-981-15-0135-7_48

somatic mutations is to analyze the identification of tumor cells. The analysis of this data mainly focused on the identification of recurred mutated cells. For evaluating and bringing out the various types of cancer categories in analyzing the data is to group the associated data values into various clusters based on the similarity concept. One among the machine learning's clustering method is expectation maximization (EM) [1] technique. It was developed by Rubin, Dempster and Laired. It is an un-supervised learning technique. The number of clusters to be projected configures the input. EM paths the repetitive structure in order to get the maximum probability of expected maximum likclihood of attributes. As it follows the iterative structure and performs the E-step [2] which approximates the likelihood of every point fitting to each and every cluster and M- step [2], which re-estimates the parameter vector of the probability distribution of every class. This process will be repeated until the maximum saturation value reached beyond knowing the details of cluster formation. Thereby, it maintains high computational cost. A variation of EM technique called improved expectation maximization (IEM) technique is proposed in this paper which estimates the attributes of a data model and calculates the extreme probability estimates for the given data values without additional computational cost. In this study, six different cancer datasets are considered and are merged to form one cancer dataset. For classification of somatic mutational data, AdaBoost algorithm is considered in this paper.

2 Related Works

Some of the clustering techniques, which are applied on somatic mutations data, are presented below.

In [3], the authors presented a model on mutational data which utilizes K-means algorithm. The K-means clustering algorithm is obtained with various values of K which can be taken from small to larger numbers. The significant value of K is selected by considering certain criterion measure. In this paper, different number of clusters was tried, and it was analyzed whether the results are biologically meaningful. To group the somatic mutations, K-means clustering algorithm was used for every cancer type in the dataset.

To evaluate the permanence of the clusters, the authors Melissa Zhao and Yushi Tang implemented K-means algorithm with training data and also on the corresponding validation datasets [4]. The K-means clustering technique originated to be associated with the training data presented in the group of predictors. The results of K-means clustering technique is very stable and also reproduces the grouping of related data elements in the worst case.

The authors Aravind Sharma, R. K. Gupta and AkileshTiwari explained the performance evaluation of the effectiveness of DBSCAN clustering technique on synthetic datasets as well as on real data [5]. In this paper, the detection of clusters points is easily performed. They tried to implement the formalization of various clusters as well as outliers in a database of data points of n-dimensional region.

The aim of this proposed model is that for the data values of every cluster, the surrounding region of a given radius needs to be contained a minimum number of data points.

In [6], the authors discussed the new data approach to form clusters by considering the early objects and a given radius and minimum number of data elements. When the data values are implanting into the existed database, the clusters are restructured using DBSCAN clustering technique. The means between each object of clusters and the new coming data can be calculated and also can be inserted into a particular cluster based on the minimum mean distance.

The authors in [7] presented an algorithm based on expectation maximization. An evaluation of expectation maximization algorithm is carried out with the employment of the educational database of students. Every cluster with a particular probability is maintained, and the initial estimated values for the missing points are allocated. Then computes the maximum likelihood of the data points and iterates this process until they found a well-clustered data values. EM technique works well when the total number of clusters is known.

An Expectation Maximization Based Logistic Regression (EMLR) technique is used to identify and also to classify the breast cancer in [8]. The needed data was gathered from the cancer institution named Kuppuswamy Naidu Memorial Hospital, India. The classification was performed on that data using EMLR technique and able to classify different stages of cancer.

The combination of classification algorithms, AdaBoost classifier with random forest for building a prediction model of breast cancer data [9], is used. Random forest is used as a weak learner of the classifier AdaBoost, which selects the more weighted data values in the process to maximize the accuracy with the reduction of overfitting difficulties. The experimental results proved that the random forest with AdaBoost classifier outperforms the classification in terms of accuracy.

R. Senkamalavalli and Dr. T. Bhuvaneswari proposed a paper to diagnosis the breast cancer disease using AdaBooast, SVM and k-means classifiers to enhance the accuracy of classification [10]. The dataset used for the proposed innovative classifier was breast cancer data. The final results in the classification of breast cancer data showed that the classification achieved a maximum accuracy than the existing techniques.

3 Somatic Mutations Cancer Dataset and Its Description

In this study, the general datasets consist of related six different cancer datasets namely Breast Invasive Carcinoma, Colon Adenocarcinoma, Esophageal Carcinoma, Kidney Renal Clear Cell Carcinoma, Pancreatic adenocarcinoma, Uterine Corpus Endometrial Carcinoma, are considered and these data sets are abbreviated as BRCA, ESCA, KIRC, PAAD, UCEC, COAD. All six datasets consist of somatic mutations patterns data as well as normal patient's data. These six cancer datasets are merged to form the entire cancer dataset which is given as

input to the IEM clustering technique. These datasets are obtained from https://
github.com/ikalatskaya/ISOWN, and datasets are prepared from COSMIC repository. The numbers of instances of each cancer type dataset which are considered in
this paper are shown in Table 1.

The various attributes in cancer dataset are presented in Table 2.

The description of each attribute [11] in this table is presented in Table 3.

The composition of each attribute in the merged dataset which is considered in
this paper is shown in Fig. 1.

Table 1 Instance count of cancer dataset

SI. No	Label	Count
1	BRCA	2478
2	COAD	47,134
3	ESCA	35,524
4	KIRC	6542
5	PAAD	6832
6	UCEC	18,142

Table 2 Attributes information in cancer dataset

@relation SomaticVsGerm line
@attribute ExAc {true, false}
@attribute dbSNP {true, false}
@attribute CNT numeric
@attribute fre numeric
@attribute VAF numeric
@attribute mutAss {'neutral','low','medium','high','stopgain','stoploss'}
@attribute pattern {'CG', 'CA', 'CT', 'TA', 'TC', 'TG'}
@attributeSeqContext
{'ATT','CTT','GTT','TAT','AAA','CAA','AAC','CAC','GAA','AAG','CAG', 'GAC','GAG','TGA','TGC','TCA','AAT','TCC','TGG','CAT','TCG','GAT','TGT', 'TTA','TTC','TCT','TTG','TTT','AGA','CGA','AGC','CGC','ACA','CCA','GGA','ACC',' AGG','CCC','CGG','GGC','GCA','ACG','CCG','GCC','GGG','GCG','AGT', 'ATA','ATC','CGT','CTA','ACT','CTC','ATG','CCT', 'GGT','GTA', 'TAA','CTG','GTC','TAC','GCT','GTG','TAG',}
@attribute isFlanking numeric
@attribute polyphen {'benign', 'probably', 'possibly'}
@attribute isSomatic {true, false}

Table 3 Each attribute description

Attributes	Description
Exac	The Exome Aggregation Consortium (ExAC) is a collection of germline as well as somatic variants from different individuals. It is a database which consists of germline variants collected from individual variants. Each variant in validation sets is given a boolean value based on the existence in ExAC. This can be used as an independent feature
dbSNP	The single nucleotide polymorphism database (dbSNP) resource classifies given variants into normalpolymorphisms and abnormal polymorphisms
CNT	CNT is an attribute given to each coding mutation classified by COSMIC and signifies the data values with a mutation against all tumor categories. If the given mutation was not cataloged by COSMIC, then a zero will be given to CNT. So, CNT values vary from 0 to 19,966
Fre	fre (frequency) samples are computed as the fraction of the total number of samples to the samples which carry a specific mutations in the dataset
VAF	Variant allele frequency (VAF) can be computed as the ratio of quantity of reads associating to the variant allele (VA) over the sum of the entire quantity of reads
mutAss	It forecasts the functional influence of amino acid replacements actually based on conservation of the affected amino acids. Categorical outputs from mutAssare: high, low, medium, stop gain, stop loss
Pattern	Substitution pattern is defined as a two-base gene sequence which constitutes the mutations and also the recently introduced mutational base. Every mutations' categorical patterns are grouped into six varieties of subtypes: "TG", "CG","TA, "TC", "CA" and "CT"
isFlanking	isFlanking is a region which measures the features and checks whether the VAF of an unknown mutation is matched with the VAF of flanking known mutations. This attribute mainly depends on the presence of known mutational polymorphisms
polyphen	The value assigned to this attribute is the outcome of a tool. The probable values that can be assigned to polyphen attribute are benign, probably and possibly
isSomatic	isSomatic specifies whether the patient is having somatic mutations or not

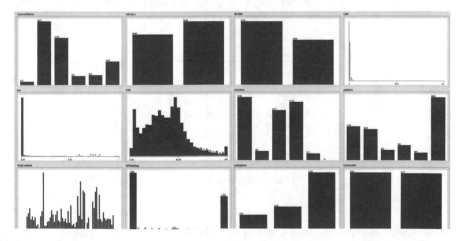

Fig. 1 Graphical representation of instance distribution of each attribute in merged cancer dataset

4   Proposed Model

The six types of cancer dataset, BRCA, ESCA, KIRC, PAAD, UCEC, COAD, are merged to form one cancer dataset which consists of normal data and also somatic mutational data related to six types of cancers. This merged cancer dataset is considered for further analysis. The main aim of proposed model is to form different clusters from the merged dataset for the analysis of somatic mutations data, and for this purpose, an algorithm is proposed in this paper which is an improved version of expectation maximization method. The proposed clustering algorithm is named as improved expectation maximization (IEM) method. The improved expectation maximization performs well with cluster validity checking and minimum number of iterations performed and avoids unnecessary iterations. During clustering formation process, highly significant instances are identified and included in the appropriate clusters, and less significant instances which are having low probability values are ignored and not included in any clusters. There by achieved dimensionality reduction of dataset which improves the efficiency of the clustering algorithm.

The flowchart representation of the proposed model is shown in Fig. 2. The entire cancer dataset is given as input to the improved expectation maximization algorithm in order to form clusters.

The proposed improved expectation maximization algorithm is presented in the next section.

4.1   Improved expectation maximization clustering algorithm

The algorithmic representation of expectation maximization follows an iterative three-step process. The various steps in proposed improved expectation maximization (IEM) algorithm are as follows:

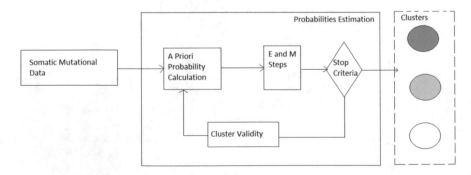

Fig. 2 Representation of flow process in improved expectation maximization method

Step 1:

E-step: In this step, based on the model parameters, proposed model computes the probability of each data point as a cluster.

Let $\hat{\theta}$ be the model parameter to be estimated using the posterior probability computation. $\mathrm{Pr}\,ob(\hat{\theta}/c)$
is the probability of the occurrence of the somatic gene in the given gene classes. In the improved maximization step, the parameters of the expectation step are optimized using the following equation

$$\mathrm{Prob}(\hat{\theta}/c) = \frac{1 + \max\{\sigma_k(A_i, T), \sigma_k(A_i, F)\} \sum_{i=1}^{N} \mathrm{Prob}(D_{i,m} = c/D_i)}{|N|.\mathrm{Prob}(\hat{\theta}/c_k)} \quad (1)$$

where A_i = ith attribute, T = True somatic cancer cases, σ_k = Standard deviation F = False somatic cancer cases, D_i: ith partitional data, c: class label, m = class index,
Prob: Probability estimator, be the model parameter to be estimated using the posterior probability computation

Step 2: For M step, update the model parameters using the assignments in the first step. For the next iteration, the covariance matrix is calculated by using Bayes theorem [12]. Grouping the somatic gene instances using the naive parameter estimation computation in proposed EM model based on the conditional probability of class occurrence is

$$\mathrm{Prob}(c/D_i, \hat{\theta}) = \frac{\mathrm{Prob}(D_i/c, \hat{\theta}).\mathrm{Prob}(c/\hat{\theta})}{\mathrm{Prob}(D_i/\hat{\theta})} \quad (2)$$

The covariance matrix for the next iteration is given as

$$P_q(t+1) = \frac{1}{N} \sum_{k=1}^{n} \mathrm{Prob}(\hat{\theta}/c)$$

Step 3: Iterate the process until all the parameters are updated or convergence or iterations completed.

In the IEM algorithm, model parameters in the E-step are computed using Eq. (1). Similarly, each model parameter is updated using Eq. (2). The probability estimator is used to improve the occurrence of somatic gene patterns in the given dataset.

5   Implementation of the Proposed Model

The IEM algorithm is applied on merged cancer dataset which consists of instances related to six cancer datasets. During the clustering process, the significant instances are included in the clusters, and very low probability instances are eliminated. Thereby, we get the instances which have high probability values based on the specified threshold probability value, and these instances fall into six clusters. The distribution of each type of cancer instances in each cluster is shown in Table 4.

After completion of clustering process, the analysis of each cluster is performed for each attribute and is shown in Table 5.

The percentage of instance distribution in each cluster is shown in Table 6.

The efficiency of the proposed IEM algorithm is compared with the existing clustering techniques namely K-means, DBSCAN and expectation maximization. The algorithms are compared in terms of error clustering rate, and the results are shown in Table 7.

The experimental results shown in above table proved that the proposed IEM cluster algorithm outperforms than the existing cluster algorithms.

6   Application of AdaBoost Algorithm on Each Cluster

The merged cancer dataset instances are grouped into six clusters by applying proposed improved expectation maximization algorithm. For classification of somatic mutational data, AdaBoost classifiers are developed for each cluster. Some of the class-based rules, which are generated for each cluster, are shown in Table 8.

The performance of each AdaBoost classifier developed for each cluster is analyzed in terms of accuracy; True Positive Rate and Recall and the results are shown in Table 9.

The results shown in Table 9 proved that more than 95% performance is achieved in terms of accuracy, TP-Rate and Recall, if the clustering is performed initially and then classification on each cluster.

Table 4 Distribution of cancer types in every cluster

Type of cancer	Clusters					
	0	1	2	3	4	5
BRCA	164	54	3	10	200	812
COAD	4438	2200	1467	596	2716	12231
ESCA	2837	1338	468	442	2119	10625
KIRC	608	259	70	59	349	1941
PAAD	625	301	216	132	390	1764
UCEC	1462	458	37	43	1309	5798

Table 5 Attribute values for each cluster results

Name of the attribute	Value	Clusters					
		0	1	2	3	4	5
Inexact: (Nominal)	True	9562.151	4409.260	1469.707	8688.724	8393.625	648.5305
	False	176.56	63.789	1257.432	9977.909	321.727	13583.580
dbSNP: nominal	True	8143.8345	3694.132	838.547	3714.988	8283.121	131.375
	False	1594.877	778.917	1888.592	14951.645	432.231	14100.7356
CNT: numerica	Mean	0.0303	0.0158	0.0107	0.01	0.2439	0.01
	std. dev	0.0571	0.0217	0.0033	0.1468	0.3093	0
fre : numerical	Mean	0.0303	0.0158	0.0107	0.01	0.2439	0.01
	std. dev	0.0571	0.0217	0.0033	0.1468	0.3093	0
VAF: numerical	Mean	49.65	48.2568	24.4222	32.3432	53.1323	24.7589
	std. dev	16.2743	16.5708	14.467	16.7373	20.843	15.0549
isFlanking: numerical	Mean	0.2661	0.259	0.6007	0.3903	0.2344	0.4487
	std. dev	0.2671	0.2679	0.3608	0.3297	0.2242	0.3595
isSomatic: nominal	True	85.6207	124.3343	2439.9486	12936.2326	18.009	13493.8548
	False	9653.0912	4348.7157	287.1916	5730.4012	8697.344	738.2563
polyphen: nominal	Possibly	766.386	620.9028	272.1098	1197.7358	197.8234	1309.0422
	Probably	381.4678	1671.368	415.2558	2616.7893	107.4619	2221.6572
	benign	8591.8581	2181.7793	2040.7745	14853.1087	8411.0677	10702.4118
mutAss:	Low	7503.930	3 751.3531	1937.8499	3215.8637	907.7804	6329.2226
	Stop gain	205.4006	9.4537	126.8292	943.3901	45.1212	1410.8051
	Medium	1188.4757	3039.5402	493.4992	6149.1195	566.7483	3697.6172
	Neutral	705.373	340.52	96.63	7306.62	7146.5714	2080.27
	High	118.769	334.0513	75.2777	1027.5125	47.8395	690.5496
	Stop loss	16.7628	2.1317	1.0469	28.1251	5.2922	27.6413

(continued)

Table 5 (continued)

Name of the attribute	Value	Clusters					
		0	1	2	3	4	5
Pattern	TC	2423.877	832.4465	66.3158	3333.7009	3055.6126	2155.0471
	CA	1000.7768	438.0442	91.1976	845.9574	724.6521	8093.3718
	TG	545.867	218.6426	21.7765	1300.9975	647.9144	1243.8019
	CG	1328.2569	546.2055	58.1529	1019.2482	1037.3854	1506.7511
	TA	474.9016	176.6491	27.5374	917.8279	464.7956	751.2884
	CT	3969.0325	2265.0621	2466.1599	11252.9019	2788.9928	485.8508
Sample SeqContent	CAG	191.9132	41.8476	7.6792	204.5159	143.9061	217.1381
	GGG	156.8667	87.0468	18.0204	235.9181	133.9745	278.1735
	GGA	208.281	137.3368	30.7864	255.6597	103.4345	418.5016
	AAC	104.1688	35.3942	2.7284	177.9406	144.7305	143.0375
	AAT	179.7241	56.3783	4.1311	287.0815	242.5452	160.1399
	CTT	99.0916	21.608	10.6743	226.4981	94.919	210.2091
isSomatic	True	85.6207	124.3343	2439.9486	12936.2326	18.009	13493.8548
	False	9653.0912	4348.7157	287.1916	5730.4012	8697.344	738.2563

Table 6 Percentage of instance distribution in each cluster

Clusters	Instances	Instance distribution in each cluster (in %)
0	10134	17
1	4610	8
2	2261	4
3	1282	2
4	7083	12
5	33171	57

Table 7 Comparison of proposed clustering model to the existing clustering approaches

Cluster algorithm	Error rate (%)
K-Means	5.64
DBSCAN	4.97
EM	3.78
Improved EM	3.17

Table 8 Generated class-based rules for each cluster

Cluster names	Description
Cluster 0	(pattern = TC) and (fre $\geq$ 0.02) and (VAF $\leq$ 44.66) and (VAF $\geq$ 43.157) $\geq$ CancerName = **BRCA** (32.0/0.0)(VAF $\leq$ 24.5) and (polyphen = probably) and (pattern = CG) $\geq$ CancerName = **PAAD** (14.0/0.0) (SeqContent = ACG) and (mutAss = low) $\geq$ CancerName = **KIRC** (28.0/8.0) (pattern = TC) and (isFlanking $\leq$ 0) and (VAF $\leq$ 53.06) and (VAF $\geq$ 46.03) $\geq$ CancerName = **UCEC** (58.0/0.0)(VAF $\leq$ 7.84) and (isSomatic = TRUE) $\geq$ CancerName = **ESCA** (143.0/2.0) (isFlanking $\leq$ 0.3) and (VAF $\leq$ 19.94) and (SeqContent = CCG) $\geq$ CancerName = **COAD** (7.0/0.0)
Cluster 1	(isFlanking $\leq$ 0.048) and (pattern = TA) and (VAF $\leq$ 31.108) and (VAF $\geq$ 19.63) $\geq$ CancerName = **BRCA** (13.0/1.0) (fre $\geq$ 0.6) and (pattern = CT) and (fre $\leq$ 0.66) $\geq$ CancerName = **KIRC** (14.0/0.0) (CNT $\leq$ 0) and (VAF $\leq$ 20.37) and (SeqContent = AGT) $\geq$ CancerName = **PAAD** (10.0/1.0) (pattern = TC) and (VAF $\leq$ 49.163)and(VAF $\geq$ 41.153)and (fre $\leq$ 0.03) $\geq$ CancerName = **UCEC** (92.0/11.0) (CNT $\leq$ 0) and (inExAct = FALSE) and (VAF $\leq$ 8.11) $\geq$ CancerName = **ESCA** (26.0/0.0) (mutAss = neutral) and (SeqContent = CCA) and (inExAct = FALSE) $\geq$ CancerName = **COAD** (4.0/0.0)

(continued)

Table 8 (continued)

Cluster names	Description
Cluster 2	(isFlanking $\leq$ 0.048) and (pattern = TG) and (VAF $\geq$ 56.4) $\geq$ CancerName = **BRCA** (7.0/1.0) (polyphen = possibly) and (mutAss = neutral) $\geq$ CancerName = **PAAD** (25.0/4.0) (pattern = CT) and (mutAss = neutral) and (dbSNP = TRUE) and (SeqContent = CCG) and (inExAct = TRUE) CancerName = **KIRC** (10.0/1.0) (CNT $\geq$ 1) and (pattern = TG) and (SeqContent = AAT) $\geq$ CancerName = **UCEC** (6.0/0.0) VAF $\leq$ 7.55) and (isSomatic = TRUE) $\geq$ CancerName = **ESCA** (133.0/1.0) (isSomatic = FALSE) and (SeqContent = TCT) $\geq$ CancerName = **COAD** (5.0/0.0)
Cluster 3	(pattern = TC) and (VAF $\leq$ 44.66) and (VAF $\geq$ 43.157) and (fre $\geq$ 0.02) $\geq$ CancerName = **BRCA** (18.0/0.0) (polyphen = possibly) and (SeqContent = TCT) $\geq$ CancerName = **PAAD** (10.0/0.0) (SeqContent = ACG) and (mutAss = low) $\geq$ CancerName = **KIRC** (10.0/0.0) (CNT $\geq$ 1) and (pattern = CA) and (SeqContent = AGG) $\geq$ CancerName = **UCEC** (19.0/6.0) (CNT $\leq$ 0) and (inExAct = FALSE) $\geq$ CancerName = **ESCA** (151.0/22.0) (dbSNP = TRUE) and (VAF $\leq$ 10.26) $\geq$ CancerName = **COAD** (3.0/0.0)
Cluster 4	(pattern = TC) and (VAF $\leq$ 45.95) and (VAF $\geq$ 43.157) $\geq$ CancerName = **BRCA** (26.0/6.0) (polyphen = possibly) and (mutAss = neutral) $\geq$ CancerName = **PAAD** (23.0/5.0) SeqContent = ACA) and (VAF $\leq$ 37.8) $\geq$ CancerName = **KIRC** (5.0/0.0) (isFlanking $\leq$ 0) and (fre $\geq$ 0.03) and (pattern = CA) $\geq$ CancerName = **UCEC** (9.0/0.0) isSomatic = FALSE) and (isFlanking $\geq$ 0) and (VAF $\leq$ 40.755) $\geq$ CancerName = **ESCA** (23.0/1.0) (VAF $\geq$ 63.04) and (mutAss = low) and (VAF $\leq$ 68.54) $\geq$ CancerName = **COAD**(7.0/0.0)
Cluster 5	(SeqContent = CAT) and (pattern = TG) $\geq$ CancerName = **BRCA** (5.0/0.0) (VAF $\leq$ 22.554) and (VAF $\geq$ 21.93) and (inExAct = TRUE) $\geq$ CancerName = **PAAD** (12.0/0.0) (polyphen = probably) and (VAF $\leq$ 20.33) and (VAF $\geq$ 13.56) $\geq$ CancerName = **KIRC** (10.0/2.0) (SeqContent = ATG) and (VAF $\leq$ 55.26) and (CNT $\leq$ 0) $\geq$ CancerName = **UCEC**(4.0/0.0) (isFlanking $\leq$ 0) and (VAF $\geq$ 59.108) and (pattern = CT) $\geq$ CancerName = ESCA (7.0/0.0) (CNT $\geq$ 3) and (SeqContent = GGC) $\geq$ CancerName = COAD (5.0/0.0)

Table 9 Each cluster and its classification performance measures

Clusters	Accuracy	TP-rate	Recall
Cluster 0	0.972	0.972	0.981
Cluster 1	0.986	0.985	0.983
Cluster 2	0.973	0.982	0.986
Cluster 3	0.968	0.975	0.989
Cluster 4	0.989	0.974	0.975
Cluster 5	0.9810	0.969	0.987

Table 10 Performance of classifiers for each cancer type in each cluster

Cancer type	Cluster-0	Cluster-1	Cluster-2	Cluster-3	Cluster-4	Cluster-5
BRCA	0.983	0.975	0.982	0.979	0.988	0.974
COAD	0.974	0.974	0.98	0.985	0.985	0.98
ESCA	0.976	0.979	0.973	0.981	0.982	0.982
KIRC	0.973	0.988	0.984	0.978	0.985	0.977
PAAD	0.984	0.979	0.985	0.988	0.989	0.975
UCEC	0.968	0.971	0.987	0.969	0.983	0.982

Each cluster consists of six types of cancer data, and the classification accuracy of each cancer type in all six clusters is also performed and analyzed. These results are shown in Table 10.

The experimental results shown in Table 10 proved that more than 95% of accuracy is achieved for each cancer type in each cluster. Finally, in this paper, it is proved that the proposed IEM clustering method is better than existing methods. The results in various experiments also proved that more than 95% performance measures for classifiers are obtained with the combination of clustering and then classification on each cluster.

7 Conclusion

It is difficult to process the classification of somatic mutational data if the size of the data and dimensions increases. Because of this reason, in this paper, initially, clustering is performed in order to reduce the dimensionality of the dataset and then applied classification process on each cluster. To perform clustering process in an efficient manner, an algorithm called improved expectation maximization is proposed and proved this algorithm performed better than existing standard algorithms. AdaBoost classifiers are developed for each cluster to classify the somatic mutational data and achieved more than 95% of classification accuracy. Finally, it is proved that a combination of clustering and classification yields good performance measures.

References

1. Joshi, J., Doshi, R., Patel, J.: Diagnosis of breast cancer using clustering data mining approach. Int. J. Comput. Appl. **101**(10), 0975–8887(2014)
2. Krishnamoorthy, I. Aroquiaraj, L.: A comparative study of clustering algorithm for lung cancer data. Int. J. Sci. Eng. Res. **7**(9) (2016)
3. He, B., Torkey, H., Azam, S.Z.M., Zhang, L.: Analysis of cancer somatic mutations taking into consideration human genetic variations. In: Conference on Bioinformatics and Computational Biology, Mar (2014)
4. Zhao, M., Tang, Y., Kim, H., Hasegawa, K.: Machine learning with K-Means dimensional reduction for predicting survival outcomes in patients with breast cancer **17**, 1–7 (2018)
5. Sharma, A., Gupta, R.K., Tiwari, A.: Improved density based spatial clustering of applications of noise clustering algorithm for knowledge discovery in spatial data. Math. Prob. Eng. **2016**, 9 (2016). Article ID 1564516
6. Chakraborty, S., Nagwani, N.K.: Analysis and study of Incremental DBSCAN Clustering algorithm. Int. J. Enterp. Comput. Bus. Syst. **1**(2) (2011)
7. Adebisi1, A.A., Omidiora O.E., Olabiyisi S.O.: An exploratory study of K-Means and expectation maximization algorithms. British J. Math. Comput. Sci. **2**(2), 62–71 (2012)
8. Rajaguru, H., Prabhakar, S.K.: Expectation maximization based logistic regression for breast cancer classification. International Conference on Electronics Communication and Aerospace 20–22 April 2017, Coimbatore, India (2017)
9. Thongkam, J., Xu, G., Zhang, Y.: AdaBoost algorithm with random forests for predicting breast cancer survivability. IEEE Int. Joint Conf. Neural Network, June 1–8, 2008, Honkong, Chaina (2008)
10. Senkamalavalli, R., Bhuvaneswari, T.: Improved classification of breast cancer data using hybrid techniques. Int. J. Adv. Eng. Res. Sci. (IJERS) **8**(8) (2017)
11. Trinh, Q.M., Spears, M., McPherson, J.D.: ISOWN: accurate somatic mutation identification in the absence of normal tissue controls Irina Kalatskaya. Genome Med. **9**(1), 59 (2017)
12. Kharya, S., Agrawal, S., Soni, S.: Naive bayes classifiers: a probabilistic detection model for breast cancer. Int. J. Comput. Appl. **92**(10), 0975–8887 (2014)

Identification of Ontologies of Prediabetes Using SVM Sentiment Analysis

V. Vasudha Rani and K. Sandhya Rani

Abstract Sentiment analysis is considered as a classification task as it classifies the polarity of a text into positive or negative. Different methods of sentiment analysis can be applied for the health domain, especially for prediabetes domain which has not been completely explored yet. And there is a lack of approaches for analyzing positive and negative tweets separately to identify the positive and negative ontologies for modeling the features in a domain of interest. Here in my work, proposed domain and sub-domains are Health and Prediabetes, respectively. Prediabetes defines the condition of blood sugar levels that are higher than normal but not high enough to be diabetes like a pre-warning call for diabetes. The proposed methodology is the deployment of original ontology-based techniques toward a more efficient sentiment analysis of Twitter posts on prediabetes. As part of experimentation, sentiment analysis uses the SVM algorithm with term frequency as a feature extraction method to train and test a large and sub-data set of tweet text. Negative ontologies are constructed for a better understanding of the aspects identified through semantic annotations. The results of the classification method are evaluated using the performance metrics accuracy, precision, recall, and F-measure for effective evaluation of the proposed method.

Keywords Prediabetes · Sentiment analysis · Support vector machine classification technique · Ontologies

V. Vasudha Rani
IT Department, Sri Padmavathi Mahila Viswa Vidhyalayam University, Tirupati, India
e-mail: vasudharani.v@gmrit.org

GMRIT, Rajam, India

K. Sandhya Rani (✉)
Department of Computer Science, Sri Padmavathi Mahila Viswa Vidhyalayam University, Tirupati, India
e-mail: Sandhyaranikasireddy@yahoo.co.in

© Springer Nature Singapore Pte Ltd. 2020
P. Venkata Krishna and M. S. Obaidat (eds.), *Emerging Research in Data Engineering Systems and Computer Communications*, Advances in Intelligent Systems and Computing 1054, https://doi.org/10.1007/978-981-15-0135-7_49

1 Introduction

1.1 Social Media and Prediabetes

Social media mining is defined as the extraction of useful information from the social network. Primary objective of the social data mining process is to commendably handle large-scale data, gain insightful knowledge, and extract actionable patterns. Social media blogs such as Twitter is one of the growing multidisciplinary areas where there is a lot of scope for social media research and development for the researchers of different backgrounds can make an important contribution. Tweet mining [1] has got its so many applications like sentiment analysis, opinion mining, community analysis, influence modeling, social recommendations, privacy, security and trust, information diffusion and provenance. Health data-related mining has got its extended applications *(i) Trend and Event Detection, (ii) Patience Insight: (Understand your Patients), (iii) Making Sentiment Decisions.* Among which sentiment analysis and opinion mining has become the field of ongoing research in text mining. Text mining research [2] is extended to even health domain because there are a lot of tweets getting popped into Internet regarding health such as people suffering from diseases info, their experiences, other problems they are facing, and suggestible treatments. And these health-related tweets are useful to extract actionable patterns and insightful knowledge about any health aspect such as diseases, treatments, causes, effects, food diets, symptoms, death issues, and general suggestions to be useful for people.

Here, the proposed method experiments on tweets retrieved on prediabetes because of its significance facts that (i) prediabetes is a serious health condition where blood sugar levels are higher than normal, (ii) approximately 84 million American adults—more than 1 out of 3—have prediabetes. Prediabetes is a metabolic disorder that is considered a precursor for the development of diabetes mellitus. It is characterized by higher than normal blood glucose levels that have not yet reached diabetic levels. Prediabetes is an indication that you could develop type 2 diabetes if you do not make some lifestyle changes. People getting awareness over prediabetes definitely reduce the risk of conversion rate into diabetes.

1.2 SA and Ontologies

Sentiment analysis and opinion mining is the computational study of people's emotions, opinions, feelings, difficulties, experiences, solutions, and every good and bad happening in their daily lives toward any issue. The entity can be an individual, event, new trend, status of a disease, status of a politician, side effects of a decision, and new scheme by the government. This gives a more elaborate analysis of posted opinions regarding a specific topic. Opinion mining extracts and analyzes textual opinions of people to derive facts, and sentiment analysis implicates in defining a sentiment score for each fact as positive, negative, or neutral.

It is a powerful tool for classification of textual data. There are three different levels at which sentiment analysis can be applied. They are aspect-level sentiment analysis, sentence-level sentiment analysis, and document-level sentiment analysis. Document-level SA classifies a document as positive or negative as expressed by the user. Sentence level aims to classify a sentence as user expressed it positive or negative. Aspect level is also known as word-level SA process involves in identifying the semantic annotation of the word felt by the user. Now with the explosion of data in the social media, it obviously becomes very difficult to keep track of everything, so you would need to have machines do this analysis, machine should be parsing these reviews, comment emails, tweet, and tell you information about these things, what the opinion or feeling that these pieces of documents expressing are also called subjectivity analysis, basically it is a field of natural language processing, and the objective is to try and extract subjective information.

Ontologies can be a supportive technique for describing subjective information. Ontology can be defined as an "explicit, machine-readable specification of a shared conceptualization." Ontologies serve as the primary means of knowledge representation in the semantic web. Ontologies are used for modeling the terms in a domain of interest as well as the relations among these terms and are now applied in various fields. So, here we propose a method for sentence-level SA that uses support vector machine classification—machine learning algorithm and an ontology technique, with the objective of deriving positive and negative tweets and describing the relations among the aspects in the domain of prediabetes, respectively. SVM is a supervised classification technique that takes labeled data, trains the classifier, and applies the classifier to categorize the new test tweet instances into positive and negative.

The ontologies work explains the modeling of the features of positive and negative tweets of prediabetes and the relations among the features, derived from SVM sentiment analysis. Performing analysis on the current prediabetes tweets explores detailed awareness over the current status of it and helps in reducing the risk of transforming into diabetes as a support of the Diabetes Prevention Program (DPP).

2 Description of the Proposed Methodology

The proposed method of sentiment analysis and deriving ontologies from prediabetes data set involves mainly four phases of work. They are (i) prediabetes tweets gathering, (ii) normalization of prediabetes tweets, (iii) sentiment analysis through SVM classification, and (iv) deriving positive and negative ontologies. It is illustrated through overall architecture for the sequence of flow operations. Figure 1 refers overall architecture diagram. The process is described as follows.

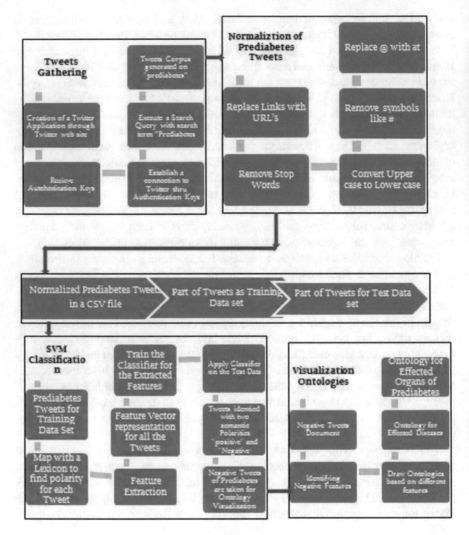

Fig. 1 Overall architecture of the proposed approach

Phase 1. Tweets Gathering

Tweets gathering process starts with the creation of Twitter application through, Twitter developers web page. Having a Twitter account is a prerequisite for this process. This phase involves four-step process defined with. They (i) create a Twitter application, (ii) install and load required R packages, (iii) create and store Twitter authenticated credential object, and (iv) extract tweets. Twitter application creation requires details like application name, its description, web site, and callback URL to be filled by the user. This process is called as Twitter registration to access Twitter through API. After the registration process, it continues to providing

API_key, API_SecretKey which are used to generate access_token and access_-token_secret. All the four keys are required to be used as part of the user authentication process through the R console and generate a PIN for verification. Once this handshake process completes with Twitter, next is to execute search Twitter function with the minimum parameters such as search word and number of tweets required.

Here, our search word is "prediabetes" and the count $n = 100,000$ number of tweets to download from Twitter. The downloaded tweets are in the form of a list and need to be converted into a data frame to perform operations. The data frame of tweets is written into a .CSV file. This .CSV file will be used as a prediabetes tweet corpus for further disease analysis.

Phase 2. Preprocessing the downloaded tweets

The downloaded prediabetes tweets data need to be cleaned so that it can be used for analysis. This preprocessing phase [3] further refines our data and also filters the data which is unimportant to us. This phase also removes the unnecessary columns and also having functions to get only the unique data because the duplicates do not help us much. This includes removing of rid of URLs, retweet headers, and references to other screen names, removing non-English characters and unnecessary spaces. The preprocessed tweets file results in one column, i.e., tweet text. An example of this can be seen here. The tweet text before preprocessing was "*An early diagnosis is crucial to stop the progression of #prediabetes to #type2diabetes. Take the risk test now:*" and now after preprocessing this may look like "*An early diagnosis is crucial to stop progression of prediabetes to type2diabetes. Take risk test now.*"

In the normalization process of tweets, #removing retweet headers, #removing links, #removing @, #removing alphanumeric words, #converting to lowercase, # to remove non-English characters, and #to get rid of unnecessary spaces are the major preprocessing tasks applied to the tweets data. After removing unnecessary spaces and symbols, another two key steps to be performed are (i) removing all the unnecessary columns except tweet text column for sentiment analysis, and (ii) remove the duplicate tweets to get only unique data as duplicates do not help us for analysis. Now the prediabetes tweet data set is ready for any machine learning-based sentiment analysis process.

Phase 3. SVM Classification

This is the third phase of the proposed model which is shown in the overall architecture in Fig. 1. The normalized prediabetes tweets from phase 2 are the intake for SVM classification. The data set is divided into two parts, more than half of the tweets as training data and the remaining tweets for test data. This approach mainly consists of eight modules of work (i) preprocessing training data set, (ii) classifying training data set using sentiment() function in R, (iii) feature extraction through unigram-based term-tweet matrix, iv) train the SVM classifier for the extracted features, (v) preprocessing of test data, (vi) apply the classifier on the test prediabetes data, (vii) derive prediabetes tweets classified into positive and negative with their corresponding sentiment scores, (viii) deriving domain features with high weighted sentiment scores for ontology preparation.

Preparation of training data set: Applying sentiment analysis [4] for the prediabetes data set starts with the questions: (i) What do you use as training data and (ii) what features do you choose. Starts with answering the first question, i.e., part of the data set is used as training data. In order to classify a document as positive or negative, we need documents that are already marked as positive or negative, i.e., that which data acts as training data. For the process of classification of training data into positive, negative, and neutral, a predefined lexicon resource is used. Lexicon is a resource with information about words like a dictionary of polarized words. The lexicon used here is the Jockers' (2017) dictionary found in the lexicon package. More of positive connotations over negative connotations lead to a more positive score or vice versa.

Here, the process of **training** data set classifications starts with the first step of (i) creating a document-term matrix for the training tweets and the second (ii) putting the document-term matrix in a container so that the resultant word corpus is annotated by using an online corpora, i.e., Jockers' dictionary which has been hand annotated by researchers. The result is a corpus of tweets which are marked into either positive, negative, and neutral. Now the trained data is ready to be given for SVM classification algorithm. The SVM classifier uses the labeled data to train the SVM classifier.

Algorithm 1 Prediabetes Sentiment Classification

Input : Tweets on Prediabetes T_p
Output: Derived Ontologies O_p, O_n
Classification Process
Step1: Preprocessing Training Data T_{tr}
Step2: Classifying Training data set using Sentiment() fun
Step3: Feature Extraction
Step4: Train the classifier
Step5: Preprocessing Test Data T_t
Step6: Classification on Test Data T_t
Step7: Derive positive and negative tweets P_t, N_t
Step8: Ontologies preparation O_p, O_n from P_t, P_n
Step9: End

Feature Extraction: From the labeled tweets, data features need to be identified to train the SVM classifier. The answer for the second question "What features do you choose for classification" is the simple way of looking at the problem and also at the individual words which are the identified unigrams that attains a frequency threshold value.

To do classification of **test** tweets, we need to perform the operations as (i) preprocess the test tweets, (ii) create a document-term matrix, (iii) put it in a container, and (iv) execute SVM classification algorithm sending two input parameters as trained data in a container, i.e., model.svm and test data in container, i.e., testcontainer.svm. This is where the testing happens, and the classification is done through SVM[5] classification technique. The resultant tweet file is attached with the additional columns of SVM_LABEL {positive, negative, and neutral} and SVM_PROB {between 0 and 1 for positive and −1 to 0 for negative tweets} for the existing column of TWEET_TXT from results.svm. The tweets data set identified with semantic polarities will be the input for the last phase of work, i.e., identifying ontologies of prediabetes.

Phase 4. Identifying Ontologies of Prediabetes

Ontology is defined as an unambiguous specification of conceptualization. Ontologies are an old but effective technique to represent a specific domain-related information. Ontologies can be used as an information-sharing mechanism to share among researchers and can be reused also. In practice, ontologies [6] can be very complex with several thousands of terms or very simple describing one or two concepts only. Ontologies are used to make some commitments in the domain.

Prediabetes Ontologies: Positive and negative tweets which are derived from previous phase are the input for phase 4. Positive prediabetes tweets file has the columns tweet text, SVM label as positive and SVM score. SVM positive score will be >0 and up to max value of 1. Negative prediabetes tweets file has the columns tweet text, SVM label as negative and SVM score. SVM negative score will be <0 and up to min value of −1. However, we prepare our data for SVM. Similarly, like classification, we need to prepare either positive tweets or negative tweets data for designing ontologies.

Algorithm 2. Prediabetes Ontology Designing

Input: Negative Prediabetes Tweet Text
Output: Negative Ontologies for a specific aspect
Identification of Ontology
Step1. Negative Tweets in a .CSV file
Step2. Take Tweet text
Step3. Find the highest scored negative Tweets above a threshold
Step4. Create a Term-Tweet Matrix of High ranked Tweets
Step5. Find Unigrams W_u for each tweet
Step6. Map each Unigram for its Tweet score T_s
Step7. Design ontologies for Unigram Features
Step8. 1.Ontologies for effected Diseases of Prediabetes
       2.Ontologies for Effected Organs of Prediabetes
Step9: End

Let us consider negative tweets for negative ontologies preparation for prediabetes. The pseudocode for prediabetes ontology designing approach is given by **Algorithm 2**. The process is described as follows. The negative tweets text file with a single column of tweet text is the input for this phase. Firstly, tweets are filtered for getting only the highest scored tweets by specifying a threshold value score of 5. The resultant filtered tweets will be input for the next step. In order to create a term-tweet matrix of these tweets, two steps need to be done (i) install RTextTools package and (ii) run a command create_matrix(). Code is given as follows.

```
library(RTextTools)
dtMatrix.svm <- create_matrix(Negative_tweets$Text)
```

Find the unigrams with more frequency and map them to their corresponding tweet scores. And based on these findings, different visualizations are created such as histograms, bar chart and ontology diagrams (visualizations).

3 Experimental Setup

In this work, we experimented with prediabetes tweets text data and classified them into positive and negative. Experimentation was done through R coding. Experimentation process results in positive and negative features of prediabetes data set. To do this, R coding requires many packages to be installed and many commands to be executed. Here for this paperwork, the packages utilized are (i) library(tm), (ii) library(RTextTools), (iii) library(e1071), (iv) library(dplyr), and (v) library(caret). The different methods executed are

```
dtMatrix.svm <- create_matrix(data_to_train$Text)
container.svm <-create_container(dtMatrix.svm, data_to_train$Type,
trainSize = 1:8721, virgin = FALSE)
model.svm <- train_model(container.svm, "SVM," kernel = "linear," cost = 1)
Similarly for test tweets data, code executed is
testMatrix.svm <-create_matrix(testing_data$Text, originalMatrix = dtMatrix.svm)
testContainer.svm <-create_container(testMatrix.svm, labels = rep(0, testingSize.
svm), testSize = 1: testingSize.svm,
virgin = FALSE)
results.svm <- classify_model(testContainer.svm, model.svm)
```

The process results in histograms with different colors represent different features that were developed in Fig. 2 for negative tweets and Fig. 3 for positive tweets. The identified positive and negative features with their corresponding scores are listed in Tables 1 and 3. The highest ranked tweets, both positive and negative, are mentioned in Tables 2 and 4.

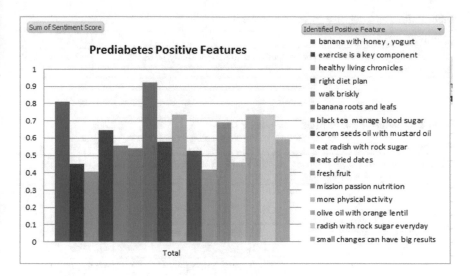

Fig. 2 Comparative analysis among positive features

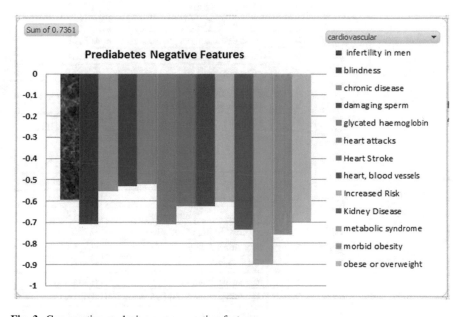

Fig. 3 Comparative analysis among negative features

Table 1 Identified negative tweet with highest score

Negative tweet	High score
"Adults with *prediabetes metabolic syndrome* are more likely to die from heart disease"	−0.901387819

Table 2 Identified positive tweet with highest score

Positive tweet	High score
"Diabetes management ***prediabetes*** healthy life black tea helps manage blood sugar new study an exciting new study has *p*"	0.921635375

Table 3 Negative features

	Identified negative feature	Sentiment Score
1	Cardiovascular	−0.7361
3	Glycated hemoglobin	−0.5214
4	Damaging sperm	−0.5303
5	Infertility in men	−0.5942
6	Morbid obesity	−0.7602
7	Heart attacks	−0.7111
8	Blindness	−0.7111
9	Kidney disease	−0.7361
10	Chronic disease	−0.5547
11	Heart Stroke	−0.625
12	Metabolic syndrome	−0.9013
13	Obese or overweight	−0.6992

Table 4 Positive features

	Identified positive feature	Sentiment score
1	Banana honey yogurt	0.8124
2	Exercise key component	0.4518
3	Healthy living chronicles	0.4065
4	Eats dried dates	0.5269
5	Fresh fruit	0.4166
6	Radish rock sugar	0.7349
7	Black tea manage blood sugar	0.9216
8	Right diet plan	0.6444
9	More physical activity	0.4588
10	Olive oil and orange lentil	0.735
11	Banana roots and leafs	0.5422
12	Eat radish and rock sugar	0.7349
13	Mission passion nutrition	0.6928

4 Two-Level Performance Evaluation

The proposed model has this evaluation phase at the end. The proposed model uses a two-level performance evaluation method. The proposed method of SVM classification has two key phases in it: (i) preparation of training data and (ii) classification of test data. Performance evaluation can be carried out for the two key phases. The process of evaluation of the classification methods at the two levels is depicted in Fig. 4.

First-level evaluation: The first level of verification starts with evaluating the resultant prepared training data set. This evaluation process needs a part of the training data to be taken as validation data. Firstly, the training data is classified using the predefined method of classification. Then to test the process, using the features from training data is used to classify the validation data. Then these predicted tweets polarities of validation data are compared with actual tweet polarities to know the correctness. Accuracy measure is used to evaluate the level of correctness of this process.

The second level of evaluation starts using the tested training data to identify features for training the SVM classifier. Using SVM classifier, test data is classified to find the polarities. The SVM classifier needs to be evaluated using some of the metrics.

There are four effective performance evaluation metrics are identified for text classification in my study. They are accuracy, precision, recall, and F-measure.

In the field of machine learning, a confusion matrix also called an error matrix is used to describe the performance evaluation of a classifier or any classification method. Here, we use this to verify the resultant positive negative tweets of SVM classification. Predicted results are compared to its actual. Both positive and negative tweets put together are classified into four categories. They are (i) actual positive tweets are the predicted positive - > Tp, (ii) actual positive tweets are predicted negative - > Fn, (iii) actual negative tweets are predicted negative- > Tn, and (iv) actual negative tweets are predicted positive − > Fp. These four parameters are used in the calculation of performance metrics. The computed values of the four

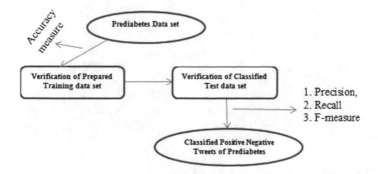

Fig. 4 Two-level performance evaluation method

Table 5 Comparison of related document level works

Author	Level	Domain	P	R	F
Bindal [7]	Document level	Reviews	80.00	85.71	82.76
Proposed SVM Model	Document level	Prediabetes posts	90.95	84.7	88.03

metrics are precision 90.95%, recall 84.70%, F-measure 88.03%, and accuracy as 82.69%. *Comparison:* The resultant values of the metrics of the proposed model of classification are compared with other paperworks and are shown in Table 5.

5 Results and Discussion

The proposed method of "classification of prediabetes tweets and designing ontology" was executed in R studio environment, and the results were obtained. The SVM classification was done on one Lakh prediabetes tweets. The resultant positive and negative tweets of prediabetes are taken into separate .CSV files for further analysis. After that each .CSV file was taken individually for developing ontologies. Positive features are extracted through the implementation of Algorithm 2, on positive tweets text data.

Negative features are extracted through the implementation of Algorithm 2, on negative tweets text data. Positive and negative features and their tweet scores are stored in Tables 2 and 4. Highest scored positive and negative tweets are identified (Figs. 5, 6, and 7).

The results make use of different visualization techniques, basis visualizations such as histograms, advanced visualizations such as bar/line charts, and also

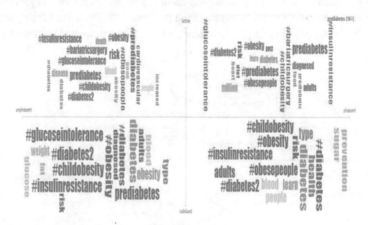

Fig. 5 Tagged cloud for prediabetes-associated aspects

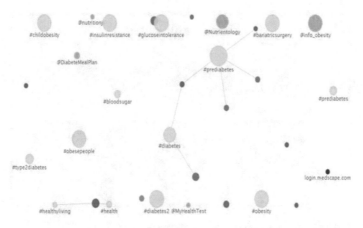

Fig. 6 WordNet graph for prediabetes-related aspects

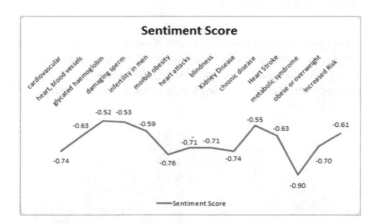

Fig. 7 Line chart of negative features with tweet scores

ontology visualizations. As a result of traditional/basic visualization methods, line charts for prediabetes features analysis for both negative and positive features are shown in Figs. 7 and 8.

5.1 Advanced Visualizations in R

There are so many advanced visualizations can be found in R. Here for the visualizations of my results, methods of tagged cloud and word net graph have been used to display the related neighborhood features for the prediabetes data domain. And different ontologies are drawn for prediabetes.

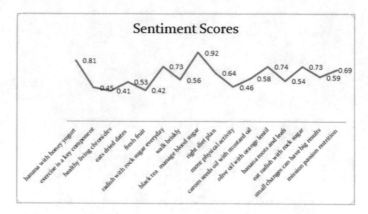

Fig. 8 Line chart of positive features with tweet scores

5.2 Ontology Visualizations

Ontologies designed are conceptualized facts mined from negative tweets identified. Figures 9 and 10 are shown to visualize negative ontologies of prediabetes. In Fig 9, ontology visualization is to depict the negative impacts of prediabetes and in Fig 10, ontology visualization is to depict the organs affected through prediabetes.

Fig. 9 Ontology visualization for all the negative impacts of prediabetes

Fig. 10 Ontology
visualization for all the
diseases effected of
prediabetes

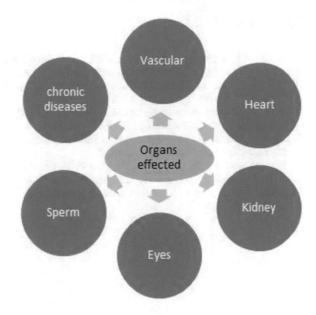

6　Beneficial Features of the Proposed Model

The proposed model for designing positive negative ontologies of prediabetes disease from tweets data has some advantages over other methods. They are listed here (i) fresh data -> work progressed on recent tweets data, (ii) availability of potential Twitter APIs, (iii) abundant preprocessing methods in R, (iv) selection of SVM classification, (v) conceptualization through ontology, and (vi) two-step evaluation of results.

SVM classification method has the potential to handle huge no of features, robust in handling large data with more of sparse columns, and also has proven results in sentiment analysis.

7　Conclusion

This paper has presented an ontology classification model in the prediabetes domain. Among so many health issues, prediabetes is one of the non-communicable diseases in India. We have presented experiments aiming to validate our approach of tweets classification at two levels as training data classification and testing data classification. The second phase of work used the effective concept of ontology. The approach of ontology is a proven effective method for mining tweets in a better way. The results show that (i) validation of training data results with accuracy of 82.69% and (ii) SVM classification results with a precision of 90.95%, a recall of 84.7%, and F-measure of 88.03%.

References

1. Introduction to the tm Package, Text Mining in R, Ingo Feinerer, 6 Dec (2017)
2. An Article on "Big Data is the Future of Healthcare", by cognizant 20-20 insights | Sep (2012)
3. Rani, V.V., Sandhya Rani, K.: Twitter Streaming and Analysis through R". Indian J. Sci. Technol. Dec 2016
4. Fikri, M.: A comparative study of sentiment analysis using SVM and SentiWordNet. In: The 2nd International Conference on Informatics for Development (2018)
5. Go, A., Bhayani, R., Huang, L.: Twitter Sentiment Classification Using Distant Supervision. Stanford University, Stanford, CA (2017)
6. Kontopoulosa, E., Berberidisa, C., Dergiadesa, T., Bassiliadesb, N.: Ontology-based sentiment analysis of twitter posts. An Int. J., Expert Syst. Appl. Jan (2013)
7. Bindal, N., Chatterjee, N.: A two method for sentiment analysis of tweets. Int. Conf. Inf. Technol. IEEE (2016)

A Systematic Survey on Software-Defined Networks, Routing Protocols and Security Infrastructure for Underwater Wireless Sensor Networks (UWSNs)

P. V. Venkateswara Rao, N. Mohan Krishna Varma and R. Sudhakar

Abstract The vast evolution and development of routing algorithms in "Underwater Wireless Sensor Networks (UWSNs)" is to monitor and collects the data for environmental studies and pollution monitoring applications. The UWSNs sensing range from the oil industry to aquaculture. It includes "instrument monitoring, pollution control, climate recording, prediction of natural disturbances, search and survey missions, and study of marine life." The sensed data can be utilized for various applications and benefit for the humans. The sensor nodes, such as stationary and mobile, are connected through wireless communication modules to transfer various events of interest. Energy efficiency plays a vital role in UWSNs. Sensor nodes are powered by batteries which are difficult to replace or charge once the node is deployed. In this paper, we focus on the survey based on UWSNs routing algorithms, metrics used, critical challenges, and provide recommendations that will shape the focus of future research efforts.

1 Introduction

UWSNs have been significant research community from the past of 2000, and notable advancements have been made in communication and routing protocols, software-defined architectures and modems [1–5]. "UWNSs have the potential to enable unexplored applications and to enhance our ability to observe and predict the ocean. Unmanned or "Autonomous Underwater vehicles (UUVs, AUVs)," equip-

P. V. Venkateswara Rao · N. Mohan Krishna Varma (✉) · R. Sudhakar
Department of Computer Science & Engineering (CSE), Madanapalle Institute
of Technology & Science, Madanapalle, India
e-mail: drmohankvn@mits.ac.in

P. V. Venkateswara Rao
e-mail: raovenkat21@gmail.com

R. Sudhakar
e-mail: sudhavk1983@gmail.com

© Springer Nature Singapore Pte Ltd. 2020
P. Venkata Krishna and M. S. Obaidat (eds.), *Emerging Research in Data
Engineering Systems and Computer Communications*, Advances in Intelligent
Systems and Computing 1054, https://doi.org/10.1007/978-981-15-0135-7_50

ped with underwater sensors are also envisioned to find application in an exploration of natural underwater resources and gathering of scientific data in collaborative monitoring missions. The main applications of UWSNs are assisted
navigation, undersea explorations, disaster prevention, seismic monitoring, equipment monitoring, ocean sampling network, distributed tactical surveillance, mine
reconnaissance, and environmental monitoring. The data collected by the sensor
devices from the bottom level of the seawater and transfer that data to the sink
nodes which are deployed on the surface of the water and sink nodes further its
transfer to the coastal (onshore) data center [6–8]." The UWSNs resemble with the
terrestrial networks when we encounter the distinguish between UWSN and terrestrial wireless sensor network such as UWSNs reveal some unique characteristics
like high bit error rate, acoustic communication, low bandwidth, and limited storage
power of sensor devices.

Underwater acoustic networking is one of the technologies for the above
applications which are mentioned in the previous paragraph. "Underwater Acoustic
Sensor Networks (UW-ASN)" consist of a different number of sensors and vehicles
that are used to perform collaborative tasks over beneath the sea level. To achieve
the primary objective, the sensors and vehicles working in the way of an autonomous network can directly adapt to the characteristics of the sea environment, in
other words, biological effects.

2 Underwater Communication Techniques

There are different techniques regarding underwater communication, some work
better than others depending on the application and what it should be able to handle
when it comes to data transfer and size. There are a lot of factors that affect
underwater communication in negative ways; some of these are signal attenuation,
multipath propagation and time variations in the channels and more [5].

1. **Acoustic communication**: Acoustic communication (ACOMM) is a method on
   how to communicate underwater through acoustic sound waves. Underwater
   acoustic communication (UAC) is the most used underwater communication
   technique due to its low signal attenuation in the water. The most common way
   to use ACOMM is by using hydrophones. Hydrophones are microphones that
   are designed to be applied underwater for listening and recording underwater
   sounds. If several hydrophones are placed as an array, they can be used to add
   signals from the direction that the array is pointing, while subtracting signals
   from other directions. Some severe limiting factors for ACOMM are the slow
   speed of acoustics in water which is about 1500 m/s and the low data rate due to
   acoustic waves [5].
2. **Electromagnetic waves**: Compared with ACOMM, electromagnetic waves
   (EM) in radio frequency bands have higher speeds and higher operating frequency which leads to higher bandwidth. Although in water, EM waves do not

work well due to the conducting nature of the medium. Since the medium is not the same in freshwater and seawater, EM works differently depending on the water medium. An EM implementation is attractive when it comes to using it as a communication carrier in freshwater. However, the most significant problem with using EM comes with the size of the antenna in an EM transmitter. One would need a couple of meters for a 50 MHz EM antenna which makes it unpractical when the size of the product is essential. Using EM in seawater is difficult due to high signal attenuation [5].

3. **Free-space optical waves**: Using optical waves for communication can be a great choice when it comes to high data rates since they potentially can exceed 1 Gbps. However, like the other mentioned communication methods, optical communication behaves differently in water as well. The most significant problem that occurs when using optical signals in water is that the signals rapidly get absorbed in water and that optical scattering occurs by suspending particles and planktons. Ambient light from the surface may also affect optical communication negatively.

The relationship between the achievable data rate and distance when various technologies are used for UWSNs is shown in Table 1. And it also exhibits the relation between proximity "range communications and magnetic induction is a better candidate (in terms of transmission delay and data rate) than acoustic signals range (below 10 m) and its communication range can be extended up to a (few 100 m) through a waveguide technique; still, for large distances and telemetry applications. Low-frequency acoustic signaling is most versatile and applicable for the physical layer communication. However, underwater acoustic communications are quite challenging [1]. Below, we have listed the well-known challenges that engineers are confronted with, in designing an underwater communication system."

Table 1 The technologies used for underwater digital communication and its comparison

Technology	Working frequency	Modulation	Distance	Data rates	Reference
Optical waves	–	PPM	1.8	100 Kbps	NA
	–	–	10	10 Mbps	NA
	–	–	11 M	9.69 Kbps	NA
Electromagnetic waves	2.4	CCK	0:16 m	11 Mbps	[19]
	2.4	QPSK	0.17 m	2 Mbps	[19]
	1 kHz	BPSK	2 m	1 Kbps	NA
	10 kHz	BPSK	16 m	1 Kbps	NA
	3 kHz	–	40 m	100 bps	NA
	5 kHz	–	90 m	500 Kbps	NA
Acoustic waves	800 kHz	BPSK	1 m	80 Kbps	NA
	70 kHz	ASK	70 m	0.2 Kbps	NA
	24 kHz	QPSK	2500 m	30 Kbps	NA
	12 kHz	MIMO-OFDM	–	24.36 Kbps	NA

Multi-path In underwater sensor networks, multipath is one of the primary issues which is governed by two facts such as:

- Sound refraction in the water
- Sound reflection at the bottom and surface of the sea level.

"The sound reflection at the bottom and surface of the sea level is a consequence of the spatial variability of the sound speed. In the depth of the sea level, the delay of a channel can be more to tens or even sometimes hundreds of milliseconds [9]." At a depth of the sea level, the motion depends on wind, marine creatures, ships, and so on. At a depth of the sea level, the motion depends on wind, marine creatures, ships, and so on, owing to the underwater channel to differ fast, and follows the precise characteristics of a so-called doubly selective channel which results in a frequency-selective as well as time-selective signal distortion [10].

Attenuation The acoustic signal attenuated primarily depends on two reasons; first, due to the spherical (deep sea) or cylindrical (shallow waters) geometric spreading; second, due to the frequency-dependent absorption caused by scattering and reverberation.

Variable Sound Speed The most critical challenge in the UWSNs is the "underwater sound speed is not constant. Sometimes, it may vary with cloud temperature and salinity of the medium and pressure. In the depth of the water (beneath the sea level), the salinity and temperature are constant. The water sound speed is directly proportional to the depth; while in shallow water, it may differ with day by day and has behaviors in different monsoon conditions of the year."

Doppler "The motion of the sea surface and the low speed of sound penetration introduces a large Doppler spreads and results in a fast-fading frequency-selective behavior (or temporal and spatial variability) of the underwater acoustic channel." UWSNs use the different layers for achieving communication; for example, physical layer and medium access control (MAC) used for acoustic technology. Low bandwidth, variable delays, and capacity are the essential characteristics of acoustic networks. The network topology is, in general, a crucial factor because the abovementioned factors are applicable. The efficiency of network topology depends on network reliability and energy consumption. Design of topology for an underwater sensor network is an open area for research which is already discussed in [1, 11].

3 Underwater Wireless Sensor Network Routing Protocols

Routing is one of the fundamental techniques over the network. The routing protocols are considered to be as not only the establishing the path between source and destination but also maintaining the routes. Most of the existing approaches on underwater sensor networks are a concern on the physical and MAC layer issues but routing also a significant concern. Because of the issues raised by the network later such as "routing, mechanisms are a relatively new area, thus providing an efficient algorithm for routing. In this section, we discussed various routing techniques proposed to date for UWSNs."

Depth-Based Routing Protocol Yan et al. proposed a "depth-based routing (DBR)" in 2008 [12]. Yan et al. approach of DBR uses the advantage of the greedy method; it involves the packets delivered at the destination which sinks at the water surface. DBR employs the general underwater sensor network architecture: information stored at the sinks is commonly located at the water surface. Because of this reason, DBR forwards data packets greedily toward the water surface (i.e., the plane of data sinks) based on depth information of each sensor. The routing information of the packet has a field that stores and identifies depth information of its most recent forwarder and is updated at every hop. The primary objective of DBR is: "when a node receives a packet, it forwards the packet if its depth is smaller than that embedded in the packet. Otherwise, it discards the packet."

Vector-Based Forwarding Protocol (VBF) One of the essential properties of routing vectors is continuous node movements observation requires often preserve and recovery of routing paths. But it leads to more cost to maintain in 2D and 3D underwater network issues. To address this problem, a position-based routing method is commonly called as VBF. Xie et al. (in 2006) proposed a "vector-based forwarding (VBF)" [13]. VBF protocol solves "node mobility issue in a scalable and energy-efficient way. In VBF, each packet carries the positions of the sender, the target, and the forwarder. For this, state information of the sensor node is not needed because of only a small number of nodes are actively participated during the packet forwarding." The routing vector identifies the path during transmission from source to destination. VBF also called a geographic routing protocol. It is the first effort to apply the geo-routing approach in underwater sensor networks.

Directional Flooding-Based Routing Protocol (DFR) "Hwang et al. [14] developed a DFR protocol in 2008. Path establishment between the source and destination requires much communication overhead in the form of control messages. Sometimes, the network overhead on dynamic conditions, network topology, and high packet loss degrade reliability." It leads to packet retransmission. Existing routing protocols cannot address the issue on link quality. Finally, most of the approaches do not give a guarantee that is 100% packet delivery. To address the above issue by DFR, DFR applies the concept of packet flooding to increase link quality and reliability. The number of nearest nodes is identifying during transmission, and the packet floods over the network controlled to prevent a packet from flooding are decided by the link quality. Apart from that, DFR also solves a well-known void problem by allowing at least one node to participate in transmitting a packet. The complete simulation study observed by using a tool of an ns-2 simulator. It proves that DFR is more fit for UWSNs especially when links are prone to packet loss.

Hop-by-Hop Vector-Based Forwarding (HH-VBF) Nicolas et al. proposed an H2-VBF protocol in 2007 [15]. The routing protocol which was mentioned earlier VBF is similar to the H2-VBF. The main difference between these two routing protocols is the VBF used "single virtual pipe from the source to the sink," whereas others used "different virtual pipe around the per-hop vector from each forwarded to the sink." Because of this reason, H2-VBF routing algorithm makes use packets adaptively and forwarding the respective destination based on its current location. The routing table gives this updated information to the nodes. Hence, the robustness of the packet delivery in sparse networks gives the significant improvement.

4 Security Issues in UWSNs

Underwater wireless sensor network has emerged and cutting edge technology for the application of ad hoc networks paradigm because of authenticating source and destination over the network it encounters security issues. Any untrusted third party receives the messages at that time, be aware of the security threats. And, on the other hand, the wireless sensor networks having the limitation of low battery power, less processing speed, and communication range. With these limitations, sometimes the network facing with various such as:

1. Sybil Attack
2. Hello Flood Attacks
3. Sinkhole Attacks
4. Selective Forwarding
5. Acknowledgment Spoofing
6. Wormhole Attack
7. Jamming
1. **Sybil Attack**: The adversary subverts the organization reputation of a (Peer-to-Peer) P2P network by sending the maximum number of anonymous credentials and uses them to obtain valid data. How reputation systems can fall into the vulnerability means the untrusted parties can send randomly identities which do not form a chain of trust linking between legitimate users and organization finally it vulnerable to the Sybil attacks. According to a survey made by the papers, Sybil attacks could be carried out in a cheap and planned way in existing realistic systems.
2. **Hello Flood**: "The next attack is hello flood attack, in which the attacker is not a legitimate node in the wireless sensor network. The anonymous user can flood hello message to any original node and violate the security issues in underwater sensor networks. The next attack is hello flood attack, in which the attacker is not a legitimate node in the wireless sensor network. The anonymous user can flood hello message to any original node and violate the security issues in underwater sensor networks. To reason behind the hello flood attack is mainly by the cryptographic problems because these nodes may suffer from massive

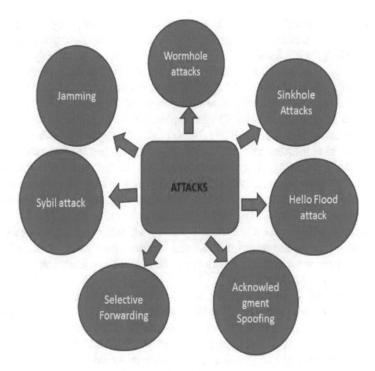

Fig. 1 Attacks in UWSNs

computational complexity. Hence, they are less suitable for WSN. To resolve the hello flood attack based on signal strength, it has been proposed to detect and prevent hello flood attack as shown in Fig. 1."

3. **Sinkhole Attack**: One of the most dangerous attacks in the wireless sensor network is the sinkhole attack. Because the once adversary implemented the sinkhole attack, other associated attacks also possible. Example of associated attacks is (i) Selective forwarding attack and (ii) Acknowledge spoofing attack, etc. The attack is primarily applicable at the routing table because to attract network traffic by advertising it is fake routing update.

4. **Selective Forwarding**: WSN is a cutting edge technology and is being emerged as an upcoming technology shortly due to its variety applications of private domains, military, and UWSNs. The selective forwarding attack is mainly applicable at the router level so prone to network security attacks. Because the sensors were having the limitation of battery degradation and processing and computational speed also lower these facts make sensor networks infeasible or inappropriate to utilize the conventional security solutions, which needs complex operations and huge memory the similar attacks on these networks which can be classified as data traffic and routing attacks. In a WSN, a node can send all the packets like the flood to all nodes. A particular case of black hole attack is

a selective forwarding attack. For this case, a node can select the target and drops packets, which may degrade the efficiency of the network [16].

5. **Acknowledgment Spoofing**: Generally, this attack is mainly caused by improper session handling at the transport layer. Whenever a packet is sent to the WSN, at that moment the copy of the valid data is stored at the local nodes. The nodes which were received acknowledgment that the data sent by the other end. The Internet Protocol (IP) not responsible for performing this operation, only transport layer is responsible. Because the sliding window deals with what are the waiting packets (not acknowledged) at the current buffer and what are the acknowledged packets. If an acknowledge may be delayed for any reason like packet loss, then the transmitting side must wait until the window opens again. Meanwhile, the system may not wait more time for getting acknowledged. So, the transmitting widow says about "Windowing out."

6. **Wormhole attack**: In a wormhole attack, an adversary may receive large number of packets at one point in the wireless sensor network, "tunnels" them to individual point over the network, and then resending them into the network from that point [16].

7. **Jamming**: The jamming attack can be referred to as "intentional interference attacks on wireless networks." It is an attempt at making legitimate users not possible to utilize available network resources. In other words, these attacks are similar to the "Denial-of-service (Dos)" attacks against in the wireless medium. If the adversary is able to implement the Man-in-the-Middle (MitM), then DoS is also possible to occur over the network [17, 18].

5 Conclusions

WSN is a cutting edge technology being emerged as an upcoming technology shortly due to its various applications of private domains such as military and UWSNs. The goal of the paper is we performed a systematic literature survey on three areas: (i) Software-defined networks, (ii) Routing protocols, and (iii) Security infrastructure over the Underwater Wireless Sensor Networks. These primary studies have contributed to technology frameworks, protocols such as routing methods, tools, approaches, and others. We have identified the criteria based on current research and trends in underwater sensor networks. In this study, we have identified various routing protocols which are currently used in UWSNs and provide recommendations that will shape the focus of future research efforts.

References

1. Akyildiz, I.F., Pompili, D., Melodia, T.: Underwater acoustic sensor networks: research challenges. Ad Hoc Netw. **3**(3), 257–279 (2005)
2. Akyildiz, I.F., Wang, P., Lin, S.C.: Softwater: software-defined networking for next-generation underwater communication systems. Ad Hoc Netw. **46**, 1–11 (2016)
3. Chitre, M., Shahabudeen, S., Stojanovic, M.: Underwater acoustic communications and networking: recent advances and future challenges. Mar. Technol. Soc. J. **42**(1), 103–116 (2008)
4. Heidemann, J., Ye, W., Wills, J., Syed, A., Li, Y.: Research challenges and applications for underwater sensor networking. In: IEEE WCNC 2006 Wireless Communications and Networking Conference, 2006, vol. 1, pp. 228–235. IEEE (2006)
5. Lanbo, L., Shengli, Z., Jun-Hong, C.: Prospects and problems of wireless communication for underwater sensor networks. Wirel. Comm. Mob. Comput. **8**(8), 977–994 (2008)
6. Diao, B., Xu, Y., An, Z., Wang, F., Li, C.: Improving both energy and time efficiency of depth-based routing for underwater sensor networks. Int. J. Distrib. Sens. Netw. **11**(10), 781932 (2015)
7. Jadhao, P., Ghonge, M.: Energy efficient routing protocols for underwater sensor networks-a survey. Energy **1**(1) (2015)
8. Jain, S., Mishra, J.P., Talange, D.: A robust control approach for magnetic levitation system based on super-twisting algorithm. In: 2015 10th Asian Control Conference (ASCC), pp. 1–6. IEEE (2015)
9. Stojanovic, M., Preisig, J.: Underwater acoustic communication channels: propagation models and statistical characterization. IEEE Commun. Mag. **47**(1), 84–89 (2009)
10. Fang, K., Rugini, L., Leus, G.: Block transmissions over doubly selective channels: iterative channel estimation and turbo equalization. EURASIP J. Adv. Signal Process. **2010**(1), 974652 (2010)
11. Akyildiz, I.F., Pompili, D., Melodia, T.: Challenges for efficient communication in underwater acoustic sensor networks. ACM Sigbed Rev. **1**(2), 3–8 (2004)
12. Yan, H., Shi, Z.J., Cui, J.H.: DBR: depth-based routing for underwater sensor networks. In: International Conference on Research in Networking, pp. 72–86. Springer, Berlin (2008)
13. Xie, P., Cui, J.H., Lao, L.: VBF: vector-based forwarding protocol for underwater sensor networks. In: International Conference on Research in Networking, pp. 1216–1221. Springer, Berlin (2006)
14. Hwang, D., Kim, D.: DFR: directional flooding-based routing protocol for underwater sensor networks. In: OCEANS 2008, pp. 1–7. IEEE (2008)
15. Nicolaou, N., See, A., Xie, P., Cui, J.H., Maggiorini, D.: Improving the robustness of location-based routing for underwater sensor networks. Oceans 2007-Eur. **18** (2007)
16. Bysani, L.K., Turuk, A.K.: A survey on selective forwarding attack in wireless sensor networks. In: 2011 International Conference on Devices and Communications (ICDeCom), pp. 1–5. (2011)
17. Bojjagani, S., Sastry, V.: Stamba: security testing for android mobile banking apps. In: Advances in Signal Processing and Intelligent Recognition Systems, pp. 671–683. Springer, Berlin (2016)
18. Bojjagani, S., Sastry, V.: Vaptai: a threat model for vulnerability assessment and penetration testing of android and ios mobile banking apps. In: 2017 IEEE 3rd International Conference on Collaboration and Internet Computing (CIC), pp. 77–86. IEEE (2017)
19. Lloret, J., Sendra, S., Ardid, M., Rodrigues, J.J.: Underwater wireless sensor communications in the 2.4 GHz ISM frequency band. Sens. **12**(4), 4237–4264 (2012)

Novel Probabilistic Clustering with Adaptive Actor Critic Neural Network (AACN) for Intrusion Detection Techniques

R. Sudhakar, P. V. Venkateswara Rao and N. Mohan Krishna Varma

Abstract Interruption detection is the procedure of assault distinguishing proof in the PC frameworks and it clears path for the recognizable proof of entrances, breakings, and other PC-related maltreatment. However, the development of the web-based gadgets makes the discovery procedure a confused strategy, representing the requirement for the robotized framework to recognize the assaults. In view of this, the paper proposes method of intrusion detection using the Novel Brainstorm-Crow Search-based Adaptive Actor Critic Neural Network. Clustering is the way toward making a gathering of conceptual objects into classes of comparative items. The clusters are subjected to the different-advance arrangement that is advanced utilizing the proposed enhancement calculation, and in the second dimension of characterization, the interruption in the information is distinguished. The experimentation of the proposed strategy utilizing the KDD cup dataset yields a precision of 0.69, True Positive Rate of 0.68, and False Positive Rate of 0.55.

1 Introduction

The intrusion detection may be a technique that provides security to the PC system model from the cyber attacks mechanism through serious observation that will increase because of the multiple connected devices over the network. The advance detection of the intrusion method termed as Intrusion Detection System within the network [1, 2] and Intrusion Detection System is capable of police work the violations of principles within the network specified the abnormal patterns of the

R. Sudhakar · P. V. Venkateswara Rao · N. Mohan Krishna Varma (✉)
Department of Computer Science & Engineering (CSE), Madanapalle Institute of Technology
& Science, Madanapalle, India
e-mail: drmohankvn@mits.ac.in

R. Sudhakar
e-mail: sudhavk1983@gmail.com

P. V. Venkateswara Rao
e-mail: drvenkateswararaopv@mits.ac.in

© Springer Nature Singapore Pte Ltd. 2020
P. Venkata Krishna and M. S. Obaidat (eds.), *Emerging Research in Data
Engineering Systems and Computer Communications*, Advances in Intelligent
Systems and Computing 1054, https://doi.org/10.1007/978-981-15-0135-7_51

system are extracted [3]. Thus, Intrusion Detection System is an intensive line of defense in any network that resists and degrades the consequences of security risks within the network. Intrusion Detection System possesses the power to notice the intrusion activities within the network through system alerts that are analyzed by the protection analyst for designing the security response of the network. The alternative techniques, the method of analyzing huge number of attacks in the networks raises a huge network burden to the analyst and may sometimes lead to faults in analyzing the networks [4]. The issue faced by Intrusion Detection System in proceeding the intrusion in a network is the high network dimensionality problem, selecting the classifier, and distance measurement techniques [5]. In the starting of the detection system, there are plenty of rule primarily based approaches engaged in detective work the intrusion within the network and therefore the rule-based approaches define rules that outline the conventional and status of the network specified the best-known attacks are detected, whereas the rule-based strategies.

The same terms raise the requirement for developing the economical and correct intrusion models [6, 7]. Clustering algorithms play a significant role within the specific applications of intrusion model and cluster validation could be a technique that assesses the clump quality and determines a more robust cluster that is applicable for any application. The most aim of the clump technique is to seek out the optimum cluster theme for extracting and analyzing the cluster patterns [8]. Cluster analysis is straightforward that deals with the study of the algorithms and strategies that aim at ripping the objects as homogenous teams, named as clusters. The most aim of clump is to unravel the matter of ripping the objects in such some way that the objects in a very cluster gibe one another, whereas the objects in two totally different clusters vary among one another. The foremost common clump formula used for determination the IDS is K-means clustering algorithm that finds the intruders within the network [9, 10]. The anomalies effectively, a clump formula is used in such some way that the anomalies are known through an unattended detector dubbed 'IKD' at the time of clustering and therefore the clustering algorithm forms the clusters supported the cluster breadth. Brainstorming could be a cluster ability technique by that efforts are created to search out a conclusion for a particular downside by gathering an inventory of ideas impromptu contributed by its members.

2 Proposed Method

A significant analysis space is that the usage of machine learning that assists the network administrator in taking effective measures to forestall intrusions within the network. The main aim of the analysis is to model associated establish an intrusion system by proposing a unique classifier, known Adaptive Actor Critic Neural Network (AACNN) classifier for the classification. Initially, the input file is clustered exploitation the projected probabilistic thin FCM that is developed by modifying the prevailing sparse FCM. The projected model offers the clusters and also the sorted information is fed to the proposed AACNN classifier, wherever the weights within the actor critic neural network are chosen optimally exploitation the algorithmic rule. Based on the classifier output, shrink feature is made associated is employed to coach the categoryifier exploitation the projected AACNN classifier to see the ultimate class that defines whether or not the user is a trespasser or traditional user. The projected probabilistic thin fuzzy C-means bunch is that the integration of the probabilistic theory within the sparse FCM and also the intention of the proposed technique are to perform optimum information clustering. The paper is organized as one deliberates the background of intrusion detection within the information and sections two discuss the literature survey of the intrusion detection ways. The projected information bunch technique is mentioned. The major concern of the prevailing automated clustering algorithms is to handle the randomly formed information distribution of the datasets. Moreover, the method of evaluating the bunch quality mistreatment the statistics-based ways was time intense once just in case of the large-sized information that was terribly arduous to include. The main aim of proposing the probabilistic Adaptive Fuzzy C-Means clustering algorithm is to alter the effective clustering of the info to make optimum clusters. The planned bunch rule is that the integration of probabilistic theory [11] and thin Fuzzy C suggests that clustering algorithm [12] and inherits the benefits of each the ideas. The planned rule performs well even within the presence of style of options during a cluster. The prevailing thin Fuzzy C suggests that bunch rule [12]. This is not capable of handling the large information and it is incapable of generalization. Meanwhile, the planned rule is capable of choosing the relevant options and capable of removing the noise from the info. Thus, the planned rule overcomes the demerits of the prevailing thin Fuzzy C suggests that bunch algorithm. The planned rule allows the higher applied mathematics understanding and insures the wide in operation ranges. Intrusion detection may be a vital space of analysis that is primarily engaged in distinctive the attacks within the automatic information processing system as a result of the technology changes and because of the presence of the huge association of the devices in the network. The attackers are capable of deactivating the intrusion system and thus, most of the developed ways assure the strength and supply the power to handle the uncertainty problems. However, most of the ways fail to perform effectively in detective work the cyber attacks in adversarial environments. Thus, the analysis concentrates on the effective intrusion detection mechanism victimization the AACNN classifier that provides a platform to spot the interloper within the knowledge. The projected intrusion detection systems realize the traditional user and therefore the interloper in efficient and optimum approach victimization the proposed algorithmic rule. The projected algorithmic rule trains the AACNN classifier thus on verifies the optimum weights of AACNN through determination the minimum error perform.

Algorithm

The major steps within the intrusion detection are: within the opening, the computer file is clustered victimization the projected probabilistic distributed algorithmic rule to create the optimum clusters that enter the subsequent step of intrusion detection. For the generation of classification, the projected BCS-based AACNN is used. The optimum clusters that are given as input to the projected BCS-based AACNN type the shrink feature that is fed because the input of the second level of AACNN for the classification of the attackers and therefore the intruders BSO suffers from native convergence problems and there is no balance within the exploration and exploitation phases. Hence, the interbreeding of BSO and CSA balances the switch-over between exploration and exploitation because of the advantage of CSA. BSO possess quicker world convergence that forms the foremost advantage of the projected algorithmic rule. Thus, the most aim of the projected algorithmic rule is to work out the world optimum weights to coach the AACNN classifier.

(a) Initialization: within the data formatting step, the adjustable parameters are initialized.
(b) Cipher the target performs: the target function for computing the world optimum weight to tune the AACNN classifier relies on the minimum error function.
(c) Location modification victimization the projected algorithm: The position of the crow is updated supported to conditions, one relies on the possible resolution and therefore the different is based on the non-feasible solution.
(d) Calculate the target performs: the target function is evaluated for the new position of the crow so as to pick out the simplest resolution.
(e) Final position: the simplest position of the crow is set looking on the target perform.

3   Results

The implementation of the projected system is going to be in MATLAB. The experimentation is performed and therefore the results are compared with the present works supported True Positive Rate, True Negative Rate, and Accuracy.

The experimentation is performed victimization the DARPA's KDD cup dataset the metrics used for the effective comparison of the intrusion detection ways embrace the following:

Accuracy The term accuracy defines the accurate detection and it is formulated as,

$$\text{Accuracy} = \frac{\text{TP} + \text{TN}}{\text{TP} + \text{TN} + \text{FP} + \text{FN}} \tag{1}$$

where, TP, TN, FP, and FN are true positive, true negative, false positive, and false negative, respectively.

True Positive Rate (TPR) TPR is the ratio of true positive (TP) to the sum of true positive (FP) and false negative (FN).

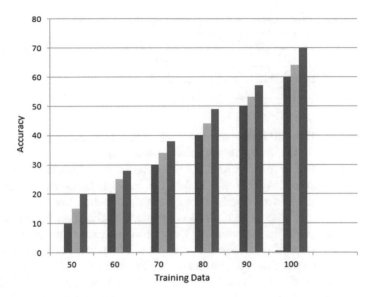

Fig. 1 Analyzing different cluster sizes

TPR is computed as,

$$TPR = \frac{TP}{TP + FN} \qquad (2)$$

The ways taken for the comparison embrace K-means-K-Nearest Neighbor (K-means_KNN), Fuzzy C-Means K-Nearest Neighbor (FCM_KNN). The results of the present ways are compared with the projected technique. The analysis is progressed supported numerous cluster sizes and therefore the comparison of the ways is created based on the performance metrics. The analysis of the intrusion detection ways supported the cluster size five. The accuracy, TPR, and FPR of the ways are delineating. Initially, the accuracy of the ways is high for the coaching proportion of fifty and upon increasing the training percentage, the accuracy of the ways reduces (Fig. 1).

4 Conclusions

The comparative discussions of the intrusion detection ways are given in Table 1. The accuracy of the ways, K-means_KNN, FCM_KNN, and AACNN methodology is 0.698, 0.688, and 0.758 and it is clear that the AACNN methodology no heritable a higher worth of accuracy. Similarly, the TPR and FPR of the AACNN methodology are higher for the remaining techniques. The paper discusses the planned intrusion detection victimization, the proposed adaptive Actor Critic Neural Network (AACNN) classifier, and that at the start, the computer file is fed to the

Table 1 Comparing the accuracy rate, true positive rate, false positive rate

Methods	K-means_KNN	FCM_KNN	BCS-AACNN
Accuracy rate	0.51	0.52	0.69
TPR	0.52	0.53	0.68
FPR	0.77	0.67	0.55

cluster module. The cluster module uses the planned Adaptive Fuzzy C-Means cluster rule that determines the optimum clusters for playing the correct and effective detection of intrusion within the information. The classification module is processed as different major levels: within the initial level of classification, the clusters are fed to the classifier to create the shrink feature that is given to the second level of classification victimization the planned classifier, for the detection of the intrusion within the information. The experimentation is performed victimization the KDD cup info and also the analysis of the intrusion detection ways is progressed by examination the performance earned thereupon of few existing methods. The planned methodology no inheritable the most accuracy of 0.69, most True Positive Rate (TPR) of 0.68, and False Positive Rate (FPR) of 0.55.

References

1. Tsai, C.F., Lin, C.Y.: A triangle area based nearest neighbors approach to intrusion detection. Pattern Recogni. **43**(1), 222–229 (2010)
2. Mukhejee, B., Heberlein, L.T.: Network intrusion detection. IEEE Netw. **8**(3), 26–41 (1994)
3. McElwee, S.: Active learning intrusion detection using K-means clustering selection. SoutheastCon **2017**, 1–7 (2017)
4. de Alvarenga, S.C., Barbon, S., Miani, R.S., Cukier, M., Zarpelão, B.B.: Process mining and hierarchical clustering to help intrusion alert visualization. Comput. Secur. **73**, 474–491 (2018)
5. Gunupudi, R.K., Nimmala, M., Gugulothu, N., Gali, S.R.: CLAPP: a self constructing feature clustering approach for anomaly detection. Futur. Gener. Comput. Syst. **74**, 417–429 (2017)
6. Peng, X., Wang, L., Wang, X., Qiao, Y.: Bag of visual words and fusion methods for action recognition: comprehensive study and good practice. Comput. Vis. Image Underst. **150**, 109–125 (2015)
7. Multitask, P., et al.: Multiple/Single-View Human Action Recognition, pp. 1–15 (2014)
8. Rahman, T.K.A., Suliman, S.I., Musirin, I.: Artificial immune-based optimization technique for solving economic dispatch in power system. Lecture Notes in Computer Science (including Subseries Lecture Notes in Artificial Intelligence, Lecture Notes in Bioinformatics), vol. 3931, pp. 338–345. LNCS (2006)
9. Nie, W., Liu, A., Li, W., Su, Y.: Cross-view action recognition by cross-domain learning. Image Vis. Comput. **55**, 109–118 (2016)
10. Harish, B.S., Kumar, S.V.A.: Anomaly based intrusion detection using modified fuzzy clustering. Int. J. Interact. Multimed. Artif. Intell. **4**(6), 54 (2017)
11. Chuan-long, Y., Yue-fei, Z., Jin-long, Z., Xin-zheng, H.: A deep learning approach for intrusion detection using recurrent neural networks. IEEE Access **5**, 1 (2017)
12. Lin, W.C., Ke, S.W., Tsai, C.F.: CANN: an intrusion detection system based on combining cluster centers and nearest neighbors. Knowl. Based Syst. **78**(1), 13–21 (2015)

Survey on Classification and Feature Selection Approaches for Disease Diagnosis

Diwakar Tripathi, I. Manoj, G. Raja Prasanth, K. Neeraja,
Mohan Krishna Varma and B. Ramachandra Reddy

Abstract Patient case similarity implies that finding and extracting a patient case have similar features in the knowledge base. The knowledge base contains data obtained through demographics, progress notes, medications, past medical history, discharge summaries and lab values. Data pre-processing is the first step and an important step in the modelling process. The aim of this step is to increase the effectiveness of the classification process by using representative and consistent data set. Pre-processing includes data cleaning, data transformation and feature selection. Further, for predicting the new cases, new sample will be submitted to trained model. In the literature, various feature selection and classification approaches are available, but it is not clear which feature selection approach may have better classification performance. So, this study presents a survey on feature selection and classification approaches applied on seven benched-marked diseases data sets obtained from the UCI repository.

Keywords Classification · Disease diagnosis · Feature selection

D. Tripathi (✉) · I. Manoj · G. Raja Prasanth · K. Neeraja · M. K. Varma · B. Ramachandra Reddy
Madanapalle Institute of Technology & Science, Madanapalle, A.P, India
e-mail: diwakarnitgoa@gmail.com

I. Manoj
e-mail: manojworkspace98@gmail.com

G. Raja Prasanth
e-mail: rajaprasanthgantasala@gmail.com

K. Neeraja
e-mail: kanugondaneeraja@gmail.com

M. K. Varma
e-mail: drmohankvn@gmail.com

B. Ramachandra Reddy
e-mail: brreddy@iiitdmj.ac.in

© Springer Nature Singapore Pte Ltd. 2020   567
P. Venkata Krishna and M. S. Obaidat (eds.), *Emerging Research in Data
Engineering Systems and Computer Communications*, Advances in Intelligent
Systems and Computing 1054, https://doi.org/10.1007/978-981-15-0135-7_52

1 Introduction

In this technological world of advancement, the healthcare industry is also getting smart and digitalized. Almost each and every hospital is maintaining and storing the patient's data specifically in its database in either structured or unstructured format. The massive hospital data cannot be entirely analysed by humans [1]. Because, the data is spread in a wide variety of formats like demographics, progress notes, lab values, etc. Moreover, the data is heterogeneous which implies the data is in the forms of images, time series, signals and many more formats [2]. The knowledge extraction is the most vital step in data analysis which cannot be done precisely and productively without the intervention of machine intelligence. So, it is highly essential to unveil the power of machine learning for the analysis and knowledge extraction from the massive hospital data. According to the analysis of the World Health Organization, the deaths are mainly caused due to the disease outbreaks primarily. In 2016, heart disease and stroke account for 15.2 million deaths followed by chronic diseases like cancers caused nearly 1.7 million deaths, diabetes caused nearly 1.6 million deaths. However, the road injuries caused only 1.4 million deaths which are a considerably smaller amount compared to the damage caused by disease outbreaks [3]. The number of death cases because of various diseases is depicted in Fig. 1 in 2016 [3]. So, the improvement in the healthcare industry dramatically leads to the decline of the death rates.

For the quick and effective treatment of a disease to a new patient, the smart approach is to search a similar case from the past data aggregated in a hospital database. That is the patient case similarity problem where the knowledge extraction of past patient data is required for the effective treatment of diseases with the help of already established, advanced machine learning classification approaches. However, the vast

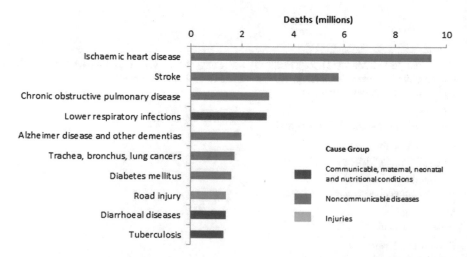

Fig. 1 Global causes of death in 2016

pool of diseases makes it difficult for the single machine learning approach for the effective analysis of the diseases and their characteristics. Because a particular approach will be efficient in a particular disease as each and every disease involves a different class of features which makes a single classification approach end-up with low accuracy in finding a similar case. So, the ensemble classification approach comes into picture where two or more classification approaches are combined and work together. This implies the building of hybrid model where it includes many classification approaches coupled with some feature selection approaches for the significant betterment of prediction. Further, the hybrid model can be used to recommend the particular medication and treatment of a disease to a new patient from the knowledge additionally extracted from the medication database of the hospitals.

As per the previous researches held, we observed that the hybrid model is the best model for the better accuracy in classification [4–6], which is a combination of two or more classifiers coupled with some feature selection approaches. It implies that the classifiers are aggregated in the model along with the application of some feature selection approaches. So, the main motive of this survey is to apply feature selection approaches with classification approaches to analyse the compatibility of these approaches in disease data sets.

Reminder of the article is structured as follows: Sect. 2 describes about various feature selection approaches, Sect. 3 describes about various classification approaches, Sect. 4 exhibits the test results investigation and comparative analysis followed by the concluding remarks based on obtained experimental results.

2 Feature Selection

This section presents a brief introduction about various feature selection approaches, namely principal component analysis (PCA), information gain (IG), OneR, short for "One Rule" (OR), ReliefF (RLF), correlation feature selection (CFS), symmetric uncertainty (SU), gain ratio (GR) and SVM attribute evaluation (SAE) are described as follows:

PCA—firstly, it calculates the covariance matrix of data points, eigen vectors and corresponding eigen values, then sorts the eigen vectors according to their eigen values in decreasing order and chooses first k eigen vectors and that will be the new k dimensions. It transforms the original n dimensional data points into k dimensions [7].

IG measures how a particular feature is associated with a target label. Highly correlated features are considered as the best features, and it considers "Entropy" for selection as feature or reduction of less relevant features [7]. GR is an extension of IG, and it considers the gain ratio of gain and intrinsic information (entropy) for consideration of impotence of that feature [7].

OneR, another way to say "One Rule", is a straightforward, yet precise, characterization calculation that creates one guideline for every indicator in the information, at that point chooses the standard with the littlest all out mistake as its "One Rule".

To make a standard for an indicator, it develops a recurrence table for every indicator against the target [7].

RLF—it calculates feature score for each feature. Further, that score is utilized to rank and features with the highest rankings are considered as the best features set. Scoring is based on the recognition of feature value differences between nearest neighbour instance pairs. It is an iterative process, and at each iteration, it takes the feature vector (X) having a place with one random instance. The feature vectors of the instance are closest to X (by Euclidean distance) from each class [7].

CFS—this approach evaluates subsets of features on the basis of the hypothesis: "Good feature subsets contain features highly correlated with the classification, yet uncorrelated to each other" [8].

SEA is an wrapper-based feature selection approach, and it considers the classification accuracy of SVM as a parameter to consider the best set of feature [9].

3 Classification

This section presents a brief introduction about various classification approaches utilized in this survey and are as follows:

Multilayer perceptron (MLP) is a feed-forward neural network with at least three layers input, hidden and output layer. Other than input layer, neurons at each layer use a nonlinear activation function with back propagation for training [10].

Radial basis function neural network (RBFN) [11] is the four-layer feed-forward architecture. First layer is the input layer, second is the pattern or hidden layer, third is the summation layer, and fourth layer is the output or decision layer. In case of RBFN, it uses the radial basis function as an activation function.

Logistic regression (LR) [12] can be considered as an extraordinary case of linear regression models. Be that as it may, with binary class classification, it violates normality assumptions of general regression models. LR indicates that a proper function of the fitted likelihood of the event is a linear function of the observed values of the available explanatory variables.

Decision tree (DT) [13] is one of the predictive modelling approaches, for learning, it makes the use of a tree as a predictive model. In this model, observations and corresponding target values are represented in the branches as conjunctions of features and leaf nodes, respectively. Tree models, if target variable is a set of discrete values are called classification trees and with real number is called regression trees. Best-first tree (BFT) [14] is similar to normal decision trees as it stops expanding a tree earlier if any splitting step appears to increase the error estimate. It is based on cross-validation. In a parallel fashion, all the trees for all the training folds are constructed.

Partial decision tree (PART) [15]—it combines two approaches C4.5 and ripper to avoid their respective problem. Unlike previous approaches, it does not require global optimization for generating the rule set and work in separate-and-conquer manner. In this approach, instances are removed which are covered. For generating a rule, it makes the use of pruned DT with current instances with leaf with largest coverage.

Random forest (RF) [16] is an approach for classification and regression problem in association with ensemble learning. In case of RF, randomness is introduced by recognizing the best split feature instead of a random selection of a subset of available features in data. Further, ensemble classifier combines the individual predictions of various trees to combine them into a final prediction, and it utilizes a majority voting approach to combine the individual predictions of tree. Random decision forests are better way for decision trees' in case of over-fitting problem with training set.

K-star is an instance-based classifier that is the class of a test instance is based upon the class of those training instances similar to it, as determined by some similarity function. It differs from other instance-based learners in that it uses an entropy-based distance function [17].

Naive Bayes (NB) [18] classifier is a probabilistic machine learning model based on the Bayes theorem. This approach can be applied with arbitrary number of independent features whether these are continuous or categorical, and it calculates the posterior probability for the event among a set of possible outcomes.

Sequential minimal optimization (SMO) is an approach to solve QP problem that arises during the training of SVM [19]. SMO solves the SVM's QP problem by decomposing the overall problem into various subproblems. Unlike SVM, it uses smallest possible optimization approach at each step with two Lagrange multipliers to find the optimal values.

4 Results and Discussion

This section presents the description about data sets along with performance measures used to validate the proposed approach and detailed result analysis on these credit scoring data sets. In this work, we have considered seven diseases data sets obtained form the UCI repository [20] and are as mentioned in Table 1. And, classification accuracy [21] is used for comparative result analysis in terms of mean of tenfold cross-validation (10-FCV).

Results on Diabetic Retinopathy data set are as tabulated in Table 2. From the results, it is observed that MLP has the best classification accuracy with all features. NB with RLF, LR with SU, MLP with PCA, SMO with SEA, DTNB with CFS, PART with CFS, BFT with CFS, K-star with PCA and RBFN with PCA have the best performance. Overall, CFS with RF has the best classification accuracy. SEA with MLP has the best classification accuracy.

Results on Breast Cancer Wisconsin (Diagnostic) data set are as tabulated in Table 3. From the results, it is observed that RF has the best classification accuracy with all features. NB with CFS, SMO with IG, LR with CFS, MLP with CFS, SMO with IG, DTNB with CFS, PART with SEA, BFT with IG, K-star with CFS and RBFN with CFS have the best performance. Overall, CFS with RF has the best classification accuracy.

Table 1 Details of data sets

Data set	Samples	Classes	Features	Class distribution
Diabetic Retinopathy	1151	2	20	611/540
Breast Cancer Wisconsin (Diagnostic)	569	2	32	212/357
Breast Cancer Wisconsin (Original)	699	2	11	241/458
Cardiotocography	2126	3	23	1655/295/176
Cervical cancer (Risk Factors)-Cytology	858	2	36	814/44
Cervical cancer (Risk Factors)-Biopsy	858	2	36	55/803
Pima Indian Diabetes	768	2	9	268/500

Table 2 Results of Diabetic Retinopathy data set

Classifier	Feature selection approaches								
	All	IG	OR	PCA	RLF	SU	CFS	GR	SEA
NB	56.82	56.92	56.83	58.38	57.01	56.11	56.47	56.57	57.14
LR	74.89	74.77	74.97	74.63	74.93	75.09	73.16	73.21	74.88
MLP	72.02	72.11	72.18	73.33	72.31	72.32	71.94	71.88	72.51
SMO	67.59	67.61	67.81	66.94	67.88	67.97	65.99	66.06	67.89
DTNB	61.86	62.06	61.93	62.58	61.97	62.11	63.59	63.51	62.11
PART	64.64	65.11	65.21	65.17	65.04	64.94	66.17	65.07	64.69
BFT	62.29	62.35	62.41	64.21	62.54	62.44	62.95	62.07	62.43
RF	68.20	68.31	68.29	70.03	68.36	68.48	68.03	68.13	68.87
k-star	61.25	61.39	61.28	66.99	61.46	61.15	64.12	61.55	61.27
RBFN	60.12	60.72	60.33	63.08	60.28	60.35	61.95	60.82	60.77

Table 3 Results of Breast Cancer Wisconsin (Diagnostic) data set

Classifier	Feature selection approaches								
	All	IG	OR	PCA	RLF	SU	CFS	GR	SEA
NB	92.97	92.91	92.89	91.98	92.42	93.12	94.55	93.22	93.06
LR	93.50	93.71	93.58	93.97	93.87	93.63	96.84	93.92	94.57
MLP	96.66	96.97	97.16	94.98	97.28	97.31	97.89	96.92	97.31
SMO	97.72	98.33	97.99	96.67	97.93	98.02	97.66	98.31	97.82
DTNB	95.43	95.72	95.78	95.91	95.93	96.03	98.20	95.73	95.56
PART	93.50	93.91	93.49	93.97	93.73	93.69	94.48	95.03	95.15
BFT	92.97	93.88	93.09	92.89	93.58	92.86	93.87	92.69	93.47
RF	96.49	96.96	96.78	94.59	96.36	96.88	97.41	96.58	96.55
k-star	94.73	94.96	94.88	94.79	94.91	94.79	96.13	95.11	95.01
RBFN	94.20	94.89	94.16	91.23	94.41	94.72	94.91	94.27	94.43

Table 4 Results of Breast Cancer Wisconsin (Original) data set

Classifier	Feature selection approaches								
	All	IG	OR	PCA	RLF	SU	CFS	GR	SEA
NB	95.99	96.12	96.25	96.94	96.32	96.41	96.01	96.15	96.14
LR	96.57	96.98	96.86	96.56	96.69	96.83	96.89	96.84	96.73
MLP	95.28	95.96	95.44	95.75	95.39	95.46	95.92	95.65	95.83
SMO	97.00	97.87	97.19	96.71	97.41	97.83	97.37	97.19	97.26
DTNB	96.85	97.36	97.15	97.69	97.02	97.65	97.91	97.85	97.85
PART	93.85	93.85	93.85	94.99	93.85	93.85	93.85	93.85	93.85
BFT	94.28	94.87	95.08	95.99	94.78	94.69	94.83	94.73	94.19
RF	96.57	96.57	96.57	96.57	96.57	96.57	96.57	96.57	96.57
k-star	95.42	96.36	96.09	96.32	96.97	96.38	97.76	96.37	97.49
RBFN	95.85	96.98	96.39	96.97	96.13	95.85	97.36	96.45	95.09

Results on Breast Cancer Wisconsin (Original) data set are as tabulated in Table 4. From the results, it is observed that MLP has the best classification accuracy with all features. NB with RLF, LR with SU, MLP with PCA, SMO with SEA, DTNB with CFS, PART with CFS, BFT with CFS, K-star with PCA and RBFN with PCA have the best performance. Overall, CFS with RF has the best classification accuracy. SEA with MLP has the best classification accuracy.

Results on Cardiotocography data set are as tabulated in Table 5. From the results, it is observed that MLP has the best classification accuracy with all features. NB with CFS, LR with CFS, MLP with OR, SMO with CFS, DTNB with CFS, PART with CFS, BFT with CFS, RF with IG, K-star with CFS, and RBFN with CFS have the best performance. Overall, OR with MLP has the best classification accuracy. SEA with MLP has the best classification accuracy.

Table 5 Results of Cardiotocography data set

Classifier	Feature selection approaches								
	All	IG	OR	PCA	RLF	SU	CFS	GR	SEA
NB	94.03	94.48	94.22	94.35	94.68	94.91	95.35	94.71	94.97
LR	98.64	98.97	98.83	98.90	98.91	98.73	99.39	98.77	98.69
MLP	99.06	99.73	99.86	98.14	99.39	99.83	99.64	99.36	99.18
SMO	98.59	98.68	98.71	96.96	98.88	98.53	98.90	98.49	98.67
DTNB	98.31	98.87	98.39	98.05	98.53	98.47	98.98	98.51	98.23
PART	98.49	99.02	98.86	98.03	98.26	98.45	98.93	98.61	98.15
BFT	98.25	98.69	98.51	92.90	98.43	98.54	98.97	98.14	98.36
RF	98.97	99.36	99.32	98.26	99.65	99.73	99.97	99.02	98.83
k-star	96.05	96.63	96.11	96.91	96.38	96.42	98.21	96.65	96.49
RBFN	97.93	97.93	97.93	96.51	98.62	98.16	98.67	98.26	98.35

Table 6 Results of Cervical cancer(Risk factors)-Cytology data set

Classifier	All	All (Deleted 2)	Feature selection approaches							
			IG	OR	PCA	RLF	SU	CFS	GR	SEA
NB	86.71	86.83	86.97	86.92	89.51	87.03	87.25	93.24	87.26	87.56
LR	93.82	93.94	94.23	94.12	94.87	93.98	93.87	95.31	95.12	93.79
MLP	93.59	93.71	93.71	93.71	94.17	93.89	93.91	94.87	93.59	93.79
SMO	94.87	94.87	94.87	94.87	95.28	94.68	94.71	95.63	94.91	94.77
DTNB	94.06	94.87	94.91	94.87	94.76	95.12	95.36	95.48	95.15	94.80
PART	93.82	93.82	93.97	93.86	93.34	93.82	93.82	94.87	94.11	94.09
BFT	94.87	94.87	94.87	94.87	94.87	94.87	94.87	94.87	94.87	94.87
RF	94.17	94.17	95.31	94.65	95.61	94.38	94.66	95.91	94.39	94.28
K-star	93.47	93.82	94.25	93.92	94.66	94.51	94.29	94.87	93.91	93.87
RBFN	94.87	94.87	95.94	95.62	95.36	94.99	95.12	95.94	94.73	94.68

Table 7 Results of Cervical cancer(Risk factors)-Biopsy data set

Classifier	All features	All features (Deleted 2)	Feature selection approaches							
			IG	OR	PCA	RLF	SU	CFS	GR	SVM
NB	88.69	89.28	89.68	89.35	87.97	89.83	89.58	91.15	89.68	89.59
LR	95.57	95.57	95.83	95.66	93.36	95.57	96.36	96.83	95.81	95.73
MLP	94.76	95.22	95.82	95.73	91.61	95.82	95.73	96.53	95.88	95.93
SMO	96.15	96.15	96.83	96.18	96.59	96.38	96.87	96.97	96.82	96.44
DTNB	95.57	95.92	95.44	96.09	95.47	96.43	96.39	96.73	95.16	95.06
PART	94.52	94.64	94.84	94.86	94.31	95.01	94.93	96.15	95.36	95.02
BFT	95.69	95.69	95.73	95.87	95.58	95.97	95.88	96.75	96.36	95.83
RF	95.69	95.92	96.36	96.37	96.71	96.08	96.28	96.82	96.63	95.03
K-star	92.89	93.59	94.05	93.97	93.96	94.28	94.25	94.61	94.25	93.43
RBFN	93.47	93.59	93.83	93.88	93.78	94.01	94.28	95.10	94.48	93.90

Results on Cervical cancer(Risk factors)-Cytology data set are as tabulated in Table 6. From the results, it is observed that SMO has the best classification accuracy with all features. NB with CFS, LR with CFS, MLP with CFS, SMO with CFS, DTNB with CFS, PART with CFS, BFT with CFS, RF with CFS, K-star with CFS and RBFN with CFS have the best performance. Overall, OR with MLP has the best classification accuracy. SMO with MLP has the best classification accuracy.

Results on Cervical cancer(Risk factors)- Biopsy data set are as tabulated in Table 7. From the results, it is observed that SMO has the best classification accuracy with all features. NB with CFS, LR with CFS, MLP with CFS, SMO with CFS, DTNB with CFS, PART with CFS, BFT with CFS, RF with CFS, K-star with CFS and RBFN with CFS have the best performance. Overall, CFS with SMO has the best classification accuracy.

Table 8 Results of Pima Indian Diabetes data set

Classifier	All	Feature selection approaches							
		IG	OR	PCA	RLF	SU	CFS	GR	SEA
NB	76.30	76.68	76.92	76.86	76.73	76.79	77.47	76.71	76.59
LR	77.21	77.81	77.91	77.84	77.73	77.88	78.13	77.43	77.37
MLP	75.13	75.63	75.58	75.86	75.92	75.88	75.91	75.82	75.26
SMO	77.34	77.56	77.82	77.86	77.91	77.96	78.08	77.56	77.81
DTNB	74.09	74.86	74.59	74.81	74.61	74.59	75.91	74.26	74.18
PART	74.48	74.83	74.49	74.61	74.49	74.83	74.83	74.19	74.28
BFT	73.57	73.83	73.19	73.82	73.46	73.19	73.73	73.76	73.58
RF	75.26	75.81	75.73	75.61	75.66	75.82	76.53	76.06	75.14
K-star	69.14	69.43	69.49	69.81	69.37	69.84	69.92	69.45	69.33
RBFN	75.39	75.44	75.37	75.63	75.48	75.52	76.56	76.14	75.72

Results on Pima Indian Diabetes data set are as tabulated in Table 8. From the results, it is observed that SMO has the best classification accuracy with all features. NB with CFS, LR with CFS, MLP with CFS, SMO with RLF, DTNB with CFS, PART with CFS and SU, BFT with PCA, RF with CFS, K-star with CFS and RBFN with CFS have the best performance. Overall, RLF with SMO has the best classification accuracy.

5 Conclusion

In this article, we have presented a survey on various feature selection approaches such as principal component analysis, information gain, OneR, short for "One Rule", ReliefF, correlation-based feature selection, symmetric uncertainty, gain ratio and SVM attribute evaluation and various classification approaches such as Naive Bayes, logistic regression, multilayer perceptron, sequential minimal optimization, decision tree (Naive Bayes), partial order tree, best-first tree, random forest, K-star and radial basis function network with seven real-world disease data sets such as Diabetic Retinopathy, Breast Cancer Wisconsin (Diagnostic), Breast Cancer Wisconsin (Original), Cardiotocography, Cervical cancer (Risk Factors)-Cytology, Cervical cancer (Risk Factors)-Biopsy and Pima Indian Diabetes. And, results are compared in terms of classification accuracy. From the experimental results, it is observed that SMO and RF are the best performers in most of the data sets, and CFS base feature selection approach improves the classification performance of these classifiers as well rest of the classifiers also. And, from results, it is also observed that other feature selection approaches are also a good way to improve the classification accuracy of other most of the classifiers.

References

1. Wang, X., Wang, Y., Gao, C., Lin, K., Li, Y.: Automatic diagnosis with efficient medical case searching based on evolving graphs. IEEE Access **6**, 53307–53318 (2018)
2. Canino, G., Guzzi, P.H., Tradigo, G., Zhang, A., Veltri, P.: On the analysis of diseases and their related geographical data. IEEE J. biomed. health Inform. **21**(1), 228–237 (2017)
3. The top 10 causes of death (Last Accessed 25 Apr 2019). https://www.who.int/news-room/fact-sheets/detail/the-top-10-causes-of-death
4. Edla, D.R., Tripathi, D., Cheruku, R., Kuppili, V.: An efficient multi-layer ensemble framework with bpsogsa-based feature selection for credit scoring data analysis. Arab. J. Sci. Eng. **43**(12), 6909–6928 (2018)
5. Tripathi, D., Cheruku, R., Bablani, A.: Relative performance evaluation of ensemble classification with feature reduction in credit scoring datasets. In: Advances in Machine Learning and Data Science, pp. 293–304. Springer (2018)
6. Tripathi, D., Edla, D.R., Cheruku, R.: Hybrid credit scoring model using neighborhood rough set and multi-layer ensemble classification. J. Intell. Fuzzy Syst. **34**(3), 1543–1549 (2018)
7. Hall, M., Frank, E., Holmes, G., Pfahringer, B., Reutemann, P., Witten, I.H.: The WEKA data mining software: an update. SIGKDD Explor. **11**(1), 10–18 (2009)
8. Hall, M.A.: Correlation-based feature selection of discrete and numeric class machine learning (2000)
9. Guyon, I., Weston, J., Barnhill, S., Vapnik, V.: Gene selection for cancer classification using support vector machines. Mach. Learn. **46**(1–3), 389–422 (2002)
10. Rosenblatt, F.: Principles of neurodynamics. perceptrons and the theory of brain mechanisms. Tech. rep., CORNELL AERONAUTICAL LAB INC BUFFALO NY (1961)
11. Broomhead, D.S., Lowe, D.: Radial basis functions, multi-variable functional interpolation and adaptive networks. Tech. rep, Royal Signals and Radar Establishment Malvern, UK (1988)
12. Le Cessie, S., Van Houwelingen, J.C.: Ridge estimators in logistic regression. Appl. Stat. 191–201 (1992)
13. Rokach, L., Maimon, O.Z.: Data mining with decision trees: theory and applications, **69**
14. Shi, H.: Best-first decision tree learning. Ph.D. thesis, The University of Waikato (2007)
15. Witten, I.H., Frank, E., Hall, M.A., Pal, C.J.: Data Mining: Practical machine learning tools and techniques. Morgan Kaufmann (2016)
16. Breiman, L.: Random forests. Mach. Learn. **45**(1), 5–32 (2001)
17. Cleary, J.G., Trigg, L.E.: K*: An instance-based learner using an entropic distance measure. In: Machine Learning Proceedings 1995, pp. 108–114. Elsevier (1995)
18. John, G.H., Langley, P.: Estimating continuous distributions in bayesian classifiers. In: Proceedings of the Eleventh conference on Uncertainty in artificial intelligence. pp. 338–345. Morgan Kaufmann Publishers Inc. (1995)
19. Platt, J.C.: 12 fast training of support vector machines using sequential minimal optimization. Adv. kernel methods 185–208 (1999)
20. UCI machine learning repository (Last Accessed 25 Apr 2019). https://archive.ics.uci.edu/ml/index.php
21. Tripathi, D., Edla, D.R., Cheruku, R., Kuppili, V.: A novel hybrid credit scoring model based on ensemble feature selection and multilayer ensemble classification. Computational Intelligence

Review on Supervised Learning Techniques

Y. C. A. Padmanabha Reddy and N. Mohan Krishna Varma

Abstract In most of the application domains, few tasks can be solved by supervised learning where they can be used to label the data patterns from the existing data. This paper reviews various supervised learning techniques like decision trees, rule-based learners, lazy learners such as NNC, and a comparison of major supervised learning mechanisms like neural networks and support vector machines. The strengths and weakness of unsupervised learning techniques are also compared. This paper reviews about various supervised learning techniques strengths and weakness, brief review of unsupervised techniques, and navigation to semi-supervised learning.

1 Introduction

One categorization of learning algorithms is to divide them into supervised methods, unsupervised methods, and semi-supervised methods. Semi-supervised learning is more recent when compared with the other two. Supervised learning is to learn from the set of given examples, where each example consists of the problem instance along with its solution. Training set, which is also called as labeled set, is useful to develop classifier. Unsupervised learning does not have labeled data, and it has only unlabeled set. Semi-supervised learning is combination of unsupervised and supervised learning. It has both unlabeled and labeled data. Literature [1] has more information about semi-supervised techniques. Labeled data is considered as training data to build classifier. This article reviews briefly about supervised and unsupervised learning methods, which are followed by the reasons to navigate for semi-supervised learning methods.

Y. C. A. Padmanabha Reddy · N. Mohan Krishna Varma (✉)
Department of CSE, Madanapalle Institute of Technology & Science,
Madanaplle 517 325, India
e-mail: drmohankvn@mits.ac.in

Y. C. A. Padmanabha Reddy
e-mail: padmanabhareddyyca@mits.ac.in

© Springer Nature Singapore Pte Ltd. 2020   577
P. Venkata Krishna and M. S. Obaidat (eds.), *Emerging Research in Data
Engineering Systems and Computer Communications*, Advances in Intelligent
Systems and Computing 1054, https://doi.org/10.1007/978-981-15-0135-7_53

2   Learning

Machine Learning [2] apprehensive with constructing adaptive computer system to improve proficiency over learning from input data or experience. Figure 1 shows classification of machine learning.

2.1   Supervised Learning

Supervision of training patterns regarding labels is called supervised learning and it is classification technique. Training set consists of patterns along with their associated labels, but test set consists of only patterns without labels. By using any classification techniques such as NNC and ID3, we can build the classifier by providing training set as input. By providing test set as input to the classifier, generation of labels for unlabeled patterns [3] is possible.

Figure 2a describes about the labeled data patterns which are binary classes where it can be categorized into diamond and circle shaped as each label. Figure 2b shows the supervised learning process where there is categorization of two class labels with boundary as margin.

This section discusses some classifiers like decision trees, rule-based classifiers, and lazy learners or instance-based learners.

2.1.1   Decision Trees

These will categorize the examples depending on features in the dataset. In tree-based classifiers, each value is a feature, which is to be classified, and each branch characterizes a value to node. The best attribute becomes the root node of decision tree. The best attribute can be found by using the numerous methods such as information such as gain ratio, Gini index [4], and information gain [5]. However, majority of the survey has concluded that there is no single best method [6]. Individual methods are compared to decide which metric should be used in

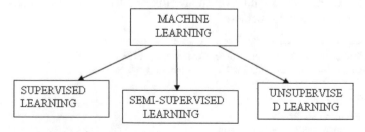

Fig. 1 Basic classification of machine learning [3]

Fig. 2 a Example of Labeled
data patterns. **b** Supervised
Learning categorize the data
w.r.t. class labels

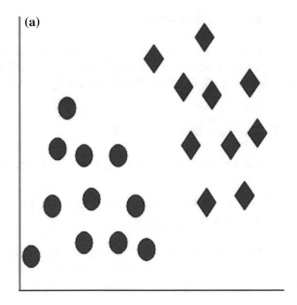

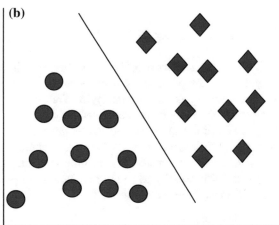

particular dataset. The same process is repeated iteratively on each partition, creating subtrees until the training data is divided into subsets of the same class.

Decision tree algorithms are suffering with over fitting. It can be solved by using
the concept pruning. There are two types of pruning methods, i.e. pre-pruning and
post-pruning. To tackle over fitting, we use the pre-pruning decision tree by not
allowing it to grow to its full size. A comparative study of well-known pruning
methods is presented in [7]. Decision trees are more a complex representation for
some concepts due to the replication problem. This can be solved by FICUS [8].
C4.5 is an extension of ID3 [9], Ruggeri presented an analytic evaluation of the
runtime behavior of C4.5; based on this evaluation, he implemented EC4.5, which
is more efficient algorithm.

2.1.2 Rule-Based Classifiers

Rules can be generated from the decision trees by considering the path from root of the tree to leaf node [9]. Rules also can be written written depending on rule-based algorithms from training set. Furnkranz [10] gives rule-based methods overview.

2.1.3 Instance-Based Classifiers

These learners come under the category of lazy learners.

Pseudo code
Procedure instance-based learner (Testing pattern)
For each test pattern
{
Find the nearest patterns of the training set according to distance
Test pattern class = k-nearest patterns which are more frequent class label
}

NNC (Nearest Neighbor Classifier)

NNC stands for nearest neighbor classifier, which comes under lazy learners category. In this method, training set acts as model or classifier. Here, we can classify test or unknown pattern by quantifying the length among the training pattern and the test pattern. There is no predefined model because it is a lazy learner. First, we can plot all training patterns and test patterns into a n-dimensional space then we compute the distance among the one test pattern to all training patterns. Among those distances, we consider the minimum value, take the label of that training pattern, and it has assigned to the test pattern.

The distance here we considered are

1. Euclidean distance
2. Manhattan distance

1. **Euclidean distance**:

$$d(x, y) = \sqrt{\sum_{i=1}^{n} (x_i - y_i)^2}$$

Here x is training set, y is test set, n represents attributes in a patterns and i represents attribute. From the above two distances, we use any one formulae to compute distance between training pattern and test pattern. But, mostly we prefer the Euclidean distance only.

Algorithm
INPUT: Training data with labels and test data without labels
  OUTPUT: Test data with labels.
  Procedure:

1. First, find all distances that one test pattern with the all training patterns.
2. Store and arrange the all distance value in sorted manner.
3. Among that the minimum distance is retrieved. Find the trained pattern label, which gives the minimum distance.
4. Assign the label for the test pattern.
5. Repeat the above process for all test patterns.

While using the NNC, if we get more than one minimum distance value that means the distance is small than the label cannot be taken as random. By taking it as random, misclassification occurs. This may lead to degrade the accuracy.

The solution to this problem is to perform voting; which pattern gets maximum votes then the pattern label is assigned to the test pattern.

Figure 3 shows NNC example. Here, the red dot mark represents the test pattern. To find distance between training pattern and test pattern, we take Euclidean distance as a metric and choose minimum distance pattern and that training set label is assigned to the test set label.

K-NNC

The above thing is the main reason to evolve the K-NNC over NNC. By using K-NNC, we may select the k-Neighbors, which are nearer to the test pattern, i.e. by

Fig. 3 Voting mechanism in NNC to give class label

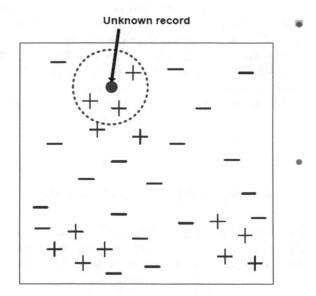

Unknown record

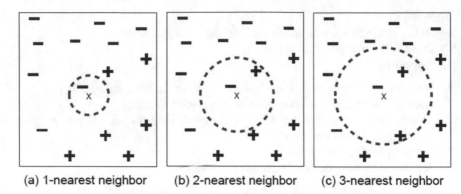

| (a) 1-nearest neighbor | (b) 2-nearest neighbor | (c) 3-nearest neighbor |

Fig. 4 The mechanism of K-NNC

Table 1 Data patterns with corresponding labels

Data in standard format

Case	Feature1	Feature2	Feature3	Class
1	***	**	*****	Excellent
2	***	**	*****	Good
3	***	**	*****	Worst

changing the k value, the result or label for the test pattern will get affected. The larger k value resolves the complexity over noise data.

Let us consider the above example. By considering if the nearest neighbor is one in Fig. 4a $k = 1$, the mark x defined the test data and negative symbol defined the trained data. If nearest neighbors are two as shown in 4b, then neighborhood comprises one positive example and the other is negative. In these two examples which is minimum distance to the test data that is assigned to the data point. Majority voting scheme is shown in Fig. 4c.

Drawbacks of these K-NNC is choosing the K value. If the K value varies then it will decrease or increases the accuracy. To avoid drawbacks of K-NNC, we used weighted K-NNC.

By observing the Table 1, KNN is very sensitive to unimportant or irrelevant features and sensitive to noise. The table gives a good comparison about the supervised learning algorithms. Naïve Bayes requires relatively small dataset.

3 Unsupervised Learning (Clustering)

Unsupervised learning is also called clustering, where data patterns are unlabeled. Appropriate suitable clustering methods to be applied for specific problems. Clustering groups are formed based on similarity among the data patterns.

Clustering deals with huge amount of unlabeled data patterns which is unsupervised learning task. The quality of unsupervised learning is measured depending on intra and inter clusters similarities [12]. Selection of specific clustering method for a particular problem is user's decision. AK Jain clustering survey categorizes the clustering methods into five types [11].

3.1 Partitioning Methods

K-means clustering is a basic partitioning method, which is proposed by Mac Queen [13]. The effectiveness depends on similarity [14]. Clustering LARge applications (CLARA), partitioning around medoids(PAM), and clustering large applications based on randomized search (CLARANS).

3.2 Hierarchical Methods

Hierarchical clustering methods classified into Divisive or Agglomerative based on way of splitting. Relationship among two clusters is quantified in dissimilar ways Viz, single-link, average-link, and Complete-link.

3.3 Density-Based Clustering Methods

It performs grouping through recognizing high-density regions in metric space. Connected dense component raises in any direction. Random-shaped clusters can be found by using density-based algorithms and also handle outliers. (density-based spatial clustering of applications with noise (DBSCAN) is one of the most popular density-based clustering algorithms. GDBSCAN can effort both numerical and categorical data. Density-based clustering methods are DBCLASD, DENCLUE, and OPTICS.

Tables 2 and 3 gives completely about the strengths and weaknesses of supervised and unsupervised learning algorithms.

Reasons to Navigate to Semi-supervised Learning:

1. Is expensive: specialists are required to execute labeling, expert want to be funded for labeling.
2. Is difficult: sometimes objects have to be accurately segmented before labeling. For example, speech signals and images.

Table 2 Comparisons of supervised learning algorithms (with ranking best, average, and worst)

Parameters	Decision trees	Neural networks	Naïve bayes classification	KNN	SVM	Rule learners
Accuracy	Average	Better	Worst	Average	Best	Average
Learning speed w.r.t no. of patterns	Better	Worst	Best	Best	worst	Average
Performance of classifier	Better	Better	Better	Worst	Better	Better
Tolerance to missing values	Better	Worst	Best	Worst	Average	Average
Tolerance to irrelevant attributes	Better	Worst	Average	Average	Best	Average
Resist to highly independent attributes	Average	Better	Worst	Worst	Better	Average
Dealing with discrete/ binary/ continuous attributes	Best	Better (not discrete)	Better (not continuous)	Better (not continuous)	Average (not discrete)	Better (not directly continuous)
Tolerance to noise	Average	Average	Better	Worst	Average	Worst
Dealing with danger of overfitting	Average	Worst	Better	Better	Average	Average
Attempts for incremental learning	Average	Better	Best	Best	Average	Worst
Transparency of knowledge	Best	Worst	Best	Average	worst	Best
Model parameter handling	Better	Worst	Best	Better	Worst	Better

4 Semi-supervised Learning (SSL)

It is one category in machine learning. SSL is mid-way among unsupervised and supervised learning. Growing of SSL is to overcome disadvantages of unsupervised and supervised learning. Supervised learning needs additional training labeled data patterns to categorize test data patterns. Building of training data is expensive and time taking. Unsupervised learning is unable to cluster unfamiliar data exactly. To overcome the problems of unsupervised and supervised learning, SSL came into existence (Fig. 5).

Table 3 Clustering algorithms comparison [15]

Category of clustering methods	Name of clustering algorithm	Time/space complexity	Ability of handling high-dimensional data
Partion-based methods	K-means	$O(NKd)$ time $O(N + K)$	No
	Fuzzy C-means	Near $O(N)$	No
	CLARA	$O(K(40 + k)^2 + k (N - K))^+$time	No
	CLARANS	Quadratic in performance	No
Hierarchical clustering methods	Average-link Complete-link Single-link	$O(N^2)$ time $O(N^2)$ space	No
	BIRCH	$O(N)$ time	No
	CURE	$O(N_{sample}^2 \log N_{sample})$ time $O(N_{sample})$ space	Yes
Density-based clustering methods	DBSCAN	$O(N \log N)$ time	No
	DENCLUE	$O(N \log N)$	Yes
Grid-based methods	Wavecluster	$O(N)$ time	No
	CLIQUE	Quadratic with no. of dimensions Linear with the no. of objects	Yes

Fig. 5 Classification of semi-supervised learning [1, 3]

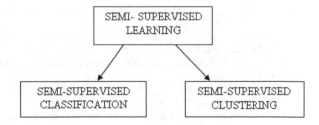

4.1 Semi-supervised Classification

This is a different type of taxonomy. Traditional classification uses additional amount of training data patterns to categorize test data patterns, whereas classification in semi-supervised fashion uses a fewer training data to categorize huge quantity of test data patterns. Semi-supervised classification reduces usage of training data.

4.2 Semi-supervised Clustering

This is one type where constraints are taken into consideration. The constraints could be in the form of instance-level, pairwise constraints such as must-link and cannot link constraints, which help in clustering the unlabeled data patterns with side information [1].

5 Conclusions and Future Work

Our contribution reviews about supervised learning techniques such as decision trees, instance classifiers, and rule-based classifiers and also given the comparison table about the strengths and weaknesses of neural networks, support vector machines, Bayes classifier with parameters as the metric such as accuracy, learning speed, noise, independent attributes, overfitting, attempt for incremental learning, and model parameter handling. Clustering which is an unsupervised learning task catergorized into five types, and are tabulated with time complexities. To solve the problems of conventional learning tasks, semi-supervised learning is introduced. In-depth review of clustering and semi-supervised learning is one of the future research directions.

References

1. Mallapragada, P.K.: A thesis on "Some contributions to semi-supervised Learning" Michigan State University (2010)
2. Mitchell, T., Machine Learning. McGrawHill (1997)
3. Kumar, K., Sreedevi, M., Padmanabha Reddy, Y.C.A.: Survey on machine learning algorithms for liver disease diagnosis and prediction. Int. J. Eng. Technol. (IJET(UAE)) **7**(18), 99–102 (2018)
4. Breiman, L., Friedman, J.H., Olshen, R.A., Stone, C.J.: Classification and Regression Trees, Wadsforth International Group (1984)
5. Hunt, E., Martin, J., Stone, P.: Experiments in Induction. Academic Press, New York (1966)
6. Murthy: Automatic construction of decision trees from data: a multi-disciplinary survey. Data Min. Knowl. Discov. **2**, 345–389 (1998)
7. Elomaa, T.: The biases of decision tree pruning strategies. In: Lecture Notes in Computer Science 1642, pp. 63–74. Springer (1999)
8. Markovitch, S., Rosenstein, D.: Feature generation using general construction functions. Mach. Learn. **49**, 59–98 (2002)
9. Quinlan, J.R.: C4.5: Programs for machine learning. Morgan Kaufmann, San Francisco (1993)
10. Furnkranz, J.: Separate-and-conquer rule learning. Artif. Intell. Rev. **13**, 3–54 (1999)
11. Jain, A.K., Murty, M.N., Flynn, P.: Data clustering: A review. ACM Comput. Surv. **31**(3), 264–323 (1999)
12. Kleinberg, J.: An impossibility theorem for clustering. In: Proceeding 2002 Conference Advances in Neural Information Processing Systems, vol. 15, pp. 463–470 (2002)

13. MacQueen, J.B.: Some methods for classification and analysis of multivariate observations. In: Proceedings of 5-th Berkeley Symposium on Mathematical Statistics and Probability, vol. 1. pp. 281–297. University of California Press, Berkeley (1967)
14. Sarma, H., Padmanabha Reddy, Y.C.A., Viswanath, P.: Recent trends in data clustering. In: Proceedings of International Conference on Nanoscience, Engineering & Advanced Computing (Icneac-2011)
15. Xu, R., Wunsch, D.: Survey of clustering algorithms. IEEE Trans. Neural Netw. **16**(3) (2005)

Wireless Technology over Internet of Things

Singaraju Srinivasulu, Modepalli Kavitha◉
and Chandra Sekhar Kolli◉

Abstract Diverse kinds of wireless technology and corresponding networks enable gadgets to address one another and to the web without cables. There are various wireless advancements out there that can be executed in hardware items for Internet of things and machine-to-machine correspondence. Wireless technology advances quickly and plays a significant role in lives of people all through the world. If we observe present-generation lifestyles, bigger number of people are depending on the technology either directly or indirectly. In IoT infrastructure, the things are to be identified or controlled wirelessly throughout existing system organization. So, many wireless technologies support objects' communication in IoT applications. But reasonable wireless technology determination for the IoT application will upgrade the performance. This paper gives analysis of various wireless technologies supported for IoT like RFID, Bluetooth, Wi-fi, Zigbee, 6LoWPAN, Z-Wave, Lora WAN, Thread, Sigfox, NFC, Cellular, and Neul. This education will be helpful to basic learners of wireless technology.

Keywords Bluetooth · Cellular · Internet of things (IoT) · Lora WAN · Neul · NFC · 6LoWPAN · Sigfox · RFID · Thread · Wi-fi · Zigbee · Z-Wave

S. Srinivasulu (✉)
Department of CSE, PACE Institute of Technology and Sciences,
Ongole, Andhra Pradesh, India
e-mail: Srinivasulu_s@pace.ac.in

M. Kavitha · C. S. Kolli
Department of CSE, Koneru Lakshmaiah Education Foundation,
Vaddeswaram, Guntur, Andhra Pradesh 522 502, India
e-mail: modepalli.kavitha@kluniversity.in

C. S. Kolli
e-mail: kcs@kluniversity.in

© Springer Nature Singapore Pte Ltd. 2020
P. Venkata Krishna and M. S. Obaidat (eds.), *Emerging Research in Data Engineering Systems and Computer Communications*, Advances in Intelligent Systems and Computing 1054, https://doi.org/10.1007/978-981-15-0135-7_54

1   Introduction

Internet of things (IoT) is the infrastructure of physical devices, structures, auto-mobiles, and diverse things embedded with hardware, software, sensors, actuators, and network accessibility which enable these articles to accumulate and exchange data. The global standards initiative on Internet of things described the IoT as "a worldwide foundation for the data society, empowering propelled administrations by interconnecting things dependent on existing and advancing interoperable data and correspondence innovations" [1]. *The IoT empowers things to be identified or controlled remotely across over existing framework structure* [2]. *It is making open entryways for increasingly direct incorporation of the physical world into PC-based structures, and achieving upgraded capability, precision and financial favorable position notwithstanding diminished human intervention* [3].

While developing IOT application, investigating wireless connectivity options is mandatory to developer for the success. For selecting better wireless technologies in our application or to introduce new wireless standard, first, we studied the existing wireless standards in this paper.

IoT can begin from our homes with basic lighting or apparatus control and venture into the domain of production lines and enterprises with mechanized machines, brilliant security frameworks, and focal administration frameworks. It has scaled up to whole urban areas with smart parking, brilliant metering, waste management, fire control, traffic control, and any comparative capacities included. That means Internet of things is extensively used in various applications counting households, medical care institutions, space, marketing, and several transports [4].

The IoT architecture is multi-layered with fragile segments unpredictably associated with one another. Figure 1 shows the structure of IoT. It begins with sensors, which are the wellspring of information being gathered. Sensors pass information onto a contiguous edge gadget, which changes over information into readable digital values and stores these incidentally. At the point when the edge detects a wireless network or the Internet, it pushes the privately put away infor-mation to a cloud server or database engaged with the application. The information is prepared, examined, put away, and sent to the end-client gadget, using an application programming.

Depends on the application, the sensors are taken. The sensors will sense the physical characteristics around. The sensed signals are sent to the coordinator device due to its less memory capacity and battery power. The coordinator sends the data to the cloud or private data base depends on application design for further processing or permanent storage. From this discussion, we found that the things are connected or coordinated using wireless technology.

Wireless advancements have increased across the board predominance in every one of the fields. Instances of wireless advances in drug incorporate versatile crisis reaction frameworks, specially appointed systems with WLANs for remote patient checking, the utilization of convenient gadgets in telemedicine applications including regular healing center utilize, for example, downloading day-by-day

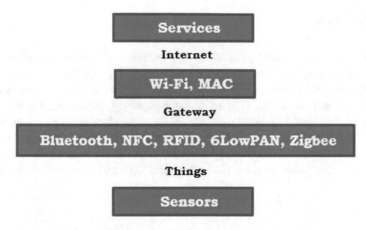

Fig. 1 IoT structure

plans, and indoor following of advantages and workforce. Different types of wireless technologies are available nowadays.

In this paper, we are going to discuss some of the widely used remote knowledge like RFID, Bluetooth, Zigbee, Wi-fi, 6LoWPAN, Lora WAN, Z-Wave Thread, Sigfox, NFC, Cellular, and Neul. This discussion will surely help the basic learners to select the suitable technology while designing application.

Section 2, we are discussed literature survey on wireless technologies with some applications. Section 3 refers frequency, range, and data rates analysis of some technologies, and Sect. 4, we are discussed one case study.

2 Literature Survey

In the previous years, we were utilized the wired innovations for the correspondence. These technologies have the best drawbacks of utilizing cable, and it is difficult to use for long distance and not a reliable one too. To conquer these disadvantages, we have been moved to the wireless one. By utilizing the wireless correspondence advancements, we make our correspondence as reliable and cable-free one [19].

Due to smartphones, tablets, and laptops' intervention in human world, the term *wireless* technology has become part of everyone's life and wireless innovations have increased broad pervasiveness in every one of the fields.

Some common wireless technologies supported to IOT are RFID, Bluetooth, Wi-fi, Zigbee, 6LoWPAN, Z-Wave, Lora WAN, Thread, Sigfox, NFC, Cellular, and Neul. But depends on application requirement, we choose the suitable technology. So, in this paper, some wireless technologies are studied and observe frequency and range capacities.

2.1 Bluetooth

Bluetooth is a low-vitality utilization correspondence innovation and a planned specialized strategy for emergency circumstance [5]. In the zones of system administration, registering, customer hardware and media transmission Bluetooth overseen has excess of 30,000-part organizations [6]. Bluetooth utilizes a radio innovation detaches transmitted data into packs and communicates each package on one of 79 assigned Bluetooth channels. Each channel has a transfer speed of 1 MHz. It is a standard wire-trade interchanges convention basically intended for low-control use, with a short-run dependent to ease receiver chips in each contraption.

Applications:

- Wireless correspondence and control between mobile and Bluetooth.
- Wireless correspondence and control with android gadget phones, compressed distant speakers, and tablets.
- Wireless Bluetooth headset and Intercom Colloquially.
- Wireless scaffold between two Industrial Ethernet (e.g., PROFINET) systems.
- Short extend broadcast of health-symptoms' related sensor data from medicinal devices to mobiles and telehealth appliances.

2.2 Zigbee

Zigbee is an unmistakable low rate correspondence paradigm essentially intended to be sent for wireless individual territory systems. It is an IEEE 802.15.4 founded common acquiring negligible measure of unpredictability, cut rate, and low-quality utilization. Zigbee bunch tree is great perceived Zigbee topologies particularly reasonable for WSN's expending low quality and continuing cut rate since it bolsters control protect activities [7]. Zigbee gadgets frequently transmit information over longer separations by going information through transitional gadgets to achieve more distant ones, making a work organize that means a system with no incorporated control or high-control transmitter/recipient ready to achieve all the arranged gadgets. It is 2.4 GHz work neighborhood (LAN) convention. It was initially intended for building robotization and control things like remote indoor regulators and lighting frameworks frequently use Zigbee.

Applications:

- Home control and entertainment
- Industrial control
- Embedded detecting
- Medical information accumulation

- Smoke and gatecrasher cautioning
- Building robotization
- Remote mouthpiece arrangement, in Shure wireless microphone systems.

2.3   Z-Wave

Z-Wave is a standout among the most well-known remote conventions for home territory systems. However, there have been few investigations going on in recent years. Z-Wave is not an IP-perfect convention like Wi-fi, so z-wave gadget cannot straightforwardly associate with the Internet or normal client gadgets (e.g., cell phone, PC). The z-wave organize utilizes controller to control and oversee all gadgets. Controller additionally goes about as a gateway and enables a client to collaborate with z-wave gadgets from cell phone or workstation by means of the Internet or local network [9]. The US-based measures body IEEE sorted out an undertaking gathering to build up a standard for low power, low information rate systems. Subsequently, the Zigbee Alliance was set up by a few organizations to cooperate to expand on 802.15.4. Be that as it may, an opponent innovation to Zigbee, called Z-Wave, has gotten the help of chip goliath Intel, just as systems administration heavyweight Cisco. Both are remote norms, use work systems, and are planned for low power. Even though, Zigbee accommodates a more information rate rather than Z-Wave [10].

Applications:

- Home and office automation
- Remotely control the electronic lock functions
- Smart hubs
- Smart lightening
- Security and alarm
- Voice-controlled application
- Water management
- Smart energy management
- Smart sensors
- Smart USB.

2.4   6LoWPAN

6LoWPAN utilizes a lightweight IP-based correspondence to go over lower information rate systems and it is proposed by IETF to adjust IPv6 into the WSNs condition. IPv6 low-power wireless personal area network convention is introduced to chop down utilization. It outfits best for battery-worked hubs, for example, temperature and light sensors [11]. IPv6 is taken as an interconnection plot for IEEE802.15.4 apparatus at framework layer, and 6LoWPAN innovation has pulled in broad consideration. 6LoWPAN working gathering takes a shot at the examination of IPv6 tradition suite subject to IEEE802.15.4 standard and constructs self-affiliation 6LoWPAN framework with course tradition.

Applications:

- Smart meters
- Smart homes
- Smart lighting
- Security
- Detecting crisis occasion, failure, caution as well as enhancing execution and lifetime of WSN.

2.5 LoRa

The LoRa innovation is a long-range low-power innovation. The LoRa innovation tends to these necessities of a battery-worked inserted gadget that offers battery lifetime, the long range, the security, vigor to obstructions. Long range wide area network (LoRa WAN) is an open-review secure standard for the IoT availability [13]. LoRa WAN is institutionalized and publicly released by the LoRa collusion.

Applications:

- Street light monitoring
- Smart parking
- Design of an automatic irrigation system
- Vehicle diagnostic system for driving safety
- Soil-propagation management
- Indoor signaling
- Remote monitoring systems
- Wastewater monitoring systems
- Smart building applications
- Tactical troops tracking systems.

2.6 Wi-fi

Wireless fidelity (Wi-fi) communication protocol is the most prominent innovation that has been as of late developed and is utilized in long-distance interchanges. Wi-fi systems give us the network that is pervasive and reasonable expense. With the ease factor, most smartphone applications play out a few foundation undertakings under Wi-fi [14]. Wi-fi development allows an electronic device to conversation of data remotely using radio waves over a PC sort out, including quick Internet affiliations.

Applications:

- Wi-fi at home
- Education institutions
- Hotspots
- Wi-fi in the industry
- Inventories
- Positioning.

2.7   NFC

Near field communication (NFC) is a remote-correspondence standard for trading information over a short distance which is institutionalized in ISO/IEC 18092 and ECMA-340 [15]. It is short-range remote innovation supports up to 10 cm or less. NFC reliably incorporates initiator and objective. The initiator viably delivers a RF field that can control an uninvolved target. This enables NFC centers to take very clear shape factors like marks and stickers.

Applications:

- Ticketing system
- Smart postal system
- Security analysis
- ERP system
- NFC antenna system
- ATM authentication system
- Mobile payments
- Data communications.

2.8   Sigfox

Sigfox is a worldwide IoT organize administrator. It utilizes differential double-stage move keying in one course and Gaussian recurrence move keying in the other heading. Sigfox and their accomplices set up radio wires on towers (like a mobilephone organization) and get information transmissions from gadgets, such as water meters or parking sensors. The Sigfox system and innovation is gone for the minimal effort, and it utilizes an ultra narrow band (UNB)-based radio innovation to associate gadgets to its worldwide system [16]. Sigfox made an UNB IoT correspondence framework intended to help IoT organizations over long ranges, for example, more than 20 km between customer gadgets and base station.

Applications:

- Outdoor fingerprinting localization
- Home and consumer goods
- Smart metering (energy consumption)
- Health care
- Transportation
- Remote monitoring and control
- Marketing
- Security
- Car-door safety system
- IoT-based monitoring systems.

2.9 RFID

A radio-frequency identification outline uses marks or tags appended to the articles to be recognized. There are two sorts of radio recurrence distinguishing proof: dynamic and latent. This convention was structured explicitly, so gadgets without batteries could send a flag. In many frameworks, one side of a RFID framework is powered, making an attractive field, which instigates an electric flow in the chip. This makes a framework with enough capacity to send information wirelessly over and again. Along these lines, RFID labels are utilized for transportation and tracking purposes. Dynamic RFIDs are being utilized in real-time location system for discovering, tracking, and checking things and individuals.

Applications:

- Home safety
- Patient health monitoring
- Interactive marketing
- Location tracking
- Equipment management
- Human activities tracking
- Race tracking
- Vehicle tracking
- Inventory tracking
- Library management.

2.10 EnOcean

EnOcean's vitality gathering wireless sensor innovation gathers vitality out of air. The EnOcean development is an imperativeness gathering remote advancement used essentially in constructing robotization framework, but at the same time is associated with various applications in smart homes, industry, coordination's and transportation. Modules subject to EnOcean advancement join scaled down scale essentialness converters with ultra-low power equipment and engage remote correspondences among battery—fewer remote sensors, switches and controllers. EO is a universal standard fundamentally utilized in private furthermore, building robotization, where discontinuous sensor centers work without the need of battery and accordingly for a long time of support free [17].

Applications:

- Intelligent lightening control system
- Building automation
- Smart home
- Environmental monitoring
- Transportation
- Industry.

These are some wireless technologies we saw in many smart applications. Some other technologies or protocols support IoT applications are Cellular, Neul, Thread, LTE, NB-IoT, Digi Mesh, Weightless-N, Weightless-P, Weightless-W, ANT & ANT+, Ingenu, EnOcean, MiWi, Dash7, Wireless HART, etc.

3 Analysis of Wireless Technology

We have so many wireless technology protocols to support IoT-based applications. To select one form, we need to compare the properties with one another. Even for each technology, we have different standards. Table 1 describes the frequency, range, data rates of some standard wireless technologies [18].

Day to day, new application requirements demand some more efficient, less cost-affordable, secure wireless technologies.

4 Case Study

Wireless technology is used in all the fields today like medical, transport, industry, agriculture, and marketing.

Industrial plants consist of large number of devices interconnected in different ways, and industry profit is depending on equipment health indirectly. Because, if

Table 1 Frequency, range, and data-rate analysis of some wireless technology

Technology	Standard	Frequency-band	Range	Data-rate	Symbol
Bluetooth	Bluetooth specification 4.2 core	2.4 GHz	1–100 m	1 Mbps	
Zigbee	ZigBee 3.0	2.4–2.483 GHz	10 m	250 Kbits/s	
Z-Wave	ZAD12837/ITU-T G.9959 Z-Wave Alliance	900 MHz (ISM)	30 m	9.6/40 Kbits/s	
6LoWPAN	RFC6282	Adjusted and used over an assortment of other networking media	N/A	N/A	
LoRaWAN	LoRaWAN	Various	2–5 km	0.3–50 kbps	
Wi-Fi	Based on 802.11N	2.4, 3.6 and 4.9/5.0 GHz bands	Basic range is up to 100 M yet can be expanded	600 Mbps maximum	
NFC	ISO/IEC 18000-3	13.56 MHz (ISM)	<0.2 m	100–420 kbps	
RFID	LF HF UHF	120–150 KHz (LF), 13.56 MHz (HF), 433 MHz (UHF), 865–868 MHz (Europe) 902–928 MHz (North America) UHF, 2450–5800 M Hz (microwave), 3.1–10 GHz (microwave)	10 cm to 200 m	Differs depends on medium	
Sigfox	Sigfox	900 MHz	3–10 km urban and 30–50 km rural environments	10–1000 bps	
EnOcean	EnOcean	868, 902 and **928** MHz	30 m Indoors and 300 m Outdoor	**125 kbit/s**	

the equipment are in good condition, the productivity also gets in proportion. If the equipment gets repair, automatically, the productivity decreases. So, continuous monitoring of equipment's health is needed in industry. Due to the advances in wireless technology, the equipment health is monitored by sensors continuously and the sensed data is sent to supervisory system through Wi-fi, Bluetooth, Zigbee, or any other technologies supporting the environment. That means the sensors like sound and vibration sensors are attached to equipment in plants to sense the working condition changes. The sensor-generated data are always checked with predefined conditions to identify the present health condition of an equipment. After analysis, the results will send to the user wirelessly. All the equipment, sensors, and supervisory systems are connected wirelessly.

With the help of technology advancements, the machine's health is predicted before, so the owner is ready with solutions. In big industries, if a machine stops one minute, it gets loss in lakhs. This type of machine breaks in industries will be predicted by sensors and its wireless data communications easily. Figure 2 refers the machine's health in industry which is observed by man [18]. Figure 3 refers the machine's health in industry plant which is monitored by sensors.

Fig. 2 Human monitoring of machine health in industry [18]

Fig. 3 Machine health monitoring using sensors and data communication using Wi-fi technology

5 Conclusion

Wireless advancements can adequately exchange the information over long distances. Sensors and actuators can be added to numerous mechanical, business, and automate entertainment applications. The fundamental favorable circumstances for utilizing a wireless solutions in these applications are more mobility and probability to move gadgets and associate with smartphones and tablets without cables, bypassing long separations and territories where cables can't physically fit, quick and simple establishment and charging, high adaptability if there is a need to change an establishment, expanded staff security by not being physically near a gadget amid design and support, adaptable human interface gadgets easy mix of gadgets into the system.

References

1. Internet of Things Global Standards Initiative. ITU. Retrieved 26 June 2015
2. Bartolomeo, M.: Internet of Things: Science fiction or business fact. A Harvard Business Review Analytic Services Report, Technical Report (2014)
3. Vermesan, O., Friess, P. (eds.): Internet of Things: Converging Technologies for Smart Environments and Integrated Ecosystems. River Publishers, Aalborg (2013)
4. Azzawi, M.A., Hassan, R., Bakar, K.A.A.: A review on Internet of Things (IoT) in healthcare. Int. J. Appl. Eng. Res. 11(20), 10216–10221 (2016)
5. Kajikawa, N., et al.: On availability and energy consumption of the fast connection establishment method by using Bluetooth Classic and Bluetooth Low Energy. In: 2016 Fourth International Symposium on Computing and Networking (CANDAR). IEEE (2016)
6. Newton, H.: Newton's Telecom Dictionary. Flatiron Publishing, New York (2007)
7. Shende, S.F., Deshmukh, R.P., Dorge, P.D.: Performance improvement in ZigBee cluster tree network. In: 2017 International Conference on Communication and Signal Processing (ICCSP). IEEE (2017)
8. Kim, T.: A study of the Z-Wave protocol: implementing your own smart home gateway. In: 2018 3rd International Conference on Computer and Communication Systems (ICCCS). IEEE (2018)
9. Knight, M.: Wireless security—how safe is Z-wave? Comput. Control Eng. J. 17(6), 18–23 (2006)
10. Le, D.H., Pora, W.: Implementation of smart meter working as IEEE1888-6LoWPAN gateway for the building energy management systems. In: 2014 11th International Conference on Electrical Engineering/Electronics, Computer, Telecommunications and Information Technology (ECTI-CON). IEEE (2014)
11. Ma, X., Luo, W.: The analysis of 6LoWPAN technology. In: Pacific-Asia Workshop on Computational Intelligence and Industrial Application, 2008. PACIIA'08, vol. 1. IEEE (2008)
12. Rattagan, E.: Wi-Fi usage monitoring and power management policy for smartphone background applications. In: Management and Innovation Technology International Conference (MITicon), 2016. IEEE (2016)
13. Saminger, C., Grünberger, S., Langer, J.: An NFC ticketing system with a new approach of an inverse reader mode. In: 2013 5th International Workshop on Near Field Communication (NFC). IEEE (2013)

14. Chung, Y., Ahn, J.Y., Du Huh, J.: Experiments of a LPWAN Tracking (TR) platform based on Sigfox test network. In: 2018 International Conference on Information and Communication Technology Convergence (ICTC). IEEE (2018)
15. Arcari, F.D.A., et al.: Development of a WirelessHART-EnOcean adapter for industrial applications. In: 2017 VII Brazilian Symposium on Computing Systems Engineering (SBESC). IEEE (2017)
16. https://www.rs-online.com/designspark/eleven-internet-of-things-iot-protocols-you-need-to-know-about
17. Abinayaa, V., Jayan, A.: Case study on comparison of wireless technologies in industrial applications. Int. J. Sci. Res. Publ. **4**(2), 2250–3153 (2014)
18. https://www.bing.com/images/search?q=machines+health+industry+image&qpvt=machineshealth+industry+image&FORM=IGRE

A Critical Review on Internet of Things to Empower the Living Style of Physically Challenged People

Chandra Sekhar Kolli⬀, V. V. Krishna Reddy⬀
and Modepalli Kavitha⬀

Abstract We are living in the world in which we can connect anything to the Internet that forms a dynamic network and reduces the human intervention in all aspects. Due to the rapid improvements in the technology, especially in Internet of things (IoT) that facilitates and empowers the way of communication, data transfer among people, between the devices. It is already proven fact that how IoT was applied in creating smart homes, providing security, and comfort to the people in day-to-day activities by enhancing the quality of the living style. By considering and integrating the power of IoT, one can empower the lifestyle of physically challenged people. We can use this IoT to unlock the new value and explore the potential to provide a better and quality lifestyle to the disable people. In this article, we perform a critical review and analyze the lifestyle of the physically challenged people and present how best we can apply IoT to empower the quality of living style of such people. IoT can offer assistance, support and empower the physically challenged people and allow them to move around in the social life with ease and comfort. In this article, different application scenarios and domains, challenging issues are identified and addressed with possible solutions.

Keywords Internet of things · Disability · Security · Comfort

C. S. Kolli (✉) · M. Kavitha
Department of Computer Science and Engineering, Koneru Lakshmaiah Education
Foundation, Vaddeswaram, Guntur, Andhra Pradesh, India
e-mail: usercsk@gmail.com

V. V. Krishna Reddy
Department of Information Technology, Lakki Reddy Bali Reddy College of Engineering,
Mylavaram, Krishna, Andhra Pradesh, India

© Springer Nature Singapore Pte Ltd. 2020
P. Venkata Krishna and M. S. Obaidat (eds.), *Emerging Research in Data
Engineering Systems and Computer Communications*, Advances in Intelligent
Systems and Computing 1054, https://doi.org/10.1007/978-981-15-0135-7_55

1 Introduction

IoT-enabled devices are acquiring massive potential as it drives our daily life toward automation and providing smart ambiance in all aspect that too with low cost. These devices impact very much in all aspects of our life starting from fitness and health, automotive and logistics, smart homes, smart cities, and industrial IoT, etc. Cultivation and agriculture field also seen many technological transformations in the last few years. IoT is nothing but grouping of embedded devices, RFID, sensors, and lightweight software which generally referred as Internet of everything [1, 2]. It is already proven fact that the IoT made a significant revolution in current hyper-competitive data-driven computing and wide variety of communication channels that enable industries, organizations, homes, business, etc., toward technology-driven. We will start with an introduction to the IoT that is taking place and entering across many industries and domains. The Internet of things (IoT) allows electronic devices to connect to the Internet in more secure way. Cisco announces that it is going to enhance more working toward to provide more security and safety solutions that offer ease management of millions of connected cameras and devices [3].

Continuing the same, IoT can be applied in health domain too because it will add great value for the physically challenged people. Physical and mental health is the most important issue in the present busy lifestyle. We strongly believe that IoT can be applied, especially for the physically challenged people to empower and overcome the difficulties in their lifestylc. Assistive IoT sensory devices and RFIDs are very influential tools to help the impaired people. In this context, we started to explore in a detailed manner to study how people with physically impaired can able to interact with and benefit it from the advancements of the IoT-assistive technologies. In this review, we are exploring the role of IoT in deeper level and adoption to get benefits out IoT for physically challenged people. IoT for the physically challenged will not only make their life easy at all the places and all the time. Technology has helped challenged people come at par with their counterparts by completely nullifying their disabilities and empowering them to do whatever it is that they would like to. Meanwhile, there has been a lot of discussion on the importance of IoT for the physically challenged as it is another step in that direction. For instance, it has helped doctors monitor physically challenged patients while they are at home. Furthermore, IoT devices like 3D printed prosthetics have helped understand those who cannot communicate through brain activity. Physically disabled people and some of the elder people usually require physical assistance either in the form of an object like hand stick or another person to meet their important personal needs. Even though they are suffering from impairments, they always prefer live independently, self-dependent, and self-managing in their daily activities either at home or workplace. The integration of technology and services with the IoT will create and provide a better quality of living [4] for disabled people and elderly.

2 Background

The root cause behind any kind of disability is the effect of an impairment that may be physical, cognitive, mental, sensory, emotional, developmental, or some mixture all the above. A disability may come from birth itself or may occur during lifetime of the person accidentally [5]. Various reasons that cause the disability can be; by inherited (genetically transmitted); by congenital, because of their parents or ancestors' jeans or other diseases effected when his/her mother conceived time and kind of medicines used, or he/she may face few special conditions with effect of illness or injury.

There are different types of disabled people, who are physically disabled, sensory disabled, mentally disabled, pervasive disabled, Mental disabled. This paper mainly focuses on physical and sensory (visual and hearing) disabled people.

2.1 Physical Disability

Physical disability is the state which significantly impacts one or more major life activities in their daily activities. The problems faced by such people may include muscle imbalance, no strength, maybe sensing or grasping the things and environment around, difficulty in interacting with others in their daily routine, difficulty in moving from one place to other indoor and outdoor places, etc. Individuals with such difficulties will respond to others very slowly.

2.2 Visual Disability

There are number of different vision conditions are possible from person to person. These conditions can affect a person's eyesight in their day-to-day life. Few people may have a minor effect and few people suffer larger effect.

Visual impairment or visual disability leads to decreased ability to see the things and surrounding. It can be affected due to disease, trauma, congenital, or by any accidental conditions. The limitations caused by such issues are glare in seeing the things, no central vision, loss of focus in seeing boards, screen, display boards, and poor night vision. They find it very difficult to handle electronic devices which displayed on the screen. They required specially designed tools to enlarge the displayed information on the screens and also required software application that will read the data on the screen in the audio format.

2.3 Hearing Disability

People with this kind of impairment usually suffer a lot in receiving audio type of information. Generally, people who from deaf and who suffers from severe hearing problems will not get any information when is presenting information in the form of audio. In majority of the cases, people who are deaf may not speak as well. They find it very difficult to use mobile phones and audio devices. There are different kinds of assistive devices are available to help them in their daily activities. These devices will help disabled people for mobility, hearing, and vision.

3 State-of-the-Art IoT

Internet of things (IoT) creating a bond between many electronic devices to the Internet. It enables the exchange of data never available before and brings information in a more secure way. It is one of the emerging technologies which is considered as the future technology of world by enhancing the quality and comfort of the people in their daily lifestyle. In these days, it is very hard to see how many devices are already impacting our lives and the degree to which they have made our life easy. However, IoT-enabled and IoT-assistive devices for the physically challenged will not only make their life easy but also add integrity to it. Technology has helped challenged people come at par with their counterparts by completely nullifying their disabilities and empowering them in all aspects. For instance, it is very useful for doctors, and they can monitor physically challenged patients while they are at home. Furthermore, IoT devices like 3D printed prosthetics have helped understand those who cannot communicate through brain activity.

3.1 IoT for the Physically Challenged Will Help Build Smart Homes

In the early days, people must depend on switches and other traditional physical ways to interact with household items. With the introduction of Amazon Echo and Google Home which adopts IoT in people's homes. In order to use household utilities and other basic services within a house, people now no need to depend on switches and conventional physical methods of interaction. For example, IoT devices allow a visually impaired user to change the heat settings without needing to program a controller. These devices work by using a voice recognition system, without any kind of special setup. For people suffering from paralysis or those who are completely bed-ridden, such technologies are no less than a boon as they perform functions like unlocking a door without having to move.

Old-age people, handicapped patients, and people with other such kind of disabilities can be benefited from smart home, in which all appliances and devices of the house can be managed and controlled with IoT-enabled smart devices. When a resident is living alone, the ubiquitous [6] and seamless access becomes very important in providing security.

3.2 IoT for the Physically Challenged Will Help Increase Mobility

It is really very good and impressive thought of self-driving cars. The self-driving cars will ensure incredible amount of mobility for the challenged people. When they are moving from one location to another by a car arises, visually impaired people find is very difficult but with the self-driving cars remove the barrier and allow challenged people to not rely on other people or public transport to move from one place to another.

3.3 IoT for the Physically Challenged Will Help Monitor Them

In general, and with respect to many cases, people with disabilities require constant monitoring which can often be challenging. All the time and constantly, one or more people must assist in their daily activities. IoT-enabled environment will surely break all such barriers. IoT-enabled smart environments [7] will create an enabling system that embodies inclusiveness, a smart ecosystem that helps challenged people to live their lives freely.

3.4 Lightweight Wearable Devices

IoT-enabled lightweight wearable or handheld devices are surely helpful for the disabled people in smarter and easiest way. This kind of devices will provide new dimensions and opportunities for the impaired individuals. IoT will surely make everyone's lives easier and smarter.

3.5 Assistive Technology (AT)

Assistive technology in the sense that it is an object, item, piece of equipment, software application, or any product that allows the impaired people to improve the functional capabilities. These products and services available either as a software product or maybe a hardware device in assisting the impaired individuals. To name a few hardware hearing aids, wheelchairs, voice recognition, etc. IoT enabled products and devices have great potential to assist and empower people in better and provide quality services. IoT surely accelerates development of assistive technologies.

For example, the Toyota Mobility Fig. 1 band helps blind people navigate [8]. Toyota is developing an IoT-enabled wearable device for the blind people that will assist the mobility of blind people. It is something like a band that can be worn around your shoulders.

It will take input from the cameras. These cameras will observe and detect the nearby environment and surrounding and senses the signs. Accordingly, the IoT band will return feedback in the form of making an alert (sound), vibrating and directs them for their navigation. This can be used both indoor as well as outdoor places. To make it smarter, few more features can be incorporated like object detection and identification, opposite people face recognition.

Another device, Philips Light Bulbs Fig. 2 can be made enabling within a smart environment [8]. The IoT-enabled Smart Light can assist people with cognitive impairments navigate through the house or reminding them about things that they still need to do.

It will take input from the cameras. These cameras will observe and detect the nearby environment and surrounding and senses the signs. Accordingly, the IoT band will return feedback in the form of making an alert (sound), vibrating and directs them for their navigation. This can be used both indoor as well as outdoor places. To make it smarter, few more features can be incorporated like object detection and identification, opposite people face recognition.

Fig. 1 Toyota Project Blade model

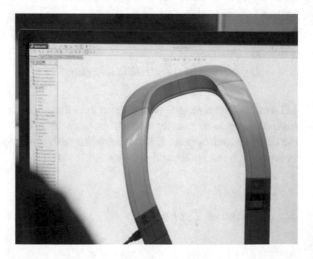

Fig. 2 Philips Hue Light
bulbs

Another device, Philips Light Bulbs Fig. 2 can be made enabling within a smart environment [8]. The IoT-enabled Smart Light can assist people with cognitive impairments navigate through the house or reminding them about things that they still need to do.

Deaf people can use this Hue Light Bulbs by operating the bulbs by making it shine in colors when there are certain noises. For instance, they will release purple light when the bell rings and emit red light when the fire alarm turned off. These are few examples that provide services when few situations If, This, Then, that. We can establish much smarter and more complex systems by considering contextual basis.

Another example decreases risks and time with Medtronic GUARDIAN® glucose monitoring [8]. IoT-enabled glucose-monitoring wearable device Fig. 3 which is embedded with a tiny sensor underneath for diabetes patients to continuously monitor and track the glucose levels of the individuals. It detects and notifies the person who wore it when a threshold level is reached. It may be high threshold value or low threshold value.

A tiny electronic glucose sensor is embedded in the skin of the patient or any individual to measure glucose levels. It is connected to a transmitter that will sense and sends the information to the display device. Even wearing the device, it is still necessary to prick your fingers for collecting blood samples, but the advantage is, it will drastically reduce the number of pricks for collecting samples in a day. This can be coupled with early insights to check, whether your glucose levels are heading in the wrong direction and maintaining at correct level or crossing more than sufficient, this will make the lives of diabetes patients easier.

Another example is, Home Automation. These devices let you control your home easier [8]. Smart Home or Home automation permits and let the users control some of the parts of their home like fans, lights, air conditioners, security like closed-circuit cameras with ease more smarter way. In some home, few items are difficult to reach, see, hear and dangerous prone. Such things can be controlled with IoT capable devices to avoid the dangers situations. Especially for the people with disabilities, these smart homes will change the way they live in the home.

Fig. 3 Medtronic's
continuous glucose
monitoring (CGM) device

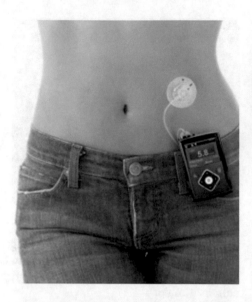

Fig. 4 Homey
voice-controlled home
automation device

Athom has just started delivering a device Fig. 4 called Homey [8]. It is a voice-controlled device which can be used to control everything in the smart home like lights, music, temperature, and TV. This device would be exactly suitable for people with disabilities. These devices can be controlled by voice itself. So, the people no need to physically go the device and no need to dependent physical touch input.

Another example, navigation without screen with connected insoles [8]. General tendency of people is wearable device means the first thing comes to our mind is gadgets like smart watches, bracelets, and glasses.

But this is the device which is integrated with our regular clothing. These insoles Fig. 5, invented by a MIT researcher, which helps people to navigate without

Fig. 5 Navigation without
screen with connected insoles

having to look at any display. Constantly looking at your smartphone screen with Google Maps showing you where to turn left is annoying and risky. But with the help of these insoles, people no need to have to look at your screen to see where to go, but it allows people to know by giving feedback and sensing through vibrations in the shoes. As you walk, it vibrates on the side that you will want to turn to. If your route continues by taking a left turn, it will automatically vibrate on the left side, and a right turn means the insoles would vibrate on the right side.

These insoles are designed in such a way to work with a mobile device and assist the user to navigate a city without looking at a smartphone continuously for getting the directions. The insoles will assist by doing this by giving feedback in the form of vibrating and let the persons know where to go, and to make best and shortest route recommendations for specific locations based on the wearer's learned behavior. People with visual or hearing impairments, this IoT enabled insoles are great invention. It increases agility and freedom by giving subtle feedback, without having voices surrounding you or having to wear weird systems.

4　Application Areas

4.1　At Shopping Malls, Hotels, and Theaters

Generally, people with blind and eye-sighted people will face many difficulties. These people often most of the time will avoid going to those public areas. They always need an assistant person if they want to go. So, IoT-enabled navigation devices will certainly enable them to visit such places on their own without having any assistance. In such places, we can implement and set up RFID system that will help them a lot. Places like restrooms and hotel rooms and other business areas have different layouts from one place to another place, so it may take time exploring to find your way around.

4.2 At Public Transport

Transportation and other conveyance places like bus terminals, railway stations, and airports are always busy with passengers and other wanderers with many activities. By familiarizing yourself with the layout of the terminals, gates, and platforms, you will be able to navigate through these spaces with more ease. With the help of IoT, we can create a smart environment that will guide them in the right direction without the need of any external physical assistance. The sounds of escalators and baggage claim carousels [9] can also help you stay oriented in the right direction.

4.3 At Schools and Academic Institutions

In school environment, if any visually impaired and low sight children are present, in that scenario, IoT-enabled environment will create great added value of designing intelligent interactive play and learning environments for toddlers and children. Learning environment can be equipped with RFID technology to create good ambiance for the toddlers. With low cost, RFID tags/readers are enough and are appropriate for setting put the educational institutions.

4.4 At Domestic Environment—Creating Smart Homes

Creating smart homes for domestic environment enables us to control household devices such as air conditioners, kitchen equipment, windows and doors, maintain indoor temperature by depending on number of people in the house along with temperature in the outside world, maintain water at a right temperature and providing security with security-enabled devices [10]. Visually disabled people require IoT-enabled gadgets like indoor navigation system [11] and an obstacle detection system [12]. These devices work based on voice-synthesized instructions and obstacle detection sensors. Physically impaired people generally require electromechanical devices for movement assistance. Such devices include powered wheelchairs and specialized lifting devices to transfer the user between the sleeping cart and the wheelchair. They require a wheelchair that can able to lift the front wheels [13] to achieve an upright position has been designed. This way, the disabled persons who cannot stand on their own can reach certain heights and are able to pick and place things on selves.

5　Case Study for Voice-Enabled IoT Device for Visually Impaired People

Navigation becomes very easy with the help of Google maps and kind of applications, but these are suited to normal people only. Visually challenged people find it very difficult using the existing navigation technologies and its features. They are the one who actually requires assistance in their daily routine at home and workplace. IoT enabled navigation assistance devices with sensors that detects all the objects or obstacles around 360° of the particular person who hold the assistance device. It should be incorporated with speaking module which converts the text or feedback from the processing module into audio format to alert when any obstacle in their way, especially in traffic road or level crossing which will be useful to avoid road accidents and other mishaps, etc. It can also include GPS-enabled navigation capability so that they can be tracked by the exact location if required by the family members, relatives, etc., in emergency. The proposed assistive device can incorporate new feature like sending SMS alert to inform the exact location of his/her. It can also include sensors for weather information alerts. The proposed model is illustrated in Fig. 6. It can also be extended, by incorporating face recognition feature in the assistive device, which will automatically recognize the face the people who stands opposed to them and informed spontaneously.

Such designed system will make use of sensors, application face recognition, and GPS navigation system. Raspberry Pi controller will be used to calculate sensor data based on that it will give signal back. This information will be converted by audio module and played to the user. Similarly, weather sensor will give information regarding weather.

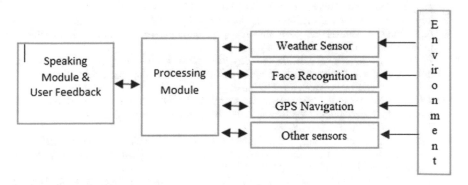

Fig. 6 Proposed IoT model for visually impairment people

6   Conclusion

It is already a proven fact that how IoT can make our lives more smart, secure, and authentic in our daily lifestyle. If we can use it in more smart and careful way, it will help, assist, and improve and empower the lifestyle of physically challenged people. In this article, we tried to gathered information regarding the different kinds of disabilities and introduced the power of IoT. I have gathered few IoT-enabled smart devices which assist the disable people. If we can use IoT in smarter way it will assist physically disabled people to move around in side home environment, office and public places, visually disabled people to reach their destination with the help of warnings, feedback and guidelines, the deaf and dumb people to communicate in better way so that the speaker and listener may comfortable with each other. In the coming future, IoT-enabled devices surely create a new world that will surely bring light into the lives of physically challenged people.

References

1. Ghosh, A.M., Halder, D., Alamgir Hossain, S.K.: Remote health monitoring system through IoT. In: 5th International Conference on Informatics, Electronics and Vision (ICIEV)
2. Fernandez, F., Pallis, G.C.: Opportunities and challenges of the Internet of Things for healthcare Systems engineering perspective. In: International Conference on Wireless Mobile Communication and Healthcare (Mobihealth), pp. 263–266 (2014)
3. https://www.cisco.com/c/en_in/solutions/internet-of-things/overview.html# ∼ stickynav=1
4. The official website of the Smart Home Association of Netherlands. http://www.smart-homes.nl/
5. https://en.wikipedia.org/wiki/Disability
6. Ghazal, B., Al-Khatib, K.: Smart home automation system for elderly, and handicapped people using XBee. Philos. Trans. R. Soc. Lond. **A247**, 529–551 (1955)
7. https://www.allerin.com/blog/this-is-why-the-physically-disabled-find-iot-promising
8. https://medium.com/@imn/5-promising-examples-of-iot-and-wearable-devices-that-enable-people-with-disabilities-f50df601e046
9. Parton, B., Hancock, R., Mihir, F.: Physical world hyperlinking: can computer-based instruction in a K-6 educational setting be easily accessed through tangible tagged objects? J. Interact. Learn. Res. **21**(1), 95–110 (2010)
10. Darianian, M., Michael, M.P.: Smart home mobile RFID-based internet-of-things systems and services. In: Proceedings of the International Conference on Advanced Computer Theory and Engineering (ICACTE '08) (2008)
11. Saaid, M.F., Ismail, I., Noor, M.Z.H.: Radio frequency identification walking stick (RFIWS): a device for the blind. In: Proceedings of the Fifth International Colloquium on Signal Processing and Its Applications (CSPA'09), March 2009
12. Martin, W., Dancer, K., Rock, K., Zeleny, C., Yelamarthi, K.: The smart cane: an electrical engineering design project. In: Proceedings of ASEE North Central Section Conference, Michigan, USA (2009)
13. Ahmad, S., Tokhi, M.O.: Linear quadratic regulator (LQR) approach for lifting and stabilizing of two-wheeled wheelchair. In: Proceedings of the Fourth International Conference on Mechatronics (ICOM), May 2011
14. https://blogs.cisco.com/video/asis-2013-dallas-area-rapid-transit-moves-safety-and-security-forward-with-cisco?dtid=osscdc000283

IoT-Cloud-Based Health Care System Framework to Detect Breast Abnormality

Modepalli Kavitha and P. Venkata Krishna

Abstract Breast cancer is utmost widely recognized cause of demise in the women all through the world from most recent 65 eons. From medical insights, identification and correct treatment of tumor at early stage will increase the lifetime of patient. According to doctor's perception, periodic breast self-examination will help the people to recognize the tumor symptoms at premature stage. But today people are highly educated toward technological revolution like smartphone and busy with their works. So, people expect non-ionic and inexpensive technological assistance for breast health monitoring. The quick technological connection amid wireless body area networks, Internet of things and cloud computing will bring innovative headway in the medicinal services by offering reasonable and quality patient consideration. Through this paper, we proposed IoT-cloud-based health care (ICHC) system framework for breast health monitoring. Using this framework, the breast cancer symptoms possible are to identify at the earliest.

Keywords Breast cancer · Cloud computing · Internet of Things · Wireless Body Area Networks

1 Introduction

Nowadays, Internet of things (IOT) assists as a worldwide stage to interconnect a huge number of physical items, RFID tags, sensors, actuators, cell phones, things, humans, and enables new ways of working, communicating, interacting, living, and entertaining. In IOT paradigm all physical objects which you wear, where you used,

M. Kavitha · P. Venkata Krishna (✉)
Department of Computer Science, Sri Padmavathi Mahila Visvavidyalayam, Tirupati, India
e-mail: dr.krishna@ieee.org; pvk@spmvv.ac.in

M. Kavitha
e-mail: modepalli.kavitha@kluniversity.in

© Springer Nature Singapore Pte Ltd. 2020
P. Venkata Krishna and M. S. Obaidat (eds.), *Emerging Research in Data Engineering Systems and Computer Communications*, Advances in Intelligent Systems and Computing 1054, https://doi.org/10.1007/978-981-15-0135-7_56

what you see, places you go, people you meet anything will be connected, addressed, and controlled remotely, i.e., anything, anyone, and any service can be linked by means of appropriate information and technologies [1].

The recent advances in communication tools and wireless sensing technology create a smart environment for extensive variety of uses in various spaces, for example, medicinal, sports, shopper gadgets, informal communication, and undertaking utilization. Among those developing applications, e-Health is perceived as the most critical and promising one because of its potential for well-being observing of lifesaving in crisis circumstances [2].

Specifically, wireless body area networks are the key empowering agents of remote and in-clinic well-being checking and are required to upset the well-being and constant body observing industry. A wireless body area sensor system comprises of various sensor hubs worn by patient that can gauge and report the patient's physiological state [3]. Also, cloud computing offers a lot of chances to administrations suppliers and clients, fundamentally encouraging calculation or storage for future usage. Wireless body area networks (WBANs) technology, Internet of things (IOT), and cloud computing mixture will advance the health nursing system effectiveness and support to achieve the goal of healthcare stakeholders.

The fundamental rule of IoT is that the objects like sensor hubs recognize, sense, process, and speak with one another [4]. With the intercession of most recent innovations wireless body area networks, Internet of things and cloud computing prompted to address the world challenges such as remote healthiness nursing, i.e., This technological combination can deliver good healthcare provision to the public.

Breast tumor is the mutual reason of death amid females throughout the world at the age of forty [5]. According to World Health Organization's cancer statistical reports more than four lacks women expire due to breast cancer [6]. Breast cancer death rate increases day to day in both developed and developing countries. The number of breast tumor cases is more in developed nations rather than developing nations, and contrary the number of incident rates is more in developing nations rather than developed nations [7]. According to Indian cancer research reports breast cancer cases increasing every year and the count may be doubled by 2025. We are designed Fig. 1 expresses how the breast cancer incident and death rate increasing year by year in India [8, 9]. The only opportunity to decrease the breast cancer death rate is early detection and effective treatment. The tumor size in breast will increase depends on number of years from attack, so the treatment may not give better results if we detect at later stages [5]. Now a day, we have so much advancement in medical treatments. So, the golden challenge in our hands is fixative the breast tumor at premature stage.

Mammography, thermography, ultrasound, MRI, and biopsy are the available medical imaging techniques to detect breast cancer. These techniques fix the scope, position, shape, organization, depth of tumor, and all other parameters which are needed to identify and treat the tumor. Doctors suggest that self-examination of breast is the best primary screening tool to detect breast cancer at premature stage

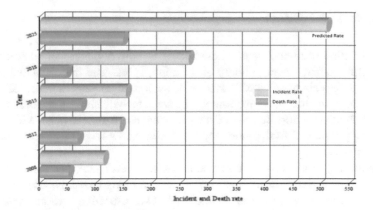

Fig. 1 Breast cancer incident and death rates in India

for patients. Breast thickness, bump, size, shape variations, skin irritation, redness, nipple scratch, retraction, discharge other than milk, and some minute calcifications are some common symptoms of tumor [5]. Continuous or periodic monitoring of breast health is significant to recognize these symptoms at initial phase.

Problem Statement:

There is a rapid technological advancement in medical services. But due to busy schedules, people are not using services properly. That means going to medical diagnostic center, taking test, showing reports to doctor, and getting results needs to spend time. Some cases the procedure will take more than one day time. Today people are well educated, so they expect remote medical assistance to monitor their health continuously.

Particularly for breast cancer, continuous or periodic self-examination of breast health will help the people to identify the disease at premature stage. And, statistical facts say that machine observation gives better results for most of the cases than manual observations. So, the situation demands to investigate the cost-effective, comfortable, portable, and non-ionic technological medical assistance to monitor breast health closely. We proposed an IOT-cloud-based health care (ICHC) expert system framework to support all categories of people through this work.

A couple of examinations have shown that the compelled access to patient-related information amid decision-making and lacking correspondence among patient consideration members are proximal explanations behind medicinal goofs in medicinal services [10]. Due to our proposed framework all these gaps will be surely possible to eliminate.

2 Associated Effort

The innovation of remote sensor networks is usually organized in different fields, for example, environment, military area, traffic, sports, and the medicinal services field [11]. Especially, in healthcare field, such a significant number of authors suggest through their examination work, sensors will show a vigorous role in constant health monitor. Because of the persistent checking, the disease identified at beginning period and furthermore conceivable to know how the patient well-being reacts for treatment.

Kumar, K., et al. monitor body temperature, breath rate, heartbeat, and body movements of patient's using Raspberry Pi. It is a Linux-based working framework fills in as a small pc processor framework. The respective sensors measure the readings and send to Raspberry Pi using amplifier. Using internet Raspberry Pi sends all the present well-being information of patient to Web database. Accessing to the Web and health monitoring of the patients is possible by doctors any time [12].

Kumar, K. M., et al. presented the solution to screen the ECG of a man utilizing an IOIO microcontroller board, which gets the bio-flag information from a man utilizing ECG electrodes and sends it to the cell phone remotely utilizing Bluetooth innovation. When observing, the client can upload it to a server private database. The doctor will access it and process the results [13].

Baba et al. developed a wearable WBAN application for well-being remote observing, utilizing ECG, SPO2, pulse, and breathing sensors. That screen patient's well-being through the persistent identification, process, and convey of human physiological parameters [14].

Lai, Xiaochen, et al. explains about Internet of Things (IoT) based health services using cloud computing that models and observes patients pervasively [11].

Nubenthan S., et al. present the design and implementation of remote observing framework contains sensor system and remote checking application. He discussed how the sensors collect the data, how the collected information will be managed and how alerts the organization in serious state of the patients [15].

Many authors examine and proven that sensors give best support in patient health monitoring. The article publications count from 1975 to 2018 we will consider as a proof for the above statement. Too many papers are available in so-called databases like IEEE, Scopus, and Springer to support the title "sensors in patient health monitoring." Figure 2 shows the graph related to how many articles published in IEEE, Scopus databases to support the title "Sensors in patient health monitoring." From this graph, we claim that sensors assume a crucial job in patient well-being checking.

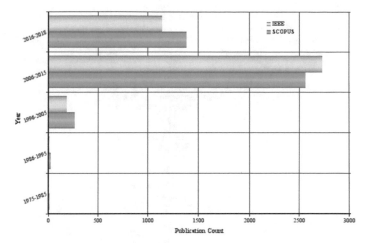

Fig. 2 IEEE and SCOPUS publication status on the title "sensors in patient health monitoring"

The data coming from sensors are processed at coordinator. The coordinator is a portable device like mobile or tablet or small laptop. The secured and permanent storage of processed data is the query now. To solve this query, after processing the results are sent to cloud for permanent storage. Because the cloud service allows to access the data anytime anywhere. Several cloud computing platforms are already available to support sensor-based data storage, either free or commercial. Some of the cloud services are Pachube, Nimbits, ThingSpeak, DropBox, iCloud, Pithos, Okeanos, Rackspace, Amazon Web Service, and Geogrid [16]. Most of them are offers free service.

3 Projected Effort

Our proposed work is in view of computerizing the strategy for gathering patient's breast physical characteristics data using sensors. And send that data to cloud for permanent storage or further processing through personal coordinator.

3.1 Proposed Model

Figure 3 depicts the proposed model. The framework includes four fundamental parts: sensors, personal coordinator, cloud server, and clients. At the patient's side, we have wearable remote sensors. The sensor hubs have the product to gather the characteristics around them; encode and transmit information over remote

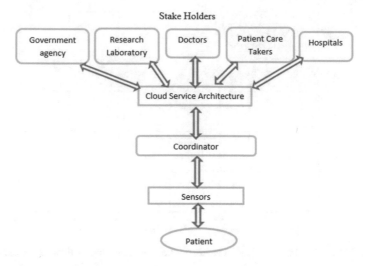

Fig. 3 Proposed model for IoT-cloud-based smart health scenario

correspondence channels. These sensors associated with coordinator, which read information from the sensors and sends it to a cloud server. The collected data is stored there and does the further processing. That processed data is accessed by different users like patients, family members, doctors, nurses, medical laboratories, hospitals, public authorities which are responsible for validating and authorizing of these doctors using Internet (stake holders).

3.2 Block Diagram of ICHC System

Figure 4 demonstrates the block diagram of proposed IOT-cloud-based health care (ICHC) expert system. It shows patient with wireless sensor jacket, coordinator, cloud server, and different type of users like doctor, hospitals, medical research laboratories, and other users. The wireless sensor nodes needed to sense the physical symptoms of breast are embedded in spandex material wearable jacket. The spandex material is stretchable, so it is comfortable to continue with all extents of people. Along with sensors, microcontroller is placed in jacket. It controls and sends signals to the sensors embedded in jacket. The sensed data is passed to coordinator using Wi-Fi, due to less memory capacity of sensors. The coordinator is a portable device. The collected resultant data are sent from the personal coordinator to cloud using Wi-Fi or Internet for permanent storage and further processing. Cloud service allows to access the data anytime anywhere. So, we are selecting cloud service to store the sensed data along with results. The results accessed from cloud server by different users depend on their need.

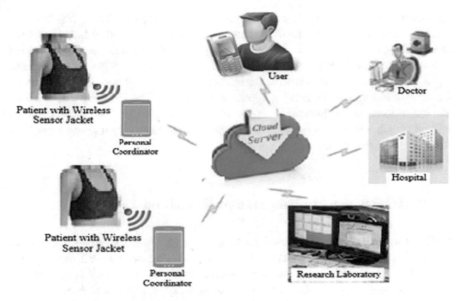

Fig. 4 Block diagram of proposed IoT-cloud-based health care (ICHC) system

3.3 Cloud Services

Software-as-a-Service, Platform-as-a-Service, and the Infrastructure-as-a-Service are the three types of facilities offered by cloud services. Reddy et al discussed clearly about these services [17]. SaaS gives the product applications as a service used to investigate the patient's information on distributed computing condition. PaaS gives the client to convey client assembled applications over the cloud foundation to analyze and store the results. IaaS provides virtualized data centers to include the storage as an administration, calculation asset as an administration, and correspondence asset as an administration. The clients can send and run the applications in these models.

3.4 Proposed Methodology

Doctors strongly suggest that the breast cancer not attack to both breast at a time. Depends on this statement, in thermography test—comparison of breasts will consider to identify breast tumor. In ICHC system, we are comparing the physical symptoms of both breasts to identify the breast abnormality. Breast thickness, bump, size, shape variations, skin irritation, redness, nipple scratch, retraction, discharge other than milk, and some minute calcifications are some common physical symptoms to fix breast abnormality [5]. Most of these physical symptoms are possible to sense by fixing suitable sensors at predetermined positions on breasts. The sensors arrangement is follows triangulation method. It helps to fix the abnormal position on breast. The sensed signals from sensors are sent to cloud through coordinator using Wi-Fi. At cloud the software service converts these signals into required data formats. These data items will be processed and generate the reports in required format by application software. The software can send the automatic message to the user depends on the report's status. The patient reports will be possible to access by doctors or other stakeholders of our system any time from cloud infrastructure using Internet.

4   ICHC System Implementation Discussion

While using ICHC system the user needs to create an account by entering personal details like name, sex, age, family history of this disease, etc. All the physical symptoms mentioned above are not common to all categories of people at one stage. It is possible to identify the tumor based on the symptoms present at ICHC. In ICHC system, we consider four common metrics to fix the abnormality of breast. All these four metrics data we can collect from both breasts using sensors. At cloud we can compare data related to both breasts.

Metric 1: Temperature
Metric 2: Red spots on breast surface
Metric 3: Nipple discharge
Metric 4: Lump.

The breasts temperature abnormality is discussed in paper [18] clearly. But the temperature of skin varies due to some health problems like fever, surgeries, etc., and from environmental changes. Using only this metric we are unable to fix breast abnormality. For better result, we are considering other three metrics which are most common symptoms at breast cancer patients.

Depends on these four metrics results our ICHC system will identify the breast abnormality and sends message along with suggestions to the patient. The given metrics in ICHC system will be represented by using M1, M2, M3, and M4. And we created files to send to the patient along with messages.

File 1: Diet plan for good health maintenance
File 2: Breast cancer symptoms case study
File 3: Breast cancer hospitals and some doctor's information.

Algorithm:

Step 1: Read four metrics results on breasts
Step 2: If all four metrics gets negative results and family history status is "no" then
Step 2.1: Patient gets message as "Your health is good" Congratulations Have a safe life

Step 3: If all four metrics gets negative and family history "Yes" then
Step 3.1: Patient gets message as "Your health is good" and follow file-1 for better health maintenance
Step 4: If M1 is positive, M2, M3, M4 are negative and family history is positive then
Step 4.1: Patient gets message as "Periodic Monitoring of breast health suggested"
Step 5: If M1 is positive, M2, M3, M4 are negative and family history is negative then
Step 5.1: Patient gets message as "Don't worry Small temperature irregularity observed"
Step 6: If M1 and M2 or M3 is positive, M4 is negative and family history is positive then
Step 6.1: Patient gets message as "Periodic Monitoring of breast health suggested and observe file-2"
Step 7: If M1 and M2 or M3 are positive, M4 is negative and family history is negative then
Step 7.1: Patient gets message as "Don't worry Small temperature irregularity and red spots on breast surface observed"
Step 8: If M1, M2, M3 are positive and M4 is negative then
Step 8.1: Patient gets message as "You are under suspicious state follow periodic monitoring and observe File-2"
Step 9: If All four metrics are positive then
Step 9.1: Patient gets message as "Immediately contact your personal doctor or consult the doctors in file-3."

Likewise, all the possible cases depend on our considered metrics can be checked carefully by the system. All three files are get accessed whenever login to our system.

5 Results Discussion

In ICHC system, using above sample algorithm, it is observed that the number of metrics gets positive results increases then health risk rate also increasing that is shown in Fig. 5. It is obvious that both breast for all categories of people are not matched for all metrics. There is a change obviously, if that change is more than predefined results for the metrics that refers positive value. If it is near to predefined range, then that is negative result. The ranges for all the metrics will be discussed in coming article.

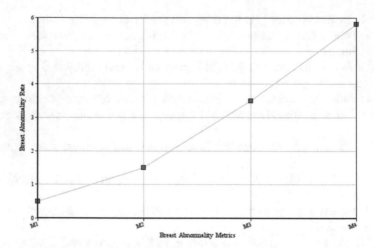

Fig. 5 Breast abnormality rate over metrics results

6 Conclusion

Breast cancer is the most mutual health problem especially amid women today. Most of the imaging technologies like mammography, thermography, ultrasound, magnetic resonance imaging, etc., support us to detect this disease. But due to high cost, most of the people are unable to take the test periodically and some people are not interested due to ionic nature of most of the suggested tests. Periodic self-examination of breast health is one of the major possibilities to identify the breast tumor. By using our proposed ICHC technological assistance breast health periodic monitoring is possible. It is comfortable, non-ionic, and supports all categories of people. This system stores the breast health information permanently at cloud, so all the authorized stake holders are possible access the breast health information any time, and also send the status of health to the patient mobile or email.

7 Future Work

As part of this proposed project we need to identify exact the sensors, wireless technologies, and cloud services to support application.

References

1. Atzori, L., Iera, A., Morabito, G.: The internet of things: a survey. Comput. Netw. **54**(15), 2787–2805 (2010)
2. Sawand, A., et al.: Toward energy-efficient and trustworthy eHealth monitoring system. China Commun. **12**, 46–65 (2015)

3. Tyagi, S., Agarwal, A., Maheshwari, P.: A conceptual framework for IoT-based healthcare system using cloud computing. In: 2016 6th International Conference Cloud System and Big Data Engineering (Confluence). IEEE (2016)
4. Liu, X., Baiocchi, O.: A comparison of the definitions for smart sensors, smart objects and Things in IoT. In: 7th IEEE Conference in Information Technology, Electronics and Mobile Communication (IEMCON), pp. 1–4, Oct 2016
5. Hela, B., et al.: Breast cancer detection: a review on mammograms analysis techniques. In: 2013 10th International Multi-Conference on Systems, Signals & Devices (SSD). IEEE (2013)
6. Mejía, T.M., et al.: Automatic segmentation and analysis of thermograms using texture descriptors for breast cancer detection. In: 2015 Asia-Pacific Conference on Computer Aided System Engineering (APCASE). IEEE (2015)
7. De Oliveira, J.P.S., et al.: Segmentation of infrared images: a new technology for early detection of breast diseases. In: 2015 IEEE International Conference on Industrial Technology (ICIT). IEEE (2015)
8. http://www.breastcancerindia.net/statistics/stat_global.html
9. https://seer.cancer.gov/statfacts/html/breast.html
10. Bringing IoT and cloud computing towards pervasive healthcare
11. Lai, X., et al.: A survey of body sensor networks. Sensors **13**(5), 5406–5447 (2013)
12. Kumar, R., Rajasekaran, M.P.: An IoT based patient monitoring system using raspberry Pi. In: IEEE International Conference in Computing Technologies and Intelligent Data Engineering (ICCTIDE), pp. 1–4, Jan 2016
13. Kumar, K.M., Venkatesan, R.S.: A design approach to smart health monitoring using android mobile device. In: IEEE International Conference in Advanced Communication Control and Computing Technologies (ICACCCT), pp. 1740–1744, May 2014
14. Baba, E., Jilbab, A., Hammouch, A.: A health remote monitoring application based on wireless body area networks. In: 2018 International Conference on Intelligent Systems and Computer Vision (ISCV). IEEE (2018)
15. Nubenthan, S., Ravichelvan, K.: A wireless continuous patient monitoring system for dengue; Wi-Mon. In: 2017 International Conference on Wireless Communications, Signal Processing and Networking (WiSPNET). IEEE (2017)
16. Doukas, C., Maglogiannis, I.: Bringing IoT and cloud computing towards pervasive healthcare. In: 2012 Sixth International Conference on Innovative Mobile and Internet Services in Ubiquitous Computing (IMIS). IEEE (2012)
17. Reddy, B.E., Suresh Kumar, T.V., Ramu, G.: An efficient cloud framework for health care monitoring system. In: 2012 International Symposium on Cloud and Services Computing (ISCOS). IEEE (2012)
18. Kavitha, M., et al.: Wireless sensor enabled breast self-examination assistance to detect abnormality. In: 2018 International Conference on Computer, Information and Telecommunication Systems (CITS). IEEE (2018)

Smart Home Monitoring and Automation Energy Efficient System Using IoT Devices

K. Suneetha and M. Sreekanth

Abstract Wastage of electricity is one of the main problems which we are facing nowadays. In home and working areas, people will be engaged with too many activities. Sometimes, there might be a chance to forget to turn off various electrical appliances like lights, fans, servers, if there were nobody in the room or area/passage. This leads to huge power consumption. Also, it is very difficult to physically challenged persons and old-age people to access the appliances every time manually. To avoid all such situations, we proposed "Smart Home System." It is the process of automating home appliances in which the various electric and electronic appliances are wired up to a central computer control system so they can be handled more efficiently. "Smart Home" helps to overcome the problems that usually people face in daily life. Behind this, it needs the involvement of multiple complex hardware components in controlling the smart home project and there is a fair amount of software architecture that is responsible for driving the hardware components. Moreover, there is a need to employ persons to monitor such simple tasks. This manual approach is time-consuming and also very costly. Thus, automation is the better alternative through which these tasks can be completed automatically without human effort and also accessed remotely from anywhere and anytime with less effort. Everyone is showing interest in making their home smarter to ease their daily activities. The proposed Smart Home system has the capability to control the home appliances based on the environment. Here, different boards like Raspberry Pi 3 and Arduino UNO are going to be deployed which can take control over the appliances and act accordingly. People can also be notified, if there are any fire/gas leakages inside the home. This security system facilitates the user to take necessary actions as soon as possible to secure their home. It also involves the deployment of different kinds of sensors which can able to detect the object motion, temperature and humidity and fire detection. This application is very simple and can be implemented in all types of homes.

K. Suneetha (✉) · M. Sreekanth
Department of MCA, SVEC, A.Rangampet, India
e-mail: umasuni.k@gmail.com

M. Sreekanth
e-mail: msreekanth53@gmail.com

© Springer Nature Singapore Pte Ltd. 2020
P. Venkata Krishna and M. S. Obaidat (eds.), *Emerging Research in Data
Engineering Systems and Computer Communications*, Advances in Intelligent
Systems and Computing 1054, https://doi.org/10.1007/978-981-15-0135-7_57

Keywords Internet of things · Raspberry Pi 3 and Arduino UNO · Sensors · MQ2

1 Introduction

With the rapid growth of population density, the services needed to supply the requirements of a citizen also grown tremendously for better quality living. Generally, in working places and at home, people are engaged with multiple activities; hence, there is a chance to forget to switch off various electrical appliances like fans, lights, which leads to huger power consumption wastage. Thus, automation is the best choice to do these activities automatically without human effort and also can be accessed remotely from anywhere and anytime with less effort, hence bring new demand and challenges.

Over the last decade, the expectations of life have been driven to a greater extent and people are enjoying much independence to live happily, hence bring new demand and challenges. Smart Home is a collaboration of technology and services through a network for better quality living. Everyone is showing interest to create a secure and smart home to ease their daily activities. The proposed Smart Home system has the capabilities to control the home appliances based on the environment. In this proposed system, different boards like Raspberry Pi 3 and Arduino UNO are going to be deployed which can take control over the appliances and act accordingly. It also involves deployment of different kinds of sensors which can able to detect the object motion, temperature and humidity and fire detection.

The proposed Smart Home automation system has the capabilities to control the home appliances based on the environment. This IoT-based smart home system consists of a combination of different components such as Raspberry Pi 3 and Arduino UNO which are to be deployed to control various appliances and act accordingly. Also with the deployment of different kinds of sensors, one can able to identify the motion of the object, temperature and humidity and fire detection.

The main idea of our proposed Smart Home automation system is

- Able to detect the room temperature automatically
- Able to monitor and automatically on and off the lights
- Able to ring doorbell automatically
- Able to identify the fire/gas accidents.

As a part of IoT, smart home systems and devices function together share consumer usage data among themselves and actions get automated based on house owner preferences. This proposed system uses sensors over IoT devices through Internet to automate home appliances. A smart home is one that equipped with lighting, heating, and electronic devices that can be controlled by a smart phone or via the Internet. Here, Raspberry Pi and Arduino microcontroller UNO R3 are used which acts as an interface between various access points.

The IoT devices [1] are connected to the Internet and send information about themselves or about their surroundings over a network or allow actuation upon the physical entities/environment around them remotely. This system has two kinds of environments. One is, the system components are integrated with other electronic devices and sensors and other is the user interaction with the appliances through cloud. Here the user can remotely access/take control over all the appliances that are connected to the system. The following are the two boards which we are going to encounter:

- Arduino UNO (Microcontroller)
- Raspberry Pi 3 (Microcomputer).

1.1 Arduino Uno

Arduino Uno [2] is an open-source electronics platform-based microcontroller easy-to-use computer. It can be easily communicated by passing instructions to the microcontroller on the board at any instant of time. Arduino also simplifies the process of working with microcontrollers, and it offers some advantages for users and other systems.

1.1.1 Inexpensive

Arduino boards are quite cheaper when compared to other microcontroller platforms.

1.1.2 Cross-Platform

The Arduino software (IDE) runs on a variety of operating systems namely Windows, Mac OS X, and Linux environment, but microcontroller systems are limited to Windows.

1.1.3 Programming Environment

An Arduino software is user-friendly for learners and adequate for advanced users It is based on processing programming environment; hence, users can able to learn the programs in that environment effectively and become expertise with the working functionality of Arduino IDE.

1.1.4 Open-Source and Extensible Software

It is an open-source tool. The user needs to upload his software to the board through Arduino Sketch.

1.1.5 Open-Source and Extensible Hardware

The Arduino board plans get publicized under Creative Commons license so that the circuit design experts can create their own version of modules with an effective enhancement and improvements.

1.2 Raspberry Pi 3

Raspberry Pi [3, 4] compared to Arduino is more powerful in terms of computation and processing. It can integrate different types of sensors and actuators, and this port is more attractive when compared to a similar kind of feature in Arduino.

1.2.1 Sensors and Other Devices

Sensors are used to identify the changes happening in the surroundings. A sensor used with an electronic system plays a vital role. It can able to measure the temperature, pressure, and so on and transform it into an electric signal. The sensor can be distinguished based on range, sensitivity, and its resolution.

The following are the sensors and devices used for Smart home system:

1.2.2 IR Sensor

IR is an electronic device which either emits and or detects infrared radiations to sense certain aspects happening in the environment. IR sensors are able to detect the motion of the object and also measure the heat emitted by an object.

1.2.3 DHT22 Sensor

An Ardunio board with the help of DHT22 sensor helps to determine the humidity and temperature in the atmosphere. It contains humidity sensor wet components and a thermostat and connected with a high-performance 8-bit microcontroller to detect the surrounding air and spits out a digital signal on the data pin.

1.2.4 MQ2 Gas Sensor

As a part of a safety system, the gas sensor detects the gas leakages. To detect the gas leakage, Grove—Gas Sensor (MQ5) is used at home or organization. This sensor is able to detect CO, H2, Smoke, Alcohol, LPG by adjusting the sensor of potentiometer. An alarm gets triggered automatically when MQ2 gas sensor detects the gas leakage.

1.2.5 Relay

It is an electromagnetic operated switch used to turn on/off a circuit by a low power signal. It uses only one signal to control a lot of circuits.

1.2.6 Transceiver

The NRF24L01 transceiver module is used to transfer the sensor data between multiple IoT boards.

The proposed system focuses on the implementation of following modules: Switch on/off the lights when person enters/exits the room, turn on/off the fans based on the room temperature, detection of fire accidents or gas leakages inside the home and implementation of automatic door alarm.

2 Literature Survey

This section explains the research related to smart home automation system.

2.1 Bluetooth-Based Home Automation System Using Cell Phones

Adhiya et al. [5] presented a paper on Bluetooth-based wireless home automation system using Arduino BT board at input–output ports using relay circuit. To control the home appliances, they used android phone and Bluetooth technology to establish a connection between Arduino board and mobile for wireless communication. It is provided with password protection so that only authorized persons are able to access and control the appliances. Here Python programming is used and installed in any of Symbian operating system environment. A circuit is developed to receive the feedback from mobile about the status of the device. This system is developed to help the old-aged and handicapped people to operate the home appliances with ease and by reducing their human efforts.

2.2 GSM-Based System Using Mobile Phones

Lamine and Abid [6] proposed a GSM-based home automation using cell phones. An appropriate SMS/GPRS/(DTMF)-based home automation communication methods can be considered based on GSM module commands. They presented how devices and sensors relate together with the network and communicate effectively using GSM and SIM. They used transducer and sensors to convert physical qualities to voltage.

2.3 Home Automation Using RF Module

Tseng et al. [7] proposed a home automation system using a RF controlled remote. Nowadays, wall switches located at various parts of the house make it difficult for the user to go nearer to control and operate. As technology grows, the homes also become smarter by replacing one switch to centralized RF control switches. On the transmitter side, they combined RF remote to the microcontroller which sends an on/off signal to the receiver where devices get connected. They used wireless technology to turn on/off globally to operate the remote switch on the transmitter. This system is developed to help the old-aged people or physically challenged people to operate the home appliances with RF technology with ease and by reducing their human efforts.

3 Implementation

IoT devices can consist of a number of models based on functional attributes, such as sensing, actuation, communication and analysis and processing. IoT device is having single board computer (SBC) including CPU, GPU, RAM, storage, and various types of interfaces. The following Arduino code is used to implement the automatic door alarm. It should be loaded into Arduino IDE in the system, and there we need to compile the sketch and upload it to the Arduino UNO board attached with an IR sensor.

3.1 Architecture of Smart Home

The devices and Arduino UNO boards are connected through jumper wires. The Arduino UNO boards will receive the data from sensors. Then these Arduino UNO boards are connected to a Raspberry PI_3, which is acting as a gateway. Transceivers are used to transfer data from Adruino UNO to Raspberry Pi 3.

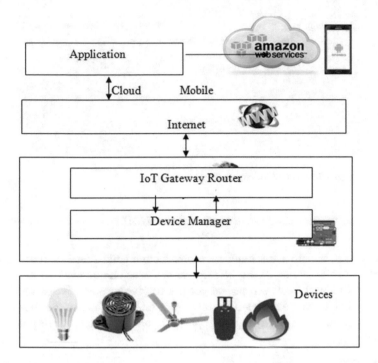

Fig. 1 Smart Home architecture

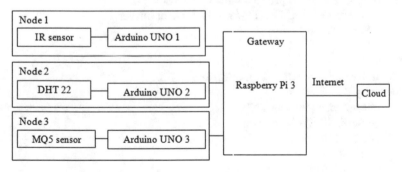

Fig. 2 Block diagram of Smart Home

The user can connect to the gateway via Internet and can control the devices remotely. The user will receive the alert message to when gas/fire accident happens inside the home. Block Diagram of Smart Home (Figs. 1 and 2).

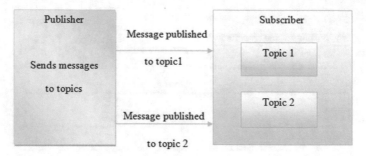

Fig. 3 Communication model of Smart Home

3.2 Communication Model of Smart Home

In this system, we are using request–response communication model and publish–subscribe model (for cloud). In the request–response communication model, client has to send request to server and server has to give response based on the client request. When server receives a request from client, it fetches the data from the local server, retrieves the resource representations, and sends a response to the client (Fig. 3).

Doorbell.ino
```
void setup()
{  pinMode(3, INPUT);// pin to connect IR sensor
        pinMode(8, OUTPUT);// pin to connect buzzer
        Serial.begin(9600);
}
void loop()
{ if(digitalRead(3) == HIGH)// motion is
  detected by the IR sensor
    {        Serial.print("buzzer on");
             tone(8, 400);// buzzer on
             delay(300);
             noTone(8);
             delay(300);
             tone(8, 400);
             delay(300);
}
else
 { noTone(8);// buzzer off
   Serial.print("buzzer off");
 }
}
```

3.3 Board Setup

The following is the Arduino sketch which is used to implement the gas/fire detection inside the home. It should be loaded into Arduino IDE in the system, and there we need to compile the sketch and upload it to the Arduino UNO board which is equipped with a MQ2 gas sensor.

The below diagram shows the light functionality when a person is inside and outside the room. Whenever a person enters into the room IR sensor detects the motion and makes light to "on" state. Otherwise, it will be in "off" state.

Screenshot of the Automation Process　　*Screenshot: Implementation of fire detection*

The above equipment consists of Arduino UNO, MQ gas sensor, and transceiver. The data which is generated by the MQ gas sensor is transmitted to Raspberry Pi through the transceiver. Whenever the fire is detected by the sensor, the owner will be notified with an alert message as well as email notification so that there is a chance to take the necessary actions at the earlier.

Screenshot 1: Alert message sent to the   *Screenshot 2: Email notification*
home owner

The above screenshot shows the email notification which is sent to the home owner when any fire/gas is being detected by the gas sensor.

4 Conclusion and Future Enhancements

"Smart Home" helps to overcome the problems that usually people face in daily life. People tend to forget switch off appliances and thus electricity will be wasted. By using this automation, electricity can be saved. People can also be notified, if there are any fire/gas leakages inside the home. This security system facilitates the user to take necessary actions as soon as possible to secure their home. This kind of products may have high demand in the coming future. This application is very simple and can be implemented in all types of homes. Some of the other areas where we can deploy IoT applications are industrial automation, healthcare monitoring, and environment monitoring, smart cities, and so on.

Limitations and Future Enhancements

- This project always needs the Internet connection for the proper functionality of this entire system.
- We cannot control the brightness of the light and also the speed of a fan. This will be addressed in the future scope.
- We cannot able to see the person who is waiting at the doorstep. So, we can use a camera to identify the person.
- To proving the security of IoT data and sensors information transmitting to cloud, change to hack the data. We are looking at the security issues in IoT and the risk of losing privacy increases. So, to avoid this loss of privacy, we can use some encryption algorithms in future developments.

References

1. Kovatsch, M., Weiss, M., Guinard, D.: Embedding internet technology for home automation. In: Proceedings IEEE Conference on Emerging Technologies and Factory Automation (ETFA) 2010, pp. 1–8 (2010)
2. ElShafee, A., Hamed, K.A.: Design and implementation of a WiFi based home automation system. Int. J. Comput. Electr. Autom. Control Inf. Eng. 6(8) (2012)
3. Suneetha, K., Sreekanth, M., Sankara, K.: An advanced method for detecting road traffic signs using IoT devices. (CCODE), SSRN: https://ssrn.com/abstract=3167212 or http://dx.doi.org/10.2139/ssrn.3167212
4. Suneetha, K., Sreekanth, M., Sankara, K.: An approach for intelligent traffic signal control system for ambulance using IoT. (CCODE), SSRN: https://ssrn.com/abstract=3167212 or http://dx.doi.org/10.2139/ssrn.3167212
5. Adhiya, Y., Ghuge, S., Gadade, H.D.: A survey on home automation system using IOT. IJRITCC 5(17). 5(3)
6. Lamine, H., Abid, H.: Remote control of a domestic equipment from an Android application based on Raspberry pi card. In: IEEE Transaction 15th International Conference on Sciences and Techniques of Automatic Control & Computer Engineering—STA'2014, Hammamet, Tunisia, 21–23 Dec 2014
7. Tseng, S.-P., Li, B.-R., Pan, J.-L., Lin, C.J.: An application of Internet of Things with motion sensing on smart house. 978-1-4799-6284-6/14 c 2014 IEEE

Realistic Sensor-Cloud Architecture-Based Traffic Data Dissemination in Novel Road Traffic Information System

S. Kavitha and P. Venkata Krishna

Abstract Transportation is a major source to carry things from one place to another place. Over many years, transportation may face many problems like accidents, traffic collisions, and congestion. Many people are killed every year due to accidents. It has a challenge to prevent these accidents. Safety applications mostly associated with users and their lives. One of the major ways is to offer traffic situations to the vehicles so that they can use them to identify the traffic situation. This can be achieved by transferring the information of the real-time traffic situation between vehicles. Besides the physical world, the integration of communication, and control and computation, CPS is identified as a new technology. Various GPS-based phones and sensor-enabled devices have been designed to provide major information like environmental information and secure data for traffic betterment, though it faces concerned constraints such as consumption of energy and the rich rate of computation. Cyber-physical system permits controlling of the physical world directly by way of the Internet. The wireless sensor network (WSN) is the integral part of CPS that has collected data related to things in nature through sensors. To meet trustworthy and security constraints, the proposed paper presents a unique approach that depends on an authentic cloud model for real-time traffic data collection and dissemination. We design a hybrid approach, which is integration of WSN and VANETs called smart road traffic information service (RTIS), which turns traditional transportation to an intelligent transportation system.

Keywords VANET's · WSN's · CPS · Sensor cloud

S. Kavitha (✉)
Ph.D. Scholar, Department of Computer Science, Sri Padmavati Mahila Visvavidyalayam, Tirupati, Andhra Pradesh, India
e-mail: Kavithabtech05@gmail.com

P. Venkata Krishna
Professor, Department of Computer Science, Sri Padmavati Mahila Visvavidyalayam, Tirupati, Andhra Pradesh, India
e-mail: pvk@spmvv.ac.in

© Springer Nature Singapore Pte Ltd. 2020 639
P. Venkata Krishna and M. S. Obaidat (eds.), *Emerging Research in Data Engineering Systems and Computer Communications*, Advances in Intelligent Systems and Computing 1054, https://doi.org/10.1007/978-981-15-0135-7_58

1 Introduction

Most people think that the advancement of transportation means developing new roads or remaking old transportation structure. The upcoming transportation system not only depends on steel and concrete but also supports in technology advancements, such as sensors and chips that gathers, organizes, and delivers data concerned to the road traffic system. Transportation systems are specifically mobile devices, and data creates more specific value to a network. VANETs contribute a major part in building an intelligent transportation system (ITS) [1] that mainly puts emphasis on road safety. The accomplishment of VANETs into ITS depends on the critical element such as routing nodes, i.e. vehicles and also the Internet-enabled mobile nodes. Vehicular network is a foundation of intelligent transportation system. In-vehicle communication is a representative for a car-to-car communication model for information dissemination [2, 3]. Traffic information dissemination in VANET focused on safety driving information [4, 5], such as early warning information, anti-collision information for vehicles which are mobile. ITS is designed to pertain real-time propagation of traffic data to drivers. Basically, three forms of communication are required for ITS design they are (1) The wireless network between the vehicles, (2) wireless network that promotes vehicles on the road interacting with the roadside elements, and (3) a wire-line link that makes connection with the infrastructure. The real-time disseminations of information via V2V and V2I are major challenges, which are mostly constrained by the physical properties of the system. For example, the powers of radio transceiver, wireless channel sharing, mobility of devices, and the node density. Most of the challenges are from the protocols like 802.11, IP addressing and routing, and address resolution protocol (ARP). Cyber-physical systems (CPS) [2] are designed to address all of these issues by integrating the physical behavior, cyber nature, and people into a single platform. Here, we introduce solutions, which are based on CPS for accurate propagation of ITS-related data. This paper introduces solutions for the real-time dissemination of ITS information, which are based on CPS mechanisms. In this, we introduce realistic sensor-cloud architecture, i.e. service-oriented cyber-physical system which offers various services like computing and communicating facilities that offer an accurate result to users. The Sensor-cloud system process collects information by deploying the sensors, which are similar and its data in the database for each type of network. Sensors of the same type are allocated to a distinct grid. Each grid has coordinator, which is responsible to control all other sensors of its grid. Here, we design a combined architecture that is a combination of WSN and VANET, which is called as a smart road traffic information system (RTIS), which provides several services to a user (i.e. speed and travel time estimation, congestion information, road condition estimation, and environment estimation).

2 Related Work

Cyber-physical system plays a vital role in designing the next generation transportation system [2]. The reliable and secure requirements for different systems and several GPS-enabled solutions have been devised. Mostly, it focuses on describing how the movable Internet is dynamically changing its nature to address the transportation system CPS. In [4], the authors proposed a design of traffic observation system, which relies on GPS deployed mobile nodes, that gather traffic-related information from mobile phones and calculate traffic condition and congestion in real time. The major pitfall of this approach is much battery consumption. In vehicular adhoc networks, sensors are placed in a vehicle for observing the traffic because of its low-cost communication. Besides VANET, recently, so many applications have designed to predict events, and to process and deliver the traffic-related information in time. Roadside access points deployed with sensors which are wireless, and more cost to create wireless sensor networks, in addition to that VANET does not guarantee to provide detection of road conditions or communication integration. VANET is disconnected due to several reasons such as they are dynamic, mobility, and inadequate distance between RSU's. So, the demanding data about road situations known by one group of vehicles cannot be shared with other groups. One possible solution is deploying more RSU's, but it may significantly increase the cost. Disconnected groups are connected by cellular networks, which depend on the communication range but it does not solve the problem of sensing road traffic situations. Detection of accident and then forwarding the message related to it in advance to other vehicles are highly essential. A vehicle enabled with sensors and GPS can detect the accidents. Both adhoc networks and WSNs have a number of applications though their characteristics are quite different in nature. For example, WSNs nodes are small, high energy constrained, and mostly static. They have much acceptable capabilities to identify events. In contrast, VANETs topologies are dynamic in nature and have much energy constraints to detect the events. However, coverage of vehicles cannot be reliable and vehicles are not present throughout the road. An efficient transmission method can be opted by using inexpensive sensors instead of deploying a much number of RSUs on the roads. The sensors can detect events, such as icy roads, digs, road traffic situations, and accidents, so they can disseminate messages to vehicles by using the roadside units.

3 Architecture of Sensor Data Storage on Cloud

Service-oriented cyber-physical system facilitates the fastest sensor data processing (i.e. data from numerous sensor nodes. Service-based CPS (SOCPS) architecture is shown in Fig. 1 has three associated things (1) The network senses and sends it to the cloud for computation. (2) It maintains separate database in a cloud for each type of network and (3) Smart road transport system services.

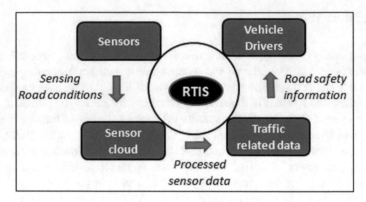

Fig. 1 SOCPS

3.1 Steps in Collecting Real-Time Traffic Data

Working of the sensor cloud [1] shown in Fig. 2 for traffic data processing is described in the following steps:

1. Different types of sensors shown in Table 1 are deployed through highways in which it is formed as grids using the method described in [4].
2. In each cluster, a coordinator is selected and that node is responsible to interact with all others of its own grid and with the cloud. The selection of coordinator is explained in [6].
3. From the base station, each coordinator in the grid of all the nodes is to be controlled.

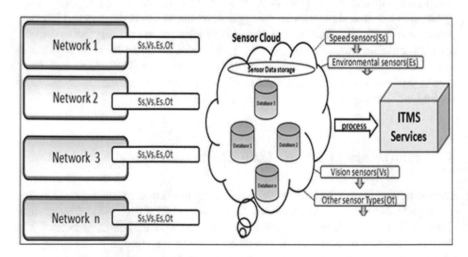

Fig. 2 Sensor data storage on cloud

Table 1 Different sensors used in proposed approach

Sensors	Purpose
Viewable sensors	Safety messages
Wheel peed sensors	Velocity estimation
Weather sensors	Weather monitoring
Crash sensors	Vehicle crash, Congestion

Table 2 Types of traffic-related data

Safeness messages	Dangerous, accident, work zone
Social messages	Emergency Vehicle notification, highway in formation
Driver messages	Road congestion information, Traffic navigation
Commercial messages Recreation messages	Advertisements Multimedia messages
Recreation messages	Multimedia messages

4. There is a separate table in a cloud database for all kinds of traffic data shown in Table 2.
5. A node that needs data from the other sensor node which is deployed in the common grid or in the different grid first it has to interact with the coordinator.
6. The sensor node (coordinator) then connects to the database for obtaining the desired result.

4   Road Traffic Information System

In this section, we define hybrid architecture [7] by taking the combination of WSN and VANET known as novel road traffic information system.

4.1   Hybrid Architecture

Sensor node in Fig. 3 deployed at pole 1 is classified as the crash sensors category shown in Table 1 whose main function is to identify the crash of the vehicle and condition of traffic. Sensor on the pole 2 called viewable sensors to sense road condition like digs on the road, merging of roads. Sensor on the pole 3 shows a vehicle type that is two-wheelers, buses, and heavy trucks. Sensor at pole 4 detects weather condition. Each vehicle is equipped with on-board unit and application units. RSUs refer roadside units, which are deployed road intersection points, parks, and so on. (OBU) on-board unit (OBU) is a mobile node and RSU is static in nature. Gateways are used to connect RSU's to the Internet. Any two RSUs can

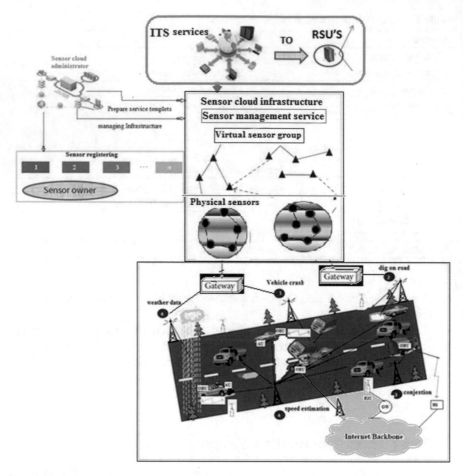

Fig. 3 Architecture of RTIS

interact directly. If any event occurs, there are several options to enter or exit from the highway roads when a vehicle is in the range of any RSU data if uploads and downloads wirelessly between them. Normally, the distances between any RSUs are 2 km or 3 km. RSU uses the control channel to inform moving vehicles that if any event occurs on the road. RSSs are placed between RSUs to receive data from sensors. Each wireless sensor can act according to its function that is sensing the physical environment to detect events. The RSSs are also referred as sinks whose function is to collect data from nodes and to transmit any event information to other RSS, and finally, to RSU near it. The on-board unit can be used to receive data from sensors for further processing. IEEE 802.11p protocol is used to connect two OBUs and broadcasting done in a multi-hop fashion. When a vehicle reaches within the range of any RSU, event information from the OBU is uploaded to the RSU, and from RSU is downloaded to the vehicle's OBU. Two RSSs communicate directly using the IEEE 802.15.4 protocol. Sensor-cloud provides service to end-users automatically, virtual sensors, which is a part from IT resources, i.e. disk storage, CPU, memory, etc. The major advantage of cloud computing is Vehicular Cloud (V-Cloud) in that all the modern cars have the permanent connection to the Internet, so that each vehicle has a considerable amount of computation, sensing, and storage capabilities. Sensor-cloud is a new generation of cloud computing. The physical sensors are to sense and transmit all data into a cloud infrastructure. In sensor-cloud infrastructure, the sensors to be utilized IT infrastructure by virtualizing all the physical sensors. These virtualized sensors are dynamic so, it is important that users may not concentrate on the actual locations of sensors. These virtual sensors are supervised using some standard methods. Sensor-cloud infrastructure provides a user interface for administering, controlling, and destroying virtual sensors, addition, and deletion of physical sensors. Sensor -cloud owners are responsible for registered or unregistered physical sensors. These resources such as physical sensors, database systems, processors, and sensor are prepared to operate in the real-time environment.

4.2 RSU Role in Road Traffic Information System

VANET has as ability to provide surrounding area information to the vehicles, for example, information related to hotel, bank, hospital, public toilet, and bus stand.

RSU provides all the above information with its exact location according to its database. Database should be updated regularly.

RSUs so far can perform different functions such as

- data disseminators,
- road traffic dictators,
- service providers
- security management, and
- service backups.

Figure 4 shows the block diagram of the roadside unit; every vehicle deployed with RFID tag that will be read by the roadside unit with its RFID reader. By sensing all the vehicles, the roadside unit will collect information about the density of traffic. The controller processed the information. Information collected by each roadside unit will be uploaded to the central cloud server database.

4.3 RSU Discovery

Vehicle identifies RSU by vehicle-to-roadside communication and vehicle-to-vehicle communication if the vehicle is out of the range of RSU. To detect RSU,

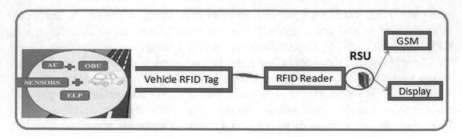

Fig. 4 RSU role in RTIS

Road Side Unit(RSU)	Verifies vehicle is in the range	Checks vehicle registration	Serve any information to vehicle

Fig. 5 Working of RSU

there is a special format for the packet proposed which is identical to ICMP format. Vehicle sends a packet and if RSU is discovered and then it replies in advertisement format. Then after vehicle and RSU can be identified each other in order to exchange messages. The registration phase is very important, in this phase RSU checks whether vehicle is registered or not. In many countries, the government collects some amount from vehicle to maintain the vehicle database. If any vehicle does not pay roadworthy, then puts in blacklist.RSU informs unregistered vehicle information to the police so that they are able to track those vehicles easily. After the registration phase, RSU is ready to serve any information. The last step of this application is data transfer, which is done unidirectional from RSU to vehicle (Fig. 5).

4.4 Abstract View of Road Traffic Information System

Figure 6 shows sensor-cloud integration for conceptualizing and communicating infrastructure called smart transport information system (STIS) services.

4.5 Steps in Sensor-Cloud Implementation for Road Transportation

Sensor-cloud is reliable, dynamic, and feasible system. Earlier technology was similar to the GPS navigation system [8] and can identify the particular location of the vehicle. When monitoring of the vehicle realized using the cloud computing

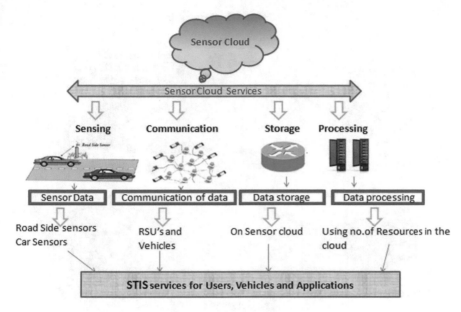

Fig. 6 An abstract view of sensor cloud

technique, it is possible to deploy centralized Web control service, GSM and GPS devices, various sensors, will facilitate the following:

(i)   Current location,
(ii) Time of arrival,
(iii) Distance traveled, and
(iv) Fuel indication.

Data is fetched and stored on the centralized control server as shown in Fig. 7 that is located in the cloud. Data pulled periodically and process [5] it by using resources in the cloud. Finally, raw sensor data is to be converted as traffic-related information (by aggregating all the data from various sensor networks) and the processed results are stored back in the cloud server to access as ITS services.

5   Simulation Results

In this, relative speed and transmission range is simulated the VANET network scheme shown in Figs. 8 and 9. Sensors form a cluster network and selection of cluster coordinator. The coordinator is responsible for collecting, organizing, and publishing all types of sensor data on the cloud.

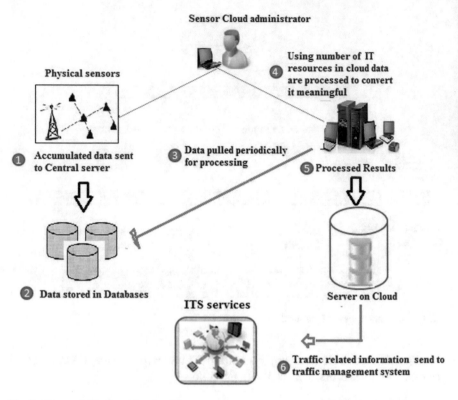

Fig. 7 Sensor data processing model

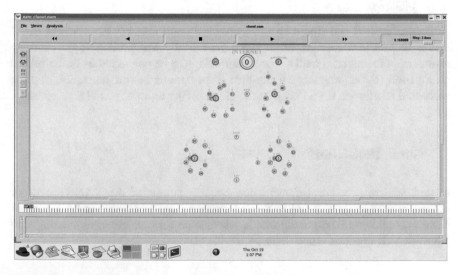

Fig. 8 Cluster network construction and the communication

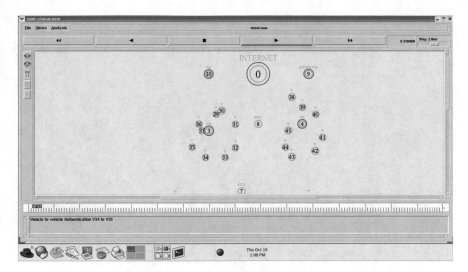

Fig. 9 Network structure of the cluster-based VANET network

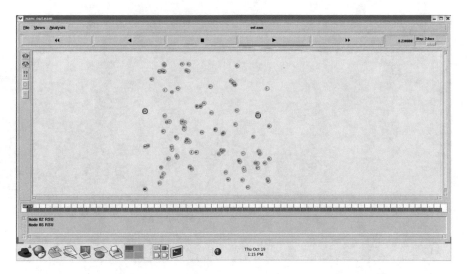

Fig. 10 RSU declarations in the VANET network

RSU plays a vital role in VANET, which is responsible to collect all data about moving vehicles on the road. Figure 10 shows the RSU announcement in the vehicular network.

Figures 11 and 12 show the major function of RSU is checking vehicle registration. If any vehicle is not registered, RSU notified it as a malicious node.

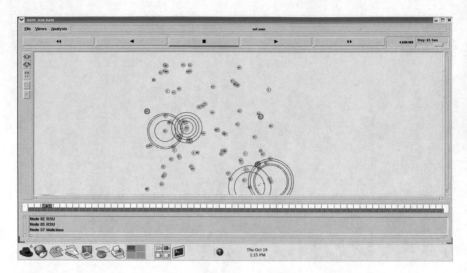

Fig. 11 Communication begins and the first malicious vehicle is detected −57 malicious

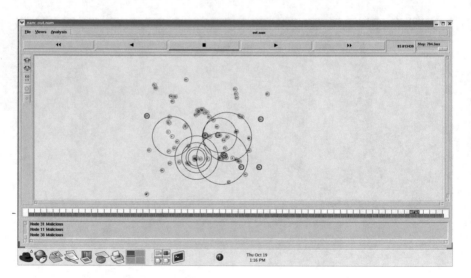

Fig. 12 Malicious vehicles 11, 36 detected

Figure 13 shows the performance parameters such as packet delivery ratio and average end-to-end delay. If the vehicles on the road are traveling in the same lane, they will have enough time to communicate. If they are in opposite directions, they have less time to interact.

The number of delivered packets and overall through put of vehicles are shown in Figs. 14 and 15. With an average interval level >50 ms and between 1 and 100 ms.

Fig. 13 Total performance analysis

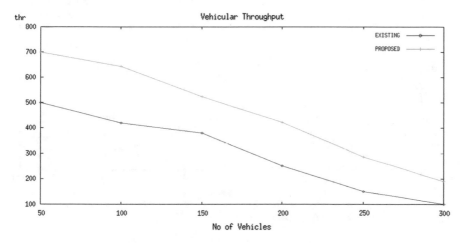

Fig. 14 Number of vehicles versus throughput of the network

We plotted a graph with total time duration 100 ms for sensing and association shown in Figs. 16 and 17. The communication between vehicles majorly associated with sensing time and setup a connection between wireless transceivers.

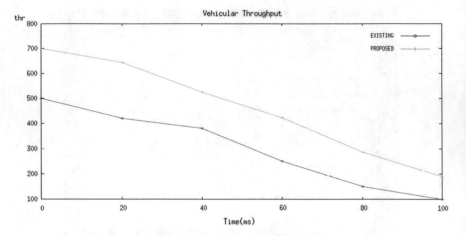

Fig. 15 Number of vehicles versus delivery ratio

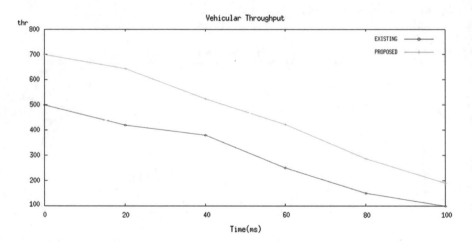

Fig. 16 Time (ms) versus throughput of the network

6 Conclusions

We have presented a realistic sensor cloud service as real-time monitoring and management system for networked sensor systems. The sensor-cloud mechanism facilitates that the data is to be efficiently and securely organized, categorized, stored, and processed. So, it is much cost-effective that data is to be available in time for easily accessing and processing. Our proposed hybrid architecture comprises cloud-service-enabled sensor pods with a variety of sensing features that measure or monitor road traffic. Finally, the sensor-cloud service framework provides real-time- and situation-aware information-sharing services to vehicles on the

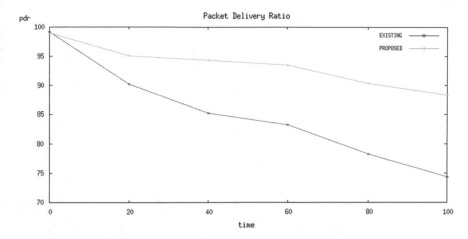

Fig. 17 Time (ms) versus packet delivery ratio of the network

road. It fills the gap between the end-users, i.e. drivers and data providers to take prior decisions before they can face any event on the road. So, we can make sensors easily a part of daily life.

References

1. Rajeesh Kumar, N.V., Kamala Kannan, R., Madhan Kumar, R.: Intelligent parking by merging cloud and sensors. ARPN J. Eng. Appl. Sci. **11**(15) (August 2016). ISSN 1819-6608
2. Rawat, D.B.: Towards intelligent transportation cyber-physical systems: real-time computing and communications perspectives. In: Proceedings of the IEEE SoutheastCon 2015, Fort Lauderdale, Florida 9–12 April 2015
3. Rajapraveen, K.N.: A trusted model for information dissemination in VANET using cloud computing technology. Int. J. Adv. Res. Comput. Sci. **8**(5) (May–June 2017)
4. Sahoo, P.K., Chiang, M.-J., Wu, S.-L.: SVANET: a smart vehicular ad hoc network for efficient data transmission with wireless sensors. Sensors **14**, 22230–22260 (2014). https://doi.org/10.3390/s141222230
5. Krutharth, C.V., Kayalvizhi, S.: Technique of data transfer between road side unit and vehicles in Vanet. IJCTA **9**(13), 6017–6027 (2016) (© International Science Press)
6. Yadav, K.A., Vijayakumar, P.: VANET and its security aspects: a review. Indian J. Sci. Technol. **9**(44) (November 2016). https://doi.org/10.17485/ijst/2016/v9i44/97105
7. Liang, W., Li, Z., Zhang, H.: Vehicular ad hoc networks: architectures, research issues, methodologies, challenges, and trends. Int. J. Distrib. Sens. Netw. **11**, 745303 (2015)
8. Azevedoa, C.L., Cardoso, J.L.: Vehicle tracking using the k-shortest paths algorithm and dual graphs. Transp. Res. Procedia **1**, 3–11 (2014)

Tech Care: An Efficient Healthcare System Using IoT

Murali Mohan Kotha

Abstract It is the dawn of new computing era—Internet of Things (IoT). It consists of seamless blending of sensors, which monitor the environment around us and actuators, that respond to the decisions taken by a strong framework of cloud computing. The lifestyle changes of people have been resulting in increase in their health problems. This demands a need for cyber-physical smart pervasive frameworks involving ubiquitous healthcare system. The health care is one of the number of application domains of Internet of Things. With Internet of Things, many medical applications can be developed such as monitoring health remotely, fitness programs, elderly care, and chronic diseases with an idea to reduce costs, increase quality of life, and enrich the user's experience. Many countries and organizations have been developing policies and guidelines for deploying the Internet of Things technology in medical field. However, the Internet of Things is in its infancy state in the area of healthcare field. Thus, this paper focuses on discussing core technologies that are shaping Internet of Things-based healthcare. Further, the paper provides challenges that must be addressed so that the Internet of Things-based healthcare system becomes robust.

Keywords The Internet of Things · Body Sensor Networks · Smart devices · Smart cities · Healthcare monitoring · Wearable devices · Human activity recognition

1 A Brief Introduction to IoT

It is the dawn of a new era, in which the physical things connect with the Internet. The things could be any device/structure. The things are embedded with sensors, software, electronics, and network connectivity. Then, these devices collect and exchange the data. Such data are analyzed using intelligent systems, and

M. M. Kotha (✉)
Department of Mathematical Sciences, Sree Vidyanikethan Degree College,
Sri Venkateswara University, Tirupati 517 102, India
e-mail: kotha_mm@yahoo.com

© Springer Nature Singapore Pte Ltd. 2020   655
P. Venkata Krishna and M. S. Obaidat (eds.), *Emerging Research in Data
Engineering Systems and Computer Communications*, Advances in Intelligent
Systems and Computing 1054, https://doi.org/10.1007/978-981-15-0135-7_59

appropriate action can be taken, without any human interaction. Thus, IoT is considered the intelligent connectivity of physical devices. It can drive massive gains in efficiency, business growth, and quality of life. The researchers have designed and developed various hardware and software platforms to support IoT applications. Figure 1 depicts the IoT architecture.

The term "Internet of Things" was first coined by Kevin Ashton in the year 1999.

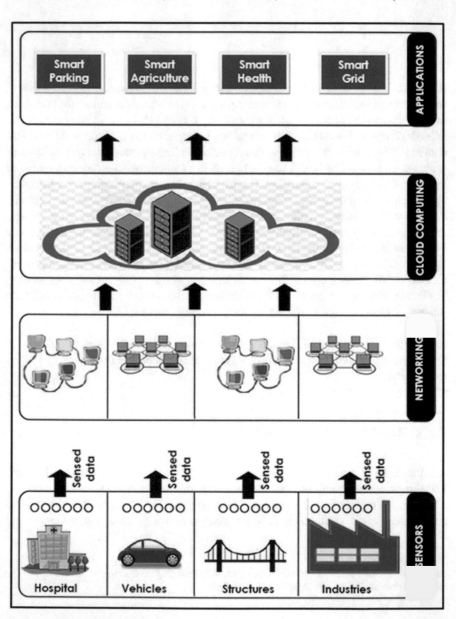

Fig. 1 IoT architecture

2 Healthcare—An Introduction

The diseases are rising largely because of lifestyles of individuals—like smoking, obesity, too much consumption of alcohol, irregular diet habits/timings, and poor physical activities/lifestyles. Cancer, which generally develops from a tumor, can be classified based on the place of tumor, tumor type, shape & size, and the way it grows, etc.

According to Word Health Organization (WHO), Ischamic heart disease and stroke are the world's biggest killers. About 15.2 million died because of these diseases in 2016. Further, three million people lost their lives due to chronic obstructive pulmonary disease, in the same year. The other diseases that are causing concern are lung cancer, diabetes, dementias, lower respiratory infections, and tuberculosis.

3 An Introduction to IoT in Healthcare

The Internet of Things can be integrated with health care. The IoT is a fusion of biological, physical and digital technologies. These features of IoT can improve the quality and effectiveness of service for patients. They provide consistent supervision, especially for chronic disease patients. The IoT devices reliably transmit urgent events, and data in real time, to a remote monitoring center for processing, where such large amount of data is intelligently processed and analyzed to achieve the smart control of objects.

Many pioneering works toward developing healthcare IoT systems are going on. The IoT in other fields already proved that remote monitoring of objects is possible with data collection and reporting. Thus, the same concept can be applied for monitoring health of people and reporting it to the same to doctors, emergency services, hospitals, or caregivers.

4 Advantages of IoT Healthcare System

4.1 Remote Monitoring

It facilitates flexible patient monitoring. Monitoring the patients remotely, when they are at their home is known as tele-monitoring system. The data are sent to remote server. After thorough analysis, it identifies any critical event.

Such remote monitoring is useful to monitor non-critical patients at home, rather than in hospital. This reduces burden on healthcare systems and can have better control over their own health at all times. Further, rehabilitation after injury can be tailored to a patient, based on symptoms.

4.2 Improved Drug Quality

- The drug production involves many critical stages. Several important parameters are to be monitored at different stages. It is required to ensure homogeneity across batches. Further, with IoT, appropriate alerts can be raised, whenever the variables exceed certain set parameters.
- All batches of drugs are compared to see how different each one is.
- Helps in standardization across all batches.
- Minimal human intervention.
- IoT in transit management ensures—safer drug transportation.

4.3 Electronic Health Records

The patients' health records can be preserved in electronic format, which will be made available to the doctors, nurses, or caregivers, when required.

4.4 Avoiding Adverse Drug Reaction

The adverse drug reaction (ADR) means unwanted, uncomfortable, or dangerous effects that a drug may have. The ADR can be considered as a form of toxicity.

At the time of giving any medicine to the patient, the IoT device can sense the compatibility of the drug using allergy profile and patient's electronic health records.

4.5 Access to Wearable Devices

Several wearable devices, are available, which continuously monitor body temperature, BP, heart rate, track runs, swims, rides, workouts, all day activity, and sleep, etc. They are extremely useful in providing accurate and reliable information on people's activities and behaviors. Such continuous monitoring of physiological parameters is significant, especially of the elderly or chronic patients. The necessary help can be provided to them, in times of dire need.

The special purpose sensors such as blood glucose monitors and joint angle sensors could also be implanted in humans to observe the problems of specific types of patients.

5 IoT Healthcare Applications

5.1 Sugar Level Monitoring

Diabetes is a group of metabolic diseases, in which there are high blood glucose (sugar) levels over a prolonged period. The sugar level of the patient is to be monitored to maintain it at specific level and also to plan food and medicine timings.

5.2 Blood Pressure Monitoring

The blood pressure is one of the important physiological parameters of the human body. The IoT device, a wireless one, displays the values. It can be synchronized with mobile health apps, which can monitor and maintain a record of values.

5.3 Body Temperature Monitoring

There are several varieties of methods to measure body temperature. The method which is more accurate one is to be chosen, to obtain precise values.

5.4 ECG—Electrocardiogram Monitoring

It is the electrical activity of the heart recorded by electrocardiography, which includes the measurement of heart rate and determination of the basic rhythm and several other key indicators. Whenever the values go beyond certain limits, an alert can be sent to doctors/caregivers, with which they can reach the patient for quick medication.

5.5 Blood Oxygen Saturation Monitoring

Using a wearable pulse oximeter, the blood oxygen saturation can be monitored.

5.6 Rehabilitation System

It is a vital branch of medicine, in which a plan can be prepared to enhance and restore the functional ability and quality of life of patients.

5.7   Medication Management

Some of the medicines may be non-compliant to the patients. It may damage the patient health and may even result in financial loss. This can be addressed using IoT. Using intelligent package methods, electronic health records, and IoT's sensing mechanisms, the adverse effects of medicines can be avoided.

5.8   Smartphone for Healthcare Solutions

Many electronic devices are coming up with smartphone-controlled sensor. With this, the smartphone can act as a driver of the IoT. A number of software and hardware are designed to make smartphone—a versatile healthcare device. A number of applications for health care, medical education, information search, etc. are developed. The diagnostic apps can be used to access diagnostic and treatment information. The drug information apps provide names of drugs with their indications, dosages, and costs. The medical education apps provide tutorials, surgical demonstrations, illustrations, etc.

A number of smartphone apps can provide healthcare solutions, like detecting asthma, chronic obstructive pulmonary disease, and several others.

5.9   Robotics in Healthcare

Robotics has been evolving as a challenging innovative field of research among industries, academia, and scientists. A robot is a machine, which is capable of doing things, autonomously. It possesses more degree of freedom than humans. The robots are considered as "things" that get integrated with IoT. Such things then use the Internet and establish connection with other things. A robot can perform cardiac, orthopedic, urology, gastroenterology, gynecology, pediatrics, and other surgical procedures. They have excellent geometric accuracy, which enables doctors to perform many types of complex surgeries, with greater precision, flexibility, and control, which is difficult/impossible with traditional methods.

The Food and Drug Administration (FDA) approved robot surgeries using da Vinci surgical system. Such a system is widely adopted by hospitals in several countries, including India. The main advantages are minimal surgical site infection, minimal blood loss, less tissue damage, low pain, quick recovery, and small scars which are less noticeable.

It is estimated that only 5% of surgeries are carried out by robots. Thus, there is a vast scope for penetration of robot surgeons.

5.10 Ambulance Telemetry

In case of emergency, the patient is shifted to hospital in an ambulance. Such ambulances, nowadays, are fitted with ambulance telemetry, which monitors and transmits vital data of patient. Such data can be made available to the doctors at hospital, with which they can plan correct treatment by the time the patient arrives. This is really useful in life-threatening emergencies.

5.11 Pills for Diagnosis

The pills which are embedded with sensors, when consumed, capture the vital health parameters and send the data to wearable devices. They, in turn, send the same to cloud, where the healthcare service providers analyze the data for further procedures/medication.

5.12 Smart Pill Bottles

They recognize and record when the patient has missed to take the medication. They even alert caregivers, by sending messages. This ensures taking medicines, at right time, which ensures well-being of patients.

5.13 Pharmacy Management

For efficient functioning of operations within a pharmacy/medical shop, RFIC technology can be used. The drug stock level can be monitored. This feature is useful for ensuring availability of rarely administered drugs—which have short shelf life and expensive too. The process includes automation of restocking process. This technology even helps in monitoring expiry dates of medicines. The climatic conditions inside the refrigerator can be monitored, where the medicines are preserved in cold conditions.

5.14 Tracking Hospital Equipment

The location of hospital equipment like stretchers, beds, ECG or X-ray machines, and ventilators can be monitored and can be made available when required. This helps in optimal utilization of the equipment, since these equipments are reused/

shared by several departments/specializations. This technology not only helps in tracking location of objects, but also hints status of them—like whether in switched on/off state, sterilized/unsterilized, and available for use or already in use. This really helps in effective use of the available resources. When equipment is out of order, the timely maintenance can even be planned.

6   Enabler Technologies in IoT Healthcare

The IoT technologies make use of the following enabling technologies, in general for it to work and succeed.

6.1   Communication Standards

The IoT healthcare network is very essential component. It acts as backbone and facilitates transmission of patients' data and requires a communication system that suits the needs of healthcare.

A number of wearable sensors deployed on human body collect large volumes of data. After edge computing, the data is sent to the cloud. The decisions taken, can be accessed by the doctors or hospitals to plan necessary action.

Thus, communications are really important in IoT health care and are classified into the following two categories:

Short-Range Communications

These are used for communication between the things (sensors) and long-range communications. The most popular and commonly used short-range communications technologies are Bluetooth Low Energy (BLE) and ZigBee.

The Bluetooth Low Energy has a range of 150 m and has a low latency of 3 ms and a high data range of 1 Mbps. The power consumption in BLE is extremely low. The BLE chip can run for 18 h, with a 180 mAH coin cell battery. Thus, if a healthcare sensor sends data at every 30 s, (i.e., 2880 times per day), then 180 mAH battery could have life for around 20 years. Even very tight security has also been implemented in BLE. Encryption, using 128-bit AES cipher, is available to protect the data.

The ZigBee has a range of just 30 m and has a data range of 250 kbps. An optional AES encryption is also available.

Long-Range Communications

Several IoT applications require long-range/distance operation. Such applications can take the advantage/capabilities of several communication technologies like GSM/3G/4G cellular. The cellular can send high quantities of data, especially 4G. However, it requires more power and expensive also. The range is about 35 km

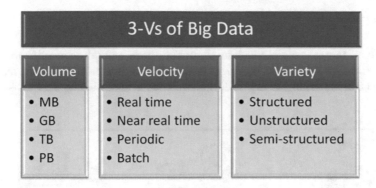

Fig. 2 Three Vs of big data

max. for global system for mobile communications (GSM). The data rates (typical download) are 3–10 Mbps in long-term evolution (LTE).

6.2 Big Data

The millions of healthcare sensors generate large volumes of data at high velocities. Further, the data may be of different varieties. The big data is generally characterized by three Vs—volume, velocity, and variety. Figure 2 depicts the same.

6.3 Cloud Computing

The sensors' computational power is very less. Thus, by following computational offloading, the complex data processing can be performed using cloud. After processing, the meaningful results are sent back to the patient/doctor/hospital/caregiver. Further, machine learning can also be applied to obtain meaningful information, from which the further course of action can be planned. The three fundamental services provided by cloud technologies are:

(1) SaaS—software as a service—which provides required applications to the intended parties.
(2) PaaS—platform as a service—which provides required tools for virtualization, database management, and operating systems.
(3) IaaS—infrastructure as a service—this provides physical infrastructure like processors (servers), storage, and several others.

6.4 Nanotechnology

It involves things' miniaturization. Scientists have started reducing the size of sensors from millimeters or microns to the nanometer scale. They are so small, that they can mingle within living bodies and can even mix directly into construction

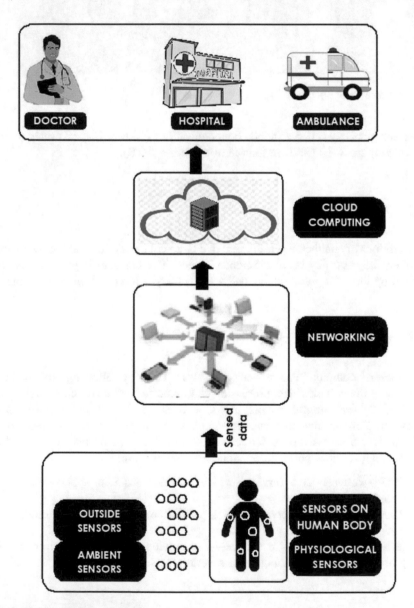

Fig. 3 Model of IoT-based healthcare system

materials. It is believed to influence many sectors—medicine, military, electronics, food processing, etc. A research is going to deliver the medicine only to the affected organ/location in the body, using nano-sized robots.

Scientists have even created DNA origami nanorobots. They cut off the supply of blood to cancerous tumors, which result in the shrinking of tumors. It is a novel way of cancer therapy. Each such nanorobot is of size 90 by 60 nm.

6.5 AI and Automation

The delivery of smart health care starts with electronic health records (EHRs). With EHRs, errors can be reduced, patient interaction can be increased, and such gathered data over a period of time provide better understanding of several aspects of patient. Further, such data can be analyzed using AI, to improve outcomes. Artificial intelligence means prediction using algorithms and data. A machine learns from the past data using intelligence.

Smart health care not only helps in curing the disease, but also aids in preventing it through early detection and prediction. The diagnosis and treatment data are stored and analyzed, using AI. Then, preventive measures can be taken. Many countries basically involve in reactive treatment but not in preventive care. Thus, several countries, including India, have been rolling out their strategies for adoption of AI in health care.

Figure 3 shows summary of all these.

7 IoT Healthcare Challenges

The IoT industry is evolving every day. Further, healthcare sector is expected to witness the widespread adoption of IoT. The IoT devices and applications contain sensitive data of patient— patient details, disease particulars, treatment offered, future course of action, etc. Since the devices are connected to the Internet, the IoT healthcare device may be a target of a hacker. Thus, a set of challenges are listed below:

7.1 Confidentiality

All the healthcare devices get connected to the Internet. Thus, such devices may be a target of hackers. Thus, confidentiality is to be maintained (anonymity).

Another promising emerging technology—blockchain—is believed to provide a solution to this problem. Several businesses want coexistence of distributed things (IoT) and distributed ledger (blockchain). By merging both of them, a secured and verifiable data recording can be achieved. Such data can be processed by smart machines; so that the interconnected devices can interact with external world and can take decisions, without any human intervention.

7.2   Empowered Users

The users must get the latest data, when required. Further, the data should not have been altered. Thus, integrity of data is to be ensured, so that the received data are not changed in transit. Further, the IoT healthcare system should be available, even under denial-of-service (DoS) attack.

7.3   Limited Resources

It is well known that, an IoT device consists of low computational power and limited memory. It just acts as a sensor or actuator. Thus, computational offloading is implemented. Further, heavyweight protocols may not run in this environment.

Further, the IoT device consists of limited battery power. Thus, the devices should be kept switched off or to be turned to power saving mode.

7.4   Cross-Platform Security

Since an IoT healthcare device moves around with patient, it needs to get connected to different types of networks. Hence, it needs security during different connections.

7.5   Standardization

Several vendors manufacture different kinds of sensors/devices. Further, new players will also enter, very often, since it is an emerging field. Then, having no standard rules and regulations, there may be issues related to protocols and interfaces. Thus, it results in interoperability issues. In addition, backward compatibility is required.

7.6   Mobility

The IoT healthcare system should permit mobility of patients. However, patients remain connected to different networks, using mobility specifications.

8 Conclusion

The Indian healthcare system is lagging behind in numerous health indicators. India is the world's second most populous country, but accounts for 20% of the global disease burden. To address these situations, one can think of IoT for healthcare purposes.

The whole idea of IoT-based health care is to take care of patient's health, at very low cost. Further, it ensures that patients move out from traditional method of visiting hospitals, for minor problems, where remote monitoring can take care of patients. It is aimed to provide seamless healthcare system to the patients. In addition, the adoption of Artificial Intelligence (AI), Internet of nano-things (IoNT) and Internet of medical things (IoMT) by healthcare providers can improve care delivery, customer experience, and operational excellence.

IoT health care suffers from inherent characteristics of IoT. Thus, lightweight protocols, interoperability, scalability, and security issues are to be addressed to. Substantial R&D efforts have to be made in IoT healthcare services and applications, so that it reaches masses in more secured way and improves their quality of life.

References

1. Gope, P., Hwang, T.: BSN-Care: a secure IoT based modern healthcare system using body sensor network. IEEE Sens. J. **16**(5), 1368–1376 (2016)
2. Yuehong, Y.I.N., Zeng, Y., Chen, X., Fan, Y.: The internet of things in healthcare: an overview. J. Ind. Inf. Integr. **1**, 3–13 (2016)
3. Zhou, J., Cao, Z., Dong, X., Vasilakos, A.V.: Security and privacy for cloud-based IoT: challenges. IEEE Commun. Mag. **55**(1), 26–33 (2017)
4. Zhu, N., Diethe, T., Camplani, M., Tao, L., Burrows, A., Twomey, N., Kaleshi, D., Mirmehdi, M., Flach, P., Craddock, I.: Bridging e-health and the internet of things: the SPHERE project. IEEE Intell. Syst. **30**(4), 39–46 (2015)
5. Kotha, M.M.: Blockchain for trusted future. Int. J. Comput. Sci. Eng. **7**(2), 655–658 (2019)
6. https://www.who.int/gho/mortality_burden_disease/en/. Accessed 23 Apr 2019

Introduction to Concept Modifiers in Fuzzy Soft Sets for Efficient Query Processing

Chandrasekhar Uddagiri and Neelu Khare

Abstract This paper presents hedges also known as concept modifiers on fuzzy soft sets. Hedges allow close ties to natural language and also allow query processing on the latest data repositories like data warehouses, big data and cloud data. Data repositories may require complex linguistic queries and at the same time efficient query processing also, while the requirements are vague and uncertain in natural languages. SQL is not able to handle such complex queries. This requires an additional application layer for computing hedges and defuzzification process. Hence, the presented framework scores a lot in terms of efficiency as well as the ability to handle linguistic queries along with aggregate operators.

Keywords Fuzzy soft sets · Hedges · Linguistic variable · Concept modifiers · SQL

1 Introduction

Database query formats are very crisp and rigid and cannot be used for practical purpose to handle vague queries as follows:

'Retrieve the details of "*young*" terrorists who were involved in "*recent*" bomb blast "*in and around*" Jammu and Kashmir'.

The parameters such as *young*, *recent* and *around* are very vague and require a soft approach to handle such queries. Natural language consists of atomic terms and adjectives to describe those terms.

C. Uddagiri
CSE Department, BVRITH, Hyderabad, India
e-mail: chandrasekhar.u@bvrithyderabad.edu.in

N. Khare (✉)
SITE, Vellore Institute of Technology, Vellore, India
e-mail: neelu.khare@vit.ac.in

© Springer Nature Singapore Pte Ltd. 2020
P. Venkata Krishna and M. S. Obaidat (eds.), *Emerging Research in Data Engineering Systems and Computer Communications*, Advances in Intelligent Systems and Computing 1054, https://doi.org/10.1007/978-981-15-0135-7_60

- The terms are called as parameters
- The adjectives that describe the terms are called as linguistic hedges.

Hedges can be used to establish the relationship between the parameters so as to minimize or maximize the choice of objects. Using linguistic hedges improves the system's objectivity [1].

Example 1 Chose parameter values such a way so as to maximize the student pass percentage.

Example 2 Minimize the choice of houses to be selected.

Le and Tran [2] show how to build linguistic fuzzy logics based on the axiomatizations and a linear hedge algebra. Many hedges are often used simultaneously to express different levels of emphasis.

1.1 Linguistic Hedges in Zadeh's Fuzzy Logic

Mathematical tools to deal with uncertainties are as follows:

1. Probability
2. Fuzzy sets
3. Interval mathematics.

In the early 1970s, Zadeh [3–7] developed and introduced the theory of approximate reasoning using fuzzy logic and linguistic variable. A linguistic variable is defined as an entity whose values are descriptions and adjectives used in a natural language. Consider a linguistic variable called 'Age'. The primary terms used to describe age are 'young' and 'old'. The linguistic terms used to describe these basic terms are 'young', 'quite young', 'more or less young', 'not very young', 'very young', 'old', 'very old', and 'not very old', etc. The goal of approximate reasoning or concept modifiers is to imitate languages as used by human beings. This is very useful while working with human interacting systems.

This paper describes fuzzy logic as described Zadeh. Its truth-values are linguistic values as applicable to that variable. The truth-values are represented using fuzzy sets where each value falls within an interval of [0, 1].

According to Zadeh's rule for truth qualification [4], a sample statement such as 'Alia is very young' is semantically equivalent to the statement 'Alia is young is very true'. Approximate reasoning needs such semantic equivalence relations. In fuzzy reasoning [4, 8], the truth-values 'True' and 'False' fall over the interval [0, 1]. The fuzzy membership values will determine the meaning of the primary terms. To compute the composite linguistic truth-values, we can use any of the two methods given below [9].

- Linguistic modifiers such as 'very', 'more or less' and 'significantly' are considered as unary operators such as concentration and dilation. Ex: CON-very, DIL-slightly, etc. Ex: $\mu_{\text{slightly } A}(x) = \mu_A(x)^{0.5}$ and $\mu_{\text{very } A}(x) = \mu_A(x)^2$
- We can use logical connectives on these terms such as 'and', 'or', 'if-then-else' using operators such as implication, negation, t-norm and t-conorm.

Figure 1 shows, diagrammatically, the relationship between true and false terms along with concept modifiers 'fairly' and 'very'. Figure 2 shows the hierarchical structure of various terms and concept modifiers for linguistic term called AGE.

Following are some of the sample linguistic hedges used:

i. Very $a = a^2$
ii. Very Very $a = a^4$
iii. Plus $a = a^{1.25}$
iv. Slightly $a = a^{0.5}$
v. Minus $a = a^{0.75}$

1.2 Fuzzy Soft Sets

The membership functions look similar to weights associated with those variables. This is one of the drawbacks of fuzzy sets. Zadeh's FUZZY SET may be considered as a special case of SOFT SET. A soft set is a parameterized family of sets. Soft set was proposed in 1999 by Molodtsov. Soft sets were able to deal with uncertainty in a parametric manner [10]. The chosen parameters determine the boundary of the soft set. Hence, it is called 'soft'.

Fig. 1 Membership functions of unitary truth-values

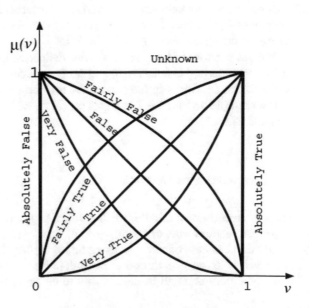

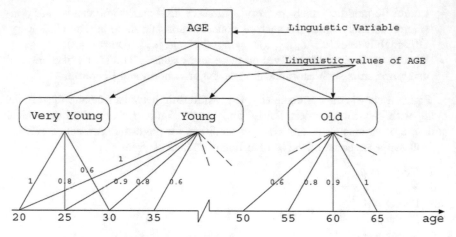

Fig. 2 Hierarchical structure for parameter called AGE

Fig. 3 Soft set describing four candidates with computer skills

$$
\begin{array}{c}
U \\
o_1 \\
o_2 \\
o_3 \\
o_4
\end{array}
\begin{array}{ccc}
low(a_1) & medium(a_2) & high(a_3) \\
\left[\begin{array}{ccc}
0.5 & 0.2 & 0.1 \\
0.1 & 0.8 & 0.1 \\
0.1 & 0.2 & 0.6 \\
0.2 & 0.25 & 0.3
\end{array}\right]
\end{array}
$$

A soft set, over a universal set U and set of parameters P is a pair (f, e) where e is a subset of P, and f is a function from e to the power set of U. For each e_i in e, the set $f(e_i)$ is called the value set of e_i in (f, e).

Various versions of fuzzy soft sets were defined to fuzzify the parameters such as fuzzy parameterized soft sets (fps) and fuzzy parameterized fuzzy soft sets (fpfs). It is possible to fuzzify both the parameters and the values of each parameter.

For example, in Fig. 3 is fuzzy soft set (F, A) which represents 'candidates with computer skills'.

2 Proposed Framework

There is an additional middle which extracts and translates the data from database as per the requirements of the fuzzy SQL query [11]. The underlying database is mostly crisp. In case of soft sets, as discussed earlier, the data can be both crisp and fuzzy values. Hence, fuzziness can be both in the soft sets as well as the query. The user in the front end will be able to choose the linguistic variable on the necessary domains, which are predefined. There is a pre-processing step involved on the database before the query can be implemented.

The first step is to identify the fuzziness that can be associated with the different attributes. For example, the previous example has illustrated that a parameter called 'technical skills' was fuzzified into terms low, medium and high, and membership functions have been defined.

Second step involves the construction of soft sets by identifying all the attributes that can be fuzzified [12]. As discussed in the previous section, various kinds of operators such as aggregation operators can be used to club different parameters to construct new parameters.

Third step involves computation of manipulated membership values based on the concept modifiers chosen. There is a metadata required to be stored to select various kinds of hedges and various kinds of defuzzification techniques that can be applied. For example, we can use Alpha, Beta, Delta or Gamma cuts for defuzzification.

Finally, defuzzification process is carried out by applying the required defuzzification cut. Mostly, Alpha (α) cut is used.

2.1 Pre-processing

As discussed earlier, there is a pre-processing step (Fig. 4) and query implementation step. The pre-processing step involves preparing the soft sets or fuzzy soft sets, preparing a table for various hedges/concept modifiers, preparation of metadata table and finally preparing a table which identifies various cuts for defuzzification (Table 2).

2.2 Meta Information Table

The following Table 1 stores the description of meta information tables. There is a table that stores various linguistic hedges discussed in Sect. 1.1, and finally Table 2 is used to store meta information. This table contains the thresholds defined for various cuts (defuzzification)

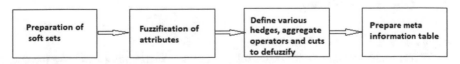

Fig. 4 Block diagram for pre-processing

Table 1 Description of meta table

Meta information	Description
Linguistic_Term	To store the name of the fuzzy parameter
Table_name	Soft set used
Column_name	Attribute associated with the fuzzy soft set
Alpha (a)	Lower range of SUPPORT of the fuzzy set
Beta (b)	Lower range of CORE of the fuzzy set
Delta (d)	Upper range of CORE of the fuzzy set
Gamma (g)	Upper range of SUPPORT of the fuzzy set

Table 2 Meta information table

Linguistic term	Table Name	Parameter	Alpha	Beta	Gamma	Delta
Cost	Houses	Expensive	0.5	0.55	0.9	0.8
Low	Houses	Expensive	0.3	0.35	0.7	0.6
Medium	Houses	Expensive	0.4	0.45	0.6	0.5
High	Houses	Expensive	0.5	0.55	0.5	0.4
Beautiful	Houses	Beautiful	0.8	0.9	0.99	0.95

$$A \quad \begin{matrix} expensive & wooden & beautiful & good-surroundings \end{matrix}$$

$$\begin{matrix}
o_1 \\ o_2 \\ o_3 \\ o_4 \\ o_5 \\ o_6
\end{matrix}
\begin{bmatrix}
0.3 & 0.4 & 0.6 & 0.9 \\
0.3 & 0.9 & 0.3 & 0.5 \\
0.4 & 0.5 & 0.8 & 0.7 \\
0.8 & 0.2 & 0.4 & 0.8 \\
0.7 & 0.3 & 0.6 & 0.5 \\
0.9 & 0.2 & 0.4 & 0.3
\end{bmatrix}$$

Fig. 5 Soft set describing six houses on various parameters

Consider the following fuzzy soft set A in Fig. 5 describing a set of houses and their properties. This soft set describes the feature of various houses in the data base. Some of those features such as expensive and beautiful can have hedges defined on them as described in the meta information.

2.3 Query Processing

A sample query can be of the following form.

'Select houses that are very beautiful and in good surroundings'. Using the data given in Fig. 5, we can compute the values for hedges and use say Alpha cut to defuzzify. Then, the sample output will look as shown in Fig. 6.

$$
\begin{array}{c|cccc}
A & expensive & wooden & beautiful & good-surroundings \\
\hline
o_1 & 1 & 1 & 1 & 1 \\
o_2 & 1 & 1 & 0 & 0 \\
o_3 & 1 & 1 & 1 & 1 \\
o_4 & 1 & 1 & 0 & 1 \\
o_5 & 1 & 1 & 1 & 0 \\
o_6 & 1 & 1 & 0 & 0 \\
\end{array}
$$

Fig. 6 Defuzzified output soft set

From the query, we can see that the aggregate operator used is 'AND' defined on the parameters 'Beautiful' and 'Good Surroundings'. There is no mention of remaining parameters. Hence, they do not matter in the selection of the houses.

It can be observed that object set {o1, o3} alone satisfies the criteria. Hence, the query retrieves only two rows in its results.

3 Conclusion

Soft sets are known for efficient query processing, especially in this case where there are lot of operators that are needed to be applied in each stage starting from pre-processing, application of hedges and defuzzification. The query can be in a fuzzy or linguistic format, and the middle tier takes care of translating the query accordingly.

Generally, the databases are crisp in nature, and there is an additional step of fuzzification that is required during pre-processing. However, it is a one time job. Hence, the framework scores a lot in terms of efficiency as well as the ability to handle linguistic queries along with aggregate operators.

References

1. Ibrahim, A.: Enhanced fuzzy system for student's academic evaluation using linguistic hedges. IEEE (2017). 978-1-5090-6034-4/17/$31.00
2. Le, V.H., Tran, D.K.: Extending fuzzy logics with many hedges. Fuzzy Sets Syst. https://doi.org/10.1016/j.fss.2018.01.014, https://doi.org/10.1016/j.fss.2018.01.0140165-0114/ ©2018 Elsevier
3. Zadeh, L.A.: A fuzzy-set-theoretic interpretation of linguistic hedges. J. Cybern. **2** (1972)
4. Zadeh, L.A.: A theory of approximate reasoning. In: Yager, R.R., Ovchinnikov, S., Tong, R. M., Nguyen, H.T. (eds.) Fuzzy Sets and Applications: Selected Papers by L.A. Zadeh, pp. 367–411. Wiley, New York (1987)

5. Zadeh, L.A.: The concept of linguistic variable and its application to approximate reasoning (I). Inf. Sci. **8**, 199–249 (1975)
6. Zadeh, L.A.: The concept of linguistic variable and its application to approximate reasoning (II). Inf. Sci. **8**, 310–357 (1975)
7. Zadeh, L.A.: The concept of linguistic variable and its application to approximate reasoning (III). Inf. Sci. **9**, 43–80 (1975)
8. Bellman, R.E., Zadeh, L.A.: Local and fuzzy logics. In: Klir, G.J., Yuan, B. (eds.) Fuzzy Sets, Fuzzy Logic, and Fuzzy Systems: Selected Papers by L.A. Zadeh, pp. 283–335. World Scientific, Singapore (1996)
9. Huynh, V.N., Ho, T.B., Nakamori, Y.: A parametric representation of linguistic hedges in Zadeh's fuzzy logic. Int. J. Approximate Reasoning **30**, 203–223 (2002)
10. Molodtsov, D.A.: Soft set theory—first results. Comput. Math Appl. **37**(4), 19–31 (1999)
11. Balamurugan, V., Senthamarai Kannan, K.: A framework for computing linguistic hedges in fuzzy queries. IJDMS **2**(1) (2010)
12. Chandrasekhar, U., Mathur, S.: Decision making using fuzzy soft set inference system. In: Smart innovations, Systems and Technologies. Springer Book Series (ISBCC-16), pp. 445–457

Smart and Efficient Health Home System

Balzhan Azibek, Shynar Zhigerova and Mohamamd S. Obaidat

Abstract Doing regular medical checkup is important for elderly population in order to stay healthy and avoid unexpected fatal consequences. Some elderly people could find health monitoring as problematical due to aging reason, therefore it is essential to develop a smart home health system. This system aims to be an affordable, cost-effective, and easy-to-use. The proposed smart health system has combined remote medical monitoring, reminding and fall detecting features. The proposed system prototype allows to observe health state of elderly people at a distance, and call an emergency by sending email notifications in case of detecting a fall or abnormal health rate vitals. Another advantage of the smart health system is safety, the design includes gas sensors and a camera that can detect occurrence of fire and unusual events (thieves entering). Moreover, the smart system includes the feature to send reminding notifications about daily routine tasks, which is crucial for elderly citizens with amnesia/memory loss. It is expected that the system saves money and time compared with in-clinic care services.

Keywords Smart homes · Intelligent systems · Efficient health homes · Microcomputer-based systems

M. S. Obaidat—Fellow of IEEE.

B. Azibek · S. Zhigerova · M. S. Obaidat (✉)
Department of ECE, Nazarbayev University, Nur-Sultan, Kazakhstan
e-mail: m.s.obaidat@ieee.org; msobaidat@gmail.com

B. Azibek
e-mail: balzhan.azibek@nu.edu.kz

S. Zhigerova
e-mail: shynar.zhigerova@nu.edu.kz

M. S. Obaidat
College of Computing and Informatics, University of Sharajah, Sharajah, UAE

KASIT, University of Jordan, Amman, Jordan

University of Science and Technology Beijing, Beijing, China

© Springer Nature Singapore Pte Ltd. 2020
P. Venkata Krishna and M. S. Obaidat (eds.), *Emerging Research in Data Engineering Systems and Computer Communications*, Advances in Intelligent Systems and Computing 1054, https://doi.org/10.1007/978-981-15-0135-7_61

677

1 Introduction

Today in most countries life expectancy has increased significantly. This improvement is achieved by advancements in medical science and technology, as well as rising awareness about personal and environmental hygiene, health, nutrition, and education [8]. On the other hand, increased life expectancy resulted in the aging of the population. Based on the statistics of the World Health Organization (WHO), the elderly population over 65 years may exceed the number of children under the age of 14 by 2050 [8]. This elderly population needs regular health-care. However the cost of these health-care services is rising and it is associated with the price of drugs prescription, diagnostic analysis, and in-clinic care services. Besides daily routine duties, even making a trip to medical centers and visiting doctors are hard for people with disabilities due to their limited mobility and independence. Therefore, it is essential to develop affordable, cost-effective and easy-to-use health-care smart home system in order to cut on expenses and offer services in a reliable and safe manner. The smart system can be defined as the implementation of the Internet of Things [1, 3, 6, 9]. Thus, this chapter aims to develop such smart home health system for elderly and people with disabilities. Utilities in this chapter are divided into 3 categories: remote health monitoring system, reminding system, and fall detection system. The remote monitoring system takes measurements of vitals such as heart rate and body temperature. Reminding system is being developed for people with problems of memory, which is also common among senior citizens. The fall detection system is based on smartphone embedded motion sensors. Furthermore, the system will try not only diagnostic health monitoring, but also consider some emergency cases and quick response for them such as notifying the ambulance and personal doctor and family. The proposed system is expected to be cost-effective as well as reliable and efficient.

The rest of the chapter is organized as follows. Section 2 provides the related work review information about smart health systems available. Section 3 presents the main architecture of proposed system and its main features, while Sect. 4 describes the hardware implementation, and Sect. 5 reveals and discusses the results. Finally, Sect. 6 concludes the work and suggests future improvements.

2 Literature Review

The development of medicine definitely will lead to an increase in life expectancy, well-being and healthy state of the population around the world. Indeed, these improvements may result in a significant rise in the cost of health-care services. Many elderly people may suffer from financial difficulties living under fixed budget conditions [2, 7]. In other words, the elderly population may not afford qualitative and expensive in-clinic health care service.

Moreover, diabetes, chronic diseases and cardiovascular diseases are the most common health problems among these elderly citizens. To illustrate, almost half of the American adults suffer from multiple heart diseases, chronic diseases, and account for 65% of mortality consequences [8]. In addition, heart diseases are the first leading cause of mortality among the aged population. Furthermore, if such diseases are not managed properly and monitored regularly on time, they may cause undesired fatal consequences.

Therefore, it is important to develop a smart health system that is affordable and adapted to elderly people.In smart home systems, the environmental and bio-sensors, actuators are embedded together through an advanced intelligent software and communication technologies [6, 9]. Such smart health systems may allow regular monitoring of elderly health at a low-cost [2, 7]. By using such smart systems, the elderly and people with disabilities can stay in their comfortable home environments instead of using expensive and limited health-care facilities.

In the e-healthcare system, every prescription, record, medical history will be automated and stored in database [2, 7, 11]. The system is useful in terms of gathering information about patients for caregivers, emergency medical services and family doctors. The infrastructure of E-health may include health-care personnel, hospitals clinics, and pharmacy. Data transfer between them is possible anywhere at any time [7]. Therefore e-health allows health-care services to be connected and perform the work with minimal error. Hence, this proposed smart system will model the concept of E-health system.

Memory and cognitive function in the older adults declines gradually with age; [13, 17] causing many elderly people to suffer from severe memory loss and dementia. This loss may have fatal consequences. For example, elderly people may forget to take medicine or takes higher doses than prescribed. In addition, the adult population may forget to turn off the oven after cooking. Such little daily routines may cause significant unexpected consequences [17]. Therefore, a reminding system would be very useful for the elderly in their daily life. It is important to note that most of elderly people may suffer from Alzheimer disease, and main symptom is associated with memory loss [10]. Therefore it is argued to include reminding system as a part of the proposed smart health system.

The proposed smart home systems is expected to allow the elderly citizens to live an independent life in their homes while ensuring maximum comfort, safety, and security. As for security it is suggested to include safety feature such as fire detection. The concept presented in [12] can be applied in smart health system. Some gas sensors can be added into design of the smart system to advance it in terms of safety. Safety purpose can also include automated activity and fall detection systems. Falls are one of the leading causes of injuries and death among the elderly [4, 5, 15]. When the system detects a fall, it will inform the corresponding personnel by triggering an alarm [15]. The literature review allows to identify the fittest design of the proposed smart health system by considering all needs of the elderly citizens.

3 Smart Health System Architecture

Based on the literature review that was done, it is argued that most elderly citizens have an issue with a medical checkup in clinics, fall cases at home, and forgetting to take medication on time. Therefore, the smart system proposed in this chapter has three main objectives/features:

1. Remote Monitoring System
2. Reminding System
3. Fall Detection System

3.1 Remote Monitoring System

The main goal of the proposed system is to give the possibility for senior citizens to check health state in their homes. From the literature review it is known that most elderly citizens have cardiovascular diseases [8]. The elderly people diagnosed with heart diseases are needed to regularly check for heart rate, heart pressure, ECG and body temperature. These vital signals can be processed using special vital sensors. The proposed smart health system uses such vital sensors as pulse sensor and body temperature sensor:

- *Pulse Sensor*. Measuring human pulse rate gives the opportunity to detect a wide range of emergency conditions, such as cardiac arrest, pulmonary embolisms, and vasovagal syncope [8]. The Pulse Sensor is a heart rate sensor that allows monitoring of the human pulse rate in real time. The front side of the sensor makes contact with skin and it has an ambient light sensor whose role is to adjust the screen brightness in different light conditions. The LED on front size will be turned on in case of contacts with fingertip or earlobe based on the brightness rate. Pulse sensors operation is based on the working principles of the LED while transmitting light into the artery, thus the sensor calculates the pulse rate by an amount of light absorbed by the blood.
- *Temperature Sensor*. The body temperature is the first thing that has to be measured in order to define the health state of individuals. To measure temperature, waterproof digital temperature signal is used. The Temperature sensor measures the temperature through an electrical signal by a thermocouple. The thermocouple can generate a voltage proportional to change in the temperature. As a result, the change in voltage can be associated with temperature value.

The health system aims to send the results of the vital sensors to physicians and/or relatives of senior citizens. If some of the vital signals will be more or less compared to the threshold values, the system will consider the case as an emergency. In such emergency cases, an emergency team (ambulance, police and/or fire department) will be called by sending email notifications.

3.2 Reminding System

As the main users are senior citizens, based on their requirements the reminding system was proposed. Recorded reminders are sent according to the schedule made by physician or relatives, for instance: (a) take medical pills, (b) go to the restroom, (c) take breakfast, lunch and dinner, (d) do exercise, (e) have a walk and (f) take a shower. The reminders are recorded in audio format, so that they can hear them anywhere in the house. In addition, the recent studies observe that the familiar voice in audio reminders has more positive effect on individuals memory. The smart health system develops reminding feature so that it is flexible to record suitable voices.

3.3 Fall Detection System

Falls can be defined as one of the causes of injury and disability among elderly people [15]. Sometimes elderly people may get fall unconscious, and to prevent any harm for senior citizens the smart health system includes fall detection feature [4, 5].

Since the main aim is to design a cost-effective smart health system, the fall detection system is basically smartphone-based. Most smartphones have built-in motion sensors as accelerometer and gyroscope. The accelerometer sensor measures the speed of the movement in all three axis, like A_x, A_y, A_z.

$$\sqrt{A_x^2 + A_y^2 + A_z^2} = A \tag{1}$$

On the other hand, the gyroscope measures angular velocity in three axis too determine orientation, as ω_x, ω_y, ω_z

$$\sqrt{\omega_x^2 + \omega_y^2 + \omega_z^2} = \omega \tag{2}$$

The final acceleration and angular frequency can be found by using Eqs. 1 and 2.

The fall event can be detected by doing experiments to define a threshold value for the acceleration of the event. Different daily activities, namely sitting, walking, laying, are associated with certain angular velocity and acceleration, so by comparison of such values, it is possible to detect the fall.

The proposed system also includes fire detecting feature, which is associated with smoke/gas detecting sensors. The feature is considered as a security and safety utility of the system, therefore was not included as the main health objective.The main goal of using gas sensors is to prevent elderly people from health injuries during fire and gas leakage cases. This feature advances the proposed system in terms of safety track.

Figure 1 shows an overall block diagram of the proposed smart system.

According to Fig. 1, the main inputs are the sensors and camera as followed:

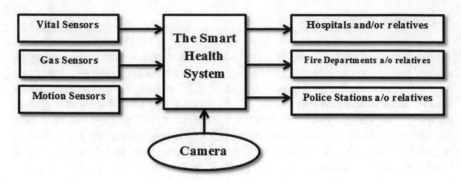

Fig. 1 The system architecture

- *The Vital Sensors* as input for remote health monitoring subsystem. The sensors process analog or digital vital signals, like pulse rate and body temperature.
- *The Gas Sensors*. To prevent any health injury, the safety of senior citizens has to be also considered. To avoid any gas leakage or fire cases it is good that the gas sensor is attached to the proposed smart system. In addition, it highlights that the system is aiming to be smart.
- *The Motions Sensors* is used in a fall detecting subsystem. Since it is argued to use smartphone-based fall recognition algorithm, the accelerometer and gyroscope are embedded in phones.
- *Camera* is used as additional device that can stream live video in real time. The main purpose of using it is double-checking for false alarms. It is always possible to check the gas leakage or falls by viewing the live streaming video from homes.

Coming to outputs of the proposed smart system, below is brief description of these components:

- *Hospitals and/or relatives*. The pulse rate, body temperature data can be sent to physician, nurse, as well as family members.
- *Fire Departments and/or relatives*. In case of gas leakage or fire, the proposed system calls the Fire Departments.
- *Police Stations and/or relatives*. Since the smart system uses a camera, it is possible to monitor for the presence of thieves. In the case of entering thieves, the system is capable to notify the police and relatives.

Based on the system architecture, a smart system model is designed as shown in Fig. 2. From the Fig. 2, it is apparent that reminding subsystem is completed in Telegram bot, while the fall detection is developed using MATLAB program. Furthermore, the remote health monitoring is implemented using a microcontroller. The microcontroller that we used is Arduino Uno that can interface both analog and digital data. According to Fig. 2, the main input is data collected from the sensors and camera. On the other hand, the main output of the smart system is email notifications

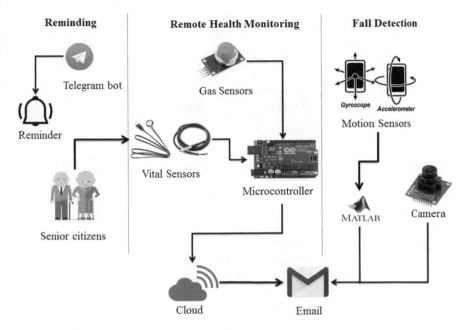

Fig. 2 The smart health system

which are sent through a Cloud platform to hospitals, fire departments, police stations, and relatives.

4   A Prototype Implementation

A Hardware Implementation. Table 1 shows the information and threshold values of the sensors used in the smart system. As it was mentioned above, there are three main groups of sensors: vital, gas and built-in motion sensors. As it is shown in Table 1, the threshold value for vital sensors is approximate typical healthy rate for the elderly people.These values may clarify whether the health state of elderly people under normal conditions. As for gas sensors, the threshold value as settled as any flammable gas or smoke in any amount. The threshold value for the motion sensors were taken from the trial experiment presented in Sect. 5.

The assembled prototype of smart system is shown in Fig. 3. For hardware implementation camera OV7670 module was used. Furthermore, the main microcontroller used in this smart health system is Arduino Uno (refer to Fig. 2).

A Software Implementation. The main software work was performed in Arduino IDE, MATLAB, and Telegram bot. In Arduino IDE the remote health monitoring was programmed, while the fall detection system was coded in MATLAB. In Telegram

Table 1 Sensors specification

Group of sensors		
Sensors name	**Technical description**	**Threshold data**
Vital sensors		
Pulse sensor	Working V = 5 V, I = 200 mA	75–120 bpm
Temperature sensor	Waterproof DS18B20 Range: −55 to 125 °C, working V = 3–5.5 V	36.1–37.3 °C
Gas Sensors		
MQ-2 Gas sensor	Working V = 5 V, I = 150 mA	H_2, CH_4, CO, Smoke in any ppms
MQ-6 Gas sensor	Working V = 5 V, I = 100 mA	LPG, iso-butane, propane, LNG in any ppms
Smarphone embedded motion sensors		
Accelerometer	BMA220 3-axis accelerometer, Iphone 5S Model	$2.01 ms^2$
Gyroscope	B329 3-axis gyroscope, Iphone 5S Model	

bot reminding system was implemented. Table 2 presents pseudo code that can be determined as an algorithm applied in coding process.

Cloud Platform: In order to send email notifications the microcontroller(Arduino Uno) has to be connected to a Cloud platform. For this purpose, ESP8622 board was used to install Wi-Fi in the microcontroller. In this smart system, Thingspeak was used as a platform that provides Cloud services, since it is cost-free and available for anyone. Thingspeak can 'react' on obtained data by posting twits in Twitter. So the alerting notifications can be seen from Twitter.

5 Results and Discussion

5.1 Remote Monitoring System

The microcontroller (Arduino) decides whether to send some notifications based on the sensor data. To illustrate the function, the temperature vital sensor was considered. It is decided that the severe threshold data for the temperature vital is around 40°C. When it exceeds the threshold, the system publishes a twit through ThingsPeak. The results are shown in Fig. 4.

For gas sensors, the same algorithm as for vital sensors is applied. The serial monitor of the microcontroller can show how the output will look like. Figure 5 shows how the gas sensors sense smoke.

Fig. 3 The hardware implementation

5.2 Reminding System

"Telegram" mobile application is used as the interface (refer to Fig. 6). This eliminates the need to develop new software and download extra applications. The user, in this case relative or close person who is setting the system should open the application; application would ask if he/she wants to set or edit the reminder time. When the time is set, the user is asked to record the reminder which will be played at specified time. It is recommended that relatives or other close people should record the audio reminder so that the voice on the record would be familiar to the senior citizen. According to Cynthia Yonan and Mitchel Sommers words have different affect depending on who said it [16]. The studies show that words spoken by the familiar voice are better perceived. Therefore it is preferred to use recorded voice of close people for reminders rather than using automatically created general audio.

Table 2 Pseudo Code Algorithm

Pseudo Code for fall detection
1: Input: Ax, Ay, Az
2: Calculate total A using formula (1)
3: if A=!threshold then
4: return exit
5: else
6: Alert!
7: sendmail (message, email address)

Pseudo Code remote health monitoring and fire detection
1: Input: data from vital/gas sensor
2: Connect ESP8622 to transfer data to Cloud through Wifi
3: if data=!threshold then
4: return exit
5: else
6: Alert!
7: connect(Thingspeak cloud platform)
8: send (data to Thingspeak)

Fig. 4 Twitter post notification

5.3 Camera

A camera is included in this smart health system as an additional device that can allow checking the warning alarms sent by the system. It is possible to monitor if the fall event detected by the system is true or false. The same concept can be applied to check for fire case.

Having a camera may allow the system to be safe and secure. By using the camera it is possible to detect entering of undesired thieves [14]. Another advantage of

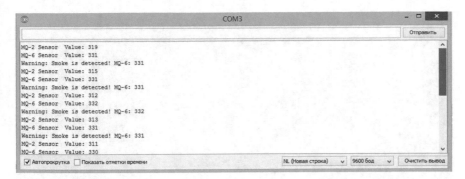

Fig. 5 The gas sensor sensing for smoke

Fig. 6 Telegram App for reminding system

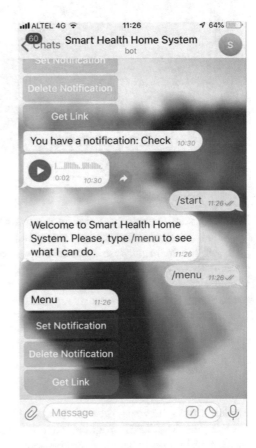

Fig. 7 Live Streaming Video screenshot

including such camera provides an opportunity to observe for any safety/security violations, namely a presence of thieves in homes of the elderly [14].

Since the main purpose is to develop a cost-effective health system, a camera module 0v7670 was used as well as ESP8622 WiFi module. Interfacing two such modules allows a live streaming video, thus it is possible to monitor activities in real time. Figure 7 demonstrates a screenshot from the live streaming video captured by the camera.

5.4 Fall Detection System

To detect the fall events, the first thing that has to be done is interfacing MATLAB Mobile with MATLAB program. Then it is possible to use the sensors data on the movement of elderly people from the phone.

The next step requires the smartphone hold with an individual or putted in pockets next to a chest. This system tests the fall detection algorithm by repeating the experiment and measuring the acceleration of the fall. The experiment results are shown in Table 3.

Table 3 The acceleration data

Trial number	The activity	The average acceleration measured, ms^2
3	Sitting	2.13
3	Falling	8.01

Smart Health System ≫

Входящие ☆

shynar.zhigero... 10:56 ↩ •••

кому: мне ˅

Fall is detected please check camera. Link is in the telegram

Fig. 8 Email notification about the fall

It can be concluded that the fall event can be detected if the acceleration is around $8.01\,ms^2$. Therefore the value is included as threshold in Table 1. When the net acceleration will exceed threshold value, the system send the email notification with attached link of live streaming video. Figure 8 presents the mail notification that was obtained when the system detects a fall event.

6 Conclusion

First of all, obviously the system is intended to help the elderly senior citizens. It is accomplished by making a useful system to help them make life easier and also by drawing attention to the problem. The topic is indeed extremely important for society. The goal of this smart health system is to make an influence in this area, which can be a start for better and bigger developments to help the elderly population around the world.

It aims an affordable, easy-to-use health system. The overall expected results are to create projects which will combine features of existing smart homes, add new components and adapt it all for the elderly people. The main criteria which are considered are price, availability, and implementation.

In this chapter a smart health system prototype is developed based on convenience, efficiency, easy to understand interface and smart interaction between a device and intelligent software. The health system includes remote health state monitoring, fall detecting, reminding features.

To conclude, the prototype version of the Health Smart Home for elderly senior citizens is designed as cost-effective. In this smart health home system the architecture of combined features that can measure vital signals, detect fire and fall, send reminders and email notifications was presented. Also the hardware model implemented and the performance as tested. It is expected the system will accomplish and investigate the further improvements.

References

1. Belghith, A., Obaidat, M.S.: Wireless sensor networks applications to smart homes and cities. In: Obaidat, M.S., Nicopolitidis, P. (eds.) Smart Cites and Homes: Key Enabling Technologies, pp. 17–36. Elsevier, Cambridge (2016)
2. Deen, M.J.: Information and communications technologies for elderly ubiquitous healthcare in a smart home. Pers. Ubiquit. Comput. **19**, 573–599 (2015). https://doi.org/10.1007/s00779-015-0856-x
3. Guelzim, T., Obaidat, M.S., Sadoun, B.: Introduction and overview of key enabling technologies for smart cities and homes. In: Obaidat, M.S., Nicopolitidis, P. (eds.) Smart Cites and Homes: Key Enabling Technologies, pp. 1–15. Elsevier, Cambridge (2016)
4. Hakim, A., Huq, M.S., Shanta, S., Ibrahim, B.: Smartphone based data mMining for fall detection: analysis and design. Procedia Comput. Sci. **105**, 46–51 (2017). https://doi.org/10.1016/j.procs.2017.01.188
5. Hsu, Y.W., Chen, K.H., Yang, J.J., Jaw, F.: Smartphone-based fall detection algorithm using feature extraction. In: CISP-BMEI, pp. 1535–1540 (2016)
6. Majumder, S., Aghayi, E., Noferesti, M., et al.: Smart homes for elderly healthcare–recent advances and research challenges. Sensors **17**(11), 2496 (2017). https://doi.org/10.3390/s17112496
7. Menachemi, N., Taleah, C.H.: Benefits and drawbacks of electronic health record systems. Risk Manag. Healthc. Policy **4**, 47–55 (2011). https://doi.org/10.2147/RMHP.S12985
8. National Center for Health Statistics: Deaths and Mortality (2017). https://www.cdc.gov/nchs/fastats/deaths.htm. Accessed 15 Sept 2018
9. Omheni, N., Obaidat, M.S., Zarai, F.: A survey on enabling wireless local area network technologies for smart cities. In: Obaidat, M.S., Nicopolitidis, P. (eds.) Smart Cites and Homes: Key Enabling Technologies, pp. 91–100. Elsevier, Cambridge (2016)
10. Patterson, C.: World Alzheimer Report 2018. In: Alzheimer's Disease International. The State of the Art of Dementia Research: New Frontier (2018). https://www.alz.co.uk/research/WorldAlzheimerReport2018.pdf?2. Accessed 15 Dec 2018
11. Poissant, L., Pereira, J., Tamblyn, R., Kawasumi, Y.: The impact of electronic health records on time efficiency of physicians and nurses: a systematic review. J. Am. Med. Inform. Assoc. **12**, 505–516 (2005). https://doi.org/10.1197/jamia.M1700
12. Roy, P., Bhattacharjee, S., Ghosh, S., Misra, S., Obaidat, M.S.: Fire monitoring in coal mines using wireless sensor network. In: Proceedings of the IEEE/SCS International Symposium on Performance Evaluation of Computer and Telecommunication Systems, The Hague, Netherlands, pp. 16–21 (2011)
13. Small, S.A., Stern, Y., Tang, M., Mayeux, R.: Selective decline in memory function among healthy elderly. Neurology **52**, 1392 (1999). https://doi.org/10.1212/WNL.52.7.1392

14. Tanwar, S., Patel, P., Patel, K., Tyagi, S., Kumar, N., Obaidat, M.S.: An advanced internet of thing based security alert system for smart home. In: Proceedings of the IEEE International Conference on Computer, Information and Telecommunication Systems, Dalian, China, pp. 25–29 (2017)
15. Tsinganos, P., Skodras, A.: A smartphone-based fall detection system for the elderly. In: Signal Processing Signal Analysis, pp. 53–58 (2017)
16. Yonan, C.A., Sommers, M.S.: The effects of talker familiarity on spoken word identification in younger and older listeners. Psychol. Aging **15**(1), 88–99 (2000). https://doi.org/10.1037/0882-7974.15.1.88
17. Zao, J.K., Wang, M.Y., Tsai, P., Liu, J.W.S.: Smartphone based medicine in-take scheduler, reminder and monitor. In: Proceedings of the 2010 12th IEEE International Conference on e-Health Networking Applications and Services, Lyon, France, pp. 162–168 (2010)

Author Index

© Springer Nature Singapore Pte Ltd. 2020
P. Venkata Krishna and M. S. Obaidat (eds.), *Emerging Research in Data
Engineering Systems and Computer Communications*, Advances in Intelligent
Systems and Computing 1054, https://doi.org/10.1007/978-981-15-0135-7

Printed in the United States
By Bookmasters